AF400736

ByDesign

A Journey to Excellence through *Science*

SEVENTH-DAY ADVENTIST® CHURCH

Kendall Hunt Religious Publishing
A Division of Kendall Hunt Publishing

Developed in Collaboration with the
Seventh-day Adventist NAD Office of Education and
Kendall Hunt Religious Publishing
A Division of Kendall Hunt Publishing Company

Acknowledgments

SDA/NAD Office of Education

Larry Blackmer, Executive Editor/Vice-President of Education

Carol Campbell, Director of Elementary Education

Debra Fryson, Director of Education, Southern Union

Program Development and Publisher

Kendall Hunt Religious Publishing, a Division of Kendall Hunt Publishing

Program Consultants Lead Team I

Dan Wyrick, Director of Nature by Design

Lee Davidson, Associate Professor of Teacher Education/Dept. Chair, Andrews University

Jerrell Gilkeson, Associate Director of Education, Atlantic Union

Ed Zinke, Biblical Researcher

Tim Standish, Geoscience Research Institute

Associates Representing Lead Team II

Betty Bayer, Associate Director of Education, SDA Church in Canada

Diane Ruff, Associate Director of Education, Southern Union

Ileana Espinosa, Associate Director of Education, Columbia Union

James Martz, Associate Director of Education, Lake Union

Keith Waters, Associate Director of Education, North Pacific Union

LouAnn Howard, Associate Director of Education, Mid-America Union

Martha Havens, Associate Director of Education, Pacific Union

Mike Furr, Associate Director of Education, Southwestern Union

Patti Revolinski, Associate Director of Education, North Pacific Union

Contributing Writers and Reviewers

Carol Raney, Freelance Writer

Katia Reinert, NAD Director of Health Ministries

Alayne Thorpe, President of Griggs University

Rick Norskov, MD, Southern Adventist University

Lucinda Hill, MD, Southern Adventist University

Raul Esperante, PhD, Geoscience Research Institute

Mitch Menzmer, PhD, Southern Adventist University

Contributing Multi-Grade Classroom Writers and Reviewers

Karen Reinke, Freelance Writer

Mark Mirek

Heidi Jorgenson

LaVona Gillham

Cindy French Puterbaugh

www.kendallhunt.com

Send all inquiries to:

4050 Westmark Drive

Dubuque, IA 52004-1840

1-800-542-6657

Science Classroom and Lab Safety

It is important to be safe when you perform hands-on investigations. Always look for the safety symbol in your textbook. It tells you how to be safe as you perform your investigations. Below are some important safety rules to follow for most investigations. Remember to be alert and listen to your teacher for any additional safety rules.

Prepare for laboratory work

- Review classroom laboratory procedures with your teacher.
- Never perform unauthorized experiments.
- Keep your lab work area organized and free of clutter.
- Know the location of the safety fire blanket and first aid kit.

Dress for laboratory work

- Tie back long hair.
- Do not wear loose sleeves.
- Wear shoes that completely cover your feet.
- Wear lab apron, goggles, and gloves as required.

Avoid contact with chemicals

- Never taste or sniff chemicals.
- Never draw materials in a pipette with your mouth.
- When heating substances in a test tube, point the mouth away from people.
- Never carry dangerous chemicals or hot equipment near other people.

Avoid hazards

- Keep combustibles away from open flames.
- Use caution when handling hot glassware.
- When diluting acid, always add acid slowly to water. Never add water to acid.
- Turn off burners when not in use.
- Keep lids on reagent containers. Never switch lids.
- Use sharp equipment and objects as they were designed to be used and in a safe manner. For example, always cut away from yourself rather than toward yourself, and be careful to study which edges and points of an object are the sharp ones.

Clean Up

- Consult the teacher for proper disposal of materials.
- Wash hands thoroughly following experiments.
- Leave laboratory work area clean and neat.

In case of accident

- Report all accidents and spills immediately.
- Place broken glass in designated containers.
- Wash all acids and bases from your skin immediately with plenty of running water.
- If chemicals get in your eyes, wash them for at least 15 minutes with an eyewash or clean water.

Scavenger Hunt

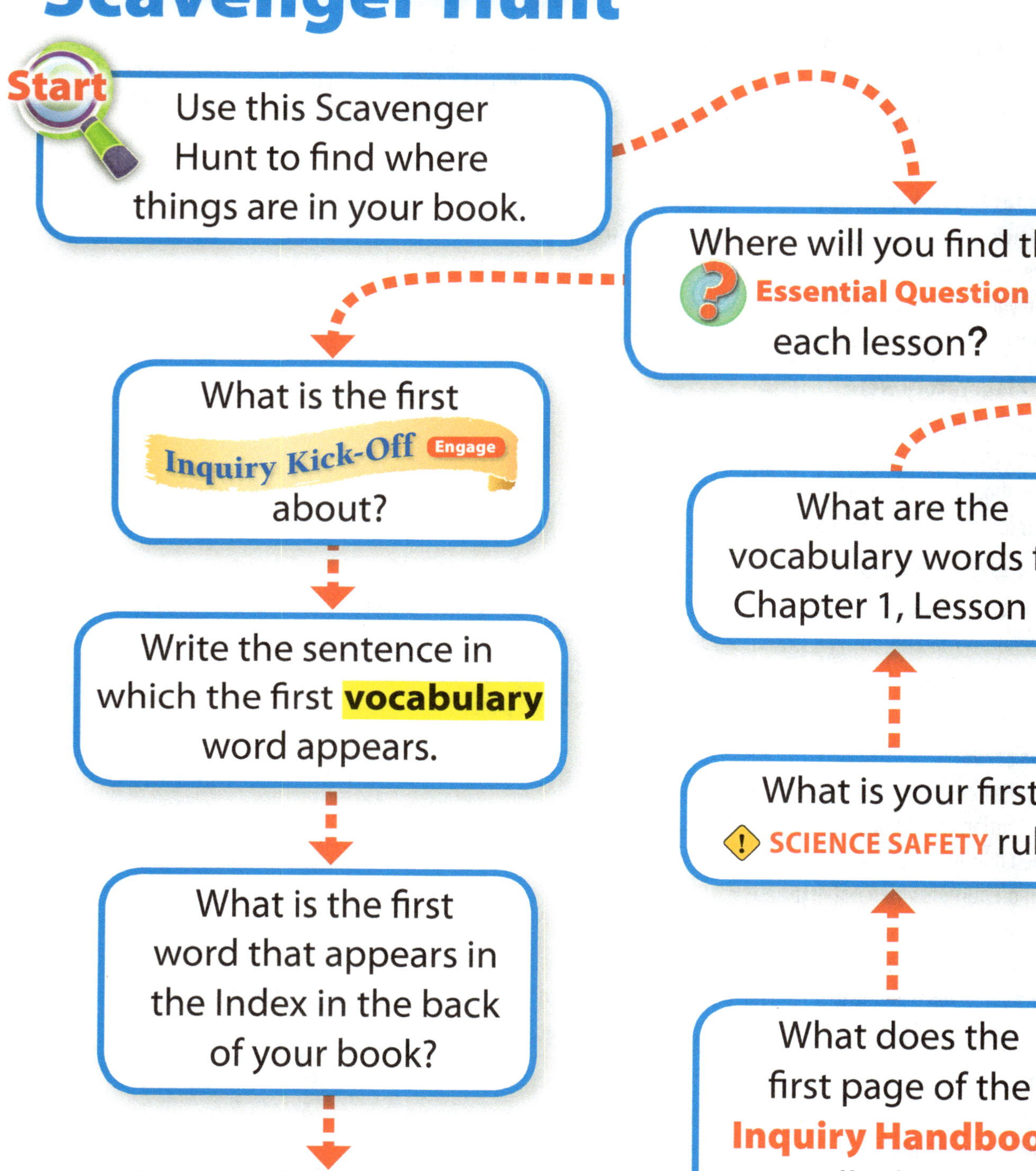

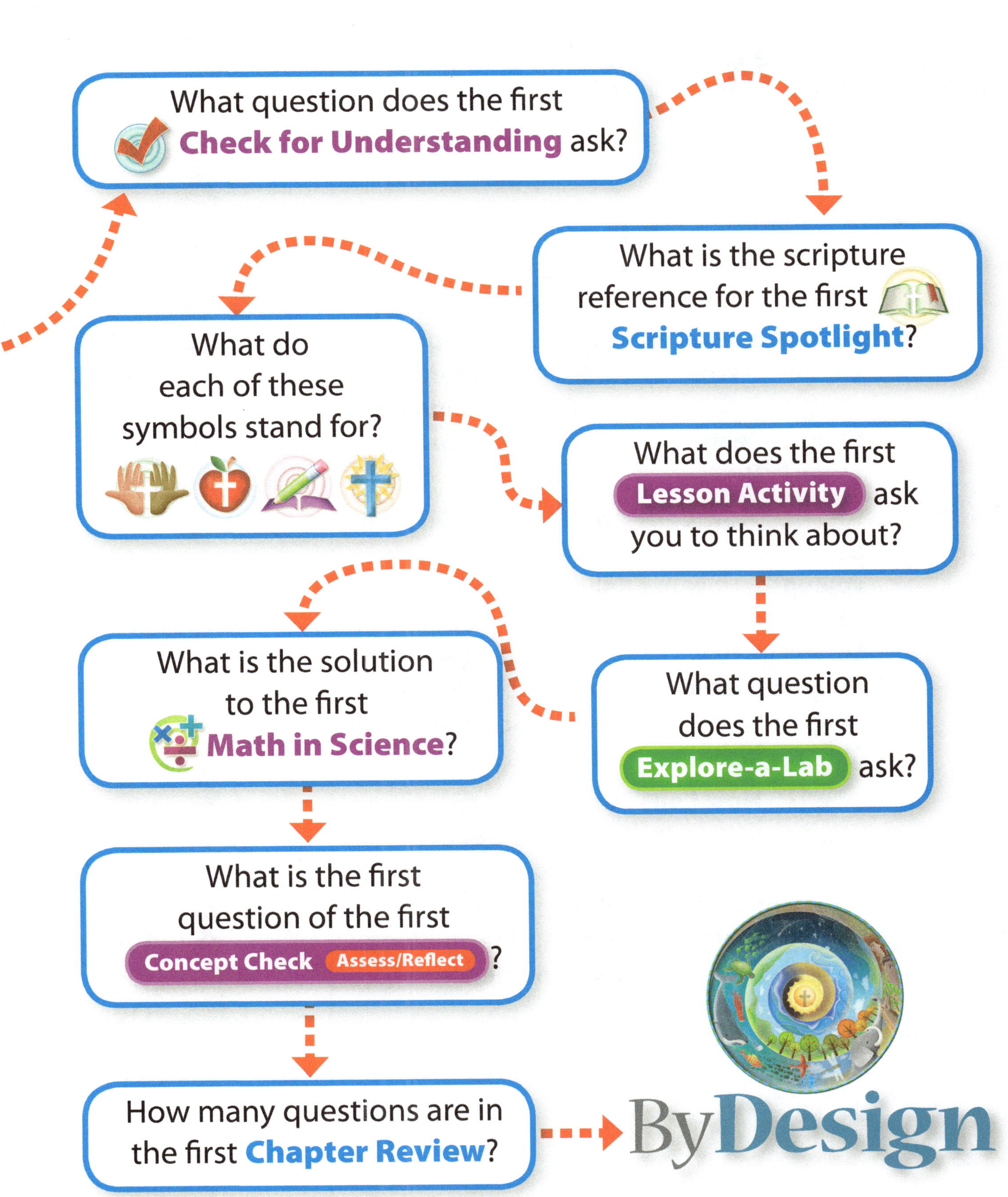

What question does the first Check for Understanding ask?
What is the scripture reference for the first Scripture Spotlight?
What do each of these symbols stand for?
What does the first Lesson Activity ask you to think about?
What is the solution to the first Math in Science?
What question does the first Explore-a-Lab ask?
What is the first question of the first Concept Check Assess/Reflect?
How many questions are in the first Chapter Review?
ByDesign

Table of Contents

Science—Understanding the Natural World

Did you ever wonder what causes a volcano to erupt, how your brain stores memories, or how water molecules interact with each other? That natural curiosity is what causes you to ask questions about the world around you. Everything, from the tiniest bug to the largest volcano to an atomic reaction, is part of the natural world. Science is a way of learning that allows you to investigate the natural world using a process based on good data. But science does not necessarily follow a single path or order.

You use the scientific process of questioning, observing, testing, and investigating to find the truth about the world. Whether you are learning about atoms, observing chemical reactions, investigating how fossils form, or researching Earth's natural resources, the testing process remains the same. In order to start understanding the natural world, you utilize the scientific process.

Science helps you answer questions in logical, meaningful ways. In the 1400s, the Europeans who explored the New World, contrary to modern myth, did not think the world was flat. In fact, the navigational techniques of Columbus's time were actually based on the fact that Earth is a sphere. They did, however, underestimate the size of the globe. You can observe the fact that Earth is round and not flat by watching the way a ship's sails emerge from below the horizon when strolling along a beach, or by observing the shadows produced by the Sun as time passes.

What are you most curious about? How can science help you find out more? Write your question, and explore more about the topic.

The Search for Truth

Using science, you can answer questions based on knowledge, not misconceptions. Just imagine if people continued to believe the Sun travels around Earth. The natural curiosity of humans has led to countless scientific discoveries about the natural world, creating advances in technology. These advancements help us understand everything from our health and the environment, to the planets and the Universe. Technology such as power tools, computers, and smartphones also make life easier and more enjoyable.

Science searches for truths about the natural world. Since science is a way of learning about the natural world, it is ever changing. The knowledge that is built by science is always open to question and revision when new facts are discovered. Ideas that are accepted today could be rejected or modified tomorrow, if new evidence calls into question those ideas. There is always something new to learn. Scientists understand that the process is not just about knowing facts; it's about understanding how the facts relate to and impact the bigger picture. How would your world be different if scientists and engineers had not conquered Earth's powerful gravity to send rockets and satellites into space?

The scientific process can guide you to truths about the natural world or to additional questions. The process helps you to better understand the natural world in which you live. The knowledge gained through the process enables you to understand how science affects life every day in a variety of ways.

The Scientific Process

The scientific process involves asking and answering questions about the natural world.

Science is a way of gathering knowledge that is accurate and reliable. It is the quest to understand things around us and how things work. The processes of science do not involve a set of predetermined steps.

However, science often begins with observations and something that the scientist wonders about those observations. For example, suppose you were on the school playground and noticed that the chains on the swings had developed a coating of rust. You might wonder what materials rust and what materials resist rusting. You might wonder what liquids promote rusting. How could you find out? What tests could you perform to find out the answers to your questions?

These are the main science processes you might follow:

- **Observe.** When you observe, you use your senses to gather information.
- **Question.** State the problem you want to answer.
- **Hypothesize.** Make an "educated" guess about the answer.
- **Experiment.** Design a test that will prove or disprove your hypothesis, which should include a prediction of expected results.
- **Data collection.** The data collected might include pH, time, temperature, distance, growth, or other measurements.
- **Interpret.** Analyze the data gathered using charts and graphs.
- **Draw conclusions** based on your results.

The data you collected during these experimental processes may lead you to further questions and to additional experiments.

These students are gathering data during a field study.

What are the advantages and disadvantages of using a field notebook to record data compared with using an electronic journal?

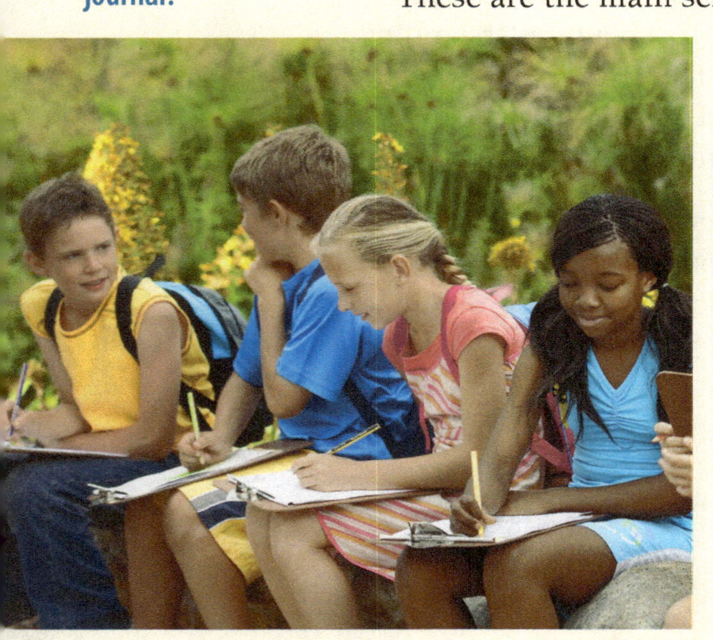

Questions in Science

The scientific process begins with observations. The observations stimulate questions. Scientists ask many, many questions throughout any scientific investigation. You probably ask questions that begin with *who, what, when, where, which, why,* and *how*. That's a good starting point, but as you expand your knowledge, the scope of your questions will grow.

Observing the natural world provides an endless source of questions. Careful observation of the natural world using your senses and scientific instruments like microscopes and probes can help you discover questions you may have never considered.

Scientists are masters at asking questions that are straightforward and can be tested. With practice you too can become a master at questioning. Scientists focus their questions so that they can identify a problem that can be tested using the processes of science. Not all questions are the same. Quick factual questions such as how many legs does an insect have require specific knowledge and build on prior learning. Experimental questions, such as do caterpillars prefer milkweed to apple leaves, that require in-depth answers, involve more observation and testing to resolve.

Read the following questions. Which one is a good scientific question?

- How does mixing chemicals that make up polymers like Gak, slime, or cornstarch putty change their physical properties?
- What is the best recipe for cookie dough?

If you are uncertain, think about which question could be answered by experimentation and which involves opinion. How could a survey help you investigate what is the best recipe for cookie dough?

Stop and think of a scientific question you can investigate and gather data about. Work with a partner to find the answer.

Good scientific questions build on what you know and lead to more questions, which, even when answered, lead to more questions. It is this type of curiosity that has led to some of the greatest discoveries. Think about scientific questions as a cycle with no beginning or end.

Look at the natural world around you. What do you observe that is yet unanswered? The Germ Theory of Disease is a great example of how observations and scientific questions and answers have led to new discoveries. The invention of the microscope led to the discovery of bacteria. This resulted in a debate about whether these tiny living things could come from nonliving matter. This led Louis Pasteur to conduct experiments that proved heat kills such organisms and prevents further contamination. Later, scientists built upon this work—the discovery that sanitation prevents the spread of disease, the invention of vaccines, and the discovery and use of antibiotics. Do you think antibiotic resistance of some bacteria will lead to new questions and answers? Look at this timeline. What new questions and discoveries do you think await us in this field?

Timeline of Germ Theory **Key Question:** What causes disease and infection?

Early Medicine	Middle Ages–1700	1700–1799	1800–1849	1850–1900	1901–1999	Since 2000
400 B.C. Hypocrates "Father of Medicine" describes immunity	Bad air, poor hygiene, or contaminated water supplies **1680** Anton Leeuwenhoek first observes tiny microbes	**1707** N. Andry proposes microbes as the cause **1766** Spallanzani observes the absence of microbes in heated and sealed containers **1798** Jenner develops the Smallpox vaccine	**1838–1839** Schleiden and Schwann develop Cell Theory **1847** Semmelweis connects hospital fevers with lack of sanitation	**1850** Virchow proves how cells reproduce **1861–1862** Louis Pasteur proves heat kills bacteria **1876–1882** Koch finds bacteria causes anthrax and TB **1879** Lister develops antiseptics	**1901** Walter Reed confirms mosquitoes transmit yellow fever **1910** Ehrlich looks for "magic bullet" **1928** Alex Flemming discovers penicillin **1983** Gallo, Montagmier and Levy identify AIDS retrovirus	**2000** First new antibiotic approved by FDA

There are many questions about the natural world yet to be asked and answered. Amazingly enough, there are many future discoveries that will be made, leading to more questions. These discoveries and questions will open doors to fields of research and study that do not exist today. By asking questions based on your observations of the natural world, you may lead the way to the next great scientific breakthrough that makes the world a better place.

Gathering Data

Suppose your parents were interested in determining which diapers are the most absorbent to use for a new baby about to join your family—cloth or disposable diapers. You might like to learn how disposable diapers work. You could conduct an investigation to test the leading brands against cloth diapers. You would want to keep certain variables constant, such as the amount and kind of liquid, the temperature, and the amount of time. You would need to gather data about how much liquid was present before and after and subtract to find out how much is absorbed by the test samples.

This is not much different from gathering data using the scientific process. Once a scientist has defined a question, formulated a hypothesis, and researched relevant information—he or she begins to experiment and make observations. Gathering or collecting data takes place during experimentation. Data can be collected in a variety of ways, such as observing, measuring, and recording. Why do you think the process of gathering data is a critical part of the scientific process?

Regardless of the techniques used when gathering data, scientists must be sure to record all information that could affect the answer to their hypothesis. So, how do they know which data to focus on? Here are some questions to guide you when gathering data:

- Have I recorded all relevant data?
- Have I recorded the data in a way that can be read and understood?
- How am I keeping track of what was done at each step?
- Can someone else understand the data?
- Is my data trustworthy and reliable?

It is important to remember that scientists often build on each other's work. If your experiment is based upon information gathered from another scientist's work and his or her data collection method was flawed, how could this affect your outcome?

Using a process that is consistent when collecting data and making sure your records are detailed helps others understand and use the data collected. This builds confidence in your conclusions and the conclusions of those who use your research.

This researcher is testing water samples to learn about the water quality.

What variables might she test to gather data for a study of the effects of pollutants on organisms living in the water?

Interpreting Data

Once all of your data has been systemically and accurately gathered, it's time to step back and ask, "What do my findings tell me?" When all of the evidence is considered, you can evaluate and draw conclusions by assessing the data. Data may be interpreted in different ways based on a person's background knowledge and perceptions. One scientist may see patterns in the data that another does not.

Let's go back to the diaper data in our earlier example. Gathering data is only the first step. Part of the process of interpreting data is to display and present it in a meaningful way so you can share your findings with others. What would be a good way to display your comparison of the absorbency of different diaper brands? The bar graph shown below is one way you could have chosen to represent the data. Which disposable diapers led the pack during the three trials? Which diaper of those tested would you recommend to your parents? Why?

To interpret data you must look at it critically. These questions can guide you when interpreting data:

- What conclusion can you reach from your results?
- Do the results answer the original questions?
- Are the results different from what you expected?
- Are there other hypotheses to explain the results?
- Did you think about all relevant data, even data that seemed odd?
- Did the data collection method change the results in any way?
- What are the implications of your conclusions?

Can you think of any additional questions that can guide you when interpreting data?

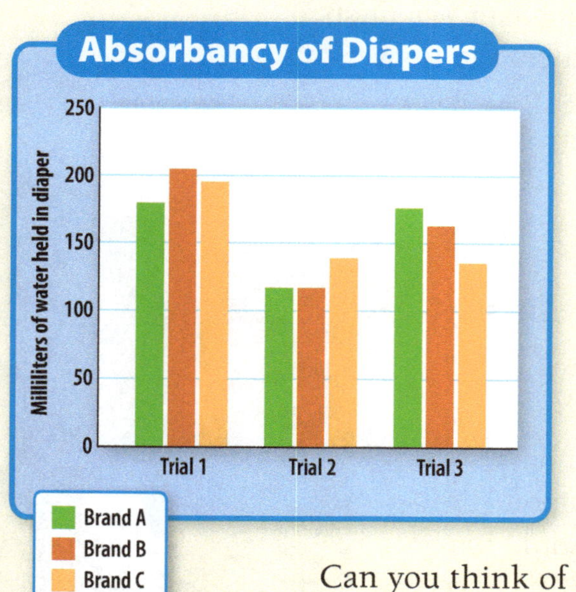

Sometimes interpreting data results in error or incorrect conclusions. The error can be systematic throughout the testing and have great effects or it can be the result of random events and have little to no effect on the outcome. What if the measuring instrument you used was inaccurate? How would that affect the results? What can scientists do to overcome making such errors and inaccuracies?

While using carefully planned methods can reduce the inaccuracy, errors still occur. Scientists indicate their level of confidence when interpreting data using a margin of error. When there is conflict about the interpretation of data, experts in that field will come to a consensus. For example, you would communicate the results of your diaper tests with another group that performed the same tests and see if you could reach a consensus on the best brand of diaper. The process of coming to consensus may require further experimentation and additional evidence.

The Scientific Process

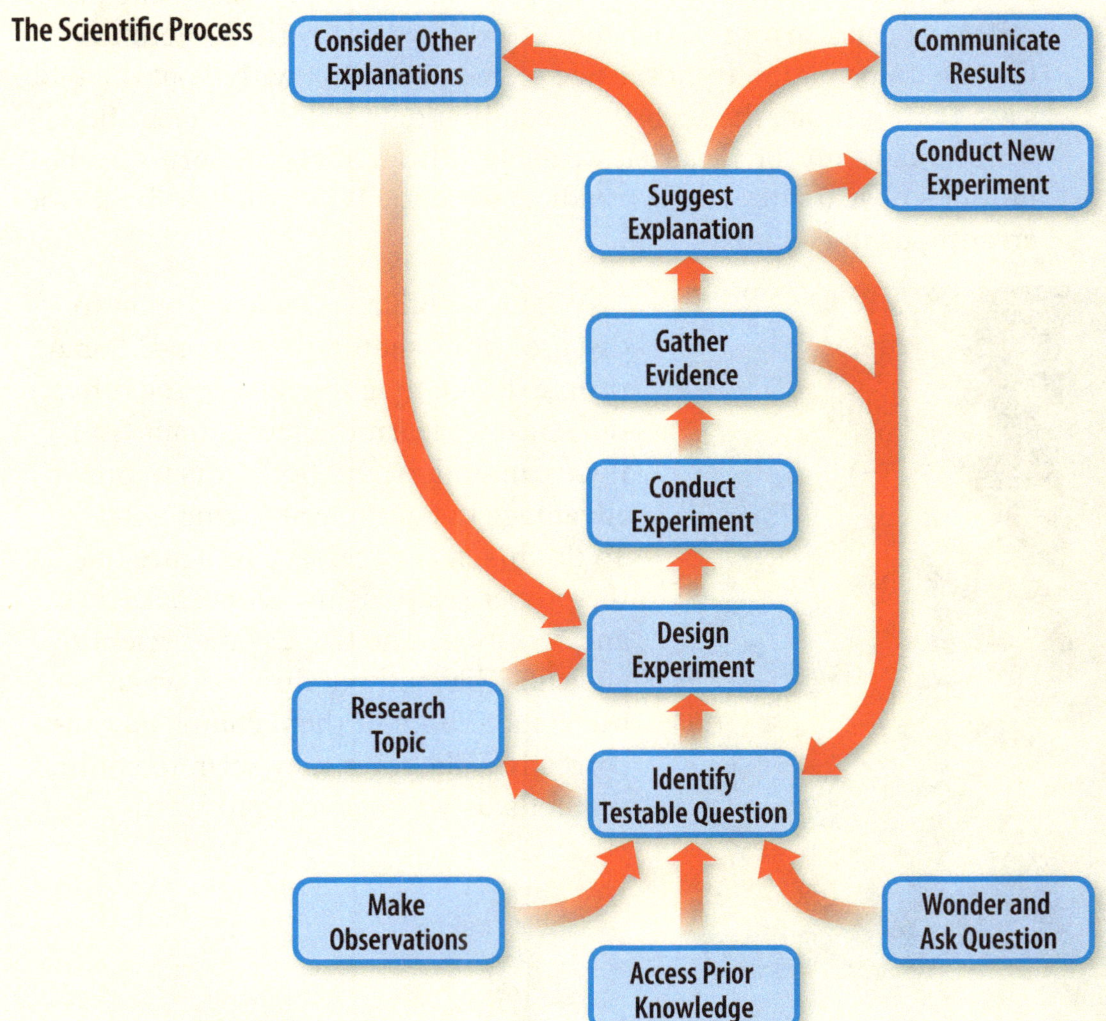

God's Word and Science

What do you like to make? Do you like to draw or paint? Build model cars or airplanes? Have you ever made up a song, created a recipe of your very own, or invented a new game? Could someone who does not know you learn anything about you by studying something you made? Another way you could tell someone about yourself would be to write the person a letter. Which way do you think someone could learn the most about you—by looking at things you made or by reading a letter you wrote to him or her? Would being able to do both help the person know you even better?

Two of the ways we can learn about God are by studying the things He made (nature—which we study in science) and reading something similar to a letter from Him (the Bible). Instead of just using one or the other, we can learn the most about God if we use both. Sometimes, though, information from science seems to conflict with what the Bible says. For example, the theory of evolution dominates the scientific world today, but the idea that we evolved from lower life forms over millions of years directly contradicts what the Bible teaches about God creating us in His image.

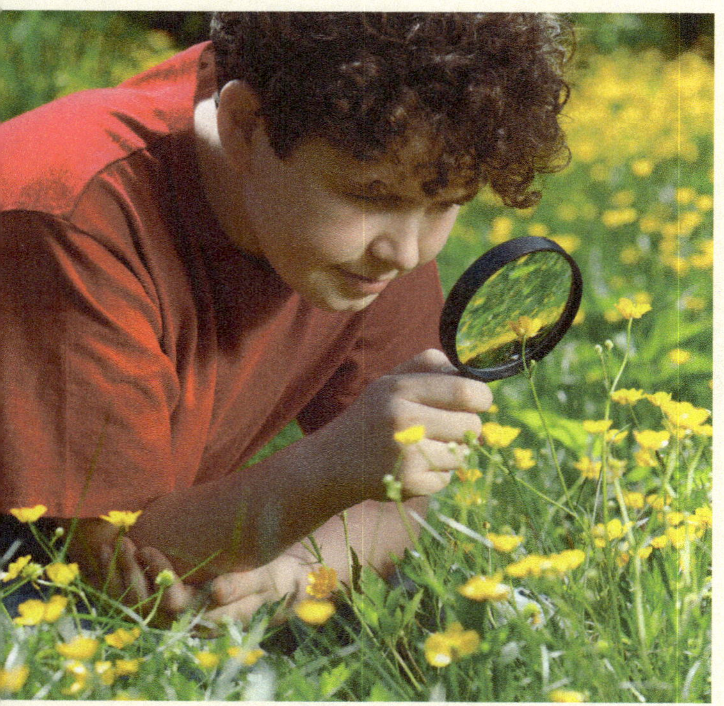

When the Bible and science seem to disagree, how should we respond? Some people decide to ignore one or the other. But since we learn the most about God if we can understand both, it is to our advantage to continue studying both. To do this, it is necessary to figure out the proper relationship between science and scripture. The truth is that science and the Bible will not disagree when we understand both of them completely and correctly. Whenever they seem to conflict, that tells us we have more to learn.

The Bible tells us in **Psalms 19:1** that "The heavens declare the glory of God; And the firmament shows His handiwork." In other words, we can learn a lot about God by studying nature. Studying the Universe can give us a glimpse of God's wisdom and power. By itself, nature reveals a conflict between good and evil but does not explain the origin or the solution to that conflict. It is limited in what it can teach us about God.

The Bible tells us of the conflict between good and evil. It explains how the conflict between good and evil began and how it will end. Seventh-day Adventists believe that the Bible is a divine revelation from God. Although the Bible does not talk about many areas of science, we can trust its accuracy when it does speak of scientific topics.

When scientific conclusions conflict with what the Bible teaches, it is appropriate for scientists with a biblical worldview to continue searching for alternate interpretations of the scientific data that are both scientifically sound and consistent with what the Bible teaches.

Can Christians Be Good Scientists?

It has been suggested by some that Christians cannot be good scientists. They suggest that allowing a biblical worldview to influence how scientific conclusions are evaluated, instead of following the evidence wherever it leads, defeats the purpose of scientific investigation.

This idea ignores the fact that *every* scientist is affected by his or her worldview. A scientist with a naturalistic worldview (which will not consider the possibility of intervention by God in history) also will evaluate conclusions based on his or her worldview and is no more likely to follow the evidence, wherever it leads, when it goes contrary to the assumptions of that worldview. Being aware of how one's worldview affects science can help all scientists be more successful in thinking about scientific ideas.

Leonardo da Vinci was an artist and scientist, best known for his painting of the Last Supper, and for his inventions.

The idea that Christians cannot be good scientists assumes that human reason should be placed above divine revelation and often reflects misconceptions about the beliefs and practices of scientists who are Christians. It fails to recognize the many contributions of people of faith in the scientific community today and in the past, who have successfully combined a biblical worldview with rigorous and productive science.

The Importance of Worldview Your worldview is like a lens through which you view the world around you or a filter you use to evaluate ideas. Using the Bible as the standard by which everything else is measured means we have a biblical worldview. For scientists with a biblical worldview, evaluating scientific conclusions through the filter of the Bible is important. When interpretations of scientific data conflict with the Bible, it is appropriate for these scientists to continue searching for interpretations that are consistent with what the Bible teaches.

As it turns out, scientists who believe in the Bible often notice things other scientists miss. A scientist who believes in the biblical Flood of Genesis might notice evidence that would point toward the rapid burial and preservation of fossils that someone who believes differently might miss. The Flood might suggest possible research questions about how massive amounts of water in a relatively short period of time might explain certain formations. If science is the search for truth about the natural world, and the Bible contains truth about the history of our world, it should not be surprising that ideas based on the truth of the Bible can generate good questions that can be answered by scientific research.

Some have even suggested that a biblical worldview gives scientists a bit of a head start in the search for truth. For example, weather data has been collected for hundreds of years. In the 1850s Mathew Maury, an officer in the U.S. Navy, established a standard system for collecting and recording information on sailing routes. Maury was a Christian who read and studied the Bible. In the Bible studies, the words of **Psalm 8** stuck in his mind: ". . . whatsoever passeth through the paths of the seas." Maury determined that if God's word said there were "paths" in the seas, there must be paths, so he set out to find them. The data Maury collected is still being interpreted more than 150 years later by scientists studying climate change—the same data, interpreted for a different purpose. Can you think of other data collected long ago that may be helpful to scientists today?

Many of the greatest scientists of all time have worked from a biblical worldview. Many scientists today still do. God calls young people to all kinds of jobs that advance His kingdom on Earth. Maybe God has plans for some of the students in your class to become scientists. Maybe He wants to use you to make the world better through scientific discoveries! Throughout this year, you will have many opportunities to see how worldview influences the interpretation of scientific data and to practice evaluating these interpretations from the perspective of your own worldview.

Unit 1
Life Science

Unit Overview

How many different species do you think exist on Earth? What features do they have in common? Do they have similar processes or go through similar cycles in their lives? God created all of the forms of life that we study. This study of life is called *biology*. In this unit, you will investigate life on Earth and the cells that make up living things.

What processes tie together all forms of life? In Chapter 1, you will study the themes that unify all living things. You will analyze the complex interactions within ecosystems and explore the various cycles that exist in nature.

In Chapter 2, you will examine cells—how all living things are made of cells, how cells carry out life processes, and how different types of cells vary. Each cell is like a tiny factory. The different parts of the cell each have jobs that God designed them to do.

Lastly, in Chapter 3, you will focus on the structure of DNA and how it is copied. You will investigate how proteins are made. These proteins respond to changes from inside and outside of the body. God designed organisms to maintain homeostasis, or balance, and this is the job of the cells and the systems in larger organisms.

Bears catch salmon as the salmon move to their spawning beds.

Your teacher may assign an Open Inquiry lab and a Lifestyle Challenge activity. Use your *Science Journal* to record your work.

An Overview of Life

Scripture Spotlight 

You can see the unity in life as you study how the species God created live and interact together. The complexity of life as seen through the study of biology reveals the design of a single master Creator.

You will read the following passages in this chapter.

John 1:1–3 (p. 22)
Colossians 1:16–17 (p. 22)
Exodus 13:21 (p. 27)
Deuteronomy 8:2–4 (p. 27)
Genesis 7–9 (p. 31)
Genesis 8:22 (p. 35)
Job 5:8–10, 28:26, 36:27–28 (p. 38)
Isaiah 11:6, 65:25 (p. 40)

Unifying themes, structures, and cycles abound in nature. These horses, the grass, and other vegetation are made of cells and must use energy. They interact in a complex ecosystem. They all play a part in different cycles of nature, such as the carbon cycle and the water cycle. As you study this picture further, you will notice that the reproductive cycle is represented by the adult horse and its offspring.

The Big Idea

There is amazing diversity among the living things that God created. The more we study them, the more we see that they also have many things in common.

Look at the living things in this picture. How are they different? How are they alike?

Inquiry Kick-Off Engage

Depending on the source, there can be thousands of microscopic organisms in a drop of water. Why are these tiny organisms important? In your *Science Journal*, you will observe and compare these creatures.

Essential Question

What Themes Repeat in Biology?

God created the creatures on Earth with great variety. At the same time, these organisms have a great deal in common. Think about the functions your body performs every day in order to survive. How are these functions similar to the functions carried out by a pet dog? What does your body and the dog's body need to survive? What does the dog have in common with an oak tree? Differences among organisms are easy to spot, but there are also many similarities. How is it possible to have so much in common when organisms look so different?

Unifying Themes Explain

Biology is the scientific study of living things and the processes that keep them alive. In addition to the fascinating diversity that is so easy to see, certain dominant ideas, or *themes,* recur throughout biology and provide a great deal of unity as well. All living things are similar in the following ways:

- They are made of cells that carry genes.
- They are part of a larger organizational structure.
- They interact with each other and their environment.
- They were created by God.

How does the structure of the leafy sea dragon help it to blend with seaweed?

16

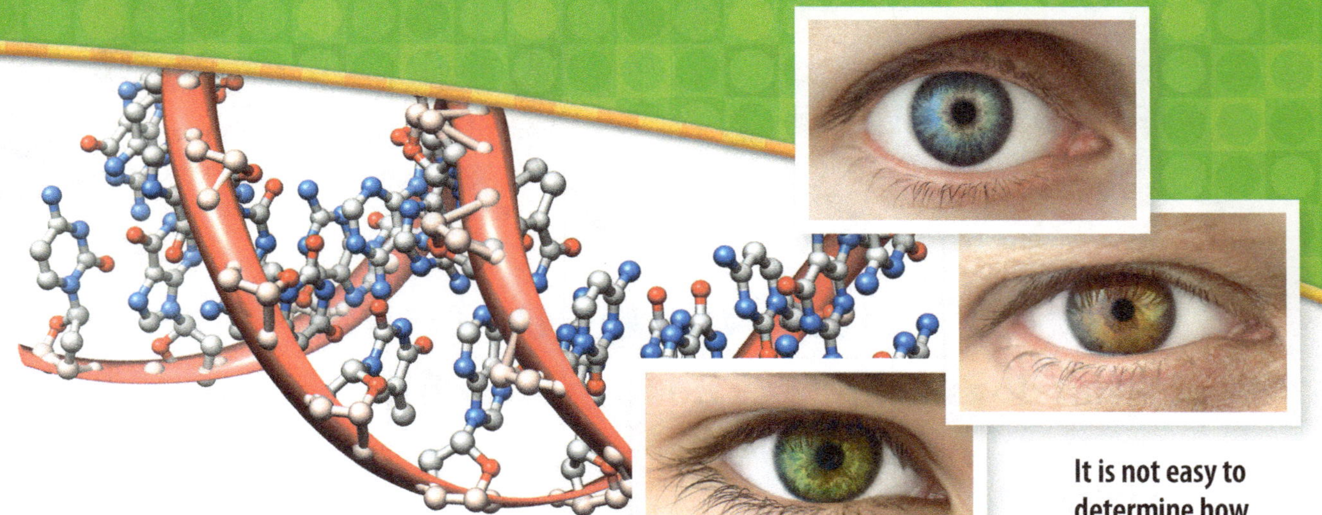

How might genetics explain the variety in eye color?

Unifying Theme 1: Cells and Genes

You know that cells are the fundamental units of life and all cells come from other cells. Even though they are tiny structures, these small structures are composed of even tinier structures. Each structure in a cell performs certain functions, enabling the cell to live. Some organisms, like *Paramecium,* are unicellular. What does the term *unicellular* mean? You are a multicellular organism. So is an onion plant. What does *multicellular* mean? How are the cells in unicellular organisms similar to cells in multicellular organisms? How are they different?

God told Adam and Eve to "be fruitful and increase in number" to create new people in His image. In this process, each new individual receives hereditary information from the parents. The parents pass on the genes in their cells to their offspring. *DNA* is the molecule that carries the hereditary information. In the cell, DNA is replicated so that each new cell produced by cell division gets a complete copy of the DNA found in the original cell. A gene is a specific part of DNA that holds the code for a specific trait. Genes contain the information necessary to create all the proteins needed to maintain and create new cells.

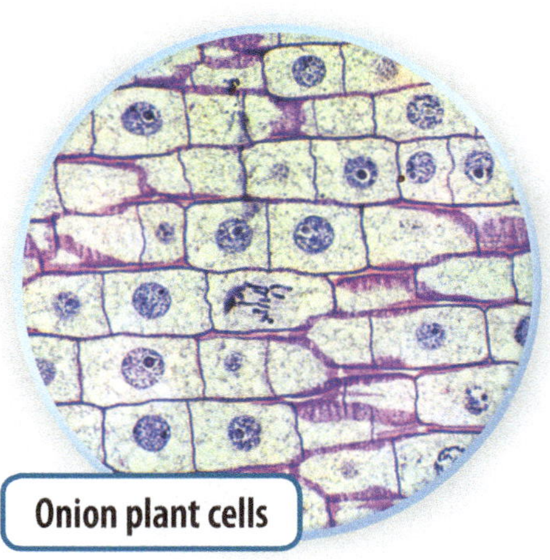

Onion plant cells

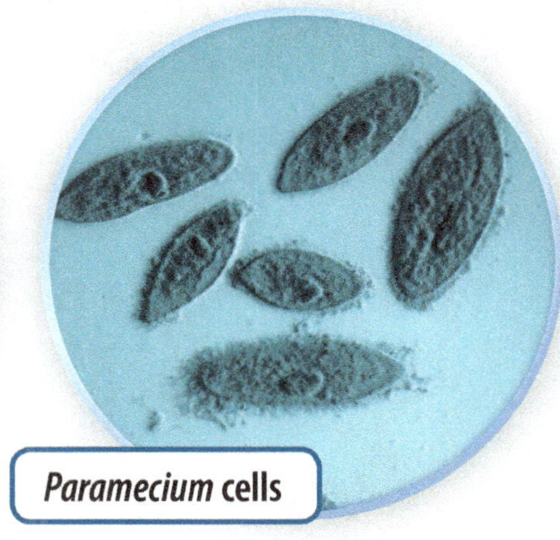

Paramecium cells

A *Paramecium* is a microscopic, unicellular organism that lives in water. An onion plant is a multicellular organism.

How are these cells alike and different?

Observing Homeostasis

What does your body do to recover from exercise?

Procedure

1. Have a partner **record** your breathing rate, heart rate, and temperature on the chart in the *Science Journal*.

2. **Predict** how each body condition might change after exercise.

3. Do jumping jacks for one minute.

4. Immediately after the exercise, your partner should **record** your breathing rate, heart rate, and temperature.

5. Rest for 3 minutes. Repeat Steps 3 and 4.

6. Repeat Step 5 one more time.

7. Switch places with your partner and repeat Steps 1–6.

Materials
- stopwatch
- thermometer strips

Analyze Results

Compare your breathing rate, heart rate, and temperature before and after exercise. **Display your data** on a graph. Note whether your predictions were accurate. **Compare** the changes in your body conditions to your partner's body conditions before and after exercise. **Develop a hypothesis** describing why your results are similar to or different than your partner's.

Create Explanations

1. What does your body do to recover from exercise?

2. How does exercise help your body recover?

3. What are some signs that a body is not in homeostasis? When do these signs signal danger?

Biology involves an organized classification of all forms of life. What things do you organize and give order to? What does **Genesis 1** tell us about God's planning and organization of Creation? Life is organized on many levels. Each level of biological organization builds on the levels below it. *Atoms* are organized into molecules. These *molecules* are used to form *organelles*, which make up the *cells* of living things. Groups of similar-functioning cells form *tissues*. Groups of tissues form *organs*. Different organs work together to form an *organ system*. The organ systems work together to form an *organism*.

Organization does not stop at the organism level. Organisms of the same species form populations. *Populations* living in the same area interact and form communities. *Communities* interact with the nonliving things in the area to form ecosystems. Finally, many smaller *ecosystems* connect to form larger ecosystems called *biomes*. In what biome do you live? What populations of plants and wildlife interact in your area?

Jaguars are one of many organisms that inhabit Central and South America.

❓ **How would you describe the population, community, and ecosystem that the jaguar is part of?**

Lesson Activity

Did you know that you can find biological organization in your schoolyard? With your partner, write the letters A–Z on a sheet of paper. Then, while outside, observe living and nonliving things in the schoolyard. Try to list something for each letter of the alphabet. The letter should be in the name of the organism or object you list. Return to the classroom and get a large sheet of drawing paper. On the paper, draw pictures and use arrows and words to show how the things you found interact.

❓ **What kinds of examples of organization can you find in the schoolyard?**

Levels of Biological Organization	
Biosphere	The part of Earth that contains all ecosystems
Biome	Connection of many smaller ecosystems
Ecosystem	Community and its nonliving surroundings
Community	Populations that live together in a defined area
Population	Group of organisms of one type that live in the same area
Organism	Individual living thing
Groups of Cells	Tissues, organs, and organ systems
Cells	Smallest functional units of life
Molecules	Groups of atoms, smallest unit of most chemical compounds
Atoms	Basic building blocks of matter

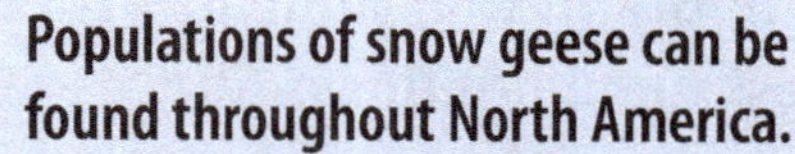

Populations of snow geese can be found throughout North America.

What levels of biological organization do these geese represent?

No matter which branch of biology we study, we find many similarities in the way living things are constructed. We find that the way they are organized at multiple levels and the way they interact with each other and their environment have similarities.

Lesson Activity

Use the Internet and other resources your teacher may provide to develop a list of different areas of study in biology. In your list, name each branch of biology and write a brief description of it. Compare your list with a partner's list. Add branches of biology to your list that you do not have. Then, choose one branch of biology and create a poster promoting that type of science as a career. Make your poster engaging and exciting!

How many branches of biology are there?

How do living things interact with each other? How do they interact with the nonliving things around them? In **Job 8:11** we are asked, "Can the papyrus grow up without a marsh? Can the reeds flourish without water?" How would removing one part of an ecosystem affect the other parts?

All living organisms interact with each other. **Interdependence** is the interaction of living things as they rely on each other in order to live successfully. What interdependence is observed between insects and flowers? Think about clover and bees. Clover needs bees and bees need clover. The clover provides nectar to bees, which in turn pollinate the clover. They are interdependent for their survival. What are some other ways in which organisms are interdependent?

How do organisms interact with their environment? In addition to interacting with each other, living organisms also respond to changes as they interact with the environment. **Homeostasis** is the ability of an organism to keep its internal environment stable in response to environmental changes. For example, the human body maintains an average temperature of about 37°C (98.6°F). When the body fails to maintain this temperature, death may occur. When you get hot or cold, how does your body respond in order to maintain a constant temperature? What examples can you see of homeostasis taking place in plants and animals living around you?

The structure of some types of leaves includes a pointed end, called a drip tip. This is especially noticeable on many rain-forest plants.

How does this structure help the plant function during heavy tropical rains? Does this design help any other organisms survive? If so, how?

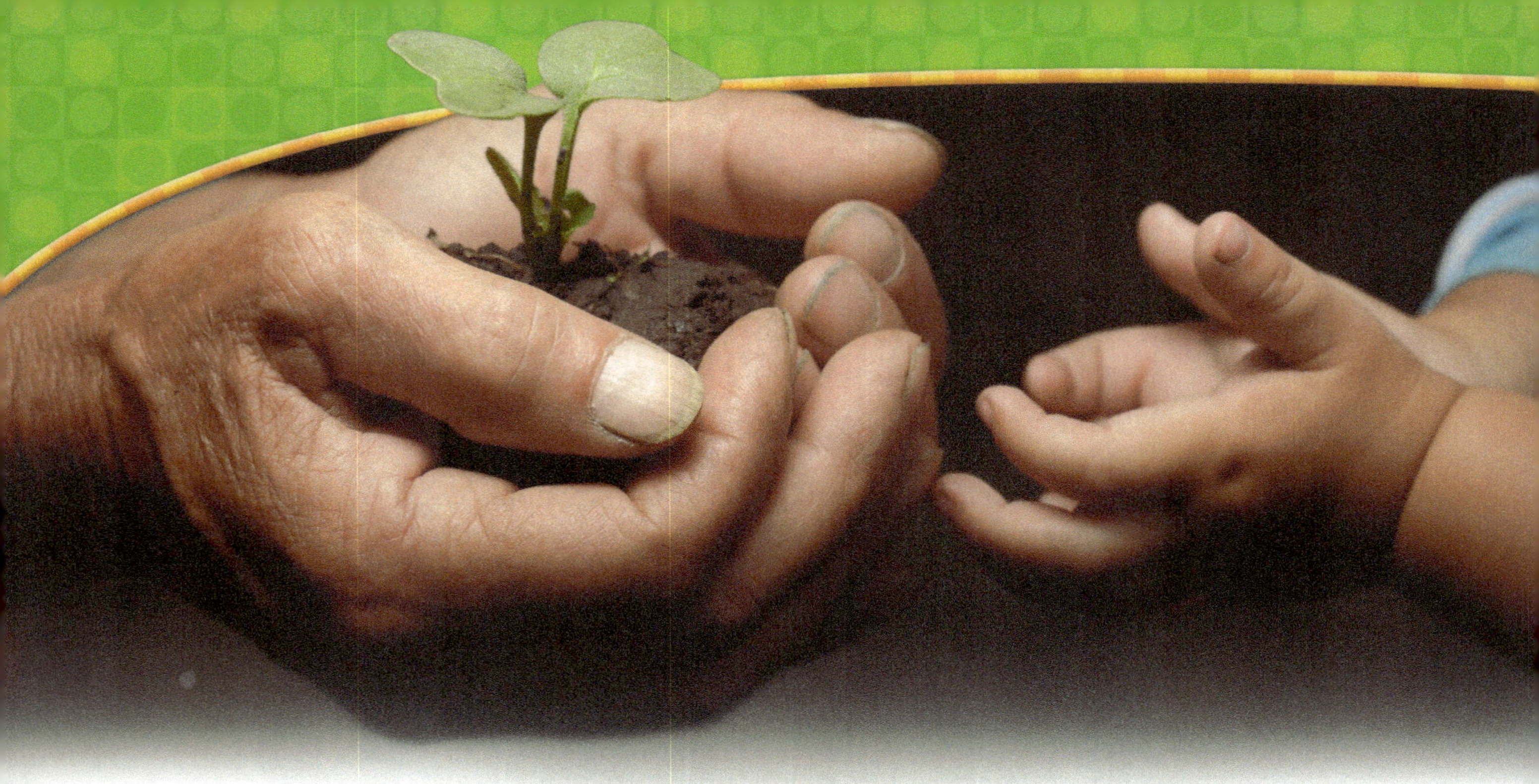

Scripture Spotlight

Compare **John 1:1–3** with **Colossians 1:16–17**. What do these verses say about where living things came from?

All the unifying themes we've discussed so far—cells and genes, biological organization, and interaction—are things found in all branches of biology. All of them have something in common. Scientists can design repeatable experiments in the present to learn more about all of them. Using the scientific process, they can study living things or the interaction of organisms in systems now and see the results of their experiments first hand. Science is very good at doing that. The next unifying theme is much different, because it deals with origins.

Unifying Theme 4: Origins

The word *origins* means "beginnings." The origin, or beginning, of all living things is another unifying theme. Scientists believe that all living organisms have a similar beginning, but they disagree on what that beginning was. Creationists believe that God created all life on Earth. Evolutionary scientists theorize that life began with a single living organism that slowly changed over time to give us the diversity we see today.

Unlike the earlier themes you studied, scientists cannot study origins with repeatable experiments. The origin of life happened a long time ago and no humans were there to see it. For this reason we say that the origin of life is part of historical science. Despite some scientists' efforts to use the scientific processes to prove how life began, we cannot re-create that exact moment. Scientists study fossils, Earth samples, and literature to develop theories and offer evidence to support their theories, but without an eyewitness, we do not have solid scientific evidence of the origin of life.

There are two basic theories about the origin of all
life. One says that life was created by God, as described in
Genesis 1–2. People have believed this for thousands of years,
and this theory is supported by scientific evidence. Much more
recently, another theory about the origin of life has become very
popular. Many scientists today believe that life came from nonliving matter and has continued to evolve to what we observe
around us today.

There are various definitions of the term *evolution*. One defines
evolution as change over time. As you may already know, *micro*
means "small." So *microevolution* refers to small changes over
time—like the size of bird beaks changing a little from generation
to generation or a flu virus changing a little from year to year.
Because *macro* means "large," *macroevolution* refers to much larger
changes—the kind and number of changes that would have been
necessary to change a microscopic cell to a fish. Similar types of
large changes would have been necessary to change a reptile to
a bird. There are many examples of microevolution in nature.
The same cannot be said for macroevolution. There is a huge
difference between microevolution and macroevolution!

Explore-a-Lab

Structured Inquiry

The only evidence we have of sharks from long ago are
the teeth they left behind. On a sheet of paper, draw
a shark tooth that is 14 cm (5.5 in.) long. This is the
length of the extinct megalodon's tooth. Next to
it, draw a modern day great white shark's tooth,
which is 6 cm (2.3 in.) long. An average great white
shark measures 6 m (20 ft) in length. Make a ratio
using the information given to find out how long a
megalodon might have been. Once everyone in class
has solved for the length of a megalodon, the class will
measure and mark the length outside in the schoolyard.

Why do you think the shark size might have changed?

All scientists, including scientists who believe in the Bible, agree that small changes happen over time. That is something scientists can learn about by using repeatable experiments today. There is nothing in the Bible that would contradict the idea of microevolution. In fact, the Bible offers evidence of microevolution. For example, after Adam and Eve sinned, plants and animals had to adapt to the changes brought on by sin. After the Flood, further adaptation was needed to survive the dramatic changes in climate and food supply.

Macroevolution, on the other hand, is something very different. The idea that life evolved randomly from nonliving matter and that humans evolved from a long line of other creatures directly contradicts the story of Creation that we read in the Bible. Genesis says that God created everything by the power of His word and that humans were lovingly formed in His image.

These organisms are very different, but they have a lot in common. They are all made of cells, each uses energy, and each stores genetic information in DNA.

What differences can you see in these organisms?

Unity and Diversity

No matter which branch of biology we study, we find many similarities in the way living things are constructed, the way they are organized at multiple levels, and the way they interact with each other and their environment. This amazing unity is evidence of a wise Creator who had a well-planned design in mind. God's plan also included an amazing variety of creatures with unique characteristics, varied environments for them to live in, and fascinating ways that they interact with each other, making them intriguing to study.

So, we see both unity and diversity in God's Creation. Even though there is an amazing amount of diversity when we look at the living things around us, the more we study them, the more we see that they have many basic things in common. This is evidence of a common Creator who loves variety!

Concept Check

Summary: What themes repeat in biology? There are many unifying themes in biology. These themes unite all living things, regardless of how diverse they are. Some themes are cells and genes, biological organization, and interaction among living things and nonliving things. The smallest atoms combine to form matter and cells, which all interact to create our biosphere. The origin of life is another unifying theme—God created all living things.

1. A zoologist is studying Seminole bats and bald eagles in the Pineywoods of East Texas. Describe some similarities, or common themes, that the zoologist will notice between the bat and eagle. What are some things that make them different?

2. Why is DNA important to living things?

3. Describe an example of homeostasis in the human body.

4. Why is the study of the origins of life considered to be historical science? What difficulty is created by historical science?

5. Living things today are different than the living things at Creation. Give an example of microevolution and explain how it is different from macroevolution.

? Essential Question

What Makes Nature Complex?

If you consider your classroom as a small ecosystem, what would be some divisions for the study of life in your classroom? Can you think of any part of your classroom that is not dependent on the whole? How is this similar to the schoolyard? The neighborhood? As we study the complexity of the world around us, we can marvel at God's design. Is there anything He created that doesn't interact with something else? What are some ways you interact with the environment you live in?

Ecological Organization **Explain**

The part of the Earth in which organisms live is called the **biosphere**. It includes the land, the water, and the atmosphere. The biosphere extends from the Earth's surface to several kilometers up into the atmosphere and down into the depths of the ocean and even deep into the rock that makes up Earth's crust.

Ecology is the scientific study of relationships between organisms and their environment. Scientists who study ecology are called **ecologists**. Ecologists study the organization within the biosphere, including biomes, ecosystems, communities, and populations. Ecology is just one of the sciences used to study the complexity of the biosphere.

The Indian leaf butterfly lives in the tropical forests of India, southern China, and Taiwan.

? If you were an ecologist, what might you want to learn about this animal?

Biomes and Ecosystems

The biosphere can be divided into biomes. A **biome** is a large area where the conditions are very similar. The plants and animal groups are adapted to the particular environment. The rain forest, for example, is a biome. Can you name some other biomes? What plants and animals live in those biomes? How are those plants and animals adapted to their environments?

The words *biome* and *ecosystem* are sometimes used interchangeably. It may be helpful to think of a biome as many similar ecosystems grouped together across the world. Ecosystems are smaller than biomes. An ecosystem is made up of the living community and the nonliving things that interact with that same community in the same area at the same time. Although ecosystems are often smaller, they can be as big as the Sahara Desert. Ecosystems can also be as small as a backyard garden. What large ecosystems have you visited? Have you visited any small ecosystems?

Communities and Populations

Biomes and ecosystems may be too large for what ecologists want to study. They may focus on something smaller, such as a community. Although smaller, communities are still full of wonder and complex interactions. Communities are different groups of organisms living and interacting with each other in the same area at the same time. Populations, or groups of individuals of the same species occupying a given area, can be as small as bacteria in the soil or as large as buffalo running across the grassland. Who are some members of your community?

The National Bison Range in Montana is home to a community consisting of many species of living things, such as bison, deer, coyotes, ground squirrels, and hawks. How do you think these populations interact with each other and with other species in the community? Do humans fit into this community?

Simple or Complex?

How do *Hydra* and *Daphnia* demonstrate the complexity of life?

Procedure

1. Use the medicine dropper to add two drops of water and *Daphnia* to the depression in the slide. Cover with a cover slip.

2. **Observe** the *Daphnia* under low power. Switch to medium power. Sketch the *Daphnia*. Label the eye, antennae, legs, eggs, and small beating heart. Write descriptions of how the *Daphnia* are interacting with each other and/or the environment. After sketching, carefully place the slide of *Daphnia* on the table.

3. Use the medicine dropper to add two drops of water and *Hydra* to another deep-well slide. Cover with a cover slip.

4. **Observe** the *Hydra* under low power. Switch to medium power. Sketch the *Hydra*. Label the mouth, tentacles, and gastrovascular cavity. If there is a bud on the *Hydra,* draw and label it. Write descriptions of how the *Hydra* are interacting with each other and/or the environment.

5. Use the medicine dropper to place *Daphnia* from the first slide onto the *Hydra* slide. **Observe** and **record** the interactions between the *Daphnia* and the *Hydra*. **Record** the time it takes for the first interaction between the two species.

6. Dispose of the specimens and clean your area as directed by your teacher.

Materials

- living *Daphnia*
- living *Hydra*
- microscope
- 2 deep-well microscope slides
- 2 cover slips
- medicine dropper

Analyze Results

Compare the structures of the two different species. **Communicate** your observations of the species' interactions with each other in a short paragraph.

Create Explanations

1. How do *Hydra* and *Daphnia* demonstrate the complexity of life?

2. *Hydra* are grouped in the same phylum as sea jellies. What did you observe about the *Hydra* that reminds you of a sea jelly?

3. Based on your observations, is one organism more complex than the other? Explain.

Complex Organisms Explain

Even the simple organisms that have been mentioned, like the *Hydra,* are actually very complex. The more scientists learn about life, the more they realize that even the "simple" cell is too complex for them to build. But there are many organisms that are even more complex and diverse. The frill neck lizard and *Titan arum* discussed below are examples of God's great ingenuity. Why do you think God made such diverse organisms to inhabit Earth?

It is difficult to think of the complexity of nature without thinking about the designs and features of the living things on Earth. The frill neck lizard, of Australia, is one of nature's fascinating designs. Most notably, this lizard has a large frill of skin around its head. The frill stays folded against the head and neck of the lizard. It is extended when the lizard is threatened. How would this frill of extended skin help the lizard when a predator threatens it? It is believed that the frill is also used in communication with other lizards. What information might they communicate? This lizard even has the ability to run on its hind legs!

When you think of a flower, what do you think of? How would you describe its fragrance or color? The scents and colors of flowers are features God designed to help plants attract pollinators. Pollinators, such as insects and bats, are required for many types of plants to reproduce. He also designed organisms with the ability to adapt to their surroundings or new circumstances. One plant with such an adaptation is the *Titan arum,* also known as the corpse lily. Imagine walking in the tropical forest of Sumatra. You smell a nauseating, rotten-meat smell. That is the odor from the *Titan arum,* a threatened species on which a single flower that can reach a height of 3 m (9.8 ft) will bloom. In addition to the terrible odor, the plant has a fleshy pink color and a temperature about the same as that of the human body. How do you think these features help the survival of the *Titan arum*?

Like all lizards, frill neck lizards need to bask in the sunlight to help them regulate body temperature.

How might the frill of the lizard help warm the lizard while it basks in the Sun?

Interactions among Species Explain

God did not create living things to exist in isolation. Ecologists have identified a variety of ways that species interact. This interaction can be in the form of *symbiosis*, or two different species living together in close proximity. Biologists identify three kinds of symbiosis: mutualism, commensalism, and parasitism.

Think about an area where you might find honeybees. What type of environment would this be? What other types of species do you think would interact with the honeybees? Some interactions of the honeybees with other organisms benefit both species. This type of symbiosis is called **mutualism**. For example, honeybees get nectar from flowers. How does this action benefit both the honeybee and the plants?

Birds nest in trees. The trees are not harmed by the birds and they are not helped either. The birds, however, benefit by gaining shelter from the tree. This is another form of symbiosis, called **commensalism**, where one species benefits and the other has no gain or loss. Can you think of other forms of commensalism?

A symbiotic interaction that benefits one species but harms the other is called **parasitism**. In parasitism, parasites harm the body of a host. Parasitic mites may invade the body of the bee. The bee is harmed because it loses nutrients and eventually dies. How does the mite benefit? **Predation** is another type of symbiosis where one species is harmed and the other benefits. In predation, one organism catches and eats another organism. How do birds and raccoons benefit when interacting with the honeybees and their hives?

Ecological Succession

What happens to an area of the forest that was burned in a forest fire? After such a major disruption, the damaged environment undergoes a natural process called ecological succession. **Ecological succession** is the gradual replacement of one type of biological community by another through a series of changes, such as when a pond fills in with grasses and becomes a meadow. There are two types of ecological succession: primary succession and secondary succession.

A beekeeper tends to honeybees housed in human-made beehives.

Which species benefit in this situation? Explain.

Primary Succession

The development of plant and animal communities in an area where living things have never existed before is called **primary succession**. What do you think must happen in a desolate area before living organisms can come back?

The first species to appear on new land are called *pioneer species*. These are usually mosses and lichens. How do you think mosses and lichens arrive on new land? These organisms grow with very few nutrients. They make soil by breaking down the rock and becoming part of the soil when they die. Gradually, grasses and other small plants, small shrubs, and trees take root. Eventually, a relatively stable community, called a **climax community**, is reached through plant succession. Different environments have different climax communities. At what point during primary succession do you think animal species appear?

Primary Succession

Primary succession begins with newly exposed rocks and progresses over many years to a climax community.

Would the organisms in a climax community be the same in all parts of North America? Why or why not?

Secondary Succession

Secondary succession is the re-growth of a community in an area that has soil and was recently inhabited by living organisms. It occurs after such disturbances as hurricanes, fires, landslides, floods, logging, farming, and grazing.

Regrowth occurs more quickly in secondary succession than new growth in primary succession. Why do you think this happens? The pioneer species of secondary succession are typically grasses and fast-growing plants. Only days after a forest fire, grasses and fast-growing wildflowers begin to sprout. How is it possible that these plants are able to regrow so quickly? Over time, longer-lived, slower-growing trees will become the dominant species to establish homeostasis as secondary succession in a climax community. What is the driving force in the changes of a climax community?

Secondary Succession

Secondary succession begins with surviving plants or seeds in existing soil and progresses over many years to a climax community.

What types of animals might inhabit the area during the early years of succession compared to those that would inhabit the climax community?

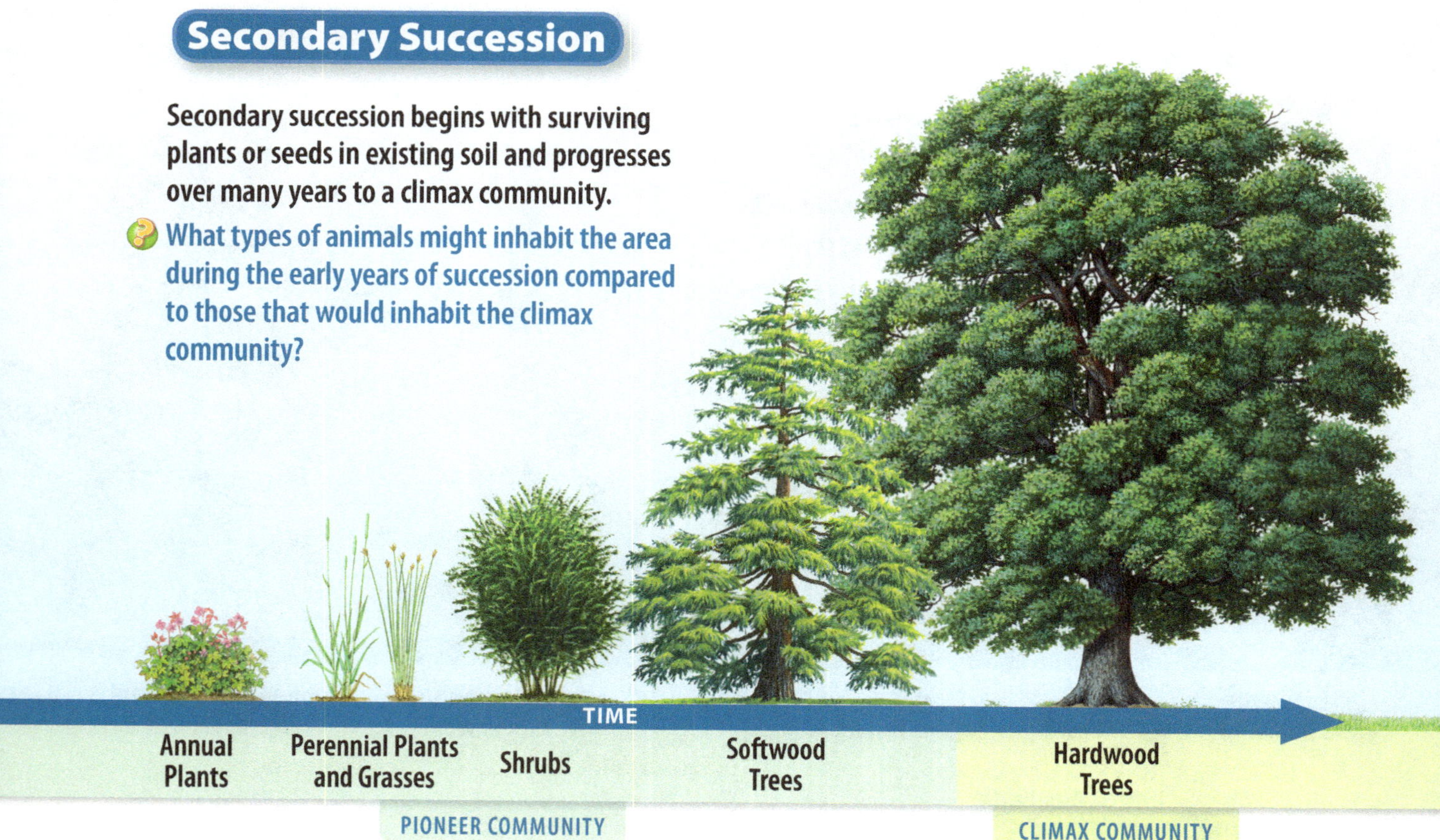

Mount St. Helens is an active volcano in Washington State. It erupted violently in 1980. It has been in a period of succession ever since. With a partner, build a model that shows what the area looked like before the eruption, right after the eruption, 1 year after the eruption, 10 years after the eruption, and today. Research and plan your model on paper first. Bring the materials you need to class. Construct your model and add labels of interesting facts you learned while researching.

 **Do you think this is an example of primary succession or secondary succession? Give reasons to support your answer.**

 **Concept Check** Assess/Reflect

Summary: What makes nature complex? The natural world shows numerous examples of complexity that give evidence of the Creator. The living organisms that make up our biosphere interact with each other on many different levels. Ecologists recognize several levels that make up the biosphere's complex organization, including biomes, ecosystems, communities, and populations. In order to live successfully, plants and animals and other living organisms have unique and complex features that allow them to thrive. When changes occur to the land, living organisms interact with each other through an ecological succession that over time brings about change, and eventually homeostasis, to an environment.

1. Describe a community and explain how organisms interact in that community.

2. The complex variety of living things is observed worldwide. How does the body structure of the frill neck lizard show evidence of God's design?

3. A remora fish can attach to the bodies of larger fish, like sharks. The remora eats scraps of food left over from the shark's meal, gets a "free ride" around the ocean, and is protected from predators. It does not harm the shark. Scientists do not think remora fish benefit the shark, either. How would you classify this specific interaction? Explain your reasoning.

4. Which do you think is a more common occurrence: primary or secondary succession? Explain.

Essential Question

What Cycles Exist in Nature?

Does the Bible describe any cycles? **Job 36:27–28** says "He draws up the drops of water, which distill as rain to the streams; the clouds pour down their moisture and abundant showers fall on mankind." You may recall learning about that cycle in previous science classes. What cycle is being described?

Cycles in Nature (Explain)

What are some cycles that you have experienced in life? How would you describe what a cycle is? Make a list of cycles that occur in nature. What do you think would happen if one of nature's cycles was disrupted?

Many processes of life follow cycles. You will explore the cycles that affect living things in this lesson.

The Carbon Cycle

Do you recall the different phases of the Moon? Every month, you can observe the change in the amount of reflection we observe in the Moon. This is an example of a **cycle**. A cycle is a series of events regularly repeated in the same order.

Another cycle involves a common element. Do you know what carbon is? What does carbon look like? How do we use carbon? How often do you think you come in contact with carbon throughout your day? Carbon, a natural element, is a basic part of life on Earth. It exists in the cells of living things. It is found in fossil fuels, which are formed from the decomposition of living organisms. Carbon can be found in rocks, soil, ocean water, and Earth's atmosphere.

Most of the carbon on Earth is stored as carbon dioxide in the air, in the bodies of living and dead organisms, and dissolved in ocean water. Carbon is exchanged between organisms and the environment in a process called the **carbon cycle**. Let us explore the carbon cycle by starting with plants.

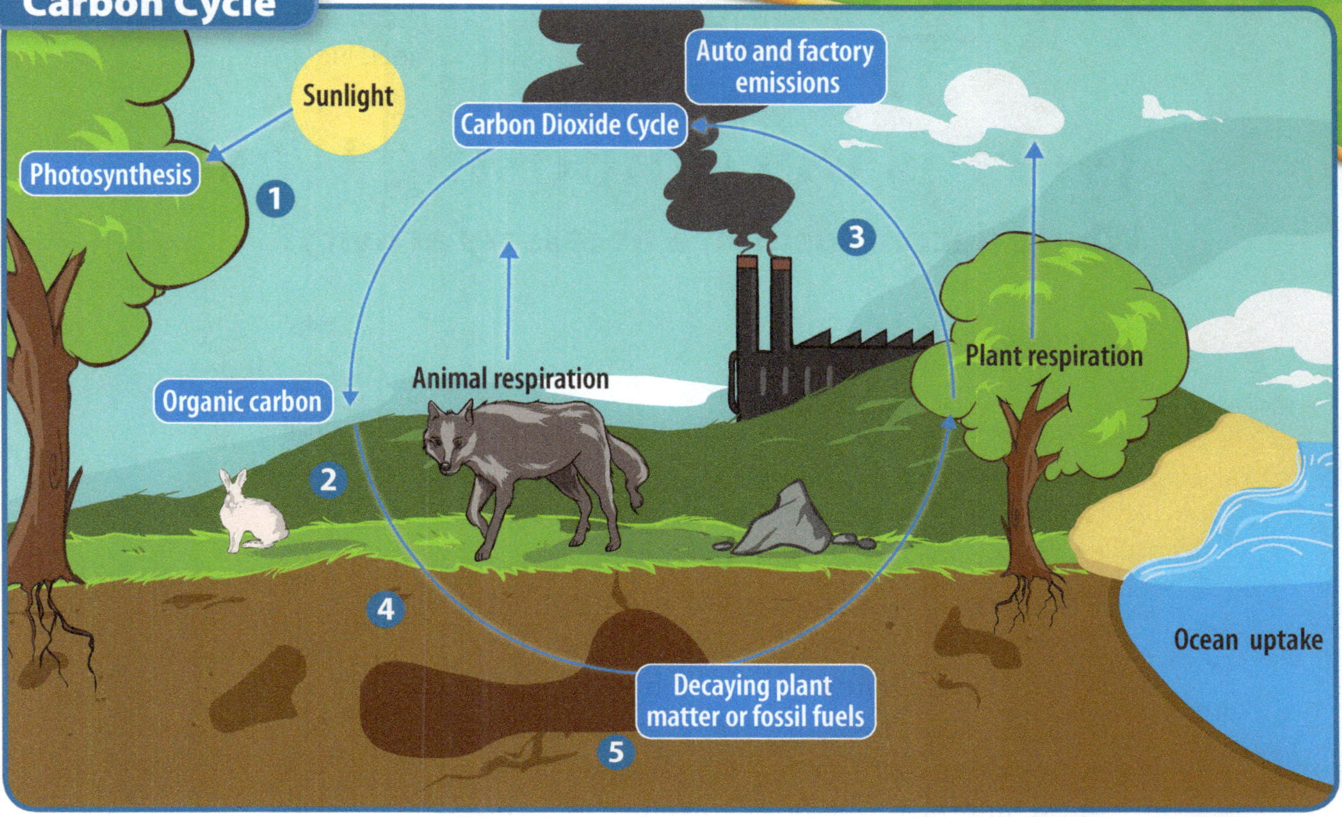

1 As land plants photosynthesize, they absorb carbon dioxide from the air and use it to make food. They release oxygen back into the environment. Ocean plants photosynthesize and absorb dissolved carbon dioxide in the water and use it to make food. The land and ocean plants' stored food contains carbon.

2 Next, carbon might travel from the plant to an animal. The animal gets the carbon by eating the plant. What else might an animal eat to get carbon?

3 During respiration, plants and animals release carbon back into the atmosphere in the form of carbon dioxide, a gas.

4 When plants and animals die and decompose, some of the carbon within their bodies is released into the soil.

5 Organisms that die and become buried deep underground for thousands of years may become fossil fuels, such as oil or coal. When humans burn fossil fuels, carbon enters Earth's atmosphere as carbon dioxide and the cycle begins again.

The carbon cycle continues at all times all over the biosphere.

Describe how two different living things contribute to the carbon cycle.

Scripture Spotlight

What nature cycles does **Genesis 8:22** describe?

Worms Decomposing Waste

Why is decomposition a necessary activity?

Procedure

1. Cut the bottle into two pieces, 9 cm (3.5 in.) from the top of the bottle. Use the awl or nail and hammer to punch ten air holes at the bottom, ten around the middle, and one in the cap.

2. Shred the paper into thin strips. Soak them in water and squeeze the excess water out of the paper. Fluff the paper and place a layer in the bottom of the bottle. Add non-protein leftovers and some potting soil. Layer paper, kitchen leftovers, and soil until it reaches the holes in the middle of the bottle. Do not pack the materials tightly.

3. Add the worms to the top of the layers. **Observe** and record descriptions of the worms' movements and how the bedding looks and smells after some time has passed.

4. Once the worms have burrowed down into the bedding, cover the bottom of the bottle with the top of the bottle. Wrap construction paper around the bottle to keep the worms in a darker area.

5. Every other day, for one month, **observe** and **record data** in a chart.

6. Once a week, dump the contents of the bottle into a bucket. Fluff the contents and add water with a spray bottle if it is dry. Layer the contents back into the bottle, adding more leftovers or yard or garden scraps.

Materials
- 2-L (68-oz) bottle (cleaned, label removed)
- nail or awl
- hammer
- scissors
- 5 full pages of newspaper
- large sheet of construction paper
- kitchen leftovers (crushed, cooked egg shells; orange peels)
- potting soil
- bucket of water
- about 75 red worms

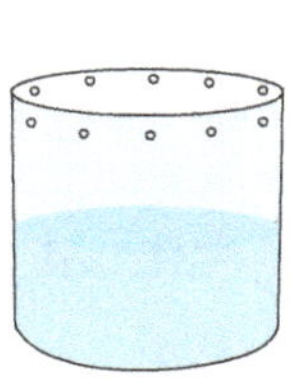

Analyze Results

Compare your original observations with the decomposed, organic matter. **Communicate** the difference between Day 1 and Day 31 in a chart.

Create Explanations

1. Why is decomposition a necessary activity?
2. Infer what would happen if the worms were removed from the process.

The Nitrogen Cycle

Like carbon, nitrogen is essential to living things. Nitrogen gas makes up about 78% of Earth's atmosphere. However, most organisms cannot use nitrogen in its gaseous form. Instead, they need nitrogen compounds. These usable compounds are formed when the nitrogen gas is changed during a process called **nitrogen fixation**. The process occurs in different ways. Nitrogen-fixing bacteria in soil or water do most of the nitrogen fixation on Earth. Some of these bacteria grow on the roots of legumes. Lightning also fixes nitrogen. The heat from lightning helps form nitrogen compounds in the air. These compounds are washed into the soil by rain. Humans can also fix nitrogen into usable compounds. These compounds are included in fertilizers.

Nitrogen is similar to carbon in another way. It is cycled between the environment and living things in a process called the **nitrogen cycle**. To explore the nitrogen cycle further, let's start with the plants.

Inside each nodule on these soybean plants' roots are thousands of nitrogen-fixing bacteria.

How do the soybean plant and the bacteria benefit each other?

1. Plants get the nitrogen they need through soil or water. Which structures of the plants do you think take in the nitrogen?

2. Animals get nitrogen from eating plants or other animals that eat plants.

3. When organisms die, decomposers change the nitrogen in the proteins of the dead organisms' bodies into ammonia compounds. Ammonia has nitrogen in it that remains in the soil or water.

4. Other bacteria change ammonia compounds into nitrites and nitrates. Then, **denitrifying bacteria** change the nitrites and nitrates to nitrogen gas. The gas returns to the atmosphere.

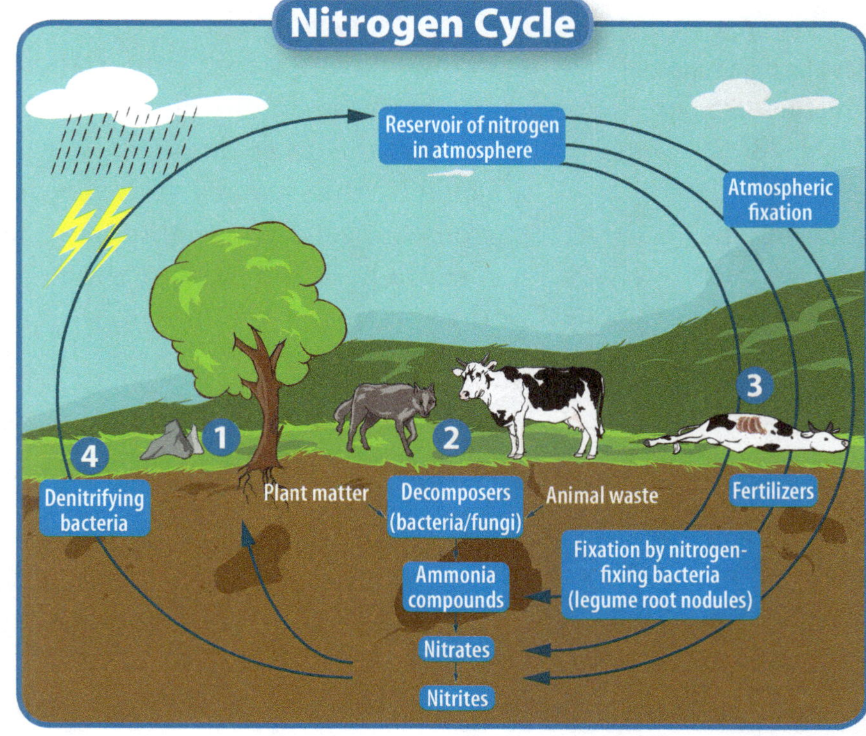

Faith Connection

Our year, month, and day come from natural occurrences related to the Sun and Moon. Where does our week come from? Compare **Genesis 2:2–3** with **Exodus 20:8–11**. See also **Revelation 14:6–7**.

Scripture Spotlight

In **Job 5:8–9**, Job says, "I would seek God, and I would place my cause before God Who does great and unsearchable things, wonders without number." What does Job say about God and rain in **5:10**, **28:26**, and **36:27–28**?

Water Cycle `Explain`

How do you use water at home and at school? Have you noticed other living things using water? If so, in what ways do they use it? Water is essential to life. It makes up about 75% of the human body. It is in the air we breathe and in the soil we work with and live on. Water is constantly recycled over and over again. Do you recall what happens in the water cycle? If you could follow a drop of water in a cloud, where might it go next?

The water cycle has an impact on life greater than any other factor in the biosphere. Since water is necessary for cells to maintain homeostasis, the availability of water is essential to all life forms on Earth. When there is a drought, a lack of rain, both plants and animals can be affected. How the water cycle works can determine how well or how poorly living things survive. Ecosystems and populations can suffer or thrive. What kind of impact can drought have on an ecosystem?

The Seasonal Cycle

Think about the seasons in your area. How are you affected by the seasons? Does it get very cold in the winter or extremely hot in the summer? Do you experience four distinct seasons? We can describe the pattern of winter, spring, summer, and fall as a *seasonal cycle*. The seasons affect plants and animals. For example, during a cold winter, there is less food for geese. What must the geese do during the winter because of a lack of food? What will the geese do once the weather warms? Plants are affected by the changes in the amount of sunlight throughout the seasons, especially in higher latitudes. How do plants respond to less sunlight in the fall and winter? What happens to these plants in the spring and summer?

Seasonal cycles affect living things as well as other cycles of nature.

How does the winter season affect how water evaporates from plants?

How does the amount of sunlight affect the rate of transpiration in plants?

Use a graduated cylinder to water two houseplants of the same size and species with the same amount of water. Place a clear plastic bag over each plant, making sure all leaves are covered. Use string to tie the bag around the main stem of the plants. Be sure that there is extra room in the bag around the stem to collect any water that may be transpired. Place one plant in the sunlight and one in a dark cabinet or closet. Check the bags in the morning. What do you observe? Collect and measure the water from each bag. Is there a difference between the amounts of water transpired by each plant? Get the data from all the groups in your class and calculate the average amounts of water.

Science Journal

Check out your *Science Journal* for a Structured Inquiry that explores the predator-prey cycle.

Extend

The Reproductive Cycle

In **Genesis 1:22** and **28**, God said, "Be fruitful and increase in number." Who or what was He speaking to? In order for a species to survive, it must reproduce. The process by which new individuals are produced is called the *reproductive cycle*. Different species have different reproductive cycles, but the individuals within a species go through the same cycle each time they reproduce. The reproductive cycle beings with the combination of a male and a female cell in a process called fertilization. The fertilized cell develops through various stages until a new individual is formed.

A sperm cell must fertilize an egg cell in order for an organism to reproduce. If the reproductive cycle stops, the species will become extinct.

Scripture Spotlight

Read **Isaiah 11:6** and **65:25**. How is the predator-prey cycle we see today different than the one described in Isaiah?

Check for Understanding

Look back to the *Structured Inquiry*. How does the function of decomposers fit in with the work of these other cycles?

The Predator-Prey Cycle

What are some predators in your area? What type of prey do they hunt? Are there ever times when you notice these predators and prey increasing or decreasing in number? The increases and decreases in the population sizes of a predator and its prey are called the predator-prey cycle. For example, foxes are predators that often prey on rabbits. What do you think would happen to the fox population if the rabbit population increases? What would happen to the environment?

Is there a point when there would be too many of a certain type of species in an environment? The **carrying capacity** of an environment is the largest number of one species that an environment can support without harmful effects. For example, if marsh rabbits reach their carrying capacity and eat all the vegetation, there won't be enough food for them to survive in that area. The lack of food would be a limiting factor. **Limiting factors** are things that keep a population from growing any larger. Limiting factors affect populations and include disease, parasites, floods, and predation. Can you think of other limiting factors that would keep a population from increasing in size?

This red fox has caught a squirrel.

If disease killed all the foxes in an area, how might that impact the population of squirrels or the population of grasses?

While your teacher scatters note cards around the play area, select two classmates to play the role of predators. Everyone else is prey. The predators stand in the middle of the area while the prey line up on one side of the space. At a given signal, the prey try to cross the area, picking up note cards, which represent their food, as they go. Anyone tagged by a predator or who does not get a note card does not survive. If weather permits, consider taking this activity outside.

 **What limiting factors did you experience?**

 **Concept Check** Assess/Reflect

Summary: What cycles exist in nature? Many different cycles exist in nature. The cycles are important for keeping balance in the biosphere. The carbon cycle and nitrogen cycle circulate much needed carbon and nitrogen between the environment and living things. The seasonal cycle and the water cycle affect the well-being of many organisms. The predator-prey cycle and the reproductive cycle maintain a balance between living things and their environment.

1. Explain the importance of decomposers to the nitrogen cycle. In what ways are the nitrogen cycle and carbon cycle similar and different?

2. Explain how the predator-prey cycle is related to the reproductive cycle. Use a specific predator-prey relationship in your response.

3. A certain plant lives in an area that has four distinct seasons. During which season would the plant have a faster transpiration rate? Explain.

4. How do the cycles of life help you to understand God's creative power?

Bears in Deep Sleep

Do you know what makes true hibernation different from deep sleep? When mammals such as ground squirrels and bats hibernate, their body temperatures drop to below water's freezing point for a long period. Their rates of *metabolism*, the chemical reactions needed for the cells and the organism to remain alive, become very low. In fact, their metabolism rates drop to about 98% less than when they are active.

Some animals go into a short-term period of sleep, which is called torpor—an animal's body cools off, but not as much as that of a hibernator. The animal can wake up and go back to sleep. Black bears, raccoons, and skunks are some animals that go into torpor.

Scientists have been able to study the metabolism of small mammals. However, that information has not helped them understand the sleeping patterns of larger animals such as bears.

A team led by Oivind Toien of the University of Alaska—Fairbanks decided to observe bears while they sleep. The team members convinced authorities to allow them to study the sleeping habits of five bears. First, they implanted sensors in the bears. Then, they provided dens for the bears where the scientists could use infrared cameras to monitor them. These cameras can record images in total darkness.

The bears were observed sleeping in a curled position, changing their position, standing, cleaning themselves, and going back to sleep. The internal monitors told the scientists that the bears' body temperatures dropped only slightly and never went as low as those of smaller hibernating animals. Their metabolism never became as low as the metabolism of smaller mammals.

Scientists hope that these bear studies may someday help humans who are seriously ill. For example, perhaps a person waiting for a transplant could be put into a state of deep sleep until the needed organ becomes available.

Concept Check

1. Why did scientists need to use infrared cameras to study sleeping bears?
2. Suppose scientists were hoping to find a cure for illness, such as cancer, AIDS, or Alzheimer's disease. How might knowledge of bears and their deep sleeping patterns help the scientists?

Ethologist

An ethologist studies animal behavior in a natural setting. This type of scientist wants to know how animals behave and why they behave the way they do. Ethologists try to understand how genetics, which is the study of heredity, and environmental factors interact in the development and expression of behaviors. Some ethologists spend months or years in the wild observing animals. They may record their observations through writing, photographs, and videos. Many use an ethogram, which is a table that describes behavior.

Ethologists collect data to discover
- how certain behaviors help an animal survive
- how the animal's environment and learning affect certain behaviors
- how behaviors change as the animal grows
- how animals that are similar to each other act in similar or different ways.

At one time, ethologists attempted to study behaviors they thought were instinctual rather than learned. Today, many ethologists believe that although certain behaviors are innate, experience and learning are the main reasons for other behaviors. For example, the suckling of the mother by newborn mammals is an instinct. The using of tools to gain food is learned. Seagulls will often drop shellfish on rocks to get them open or on the road so cars will break them open.

To become an ethologist, a person must first attend college and earn a bachelor's degree in zoology, animal science, or other related field. After that, the person will need to earn a graduate degree or a doctorate in animal science, animal biology, or ethology. Ethologists need to be skilled in math, science, biology, writing, and research.

Concept Check

1. What does an ethologist do?
2. Why is recording observations so important to an ethologist?

Study Guide

Lesson 1

1. Four themes recur throughout biology: cells and genes, biological organization, interaction, and origins.

2. Homeostasis is an important response of living things to change.

3. Living things are made of cells, which contain DNA and use energy.

4. There are two main theories of the origin of life.

Lesson 2

1. All of Earth's organisms can be organized into biomes, ecosystems, communities, and populations.

2. Living things were designed by the Creator with complex features necessary for survival.

3. Living things interact with each other. Sometimes these interactions harm a species. Sometimes the interactions are beneficial.

4. Ecological communities can change through the process of ecological succession.

Lesson 3

1. Organisms and the environment exchange carbon, a basic part of life, in the carbon cycle.

2. Organisms and the environment exchange nitrogen, which is needed by all living things, in the nitrogen cycle.

3. The water cycle impacts all life forms on Earth. Availability of water can ultimately determine the survival of an organism.

4. The seasonal cycle can affect other cycles in nature.

5. The reproductive cycle produces new individuals; the predator-prey cycle keeps populations in balance.

Show What You Know

Visualize It Complete the following concept map.

An Overview of Life

1. Themes in Biology

____
____
____
____
____
____

2. Complexities in Nature

____
____
____
____
____
____

3. Cycles in Nature

____
____
____
____
____
____

Vocabulary Check

Fill in the blank with the correct vocabulary word.

4. Parents give their offspring hereditary information through __________.
5. The internal control that keeps an organism balanced is __________.
6. A series of events that regularly repeat in the same order is called a(n) __________.
7. All parts of Earth where organisms live make up the __________.
8. The gradual replacement of one community by another is called __________.
9. A stable community of species that exist in balance is called a(n) __________.
10. The scientific study of relationships between living things and their environment is __________.

Multiple Choice

Choose the best answer.

11. What is the study of the interactions of living things and how they rely on each other?
    A. interdependence
    B. emergent property
    C. homeostasis
    D. ecology
12. When a dense forest is leveled by a tornado, what regrowth pattern takes place?
    A. primary succession
    B. secondary succession
    C. climax community
    D. carbon cycle
13. Which term describes the process in which bacteria make nitrogen gas accessible?
    A. decomposition
    B. seasonal cycle
    C. nitrogen cycle
    D. nitrogen fixation
14. Which term describes the largest number of individuals of one species that an environment can support?
    A. homeostasis
    B. carrying capacity
    C. limiting factor
    D. primary succession

Check Point

Answer the following questions.

15. Old farmland is left untended and abandoned. **Draw a conclusion** as to what will naturally happen to the abandoned farmland.
16. A backyard is home to tomato plants, oak trees, grass, squirrels, blue jays, and a variety of insects. Identify and describe three different cycles that exist in this backyard ecosystem.
17. **Compare** mutualism, parasitism, predation, and commensalism. Give examples of each.
18. Complete the chart about biological organization.

Levels of Biological Organization	
A	
B	
C	
D	
E	
F	
G	
H	
I	

Chapter 2

How Cells Are Alike and Different

Scripture Spotlight

God made the building blocks of life when He spoke life into existence. Learn more about growing your faith in the following passages in this chapter.

Colossians 1:16 (p. 50)

Philippians 4:8 (p. 53)

Ephesians 4:11 (p. 60)

1 Corinthians 12:7–10, 28 (p. 63)

Acts 1:8 (p. 65)

Luke 11:13 (p. 65)

Genesis 1:30 (p. 73)

Even the smallest living things are filled with a wonder of activity. You can observe several of these onion cells in the various stages of cell division.

The Big Idea

Cells are amazing structures and provide strong evidence of the Creator's design. They operate efficiently while carrying out the activities necessary for life.

How do cells demonstrate God's wisdom and design?

Inquiry Kick-Off Engage

You know that most cells are too small to see. What are some things around your classroom that could be used to magnify? In your *Science Journal*, you will want to check out one example that you can use to make and record your observations.

Science Journal

Essential Question
What Are Cells?

How were cells discovered? Cells are so small that their secrets remained hidden until the invention of the microscope about 400 years ago. The English scientist Robert Hooke used a microscope to look at a slice of cork. Look at the picture at the bottom of this page. If you observed what Hooke saw, what conclusions would you make?

Modern Cell Theory   Explain

Hooke observed that cork contained tiny holes. Because the matter in cork is dead, its cells are empty. They look like little boxes filled with air. Hooke named the holes "cells."

Dutch lens maker Anton van Leeuwenhoek was the first person to study living cells. He looked at pond water with his early microscope, and saw it was filled with tiny living things, which he called "animalcules." Leeuwenhoek also observed red blood cells and plant cells. What do you think he expected to see?

Why do you think so many scientists thought there were smaller living things to see? How would a microscope help them with their theories? Gradually, certain patterns about life became clear. Credit for developing the cell theory belongs to three German scientists: botanist Matthias Schleiden, zoologist Theodor Schwann, and pathologist Rudolf Virchow.

Schleiden identified that all plants are made up of cells. Schwann determined that all animals are made up of cells. About 20 years later, Virchow expanded on their observations. He stated that cells come from preexisting cells. French scientist Louis Pasteur then added the results of his research. Through experimentation, he confirmed that living things do come from other living things. You will study Pasteur's research later in this Lesson.

Osmosis Egg

How does water move across a membrane?

Procedure

1. **Measure** and **record** the mass and length of each egg. Place each egg in a cup. Cover the eggs with vinegar and soak for 24 hours.

2. Use tongs to carefully remove the eggs. Rinse the eggs. Place the eggs on paper towels and gently dry them. **Measure** and **record** the mass and length of each egg. Clean and dry the cups. Label the cups 1–4. Place the eggs back into the cups.

3. Cover the egg in cup 1 with water, the egg in cup 2 with water and a drop of food coloring, the egg in cup 3 with saltwater, and the egg in cup 4 with corn syrup. Cover each cup with plastic wrap. Let the eggs soak overnight.

4. Use tongs to carefully remove the eggs. Carefully rinse the eggs with water. Place them on paper towels and pat them dry. **Measure** and **record** the mass and length of each egg.

5. Clean and dry the cups.

6. Use tongs to carefully place the eggs back into the cups. Use the toothpick to carefully pop each of the egg membranes. **Record** what you **observe**.

Materials
- 4 large eggs
- 4 clear plastic cups
- white vinegar
- water
- dark food coloring
- saltwater
- corn syrup
- toothpick
- balance
- tongs
- paper towels
- plastic wrap

Analyze Results

Using your measurements, interpret your observations and draw a conclusion about the function of a membrane.

Create Explanations

1. How does water move across a membrane?

2. In which substance did the mass of the eggs increase? Decrease? Remain the same?

3. Explain what caused an increase in the mass of the egg.

4. If the mass of one of the eggs increased, what would need to be done to have the egg return to its original mass?

Create a time line of historical events leading to the development of the cell theory. Include the contributions of Hooke, Leeuwenhoek, Schleiden, Schwann, Virchow, and Pasteur. Include an appropriate title and illustrations.

What parts of the cell theory did these scientists contribute?

Scripture Spotlight

Read **Colossians 1:16**. Which part of this verse makes you think of tiny living things that can only be seen through a microscope?

During the Middle Ages, it was commonly believed that living organisms could come from nonliving things. If you left a pile of rags in a corner, mice would appear. Maggots would appear on meat. Tiny living organisms would appear in broth. Many scientists believed in this idea of spontaneous generation.

However, Pasteur felt another explanation was in order. He heated meat broth in special flasks with long necks bent into the shape of an S. He designed the necks this way so that air could enter the flask but living things would be trapped in the bend of the neck and never reach the broth. Then, he left the flasks alone and waited to see what would happen.

Months later, no living things had grown in the broth. When the flasks were broken, living organisms did begin to grow in the broth. Pasteur concluded that when living organisms grew in the broth it was because they had gotten in the broth from the outside. They did not form from within the broth itself. Why was this an important discovery?

Principle of Cell Theory

The discoveries of these scientists led to the development of the cell theory. This theory states:

- All living organisms are composed of one or more cells.
- The cell is the basic unit of structure in all organisms.
- All cells come from preexisting cells.

Since the development of the cell theory, scientists have learned much more about cells, their complexities, and their functions. Improved technologies help scientists to continue their study of the cell and its connection to life. Today, we understand much more about the cell and its workings. What else have we learned about the cell since the work of these early scientists?

Portrait of Louis Pasteur

How did Pasteur's work help the cell theory?

Origin of the Cell

As you know, many scientists worked hard to discover the basic unit of life and how living organisms come to be. But there is still one question that lingers. Where did the very first cell come from?

The Bible describes the creation of all life by an all-wise and all-powerful God. For thousands of years of Earth's history, most people—even those with differences of opinion about who God was or how many different gods there were—believed that life came from some intelligent, higher being. Christians, as well as Jews and Muslims, believed that the Creator was the God of the Bible. The Bible tells us the story of God's Creation. Many people today still accept the Bible story as a historical account of the origin of life.

In the nineteenth century, about the time that scientists were working to disprove the theory of spontaneous generation, Darwin's theory of evolution was gaining strength. Although his theory focused specifically on the origin of different species, scientists who did not believe God created life took the concept of evolution further. They began to ask where the very first life form came from. In 1924, a scientist named Alexander Oparin suggested that living cells arose gradually from nonliving matter millions of years ago. This idea is called **abiogenesis**. How is abiogenesis like spontaneous generation?

Through the best nineteenth-century microscopes, a cell looked like a simple blob. To think that such a simple blob could have come from something nonliving was not much of a stretch considering people had believed in spontaneous generation for years. Now, increasingly powerful microscopes, along with other tools, continue to unveil the amazing intricacies of the cell and its even tinier parts. Scientists now know that what once looked like a simple blob is actually a highly sophisticated and intricate system of parts that work together to sustain life.

Scientists in the nineteenth century may have observed cells similar to these.

How might these observations have contributed to the belief that life came from an intelligent, higher being?

Cells

You have learned that cells are the basic building blocks of life. Many of the organisms that Leeunwenhock and Pasteur examined under their microscopes were single-celled creatures. Any life form that consists of just one cell is **unicellular**. Unicellular creatures can live almost anywhere, even on or inside our bodies. Examples of these diverse creatures include bacteria, yeast, algae, and protozoa. Bacteria are among the smallest of the unicellular creatures. They can be as small as 1 to 10 µm. The unit µm stands for **micrometer**. For comparison, a human hair is about 75 µm wide! Eggs laid by birds are unicellular, so the largest single cell is an ostrich egg. If a bird egg is unicellular, how does it become a multicellular living thing?

Multicellular organisms contain more than one cell. In fact, they contain millions or even billions of cells. The human body is thought to contain 10 trillion cells. Other examples of multicellular organisms include mosses, flowering plants, and animals. Animal and plant cells are 10 to 100 µm. That is 10 times as big as most bacteria, but even at that size there's room for five of them on the period at the end of this sentence.

Explore-a-Lab

Structured Inquiry

How do materials move in and out of cells?

Predict what will happen to a clear glass of water when food coloring is added to it. Work in pairs. Fill a glass with water. Add three drops of food coloring to the water in the middle of the glass. Let the glass of water stand for 15 minutes. Observe what happens to the water during this time. Draw what you see after 5, 10, and 15 minutes.

Math in Science

Convert millimeters into micrometers. If there are 1000 micrometers per millimeter, how many micrometers are there in 30 millimeters? Extend your use of micrometers by using the same formula to measure others items in micrometers.

Cell Parts

Just like your body is made up of many parts that work together to help you function, cells are also made up of many parts that form a whole. Cells are an organized collection of material protected by a thin "skin" or **cell membrane**. The cell membrane is formed by two layers of lipid molecules with special proteins scattered throughout.

Things needed to sustain life are constantly passing in and out of cells. The cell membrane is **selectively permeable**. It controls what materials are allowed in and out of the cell. Water and gases can easily pass in and out of the cell through the cell membrane. Larger molecules do not enter and exit the cell as easily. The proteins in the cell membrane help these larger molecules pass through.

Located throughout the inside of the cell is a fluid known as **cytoplasm**. At one time, scientists believed the cytoplasm was just a jellylike fluid. Now, because of improved microscopes, they know cytoplasm is a very complex fluid. It contains many tiny structures that make, package, store, and transport everything the cell needs to carry on the functions of life. One of these structures contains the information the cell needs to repair itself and reproduce. What is this structure?

Cell membranes and cytoplasm are found in both plant and animal cells. Plant cells also contain a **cell wall**. This is what Hooke observed in the cork cells. In addition to the cell membrane, this cell wall provides additional structure to help plants hold their shape.

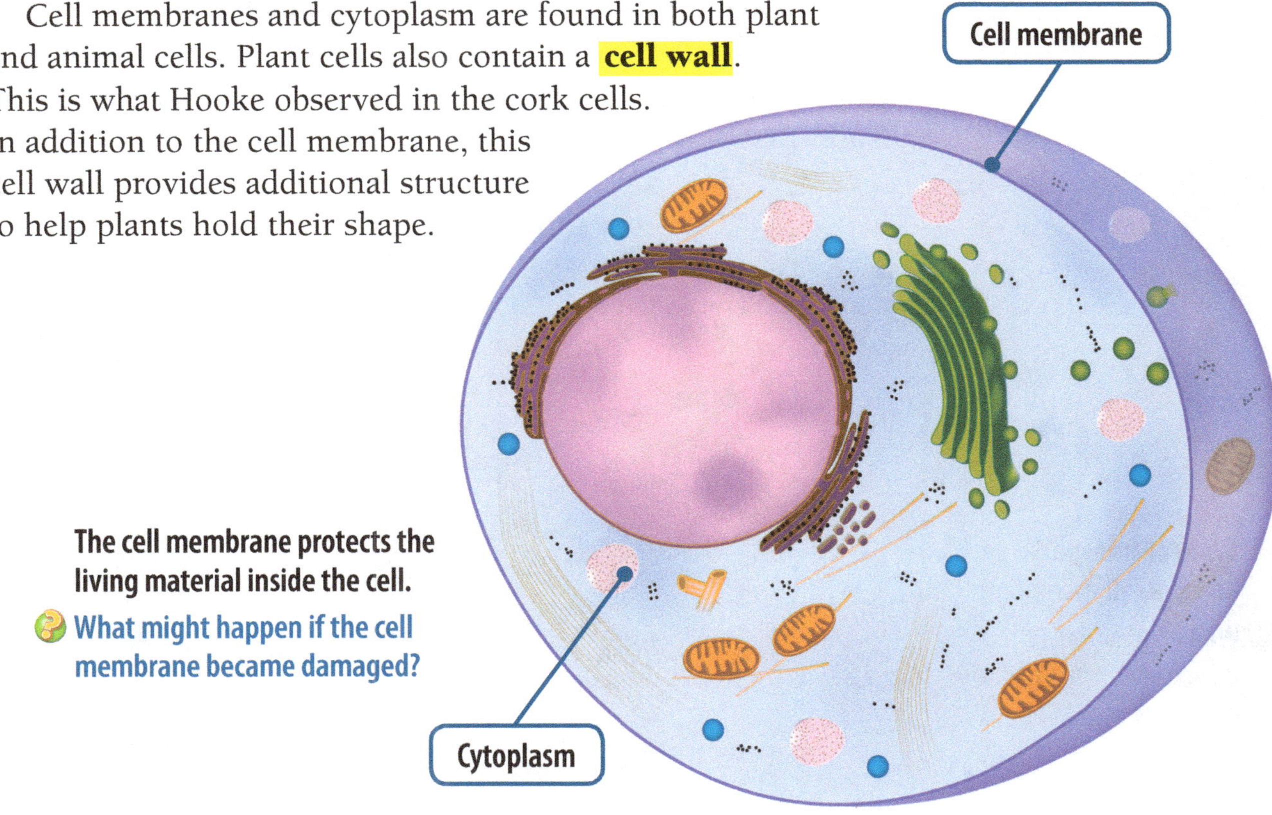

The cell membrane protects the living material inside the cell.

What might happen if the cell membrane became damaged?

Check out your *Science Journal* for a Structured Inquiry that models the function of a cell membrane.

Extend

Cell Transport **Explain**

God designed the cell with some incredibly complex transport systems. In the Explore-a-Lab, you saw the food coloring go from being all bunched together to being spread evenly throughout the water. This movement of molecules from an area of high concentration to an area of less concentration is called **diffusion**. If you have been in a kitchen when strong-smelling foods are cooking, you have observed diffusion at work. What happens when onions are cooking in the kitchen? When a pot on the stove is uncovered, molecules escape and spread out until they are evenly dispersed in the air.

Diffusion of molecules also happens in cells. Proteins embedded in the cell membrane help move molecules in and out of the cell in a process called **facilitated diffusion**. There are also tiny holes in the cell membrane that allow some small molecules, like oxygen and carbon dioxide, to enter and exit. This process happens with water molecules by osmosis. **Osmosis** refers to the diffusion of water molecules through a membrane. What happens not long after you have watered a wilted plant? Why?

Diffusion, facilitated diffusion, and osmosis occur without any use of energy by the cell. This is called **passive transport**. Sometimes energy is required in order to move molecules in and out of a cell. The molecules can move from high to low concentration, or low to high concentration. The kind of cell transport that uses energy is called **active transport**.

Types of transport across a cell membrane

🌐 What determines the direction in which water moves by osmosis?

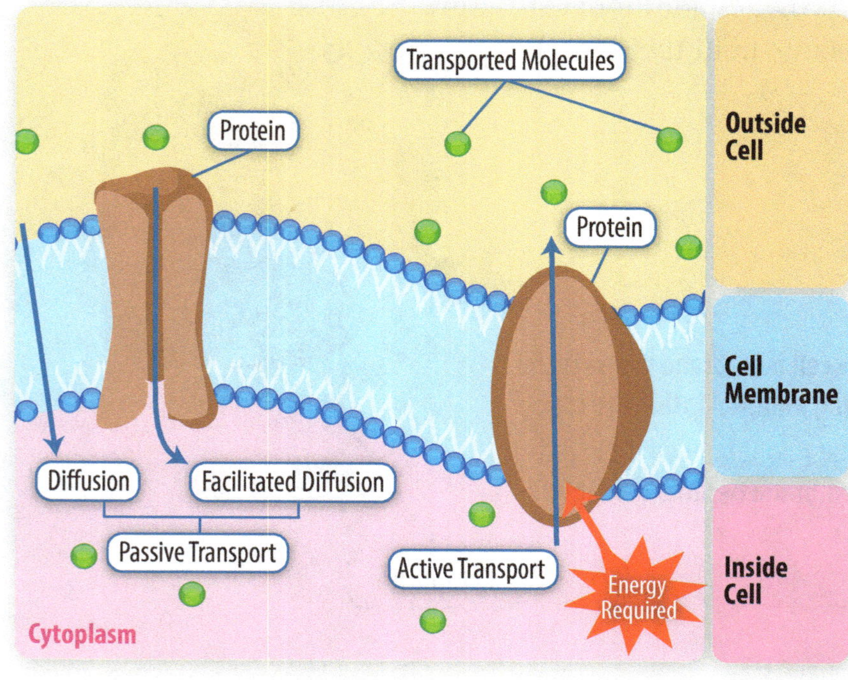

Two specific types of active transport processes are *endocytosis* and *exocytosis*. If energy is used to bring materials into the cell, it is called *endocytosis*. If energy is used to send materials out of the cell, it is called *exocytosis*. The following chart further describes ways that materials enter and exit the cell.

Cell Transport Mechanisms	
Cell Mechanism	**Function**
Diffusion	Type of passive transport requiring no energy in which a substance moves through a membrane due to a difference in concentration.
Osmosis	Type of passive transport requiring no energy in which water molecules pass through the cell membrane from an area of higher concentration to an area of lower concentration to reach equilibrium.
Facilitated diffusion	Type of passive transport requiring no energy in which proteins aid in the passing of a substance through the cell membrane.
Active transport	Type of cell transport mechanism that requires energy to function.
Endocytosis	Type of active transport requiring energy in which substances move from outside the cell to inside the cell.
Exocytosis	Type of active transport requiring energy in which substances move from inside the cell to outside the cell.

Think about the complex activities of a cell that you just learned about. Do these processes provide evidence that they were designed or that they developed by chance? Continue thinking about this in the next lessons.

Summary: What are cells? The cell theory was developed through the work of several scientists. It states that all living organisms are composed of one or more cells, the cell is the basic structure of organisms, and all cells come from other cells. The cell has parts that perform all the functions needed to keep it alive. The cell membrane controls what enters and exits the cell by way of passive transport and active transport.

1. Why was the cell theory an important discovery?

2. How does the cell membrane's structure help it perform its function?

3. How are diffusion and osmosis similar and different?

4. Use what you know about osmosis and diffusion to hypothesize what might happen if you watered a freshwater plant with saltwater.

Objectives

- Distinguish between prokaryotic and eukaryotic cells.
- Identify how the ratio of surface area to volume affects the size of cells.
- Describe how specialized cells differ in form and function.

Vocabulary

prokaryotic cell
eukaryotic cell
organelle

? Essential Question

How Are Cells Classified?

Recall that one way scientists categorize living organisms is by whether they are made of one cell or many cells. What are the words used to describe those two categories? Just like living organisms, cells can also be classified. What have you learned about the cell so far? What characteristic would you use to separate cells into two groups? One of the ways scientists divide cells into groups is by whether or not they contain a nucleus. Remember, the nucleus controls all the activities of the cell, similar to the way your brain controls your body's activities. Because of this, it is sometimes referred to as the brain or control tower of the cell.

Prokaryotes and Eukaryotes Explain

A cell that does not have a nucleus is called a **prokaryotic cell** or simply a prokaryote. Even though they have no nucleus, prokaryotes can still live and function because they have DNA. Examples of prokaryotes are bacteria and blue-green algae.

A cell that does have a nucleus is called a **eukaryotic cell** or simply a eukaryote. Examples of eukaryotes are plants, animals, fungi, and protists. The nucleus of a eukaryote is surrounded by a double membrane called the *nuclear membrane*.

Tiny holes in this membrane, called *pores,* allow molecules to pass back and forth between the nucleus and the cytoplasm. How do you think the nuclear membrane controls the movement of materials in and out of the nucleus?

Prokaryotic cells lack some of the small, specialized, organ-like structures, or **organelles**, that eukaryotic cells have. Their DNA is not contained in a nucleus like it is in eukaryotes. Instead, the DNA of a prokaryotic cell is clumped up and free-floating inside the cell. Because they do not have all of the cell organelles of a eukaryote, prokaryotic cells are small in size. Another interesting thing about prokaryotes is they do not go through mitosis or meiosis, like a eukaryotic cell. How do you think a prokaryotic cell duplicates or reproduces? What are some other characteristics of prokaryotic and eukaryotic cells?

You will often hear prokaryotes described as "simple" while eukaryotes are described as "complex." Prokaryotes are structurally different in a way that could be considered simpler. They are a little like a studio apartment compared to a house. A studio apartment consists of a single room that is set up with specific sections to eat and sleep. Eukaryotes are more like a house with separate rooms for different activities. All the functions of daily life, like eating and sleeping, take place in both the studio apartment and the house. In the same way, most of the same complex processes occur in both prokaryotic and eukaryotic cells. While the structure of the prokaryote might be considered simpler, the processes happening there are complex. Compare a computer and a mobile device. The mobile device is smaller, but it can do most of the complex processes of the computer. Calling a prokaryote simple would be like calling the mobile device simple.

Whether it is a prokaryotic or eukaryotic cell, complex cell parts and processes abound. Scientists continue to discover that even things in nature that seem simple are anything but simple. The complexities and designs are strong evidence of a wise Creator.

This is a single-celled organism called an amoeba.

Is an amoeba a prokaryote or a eukaryote? How do you know?

Characteristic	Prokaryotes	Eukaryotes
Nucleus	No	Yes
Cell membrane	Yes	Yes
Nuclear membrane	No	Yes
Cell wall	Some	Some
Perform life processes	Yes	Yes
Organelles	No	Yes

Record your work for this inquiry. Your teacher may also assign the related Guided Inquiry.

Surface Area to Volume

How do surface area and volume limit the size of cells?

Materials
- rice
- scissors
- construction paper
- tape
- metric ruler
- balance

Procedure

1. Look at the pattern in your *Science Journal*. Construct three cell models similar to the pattern with different dimensions. The dimensions of a side should be doubled each time. The sides should be 2 cm (0.8 in.), 4 cm (1.6 in.), and 8 cm (3.2 in.). Fold and tape each model into cubes with the tabs to the inside. **Record** the dimensions in your *Science Journal*.

2. **Use numbers** to calculate the surface area for each cell model. **Record** the surface area in your *Science Journal*. Use this formula:
   surface area = (length × width) × 6 sides

3. **Use numbers** to calculate the volume for each cell model. **Record** the volume in your *Science Journal*. Use this formula:
   volume = length × width × height

4. **Use numbers** to calculate the surface-area-to-volume ratio for each cell model. **Record** the ratio values in your *Science Journal*. Use this formula:
   $$\text{ratio} = \frac{\text{surface area}}{\text{volume}}$$

5. Fill each cube with rice. Use a balance to **measure** the mass of each filled cube. **Record** the results in your *Science Journal*.

Analyze Results

Compare the surface area, volume, and surface-area-to-volume ratio of the models.

Create Explanations

1. How do surface area and volume limit the size of cells?

2. Which cell model has the largest surface area? the largest volume? the largest surface-area-to-volume ratio?

3. What is the importance of this ratio? What does this suggest about God's design?

4. How does this ratio affect the amount of matter (mass) a cell can take in or remove?

Why Surface Area Matters Explain

You have learned about some of the incredibly complex processes that happen in and around the cell membrane. Think about the Structured Inquiry you just completed. Do you think a cell can carry out its life functions efficiently if it were larger? Why or why not?

Individual cells grow in size. As they grow, they obtain food and exchange gases by diffusion. Water enters and exits the cell by osmosis. Wastes are removed through the cell membrane. The larger the cell membrane, the more nutrients can enter the cell. But the larger a cell grows, the more nutrients it needs. Because the volume of the cell increases faster than the surface area of its membrane, there comes a point when the cell is too big. Its need for materials increases faster than its ability to supply those materials. There is no longer enough space on the cell membrane for all the work to happen that is needed to support the cell. Not enough nutrients can get in, and not enough waste can get out. God designed cells to be small because they are most efficient that way.

Some structures inside cells provide the additional surface area where the important life processes described above can take place. This is just one more example of God's wisdom in designing intricate complexity and efficiency in the tiny living things we call cells.

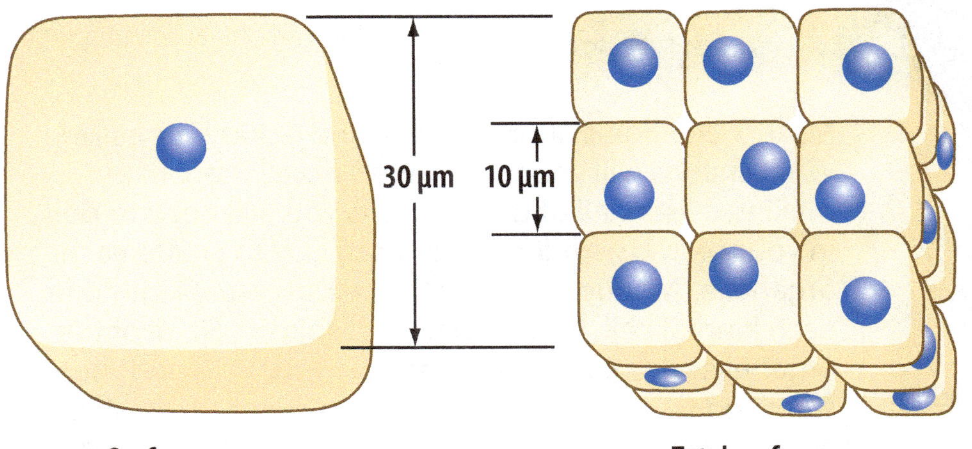

Surface area
of one large cell
= 5400 μm²

Total surface area
of 27 small cells
= 16,200 μm²

Compare the ratios of surface area to volume in these cells.

How does a higher total surface area affect the rate at which nutrients enter and wastes exit cells in an organism?

Cell Specialization Explain

Scripture Spotlight

What does **Ephesians 4:11** describe that is like cell specialization?

Red blood cells deliver oxygen and remove wastes. They have a flattened disc-like shape and no nucleus.

How does the form of these specialized cells relate to their jobs?

You have learned that cells can be classified as either prokaryotes or eukaryotes. Think about what you have learned about cells. What is another way you would classify cells? Do you think classifying cells by size would be helpful? What about classifying them based on their shape? One of the ways of classifying cells that is useful when studying them is by what function they perform. Would this method of classification be most useful in studying prokaryotes or eukaryotes? Why?

Most prokaryotic cells are unicellular and do not perform specialized functions. All life processes must be performed in that one cell. In contrast, most eukaryotic organisms are multicellular. Each of these cells has the exact same instructions—a complete set of instructions about how to maintain life—stored in its nucleus. You might think that would make all the cells identical, but that is not the case. Why?

There are hundreds of different cell types that are specialized to do different jobs. These specialized cells have different structures that are related to the task they perform. For example, red blood cells are flexible and disc-like. Both sides of a red blood cell are concave. The flexibility allows the cell to squeeze through narrow capillaries. The concave shape results in greater surface area, which permits the exchange of greater amounts of oxygen and carbon dioxide.

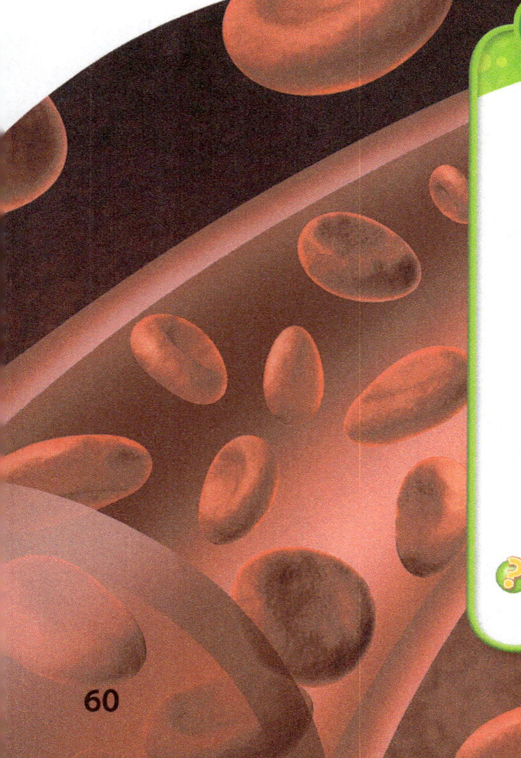

Explore-a-Lab

Structured Inquiry

Make 3-D models of prokaryotic and eukaryotic cells. What shapes will your cell models have? Prepare by making a list of all of the cell parts you will need to build in your cells. Use different colors of clay to create each organelle. Account for all of the cell structures found in each kind of cell, including the cytoplasm. Fit all of the organelles and cytoplasm inside of your cells and then surround them each with a cell membrane. What shape are your cells? Measure your cells to see how each of them compares. Then, cut the cells in half to create an inside view of each cell.

What differences do you see in models of prokaryotic and eukaryotic cells?

Root hair cells are specialized cells on a plant's roots. These cells have long, narrow, hair-like structures that extended between soil particles to reach the water and minerals. The shape of the root hair cells provide more surface area for water and minerals to move in and out of the cell.

Collar cells in sea sponges are also an example of specialized cells. These cells line the open channels in the sponge. They have whip-like structures that help push water through the channels. The cells are also sticky, which helps trap food particles moving with the water.

The table below provides a few more examples of specialized cells along with their shapes and functions.

Kind of Cell	Shape	Function
Bone cells	Osteoblasts: cubic	Make new bones
	Osteocytes: star with long branches	Direct repair and absorption of bone, by communicating with other bone cells
	Lining cells: flat or pancake shaped	Cover outside surface; control passage of calcium molecules in and out of bone
	Osteoclasts: large with more than one nucleus	Break down and reabsorb existing bone
Nematocyst	Long tubular cylinder	Stinging cell that injects poison into prey
		Found in sea jellies and sea anemone
Cheek cell	Irregular, flattened, layered	Makes a thin covering and protects underlying tissue
Palisade leaf cell	Long, rectangular block	Gives the top of a leaf a firm, flat surface

Summary: How are cells classifed? The two types of cells are prokaryotic and eukaryotic cells. Prokaryotes do not have a nucleus. Eukaryotes have a membrane-bound nucleus. Eukaryotes also have more organelles than prokaryotes. The relationship between surface area and volume limits the size of cells. Cells are small in order to be more efficient in the exchange of food, oxygen, and wastes. Specialized cells have different structures. The structures help the cells to perform their specialized functions.

1. Describe several ways in which prokaryotes and eukaryotes are different.

2. How is a cell membrane similar to a nuclear membrane? How are they different?

3. Explain how cell size is beneficial to survival.

4. What are two examples of how specialized cells differ in structure or function? Explain how each helps the organism to live.

What Do Organelles Do?

Objectives

- Explain the importance of the cell nucleus.
- Describe the functions of the cell organelles.
- Distinguish between animal and plant cells.

Vocabulary

nucleus

endoplasmic reticulum

Golgi apparatus

vacuole

mitochondria

chloroplast

vesicle

lysosome

Think about a bicycle. What parts does it have? What jobs do the different parts perform? While a bicycle is large and made up of parts that you can easily see, cells are made up of parts that you cannot see without using a microscope. Each organelle of the cell has a specific job to do. Just like the parts of the bicycle enable it to function properly, the organelles of eukaryotic cells enable the cells to function properly. The complex coordination functions performed by the cell's organelles provides further evidence of the Creator's design.

The Cell's Control Center Explain

Why do you think the nucleus is often the largest and most visible object inside a cell? The **nucleus** controls all the activities and hereditary functions of the cell, similar to the way your brain controls your body's activities. The nucleus is surrounded by a nuclear membrane that contains thousands of tiny holes, or pores, which allow materials to move back and forth between the nucleus and the cytoplasm. The nucleus sends instructions in the form of specialized chemical messengers, which are delivered to the organelles. What might happen if these instructions contain mistakes? Prokaryotic cells lack this nuclear membrane. How might this affect these kinds of cells?

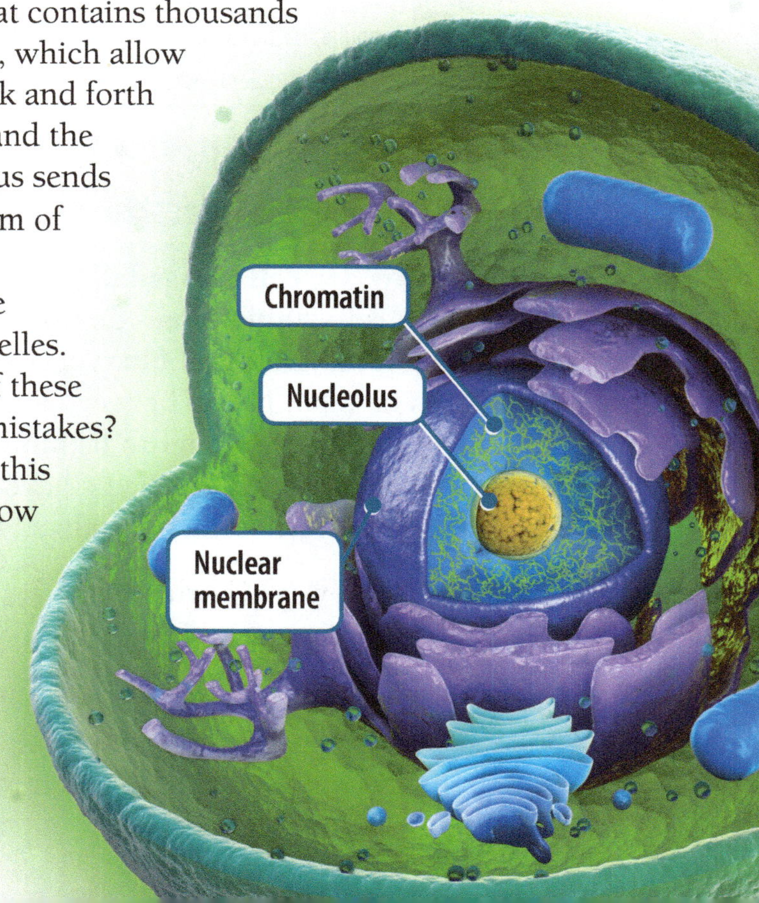

The nucleus contains the instructions for the organelles to carry out life functions.

How do you think the cell nucleus communicates with the rest of the cell?

Chromatin

Chromatin is a mass of genetic material found in the nucleus. It is composed of DNA, containing the information that directs all the functions of the cell. When the cell gets ready to divide, the chromatin condenses and forms the chromosomes.

Nucleolus

The nucleolus is a small structure inside the nucleus. This is where ribosomes are made. A cell can have up to four nucleoli. The number for each species is always the same.

Organelles and Their Function

Inside eukaryotic cells, there is a variety of tiny organelles. The work of the organelles makes it possible for cells to release energy from food, to get rid of wastes, and to do many other things that enable the cell to live and maintain and repair the body.

Ribosomes

Proteins make up most of the cell. Ribosomes are small organelles that make the proteins. Because proteins are necessary for life, there are more ribosomes than any other organelle. Free-floating ribosomes make proteins for the cell. Ribosomes attached to the ER make proteins that are moved out of the cell. What might be the advantage of having ribosomes attached to the ER?

Endoplasmic Reticulum

The largest organelle is the **endoplasmic reticulum**, or ER. It is a network of tube-like channels that winds through the cytoplasm. The ER is the cell's transport system for moving materials around the cell. There are two kinds of ER, smooth and rough. It connects to the cell membrane. Why do you think the ER connects to the cell membrane?

Smooth ER

- lacks ribosomes, giving it a smooth appearance
- transports proteins throughout the cell
- builds and stores lipids (fats)

Rough ER

- contains the ribosomes, giving it a rough appearance
- provides a surface where proteins can be built
- increases the surface area on which cell processes happen

Cell Parts

How do organelles present in plant cells differ from those present in animal cells?

Materials
- microscope
- 3 microscope slides
- 3 cover slips
- flat toothpick
- tweezers
- onion
- Elodea
- medicine dropper
- goggles
- paper towels
- iodine

Procedure

1. Place a drop of water on a clean slide. Cut a piece from the inside of the onion. Use tweezers to gently peel off the thin inner skin. Transfer the skin to the drop of water on the slide.

2. Hold one edge of the cover slip on the slide and lower the cover slip slowly. Add a drop of iodine at the edge of the cover slip to stain the sample. Use a paper towel on the opposite side of the drop of iodine to draw the stain across the slide.

3. Place your slide on the microscope and **observe** the onion under medium or high magnification.

4. Draw what you see. Label the organelles you can identify.

5. Use a medicine dropper and place a drop of water on a clean slide. Place a fresh Elodea leaf on the slide. Hold one edge of the cover slip on the slide and lower the cover slip slowly.

6. Place your Elodea slide on the microscope and **observe** the Elodea leaf under medium and high magnification.

7. Draw what you see. Label the organelles you can identify.

8. Use a clean toothpick to collect cheek cells from the inside of your cheek. Use the toothpick to spread the cells in a drop of iodine on a clean microscope slide. Put a cover slip over the cells. Repeat Steps 3–4.

Analyze Results

Compare the size, shape, and parts of both plant and animal cells.

Create Explanations

1. How do organelles present in plant cells differ from those present in animal cells?

2. How are the cells that make up your body like the cells of onion and Elodea? How are they different?

3. What other organelles would you expect to see in the Elodea cells? What other organelles would you expect to see in the cheek cells?

Golgi Apparatus

The **Golgi apparatus** is the organelle that packages the materials for the endoplasmic reticulum to transport. The Golgi apparatus is made up of a stack of three to twenty slightly curved sac-like membranes that float in the cytoplasm. The Golgi apparatus receives the proteins produced in rough ER, inspects the proteins for flaws, and makes any changes needed before sending them to the smooth ER. The proteins move off through channels in the smooth ER.

Vacuole

A **vacuole** is a fluid-filled storage container. Some vacuoles store water, other liquids, or food particles. Other vacuoles store waste products until they can be removed from the cell. A membrane keeps the vacuole contents separate from the cytoplasm. This membrane functions in a way that is similar to the cell membrane because it regulates the movement of materials between the cytoplasm and the vacuole.

Vacuoles are found in both plant and animal cells. However, those in plant cells are much larger. In green plants, a large central vacuole provides additional structure and support for the cell beyond that provided by the cell wall. These vacuoles fill up or shrink depending on how much water is available to the cell. Shrinking of the vacuoles is what causes plants to wilt.

Mitochondria

Mitochondria are organelles that release the energy the cell needs to function. Often referred to as the "powerhouses" of the cell, they may be spherical or cylindrical in shape. Mitochondria have two membranes. The outer membrane is smooth and has a structure similar to that of the cell membrane. The inner membrane has several folds. These folds, called *cristae,* provide the surface area needed for energy-producing reactions to take place.

Scripture Spotlight

Read **Acts 1:8**. What is the Christian's "powerhouse?" Read **Luke 11:13** to see how we access that power.

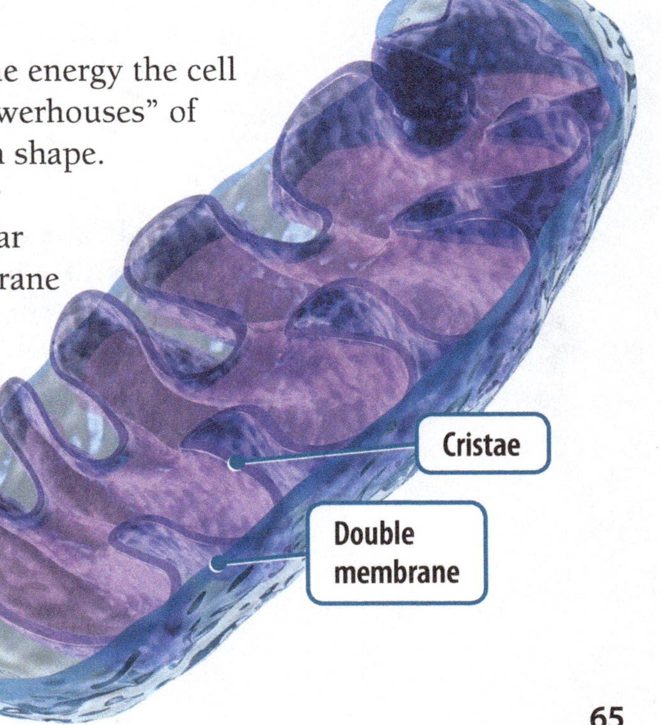

A typical mitochondria from an animal cell.

How does the structure of this cell organelle provide surface area on which reactions can take place?

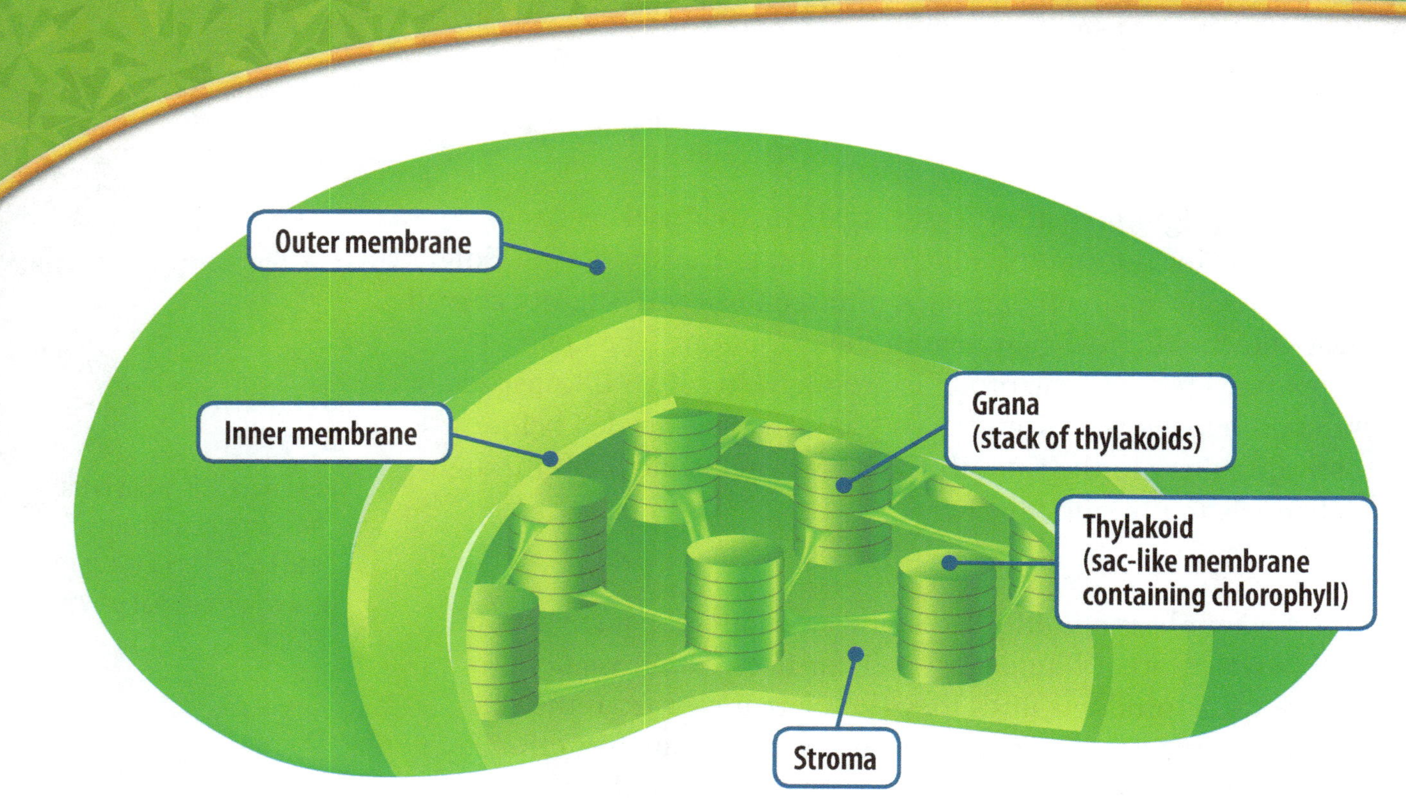

Chloroplast from a leaf.

How does the structure of grana help the process of photosynthesis?

Chloroplasts

Chloroplasts are large, green, oval-shaped organelles in which photosynthesis takes place. They are found only in plants, many protists, and some prokaryotic cells. A double membrane surrounds these organelles. Enclosed within the membranes is a gel-like fluid called *stroma*. Sugars are made inside the chloroplast. The manufacture of sugars takes place on special disc-shaped structures called *grana*. The grana contain chlorophyll-filled sacks called *thylakoids*. It is the chlorophyll that gives plants their green color. The job of the chloroplasts is to trap energy from sunlight and turn it into chemical energy.

Vesicle

The **vesicle** is a small sac-like structure that forms when part of the smooth ER buds off. The vesicles help transport proteins and other materials that the cells need to be able to do the jobs for which they are designed. When needed, the proteins and other materials contained by the vesicles are released into the cytoplasm to be used by the cell. How might the cell use these materials?

Lysosomes

Lysosomes are specialized vesicles that float freely in the cytoplasm. These organelles are formed in the Golgi apparatus. They are responsible for breaking down and digesting materials such as food and worn-out cell parts.

Lysosomes also play a role in destroying harmful bacteria. They contain powerful digestive chemicals that combine with white blood cells that surround bacteria and form a vacuole around the bacteria to destroy them.

Cilia and Flagella

Some cells have hair-like projections, called cilia and flagella, that stick out from the cell's surface. Cilia are hair-like and generally cover the cell's entire surface. Flagella are much longer, and are usually limited to one or two per cell. Both structures are often found in protozoa. These unicellular creatures live in water and wet places like seepage areas, damp soil, and leaf litter. Protozoa use cilia or flagella to swim from place to place. In your body, the movement of cilia in the trachaea and bronchioles helps clear your lungs of dust. The hair cells of the inner ear, which move in response to the vibration of sound waves are cilia that play an essential role in hearing. What causes these tiny cilia in the inner ear to move?

The chart below summarizes the function of each organelle.

Organelle	Function
Nucleus	Control center; stores hereditary information
Cell membrane	Controls what enters or leaves the cell
Cytoplasm	Contains all the other cell parts
Endoplasmic reticulum	Manufactures proteins and fats; moves chemicals around cell
Ribosomes	Build proteins
Golgi apparatus	Packages and transports proteins to rest of cell
Vesicles	Transport chemicals
Vacuoles	Store liquid, food, and waste
Mitochondria	Change sugar into usable energy
Chloroplasts	Use sunlight to make sugar
Lysosomes	Aid in digestion; get rid of wastes; fight bacteria
Cilia/Flagella	Help some cells move; clearing out airways; and responding to sound wave vibrations

Organelles in Cell Transport Explain

You have just learned that there are three organelles largely responsible for proteins moving throughout the cell. These three organelles are the ER (endoplasmic reticulum), Golgi apparatus, and vesicles.

The ER transports cellular products to the Golgi apparatus. The Golgi apparatus in turn packages proteins in other vesicles so the proteins can cross the cell membrane and leave the cell. The Golgi apparatus also transports lipids and creates lysosomes and vesicles involved in digestion. How do these organelles represent Gods' design in living things?

Explore-a-Lab

Structured Inquiry

 **How can a cell model be used to understand how a cell works?**

Think about the observations you made of plant and animal cells in the Structured Inquiry as you build your models. Use two zip-top sandwich bags as the cell membrane and syrup to act as the cytoplasm. Choose objects to represent organelles.

1. Work with a partner, but make your own cell model.

2. Choose organelles for an animal cell.

3. Add other structures and record the function of each.

4. Add syrup to your cell.

5. Repeat Steps 2–4, but this time, build a plant cell.

How are your models the same and different?

Animal and Plant Cells

Think back to the plant and animal cells you examined in the Structured Inquiry in this Lesson. What do you remember noticing that was similar? What do you remember seeing that was different? While plant and animal cells share many common features and processes, they have distinct differences. These differences allow them to carry out different jobs. Look at the illustration on the next page to see the similarities and differences between plant and animal cells.

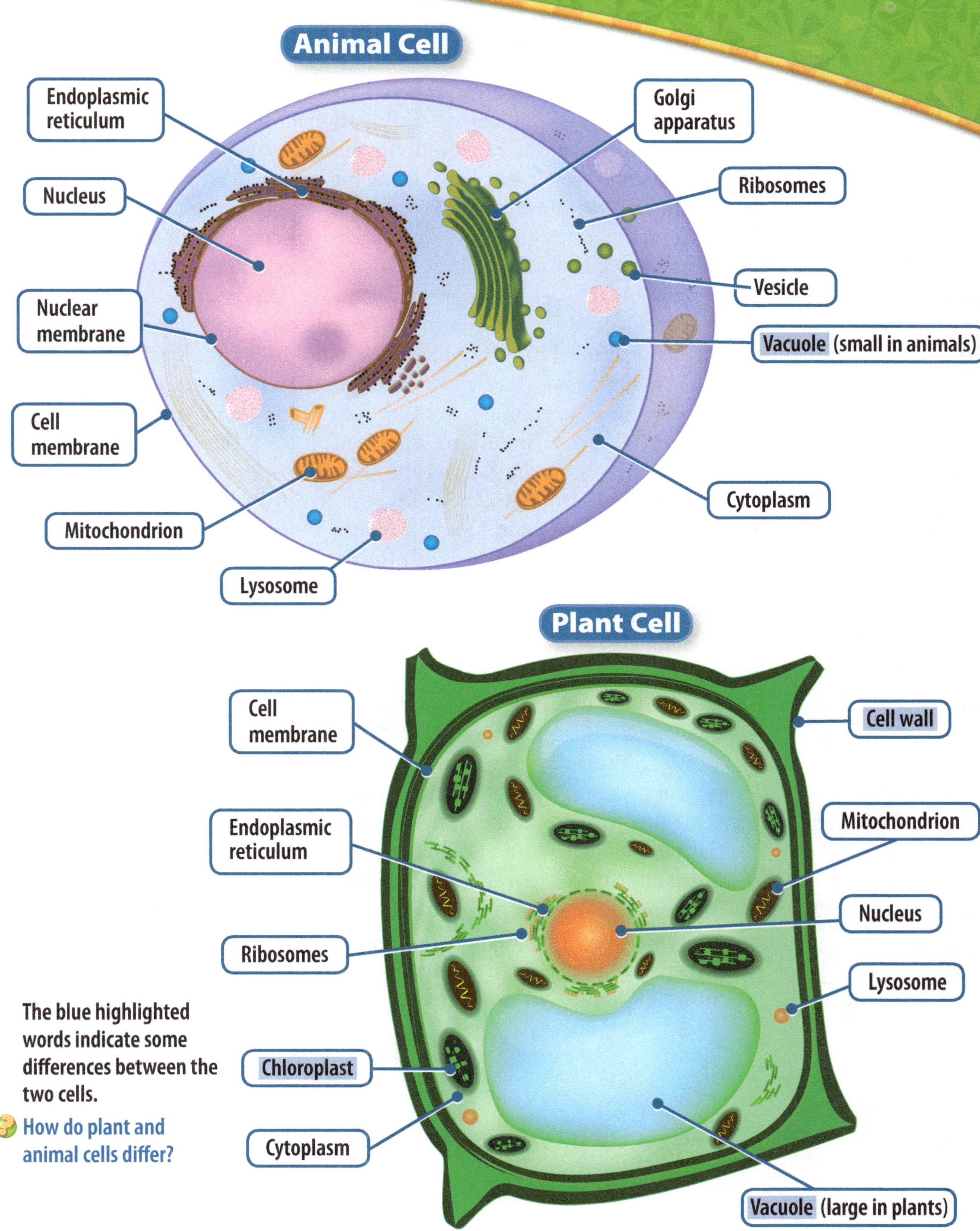

The blue highlighted words indicate some differences between the two cells.

How do plant and animal cells differ?

Machines at Work

As you have seen, the organelles are constantly busy carrying out life functions. These organelles are microscopic, but try thinking of each one as an individual factory. Each organelle includes thousands of types of tiny molecular machines—*literally* machines—with multiple moving parts that work together to perform a specific job. Like machines designed by humans, these tiny molecular machines are amazingly complex. Solar-powered machines (chloroplasts) capture light energy and store it. Tiny electrical machines in nerve cells carry messages. Tiny mechanical machines in muscle cells haul cargo and even build other machines. Every cell in every plant and animal functions because of the combined work of these complex machines. If these machines were not present, the cell would not be alive.

But the complexity doesn't end there. Most of these tiny molecular machines are made up of proteins. Proteins are required for every structure and function within the cell. There are tens of thousands of different kinds of proteins in each cell, each with a specific job to do. But before you can understand proteins, you need to know about one more thing—the building blocks of protein, which are called amino acids. Think of amino acids as differently shaped, interconnecting blocks. But instead of half a dozen different shapes like you might find in a typical set of blocks, imagine 20 different shapes with different chemical properties.

When these amino acids join together, they make proteins. A typical protein contains hundreds, or even thousands, of amino acids in a row. A specific protein is made only when exactly the right amino acids join together in just the right order. But, that is not all. The protein then folds into a 3-D shape. Exactly the right shape is necessary for the protein to function the way it is supposed to. Imagine how intricately thousands of different proteins must be folded to fit together in this way! Each time molecular biologists make new discoveries about amino acids,

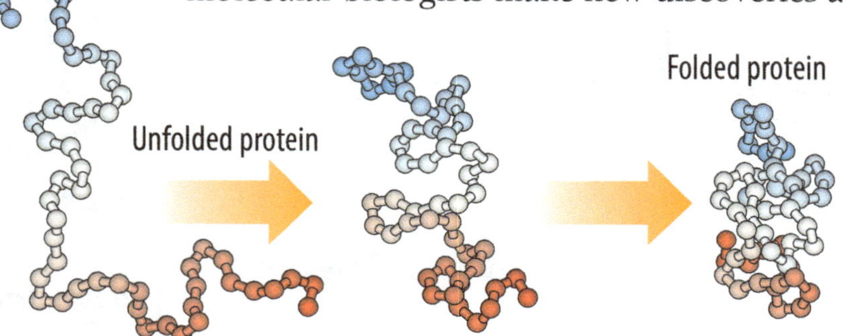

A string of amino acids folds into a 3-D shape.

proteins, or molecular machines, it becomes more challenging to imagine that such intricately complex things could have arisen spontaneously from nonliving matter.

If scientists from the nineteenth century who thought cells were simple little blobs could have known that our bodies are made up of trillions of cells, each housing thousands of organelles made up of tens of thousands of proteins, made of hundreds or thousands of amino acids, they may not have been as willing to accept the idea of abiogenesis.

Faith

When you consider the complexity of the cells and their functions, which theory about the origin of life makes most sense to you? Did such complexity happen by chance or did it require a designer? Did life result randomly or did God create the complex structures and their complex functions? Continue thinking about these questions as you study more about life functions.

Concept Check Assess/Reflect

Summary: What do organelles do? Organelles are the parts of cells that do the jobs needed to keep the cell alive. The organelles have specialized jobs and work together to perform life functions for the cell. Animal and plant cells have most of the same structures; however, animal cells do not have a cell wall and chloroplast and plant cells have larger vacuoles. Organelles are microscopic factories filled with even smaller machines that keep the factory functioning properly. Included in these tiny machines are amino acids, which join together to form proteins.

1. Name two cell organelles that are found in plant cells but not in animal cells. Describe their functions.

2. Based on the function of the mitochondria, what tissue might contain cells with a high concentration of mitochondria?

3. What would happen to a plant cell if the central vacuole lost all its water?

4. How can proteins, which are made up of chains of hundreds or thousands of amino acids linked together, fit inside the cell?

Essential Question

How Do Cells Get Energy?

You have learned how the organelles of a cell work together to carry out life functions. Many cells specialize in the jobs they perform, but there is one thing they all need—energy. As a member of the animal kingdom, you know that you eat and digest food for energy. Most plants don't get their energy in this way. How do they get energy? What is a plant's food?

How Plant Cells Make Food `Explain`

The energy source for cells is the Sun. Green plants and some other organisms are able to capture the energy in the Sun's light. How do animals benefit from this captured energy?

Needs for Photosynthesis

Photosynthesis is the process by which plants make food. The word *photosynthesis* means "putting together with light." The process occurs in green plants, in algae, and in certain kinds of bacteria. Photosynthesis requires light, carbon dioxide, water, and chlorophyll.

Chlorophyll is a pigment that traps light during photosynthesis. The energy from sunlight is used to power reactions that combine water from the soil and carbon dioxide from the air to make a simple sugar called **glucose**. A by-product of the photosynthesis reaction is oxygen.

$$\text{water} + \text{carbon dioxide} \xrightarrow{\text{light energy}} \text{sugar} + \text{oxygen}$$

What would happen if photosynthesis did not occur?

Role of the Leaf in Photosynthesis

A leaf is composed of different layers. Most photosynthesis takes place inside leaves and green stems. The bottom of a leaf contains tiny openings that open and close to allow carbon dioxide to enter and oxygen and water vapor to leave the leaf.

The leaf veins support the leaf and transport substances to various parts of the plant. What substances do the veins transport? You can better understand how a leaf functions by studying the illustration on the next page.

Steps in Photosynthesis

There are two reactions in photosynthesis: the light reaction, which happens during the day, and the dark reaction, which can happen in the day or night. These reactions are dependent on each other. Each generates a special energy molecule—ATP or ADP—that is used to keep the other running.

Light Reactions

- Occur in the grana, which are found inside the chloroplast
- Require **ADP** (adenosine diphosphate) molecules to change light energy into chemical energy
- Store the energy produced in **ATP** (adenosine triphosphate) molecules

Dark Reactions

- Occur in the stroma
- Require carbon dioxide and energy from ATP to form glucose
- Form more ADP to supply the light reactions

Energy from the Sun is converted to supply the nutrients a plant needs to perform life functions.

Where does the light reaction take place in a leaf, and how is it different than the dark reaction?

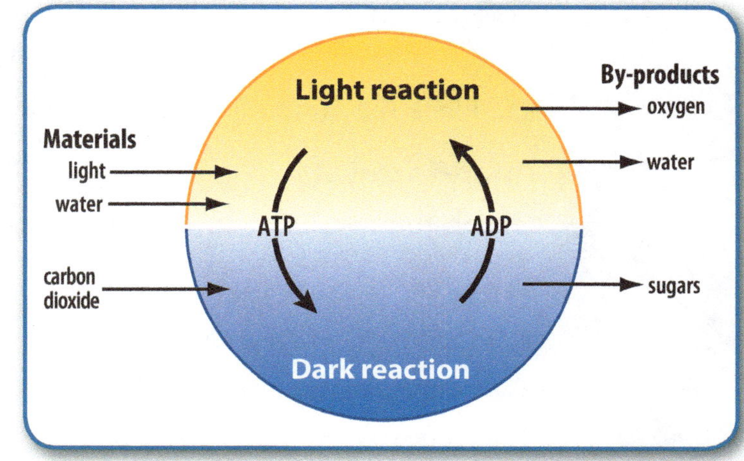

Record your work for this inquiry. Your teacher may also assign the related Guided Inquiry.

Bubbling Light Reaction

What is produced in photosynthesis?

Materials
- Elodea plants
- scissors
- 300–400-mL beaker
- clear funnel
- test tube
- balance
- distilled water
- baking soda
- graduated cylinder
- large plastic cup
- metric ruler
- table lamp

Procedure

1. Cut the stem from the Elodea plant at an angle and gently crush it.

2. Place the plant into the beaker. Invert the funnel over the plant to trap it under the funnel.

3. Use the graduated cylinder to **measure** 300 mL of distilled water and add it to a large plastic cup. Use the balance to **measure** 0.6 g of baking soda. Add the baking soda to the cup of water and stir until the baking soda dissolves. This mixture will be used in Step 8.

4. Fill the beaker with the distilled water so that the water covers the end of the funnel.

5. Fill the test tube with distilled water. Place your finger over the mouth of the test tube. Invert the test tube over the end of the funnel. Make sure no water leaks out of the test tube.

6. Place the lamp about 50 cm away from the beaker. Turn on the lamp.

7. Use the ruler to **measure** the millimeters of gas inside the test tube every 5 minutes for 30 minutes. **Record the data** in your *Science Journal*. After 30 minutes, take the test tube off the funnel and empty it. Empty the water in the beaker.

8. Set up the apparatus again, this time filling the beaker and the test tube with the baking soda-water mixture. Repeat Steps 6 and 7.

9. After 30 minutes, take the test tube off the funnel and fill it with the baking soda-water mixture and put it back over the funnel as you did in Step 5.

10. Place the lamp about 25 cm away from the beaker and repeat Step 7.

Analyze Results

Graph the results. **Compare** the data gathered.

Create Explanations

1. What is produced in photosynthesis?

2. When was the amount of gas produced by the plant the greatest?

3. What was the purpose of adding the baking soda to the water?

Cell Respiration Explain

At night the light reaction of photosynthesis shuts down. The transport of nutrients into cells, removal of wastes from cells, building of new cell parts, and other functions do not stop because it is dark. Energy is still needed to carry on these life processes. The chemical energy stored during photosynthesis is made accessible for use by the cell through **cell respiration**.

Cell respiration involves processes with and without oxygen.

Glycolysis

- **Glycolysis** is a process that breaks down glucose molecules.
- It occurs in the cytoplasm. It does not require oxygen.
- The process yields pyruvate (a type of acid necessary to convert sugar to energy) and a small amount of ATP molecules.

Aerobic respiration

- *Aerobic respiration* occurs in the mitochondria. Oxygen is required for this process.
- The pyruvate made during glycolysis enters the mitochondria.
- This process yields many ATP molecules.

Fermentation

- **Fermentation** is a type of *anaerobic respiration,* a process that does not require oxygen.
- Pyruvate is converted into water, carbon dioxide, and either lactic acid or alcohol. A small amount of ATP molecules are produced.
- Lactic acid is mainly produced in muscles and can be used as a source of energy. It is also used in food making.
- Fermentation by yeast cells is important in baking bread and making wine.

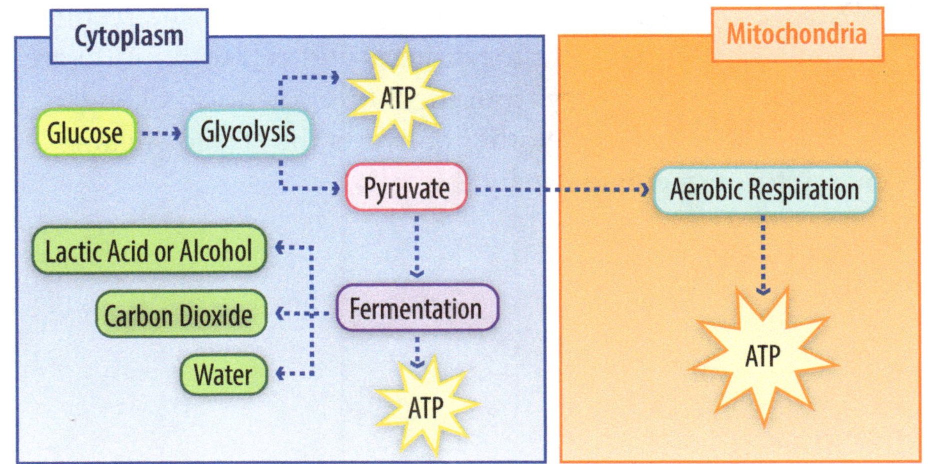

Cell respiration occurs in the cytoplasm and the mitochondria of cells.

How are the processes in the cytoplasm different from those in the mitochondria?

What happens during cellular respiration?

You will need a small-necked bottle, a balloon, a twist tie, a package of dry, active yeast, 2 tablespoons of sugar, and 118 mL ($\frac{1}{2}$ cup) of warm water.

1. Add the yeast to the bottle. Then put in the sugar and warm water.

2. Secure the balloon over the neck of the bottle with the twist tie.

3. Put the bottle in a warm location for about 60 minutes.

4. Observe what happens to the balloon.

What is the source of energy for the process? What organism performs the process?

Photosynthesis and Cell Respiration Are Related **Explain**

You can see God's design in the two processes that maintain life on Earth: photosynthesis and cell respiration. Photosynthesis builds the molecule that provides the energy for life, and cell respiration breaks this molecule down to meet the cell's current energy needs. Nearly every creature on Earth uses the food, or chemical energy, produced through photosynthesis by plant cells. Both plant and animal cells use the oxygen that plants give off during photosynthesis for cell respiration.

Photosynthesis and cell respiration go together. They involve the same basic chemical materials: carbon dioxide, water, oxygen, ATP, ADP, and glucose. Each reaction yields the opposite result of the other. As you can see, photosynthesis and cell respiration are opposites. One stores energy from the Sun in the form of sugars to be used later by cells. The other converts the stored sugars into energy for immediate use by the cells.

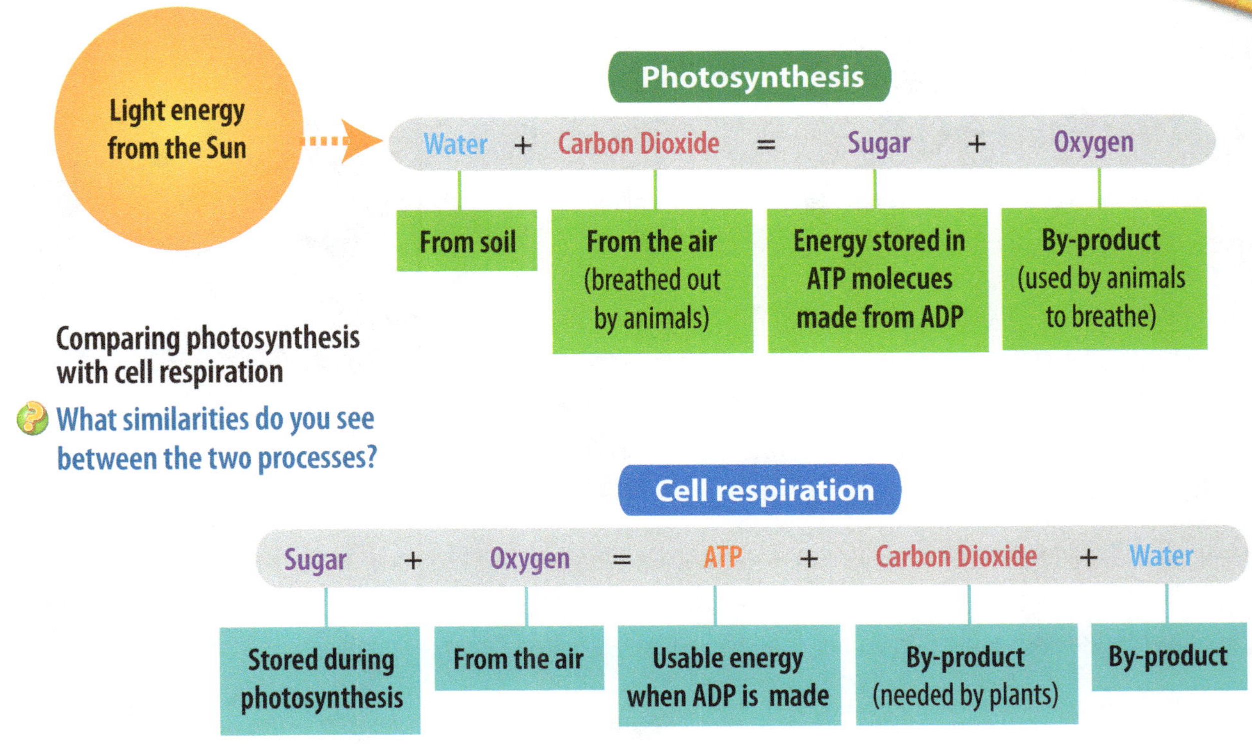

Comparing photosynthesis with cell respiration

What similarities do you see between the two processes?

Concept Check Assess/Reflect

Summary: How do cells get energy? Cells get energy through photosynthesis and cell respiration. Photosynthesis combines carbon dioxide, water, and sunlight to produce glucose and oxygen. Cell respiration uses oxygen to release energy from the glucose molecules. The cells use this energy to do the jobs they are designed to do. In the process of releasing the energy from glucose, carbon dioxide and water are produced. Photosynthesis builds sugars. Organisms break down sugars during cellular respiration to meet their energy needs.

1. Where do the oxygen and glucose used in cell respiration come from?
2. Explain the role of ATP in the energy processes of the cell.
3. How do photosynthesis and cell respiration work together?
4. Explain what cells do with ADP and ATP during cell respiration.

Get to Know
Louis Pasteur

Louis Pasteur was born in 1822 in France. He was a microbiologist who made major contributions to science, technology, and medicine.

One of Pasteur's earliest experiments disproved what some scientists of his day believed, that certain organisms came to life through spontaneous generation. Pasteur did not believe that mold, for example, just appeared when bread was left out. He believed that organisms only come into existence when other living creatures reproduce.

Pasteur also showed that tiny microorganisms cause disease. His ideas became known as the germ theory of disease. He said that if the spread of disease-causing microorganisms could be prevented, then the spread of diseases could be prevented. In addition, Pasteur created a vaccine for rabies.

Louis Pasteur also developed a way to make certain foods safe, such as milk. His method is called pasteurization and is still used today. Pasteurization kills the bacteria present in milk so that it is safe for people to drink. The milk is heated to a very high temperature to kill the harmful bacteria. This process does not change the flavor or nutritional value of the food.

In 1887, Louis Pasteur founded the Institut Pasteur in Paris, a research institute dedicated to preventing and treating infectious diseases. Pasteur was its director until his death in 1895. His work lives on through the Institut Pasteur with its more than 130 laboratories, teaching center, and medical center.

Called to Serve

As a student, Louis Pasteur was interested in studying organic crystals. He thought that their structure demonstrated God the Creator's artistic expression. He believed that this was evidence of God's wisdom and design. Pasteur observed the crystals and determined that two crystals that looked the same actually had different structures. They turned out to be mirror images of each other. His discovery gave other scientists a better understanding of crystals and their structures. Why do you think Pasteur believed the structure of crystals showed God's artistic expression?

Concept Check

1. In what way did Louis Pasteur contribute to the field of microbiology?
2. What type of research is currently being done at the Institut Pasteur?

Microbiologist

Microbiologists study microbes, which are tiny organisms such as bacteria, viruses, molds, algae, and yeasts. They study these organisms to find out where they live, how they survive, how they affect living organisms and the environment, and how humans can use them.

Some microbiologists work in universities, hospitals, or health agency laboratories. They may work alone or with other scientists. Other microbiologists travel all over the world to discover unknown microbes and ones that cannot be created in a laboratory. As they work, microbiologists may find ways to use microbes for the benefit of humankind, such as in medicines or for reducing pollution.

Because there are so many types of microbes, most microbiologists specialize, or work in one specific area, as shown in the following chart.

- Medical microbiologists study microbes that cause human disease and sickness.
- Veterinary microbiologists study microbes that cause diseases in animals.
- Environmental microbiologists study microbes in the environment, including soil, water, and air as well as animals and plants.
- To prevent food-borne illness, food microbiologists study how microbes affect food.
- Pharmaceutical microbiologists ensure medications do not contain harmful levels of microbes.
- Clinical microbiologists study microbes to prevent epidemics and to identify new infectious agents.
- Marine microbiologists study microbes found in water to control and prevent the growth of harmful bacteria in rivers and oceans.

The path to becoming a microbiologist involves many years of education and training. Most microbiologists have at least a bachelor's degree in biology, microbiology, or a related science field, and many hold a master's degree. Microbiologists who do research or work in academics also have a doctorate.

Concept Check

1. **What makes the job of a microbiologist important?**
2. **Why might a dairy plant hire a microbiologist?**

Study Guide

Lesson 1

1. The cell theory states that all living things are made up of one or more cells, the cell is the smallest unit of life, and living cells can only come from other living cells.

2. Cells are enclosed within the cell membrane.

3. Materials enter and leave the cell by diffusion, osmosis, and active transport of molecules through the cell membrane.

4. Most living things are unicellular, small in size, and unique in design.

Lesson 2

1. The nucleus contains hereditary information and controls the cell's activities.

2. Living things are divided into two main groups based on the design of their cells: prokaryotes and eukaryotes.

3. The geometry of the cell limits how big a cell can grow based on the surface-area-to-volume ratio.

4. Multicellular organisms contain many cells, each specialized to do different jobs.

Lesson 3

1. The eukaryotic cell is a complex system of many organelles that work together.

2. Animal and plant cells have many of the same cell parts. Plant cells have chloroplasts and a cell wall, which animal cells do not have. Plant cells also have larger vacuoles than animal cells.

3. Several organelles are involved in cell transport, including the vesicles, Golgi apparatus, and smooth ER.

Lesson 4

1. The source of energy that powers life on Earth is the Sun. During photosynthesis, plant cells use chlorophyll to convert light energy to stored chemical energy.

2. The process of breaking down the stored chemical energy is called cell respiration, and takes place in almost all cells.

3. Photosynthesis builds the molecule that provides the energy for life. Cell respiration breaks this molecule down so that the cell can use the energy to carry out its work.

Show What You Know

Visualize It Complete the concept map.

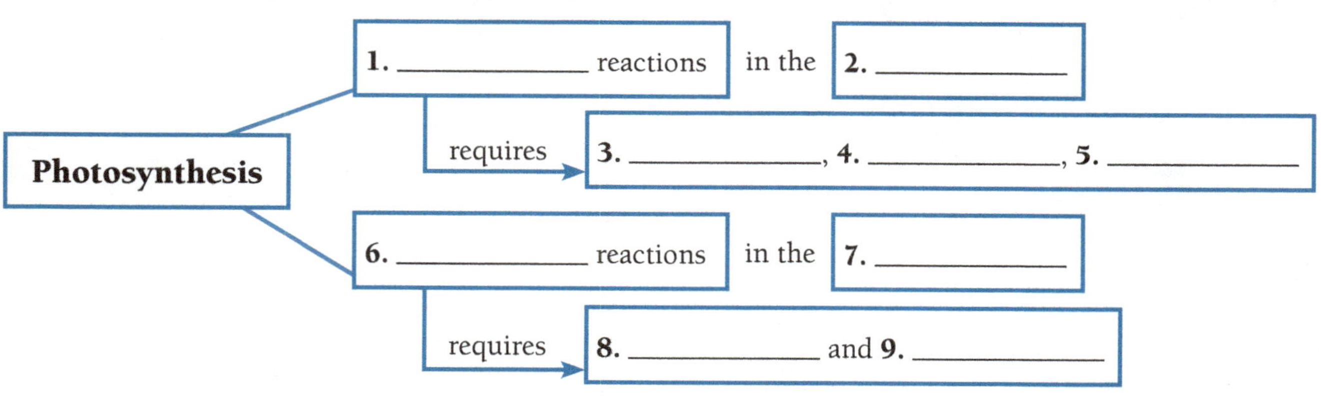

Vocabulary Check

Explain how each pair of terms is related.

10. cell membrane—osmosis
11. unicellular—multicellular
12. plant cell—cell wall
13. nucleus—organelle
14. ER—ribosome
15. chloroplast—mitochondria
16. photosynthesis—chlorophyll

Multiple Choice

Choose the best answer.

17. How did Louis Pasteur's work help to support the cell theory?
    A. He first observed living cells.
    B. He stated that cells are the basic life unit.
    C. He proved that all cells come from other cells.
    D. He invented the name "cell."

18. What process allows gases, such as oxygen, to cross the cell barrier?
    A. cell respiration
    B. diffusion
    C. photosynthesis
    D. osmosis

19. Which cell organelle helps the human ear interpret sound vibrations?
    A. lysosome   C. flagella
    B. cilia   D. rough ER

20. How do prokaryotic cells differ from eukaryotic cells?
    A. They have specialized organelles.
    B. They have free ribosomes.
    C. They lack a nucleus.
    D. They use photosynthesis to change sunlight into chemical energy.

Check Point

Answer the following questions.

21. **Compare** unicellular and multicellular life forms. Identify an advantage of each life form.

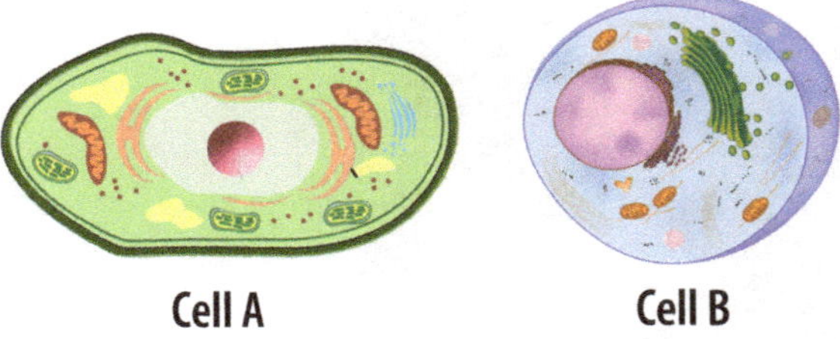

22. Look at Cells A and B. Identify each type of cell. What characteristics of the cells did you use to identify them?

23. **Infer** why many cell organelles contain folded or stacked membranes.

24. State the three main ideas of modern cell theory.

25. Describe the role of the mitochondria in cell respiration.

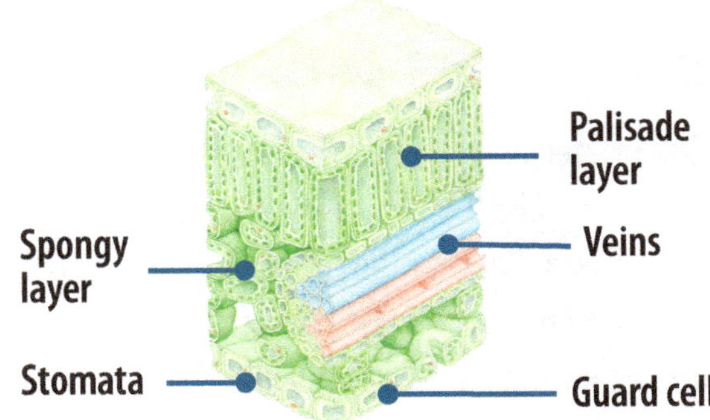

26. Look at the structure above. Where does most photosynthesis occur? Explain.

27. Explain how the structure of a red blood cell aids in its specialized function.

How Cells Function

Scripture Spotlight

God created and cares for the largest structures, like the planets, to the smallest atom. Cells are one of God's smaller creations, but they are key to living organisms functioning properly. By studying God's work you can see why it is important to take care of the largest and smallest components.

Colossians 1:17 (p. 94)
Jeremiah 31:33 (p. 95)
Malachi 3:6 (p. 104)
Proverbs 15:13 and 15 (p. 106)
Proverbs 17:22 (p. 106)

Each chromosome contains one DNA molecule. The DNA molecule in the chromosome is made of two strands twisted around each other forming a molecule shaped like a twisted ladder.

The Big Idea

Cells contain DNA, special molecules with instructions for carrying out life functions. God designed all cells with DNA so they can do amazing things.

How is DNA evidence of God's design? What is the possibility that DNA could be assembled by chance?

Inquiry Kick-Off   Engage

Have you ever written a message in code? How is it read by the recipient? Your body uses codes to carry out life functions. In your *Science Journal*, you will decipher a code, similar to how your cells decipher genes.

Science Journal

Essential Question

What Is the Structure of DNA?

Before a home is built, an architect creates a plan called a blueprint. Many workers come together to read the blueprint and put all of the components of the home in place. Without electricity, plumbing, insulation, heating, cooling, and more, the home would not function as a family would expect it to. God is the architect for our bodies. He makes the plan and brings all of the parts together to function as He planned. What parts of your body function like the construction workers? What would happen if some of these workers became sick?

The DNA Puzzle Explain

Have you ever thought about how the discovery of DNA happened? What questions do you think early scientists posed in their search for DNA? It was a long process that began many years ago and involved the work of many scientists. When research began on what was inside a cell, microscopes were not as powerful as they are today. To learn more about the cell and how it worked, scientists conducted a variety of experiments.

Some of these investigations suggested that there was something inside the cells passing on physical traits from parents to their offspring. One of the most famous experiments was conducted by Gregor Mendel who studied peas, trying to determine why some had different colored seeds, shapes, and other characteristics.

Discovering DNA and understanding its unique structure was the result of many scientists' discoveries over many years.

What type of technology do you think must have been in place in order to discover DNA?

As a class, work together to create a DNA time line. Break up into small groups. Your teacher will assign each group one or more of the scientists listed on this page. Each of the scientists contributed to the discovery of DNA. Your group's job is to research the contributions made by your designated scientist. Use the Internet or other resources to compile a brief biography of the scientist(s). Include a description of the scientific contribution and how it relates to previous discoveries.

Use images to illustrate your findings. Assemble the information on a poster board. When all of the groups have completed their work, assemble the time line in the classroom. Groups will take turns explaining their portions of the time line to the class.

Was one scientist's contribution more important than any of the others? Explain.

Contributions to the discovery of DNA

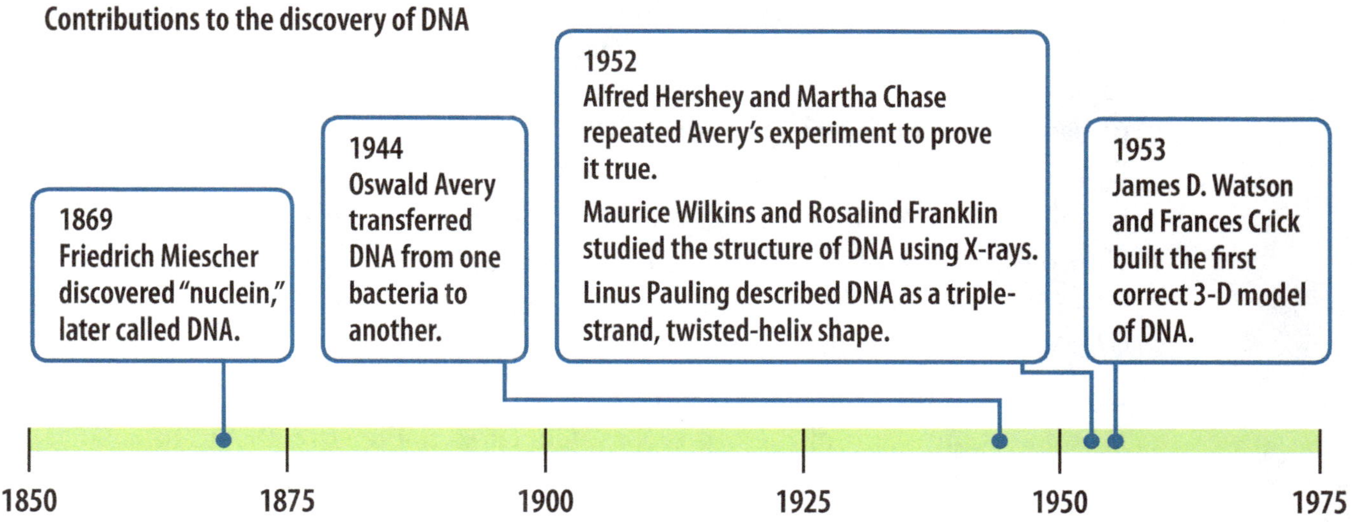

It took many researchers to unravel the mystery of DNA because it is submicroscopic and contained within the cells of living things. Why do you think so many people were involved in discovering the secrets of DNA? Do you think there are ever important scientific discoveries that can be credited to just one person?

Where are the blueprints your body uses to build new cells kept? As you may recall, DNA contains the genetic instructions that allow an organism to function and develop. In eukaryotic cells, the DNA is found within the nucleus of cells. Watson and Crick were the first to accurately model the DNA structure. It is the specific structure of DNA that allows it to hold the genetic information and pass it on to a new generation.

A Sweet DNA Model
How do DNA molecules fit together?

Procedure

1. Cut each licorice stick into six pieces 2.5 cm (1 in.) in length. Black will **model** sugar and red will **model** phosphate. Push the string through the center of the licorice pieces. Make two equal-length strings of alternating licorice colors. These will **model** the sugar-phosphate backbone of DNA.

2. **Model** the bases using four different colored marshmallows. **Record** what color stands for each base (*A*, *T*, *C*, or *G*). Connect six pairs of bases with toothpicks. *A* pairs with *T* and *C* pairs with *G*. Use toothpicks to attach the marshmallow bases to the sugar part of a backbone chain.

3. Complete the DNA by attaching the other end of each toothpick to the sugar on the second backbone chain. Carefully twist the DNA model so it looks like a double helix.

4. Use the masking tape and marking pen to label a sugar-phosphate group and one each of the four bases. Check to make sure all the base pairs in the model are correct. Draw a picture of your model.

Materials
- licorice sticks (2 red and 2 black)
- scissors
- toothpicks
- colored marshmallows, or gumdrops (4 different colors)
- string
- masking tape
- marking pen
- DNA illustration

Analyze Results

Compare your DNA model with pictures of the DNA molecule in the Chapter Opener. How is the model you built similar and different from the pictures?

Create Explanations

1. How do DNA molecules fit together?

2. You and a friend each build separate DNA models with 1000 bases. How much of your models do you think would look similar? How much would look different? How is this related to evidence of design by our Creator?

3. Why do you think the DNA in chromosomes must be folded?

In the Nucleus

Imagine peering into the nucleus of a cell from your body. Inside, you would find 46 **chromosomes**, threadlike structures of DNA. Pull the chromosomes out of the nucleus, unwind them, and you have almost 2 m (6 ft) of DNA! Imagine looking more closely at this DNA. Small sections of DNA are called **genes**, which are the basic units of hereditary information. This information is passed on from parents to offspring. These **inherited traits** are characteristics, such as height and hair color. What other inherited traits do you think are coded in DNA? Genes tell the cell when and where proteins should be formed. **Proteins** are large molecules that are the building blocks for cells. While DNA codes for the subunits called *amino acids* that make up proteins, differences in protein structure and protein activity in cells result in your traits.

Faith Connection

Do you think people in biblical times knew anything about inherited traits? Explain.

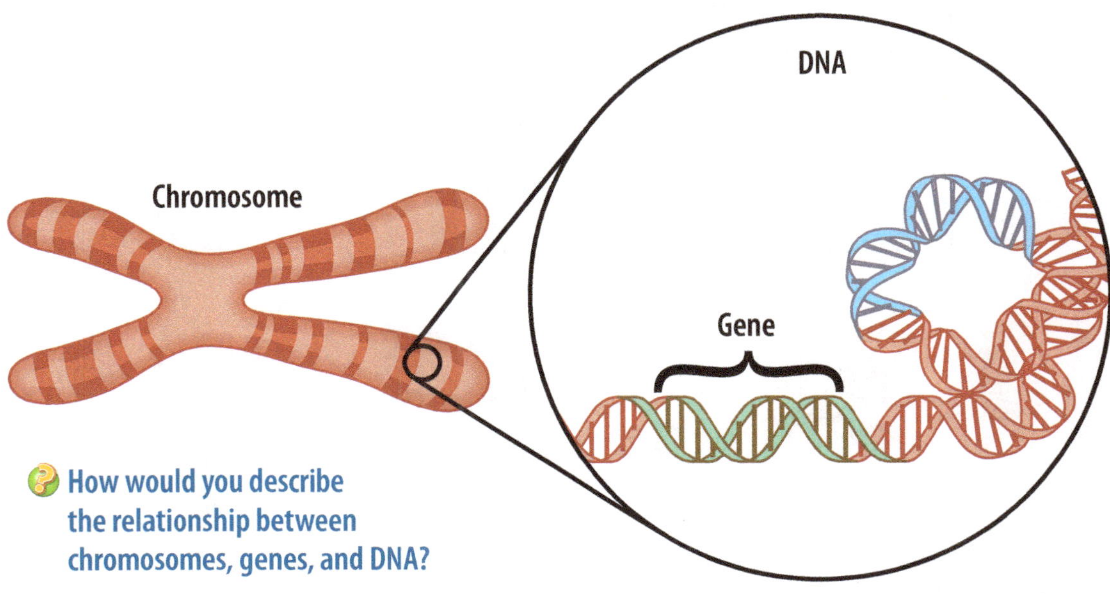

 **How would you describe the relationship between chromosomes, genes, and DNA?**

Math in Science

It is estimated that the human body may contain more than two million proteins, which code for about 25,000 genes. What is the proportion of proteins to genes?

The Twisted Ladder Explain

Workers use a blueprint to interpret what work needs to be completed. DNA is the cell's blueprint. Study the image on this page. Look carefully at the sides of the ladder. What pattern do you see? This twisted ladder is called a **double helix**. It is made of two strands of DNA twisted around each other. What do you think is the advantage of the twisted shape of the DNA molecule?

DNA is made of smaller subunits called **nucleotides**. Each nucleotide has a sugar molecule, a phosphate group, and one of four bases—*adenine, guanine, thymine,* and *cytosine.* Each base has a different shape. They can fit together like pieces of a jigsaw puzzle. Because of their shapes, when they pair up, adenine always pairs with thymine. Guanine always pairs with cytosine. These bonded base pairs hold the two strands of the DNA molecule together.

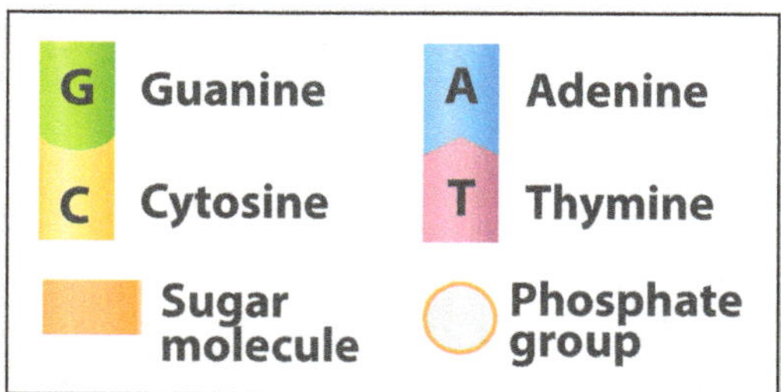

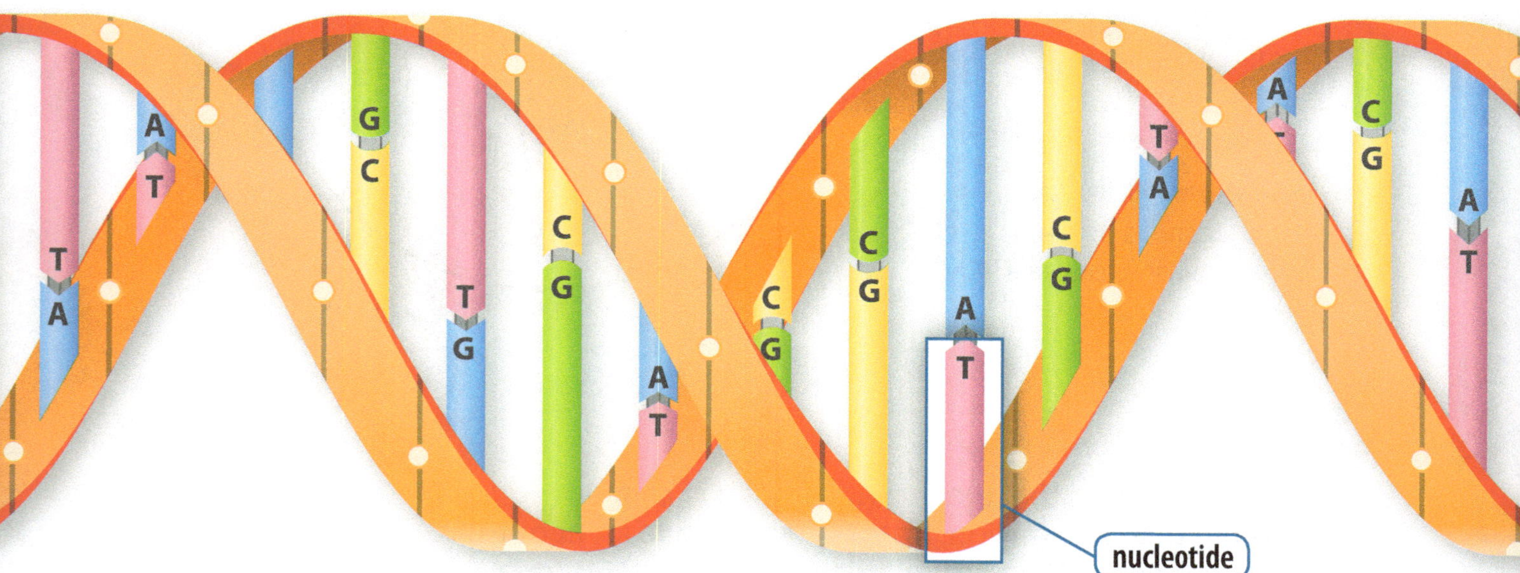

A Double Helix: The structure of the DNA molecule

What might happen if your cells misread those instructions?

The four-letter code of the nucleotides explains how chromosomes can store and use genetic information. These four pairs, AT, TA, GC, and CG, make up the letters of the DNA code. DNA's four nucleotides can be arranged in a multitude of ways. The order of the base pairs provides coded instructions. Structures in your cells read the letters of this code to make the chemicals you need to live. What do you think are some of the instructions DNA gives the cells in your body?

Check out your *Science Journal* for a Structured Inquiry that explores how DNA can be extracted from cells.

Extend

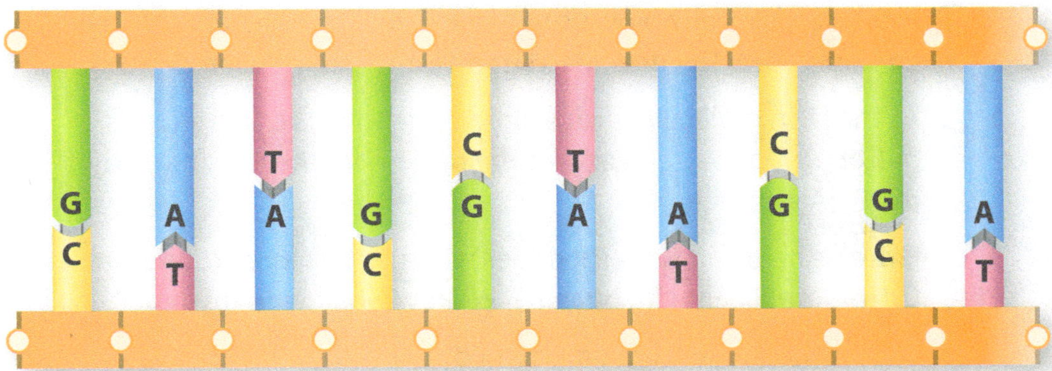

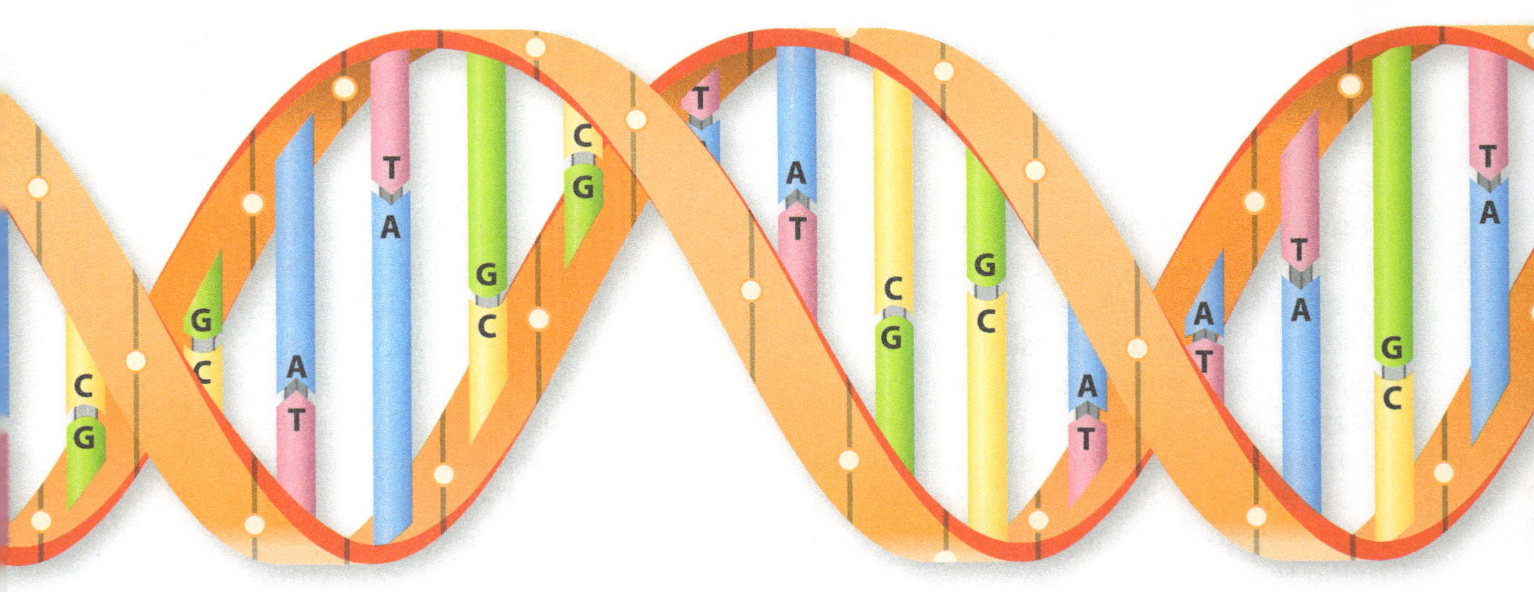

Copying DNA Explain

Before cells can divide, they must duplicate their DNA. Why do you think it is important for the DNA to be duplicated before the cell divides? Making a copy of your cell's blueprint is just the beginning of the building, or growth, process for your body. The double-helix design of the DNA molecule makes the complicated process of duplication much easier.

Genes code for the formation of the proteins. These proteins play a large part in the duplication and packaging of DNA.

1. A group of special proteins is involved in DNA synthesis.

2. These proteins unzip portions of the DNA molecule that need to be duplicated. The unzipped strand is the parent strand.

3. As the DNA molecule unzips along its length, the two strands pull apart. Each strand acts as the pattern for the missing part of the DNA molecule.

4. Proteins help pair up nucleotides with the correct partners. Cytosine will pair with guanine. What will pair with thymine?

5. The finished copies are proofread by the cell.

6. Each strand zips back up, forming two new daughter strands of DNA. The name "daughter" indicates an identical strand is produced.

Explore-a-Lab

Structured Inquiry

How can DNA fingerprinting help identify an unknown virus?

Just like humans have unique fingerprints, individuals also have unique patterns in their DNA base pairs. In fact, all organisms have their own unique DNA pattern. These patterns are called DNA fingerprints. Imagine there is an unknown virus outbreak. How would scientists identify the unknown virus? Why would this be important? You and your group members are scientists trying to identify an unknown virus. The envelope that you are given contains hand lenses and DNA fingerprints (UPC barcodes) from unknown viruses. There are also DNA fingerprints from known viruses: (A) equine flu, (B) swine flu, and (C) avian flu. Your group must decide which of the unknown viruses match the known viruses. On a sheet of drawing paper, write how you will identify the unknown viruses. Make a data table to show which unknowns match which known viruses. Did you and your group members always agree on the identification of unknowns? How are the barcodes like DNA? Besides identifying unknown viruses, for what other reasons would scientists use DNA fingerprints?

A DNA molecule consists of two parent strands that unzip during duplication.

❓ **Why is it important for proteins to check for errors in the DNA before it is zipped back up?**

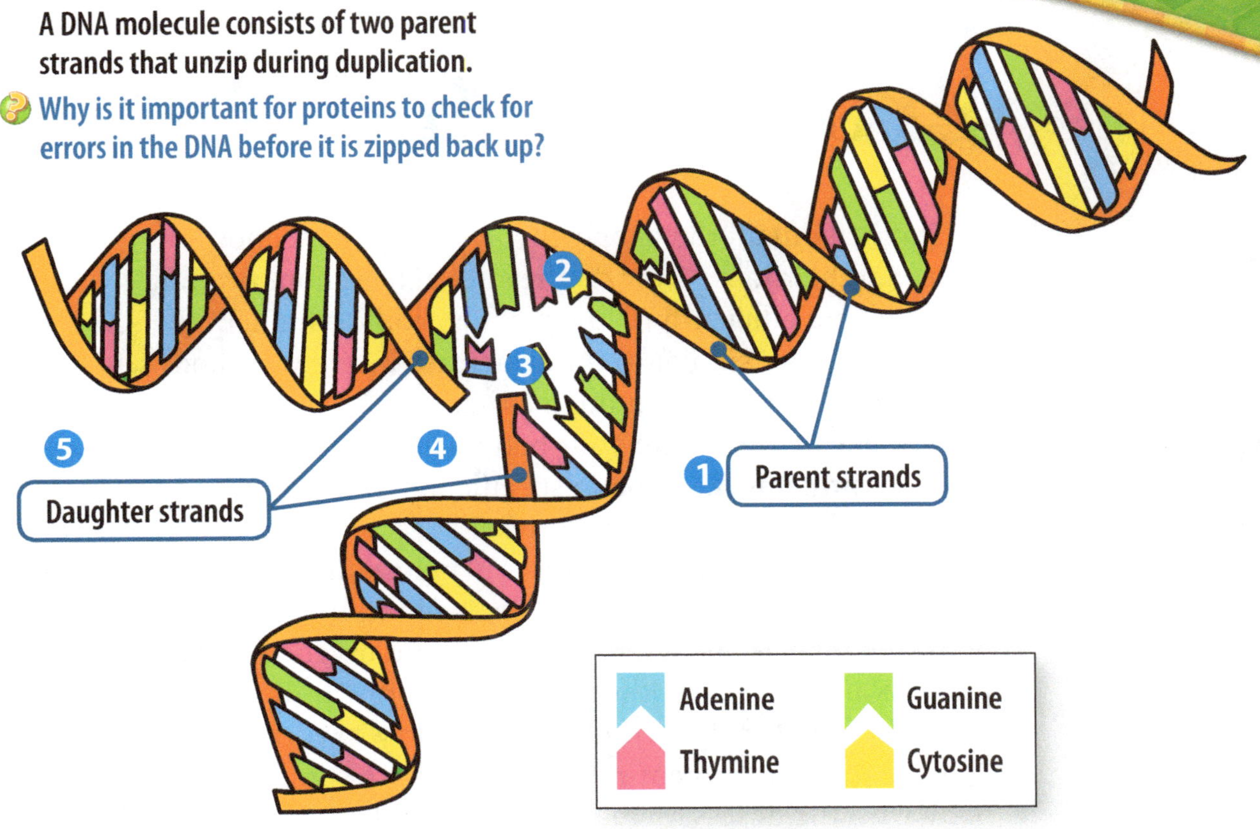

Concept Check   Assess/Reflect

Summary: What is the structure of DNA? DNA is structured as a double helix in which sugar-phosphate chains form the sides of the ladder and base pairs form the rungs that connect the ladder's sides. Its structure was discovered over several decades following multiple experiments. DNA duplicates by unraveling and zipping apart. The complementary base pairs connect to the parent strand. The daughter DNA strands, which are copies of the parent strand, recoil.

1. How many years passed between Miescher's work on DNA and Avery's work on DNA? Why do you think there were so many years between these discoveries?

2. Identify the parts of the DNA molecule. Explain how the parts of the DNA molecule fit together.

3. Suppose a molecule of DNA contains 150,000 guanine bases. How many cytosine bases would you expect the two replicated strands of this DNA molecule to contain? Explain your answer.

4. How does the structure of DNA allow the molecule to store information?

5. How does the structure of DNA allow it to copy itself?

Essential Question

How Are Proteins Made?

Think about all the people involved in your school. What are their jobs? How do they keep the school running properly and efficiently? How are the people who help to run your school similar to the structures inside your body's cells? Proteins work to build, maintain, and repair the cells and tissues that make up your body. There are thousands of different proteins. Each has a specific job to do in order to keep your body running properly.

What are some ways proteins are important for life? Specialized protein, called *collagen,* is found throughout your body. It connects and supports tissues and organs, such as muscles and lungs. Another specialized protein in your body is *hemoglobin.* It helps red blood cells carry oxygen.

When you eat protein-rich foods, you are supplying your body with the necessary materials for building proteins. What are some protein-rich foods? How much of these foods do you think you must eat each day for proper protein formation? How do you think the proteins you get in your food are made into the proteins that make up your cells?

The digestive system breaks down protein-rich foods, such as eggs, almonds, and lentils, into usable amino acids. These amino acids are used to build the proteins we need for our bodies.

Why do you think it is important to get your protein from a variety of foods?

Build a Protein

How are proteins built?

Procedure

Materials
- 4 chenille sticks (black, red, blue, green)
- colored beads (red, blue, green, yellow, orange) to represent bases
- beads (20 assorted) to represent amino acids
- codon chart
- colored pencils

1. **Model** a DNA molecule. String beads on the black chenille stick for the following sequence: TAC-GAT-CTG-TCT-CAA-ACC-TAT-CAC-GAA-TTT-AAG-ATC. Use the table in your *Science Journal* to identify the correct bead colors to **model** each nitrogen base. **Record** the gene sequence. Color-code them with the pencils.

2. Place a red chenille stick alongside the black stick. Use base-pairing rules to transcribe the DNA code into RNA. (In RNA, T codes for A, G codes for C, C codes for G, and A codes for U.) **Record** and color-code the base sequence on the RNA strand.

3. The codon, AUG, signals the start of translation. Translate the RNA code into a sequence of amino acids. Choose beads to represent the amino acids. Assemble the protein chain on the green stick. Stop assembling the protein chain when you reach UAG, which signals to stop. **Record** the amino acids in the chain.

4. The model you created represents an unfolded protein. Wrap part of it into a spiral shape and part of it into a zigzag shape. Draw a picture of this folded protein structure.

Analyze Results

Compare what you did in this activity with what happens inside a cell. How is this model limited?

Create Explanations

1. How are proteins built?

2. What do the beads and chenille sticks represent in this model? How is the model unlike what happens in the cell?

3. If errors were made while building the model, how did you fix it? What would happen to the RNA if errors are not corrected?

4. How do you think folding the protein models changes their properties?

RNA Explain

RNA, ribonucleic acid, works with DNA to build proteins. DNA is well protected in the nucleus. Why must the DNA be protected? When proteins must be formed, DNA sends messenger molecules through the nuclear membrane to the rest of the cell molecules to tell it to build proteins or control other cell activities. These messengers are in the form of RNA.

Study the chart to learn more about RNA and to compare it with DNA. Note that *uracil* substitutes for thymine in RNA.

	RNA	DNA
Full name	ribonucleic acid	deoxyribonucleic acid
Backbone	sugar and phosphate units	sugar and phosphate units
Bases	adenine, uracil, guanine, cytosine	adenine, thymine, guanine, cytosine
Structure	single strand	double strand
Function	moves genetic information from DNA in the nucleus to ribosomes in the cytoplasm to build proteins	stores genetic information; helps in protein synthesis; provides instructions to RNA

Protein Synthesis

RNA is important in the making of proteins. Protein molecules are large molecules made out of smaller units called **amino acids**. Our Creator designed 20 amino acids that combine to form more than 100,000 different proteins! What are the odds that the right combinations are made by chance? The process of building the proteins is called **protein synthesis**. This process takes place in the ribosomes of the cell.

Scripture Spotlight

One of the over 100,000 types of proteins in our bodies is called laminin. This important protein helps hold cells and tissues together. How does **Colossians 1:17** say things are held together? (Read it in NIV, NLT, or NASB.) Go online and find a picture that shows the shape of laminin.

Lesson Activity

Develop a Venn diagram that compares and contrasts DNA and RNA. Use colored beads and ribbon or chenille sticks to model the molecules. Incorporate the Venn diagram and models into a presentation style of your choice, such as poster or diorama.

Why is it important that the structures of DNA and RNA are different?

Transcription

Protein synthesis is a two-step process. The first step involves **transcription**. Have you ever transcribed classroom notes or sheets of music? How would you describe that process? Similar to you copying notes or music, during transcription in a cell, a gene that codes for a protein is copied as an RNA molecule. This takes place in the nucleus of eukaryotes. Where do you think it takes place in a prokaryote?

Steps of Transcription

1. The DNA strand unzips.

2. One of the DNA strands becomes the pattern for the RNA.

3. The RNA bases pair up with the corresponding DNA bases.

4. RNA is transcribed and released from the DNA strand.

5. The DNA zips closed.

6. The RNA leaves the nucleus.

Scripture Spotlight

DNA is copied during the transcription process. In **Jeremiah 31:33**, where does God want to write a copy of His law?

DNA acts as a pattern or template to transcribe matching base pairs. The process takes place in the nucleus.

What is the product of DNA transcription?

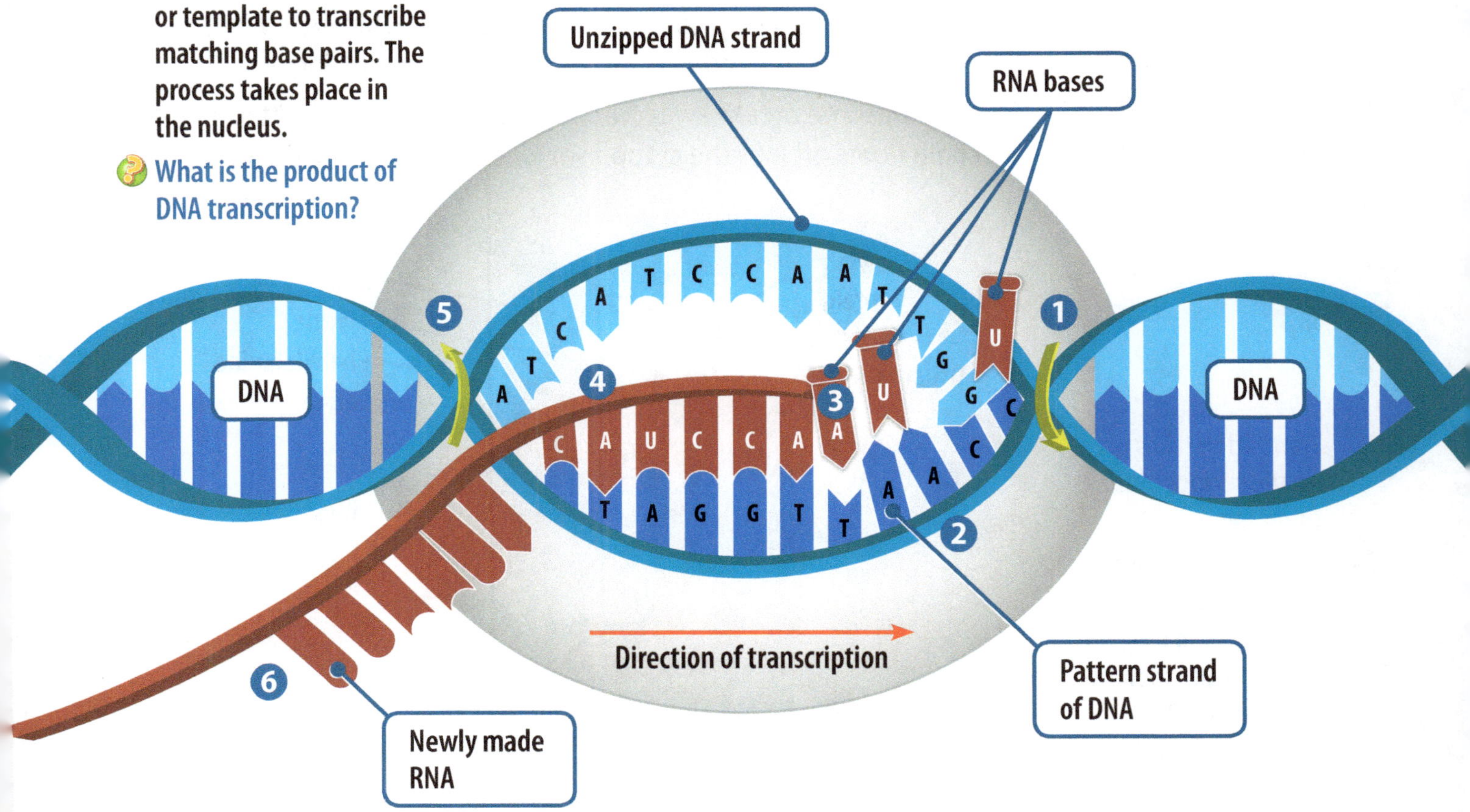

Translation

At the end of transcription, the RNA leaves the nucleus. How does it know what to do? Have you ever read the directions for building a bicycle or a model car? How did you understand the directions? How did you know where to start and stop? Similar to reading directions, the second step in protein synthesis is **translation**, which occurs outside the nucleus. During translation, the genetic code on the RNA is read and proteins are built.

Steps of Translation

1. The RNA attaches to a ribosome.

2. The ribosome reads the RNA three bases at a time. Each of these three bases is called a **codon**.

3. Protein assembly begins at a start codon. The start codon contains AUG or AUA.

4. Other RNA molecules in the cytoplasm carry anticodons. An **anticodon** has three bases that pair with a codon using base-pair matching, such as CGU and GCA.

5. The ribosome moves one codon at a time along the RNA strand until it reaches a stop codon, such as UGA, UAA, or UAG. What do you think would happen if the stop codon was missing?

6. The ribosome breaks away from the RNA. The newly built protein chain floats off into the cytoplasm where it is used by the cell.

Translation takes place on the surface of the ribosome.

How does the ribosome know which RNA and amino acid to use and where to add them to the sequence?

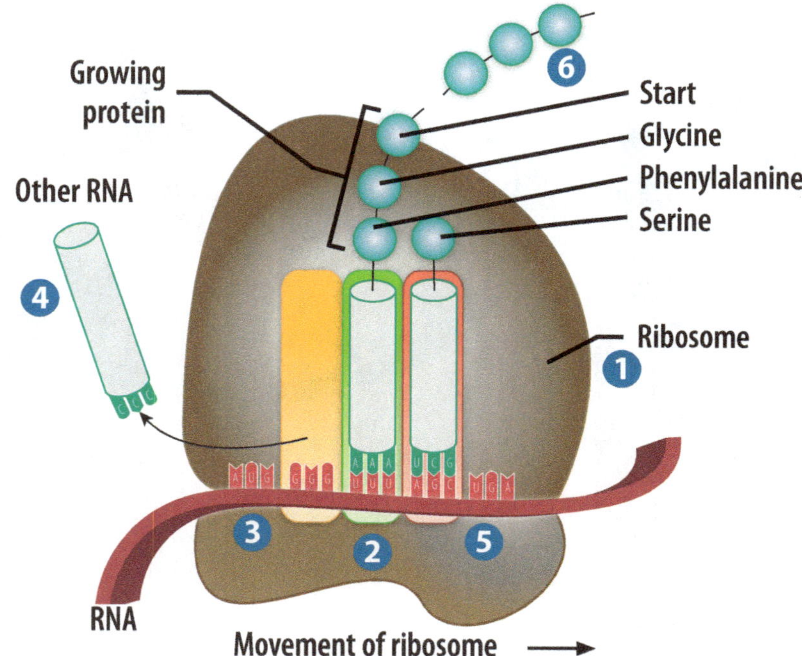

Translating the Genetic Code

Imagine that a base is like a letter of the alphabet and that a codon is a word. The "words" identify the specific amino acids needed to form each protein. Study the codon chart below. Follow these steps to find out which amino acid is coded by AUG.

1. In the first position, find "A."

2. In the second position, find "U."

3. In the third position, find "G."

4. Using your fingers or a ruler as a guide, find where the letters A, U, and G meet on the chart. They meet at the word *methionine*.

So, AUG codes for the amino acid methionine. Follow the four steps to learn which amino acid is coded by AUA. What amino acid is coded by GCG? What other codons code for this same amino acid?

SECOND Position

		U	C	A	G	
U		phenylalanine	serine	tyrosine	cysteine	U
						C
		leucine		stop	stop	A
				stop	tryptophan	G
C		leucine	proline	histidine		U
						C
				glutamine		A
						G
A		isoleucine	threonine	asparagine	serine	U
						C
		methionine		lysine	arginine	A
						G
G		valine	alanine	aspartate	glycine	U
						C
				glutamine		A
						G

FIRST Position (left axis) — **THIRD Position** (right axis)

This codon chart lists 20 amino acids active in humans.

Which three-letter code would specify the amino acid tryptophan?

Proteins are found in all living organisms. The genetic code to form these proteins is almost universal. This means that the genetic code used in humans is used in almost all other types of organisms. How does the idea of a universal genetic code give evidence of a Creator of all things?

How does the DNA code provide directions to the cell?

The order of the base pairs in DNA provides a coded message to an organism's cells, providing detailed instructions on how to produce the compounds necessary for growth and survival. Write a one-sentence message on a sheet of paper. Then, use the DNA Message Code table to translate your message into DNA code. Trade your DNA message with a partner. Using the DNA Message Code table, decipher your partner's message. Were you able to decode your partner's message? What did you find difficult about translating the message? Did you make any mistakes when translating? What would happen to an organism if its DNA was copied with mistakes?

DNA Message Code

Letter	Base Triplet	Letter	Base Triplet
a	CCC	o	GAG
b	CCG	p	GAT
c	CCT	q	GCA
d	CCA	r	GCC
e	CGA	s	GCG
f	CGC	t	ACA
g	CGG	u	ACG
h	CGT	v	ATA
i	CTA	w	ATG
j	CTC	x	TAG
k	CTG	y	TCG
l	CTT	z	TGG
m	CAA	space	TTC
n	GAC	period	AAA

Concept Check 

Assess/Reflect

Summary: How are proteins made? Proteins are large molecules found throughout your body that are needed to keep your body running properly. There are many specialized types of proteins. Proteins are created during transcription and translation. In transcription, DNA is used as a pattern to make an RNA strand. This RNA strand carries the genetic information from the nucleus into the cytoplasm of the cell. During translation, ribosomes read the genetic information from the RNA, and amino acids are assembled into long chains, forming a protein. The genetic code is based on four nucleotide bases and is almost universal in nature.

1. How are genes, DNA, and proteins related?

2. What are three ways in which an RNA molecule differs from a DNA molecule?

3. Retell the process of protein synthesis.

4. Compare how RNA and DNA use the genetic code.

How Do Cells Function?

In the last lesson, we looked at the blueprints—the DNA and RNA—needed to help your cells grow and function. We compared these blueprints to those used to build a house. When a house is built, many workers cooperate to make the house functional for the family. When the construction is complete and the family moves in, is there any work remaining to be done? What would happen if a tree fell on the house during a storm and damaged the kitchen?

An organism's cells work together like the construction crew so the organism can function as God intended. The DNA in each cell gives the instructions for the organelles to carry out. Different cells perform different jobs, yet all cells have similar needs. In multicellular organisms, the cells cooperate to perform life functions for the organism. This requires organization and a set chain of command so each worker, or cell, knows what its job is and performs it as intended. Like in a home, maintenance work needs to be done by the family or hired helpers. When cells are damaged, perhaps by a cut, the surrounding cells start to make new cells, like the construction crew that would fix the damaged kitchen. How well do you think your body would function if each cell had to perform all of its own tasks, such as obtaining its own food, repairing injuries, and protecting itself from invaders?

This is an image of pine wood magnified 200 times.

How do you think the different cells in a pine tree help it perform its life functions?

Objectives

- Describe the cell cycle.
- Explain how proteins are released into the blood.
- Explain how hormones influence body activities.
- Identify how cells help the body respond to changing conditions.
- Describe the immune response.

Vocabulary

cell cycle
daughter cells
secretion
hormone
gland
target cell
leukocyte
acquired immunity
allergic reaction

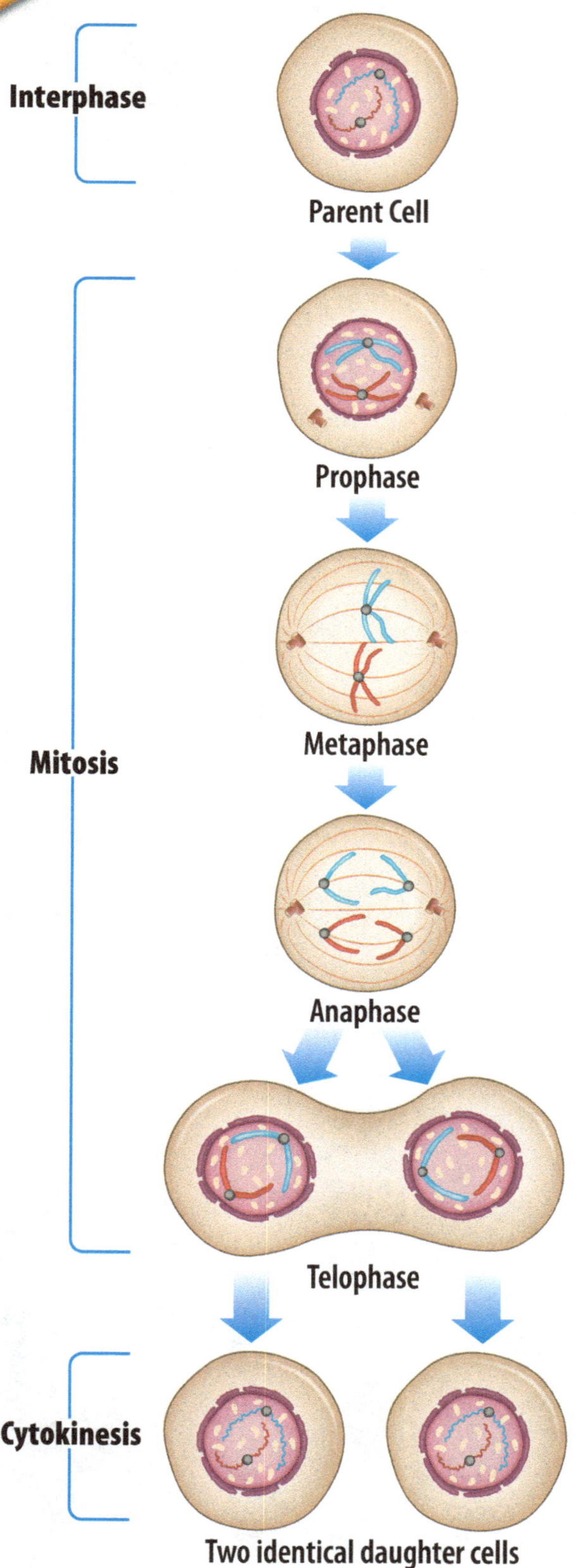

The Cell Cycle `Explain`

The **cell cycle** is a set of events in the life of a cell that leads to cell division. It results in two new cells, called **daughter cells**. Why might your body need new cells? How do you think the daughter cells compare to the original cell? The cell cycle is a controlled process, so it progresses in the correct order. There are three main parts of the cell cycle: interphase, mitosis, and cytokinesis.

The Cell Cycle
Interphase
• The cell performs its regular functions. About 90% of the cell's life is in this part.
• At the end of this part, chromosomes are duplicated. Why would the chromosomes need to be duplicated?
• There are checkpoints during this part to make sure the cell can safely divide.
Mitosis
• The cell gets ready to divide. Cell division happens in four phases: • Prophase: The membrane around the nucleus breaks down. • Metaphase: The duplicated chromosomes line up in the middle of the cell. • Anaphase: The chromosomes are pulled to opposite ends of the cell. • Telophase: New nuclear membranes form around the two sets of chromosomes.
Cytokinesis
• The cytoplasm must be divided so that two new daughter cells can be formed.
• In an animal cell, the cytoplasm pinches inward. Two identical daughter cells are formed.
• In a plant cell, a cell wall forms in order to create two identical daughter cells.

Identifying Cell Cycle Phases

What happens during different phases of the cell cycle?

SAFETY: Follow proper microscope handling techniques. Use care when using the microscope and microscope slides.

Materials
- microscope
- prepared microscope slide of plant mitosis

Procedure

1. Focus on an area of the plant specimen using low power. Take turns with your group members, **observing** the cells.

2. Then, switch to medium power. Use the fine adjustment knob on the microscope to bring the image into focus, if needed. Again, take turns carefully **observing** the cells.

3. Carefully sketch one of the cells that you see. Then, identify the phase the cell appears to be in. Write the name of the phase next to your sketch.

4. Find a cell that appears to be in a different phase and carefully sketch it. Then, identify the phase and write the name of the phase next to your sketch.

5. Repeat Step 4 with a third cell.

Analyze Results

Compare the three cells you drew. Were you more easily able to identify the phase in the cell cycle when you had one, two, or three images to compare? Discuss this with your group.

Create Explanations

1. What happens during different phases of the cell cycle?

2. Did you find the identification process challenging? Explain.

3. Did your group agree on which phases were shown in the three cells you selected? Explain.

4. Which seems to be more common: cells undergoing mitosis or cells that appear to be in interphase? Explain.

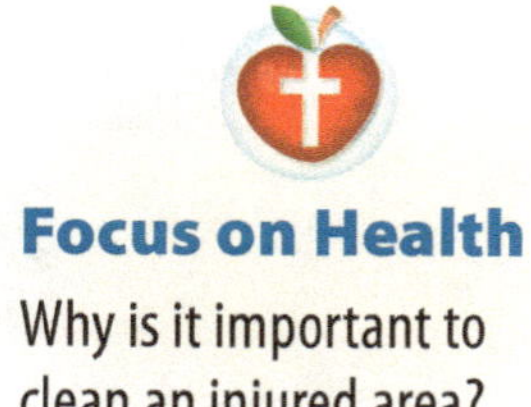

Controlling the Cell Cycle Explain

The movement of a cell through the cell cycle is controlled by proteins. These proteins work inside the cell. They allow the cell to carry out its normal job, duplicate chromosomes, or undergo mitosis. They operate during the checkpoints of the cell cycle. The proteins do not allow cell division to happen unless the chromosomes have been copied properly. If the DNA has an error, another protein may block the cell cycle. What do you think might happen to the cell if the cell cycle is blocked?

Other proteins respond from outside cells. Proteins can increase the rate of cell division during certain developmental stages, such as puberty. Proteins can also increase cell division if there is an injury to the body. Cells near a new injury begin to grow and divide rapidly to repair the tissues. How do you know when the injury is healed?

Cell proteins on neighboring cells often slow down cell division to prevent excessive cell growth. Cancer cells are an example of excessive cell growth. They are abnormal cells that grow and divide, crowding the neighboring, healthy cells. What is it about a cancer cell that is abnormal?

Protein Secretion

Recall that proteins are necessary for building, repairing, and maintaining your body's cells. How do they travel from one location to another in order to do their jobs? A cell goes through a series of steps to move proteins out of the cell during a process called **secretion**. Follow the path in the picture on the next page as you study the steps of protein secretion. What happens if the path for the protein is blocked at one of these steps?

1. A ribosome makes a protein in the endoplasmic reticulum.

2. The protein is packaged and moved to the Golgi apparatus.

3. The protein is unpacked and repackaged in the Golgi apparatus.

4. The newly packaged protein is released from the Golgi apparatus.

5. The protein is transported to the cell membrane.

6. The protein is delivered outside the cell. Once outside the cell, how does the protein identify and arrive at its new destination?

Hormone Action

Some of your cells create hormones, which are secreted by the process of protein synthesis. **Hormones** are chemical messengers produced by specialized groups of cells called **glands**. Glands make up the endocrine system, which will be studied further in Chapter 4. The cells in the glands secrete these proteins. Your brain signals the glands in your body to secrete these hormones directly into the blood.

Once in the blood, hormones travel to **target cells**. The target cells respond to the hormones. When the hormone binds with DNA in the target cell, the cell responds by directing protein synthesis or controlling specific cell activities.

Different hormones do different things for your body. Growth hormones help bones and other tissues in your body to grow and divide. They help the body to use nutrients and minerals for growth. Some hormones help regulate the amount of salt and water in your body. Still others help you respond to stress and injury.

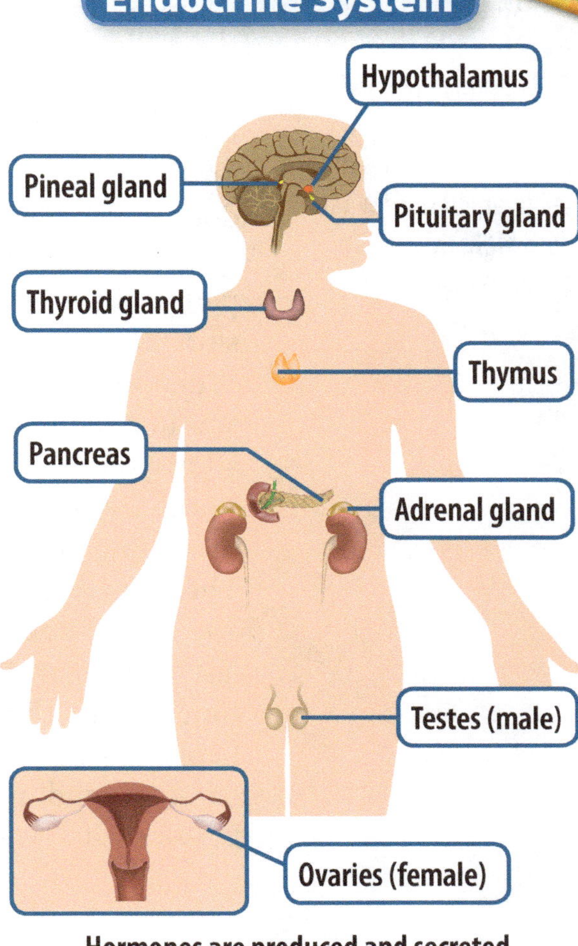

Hormones are produced and secreted from the glands of the endocrine system.

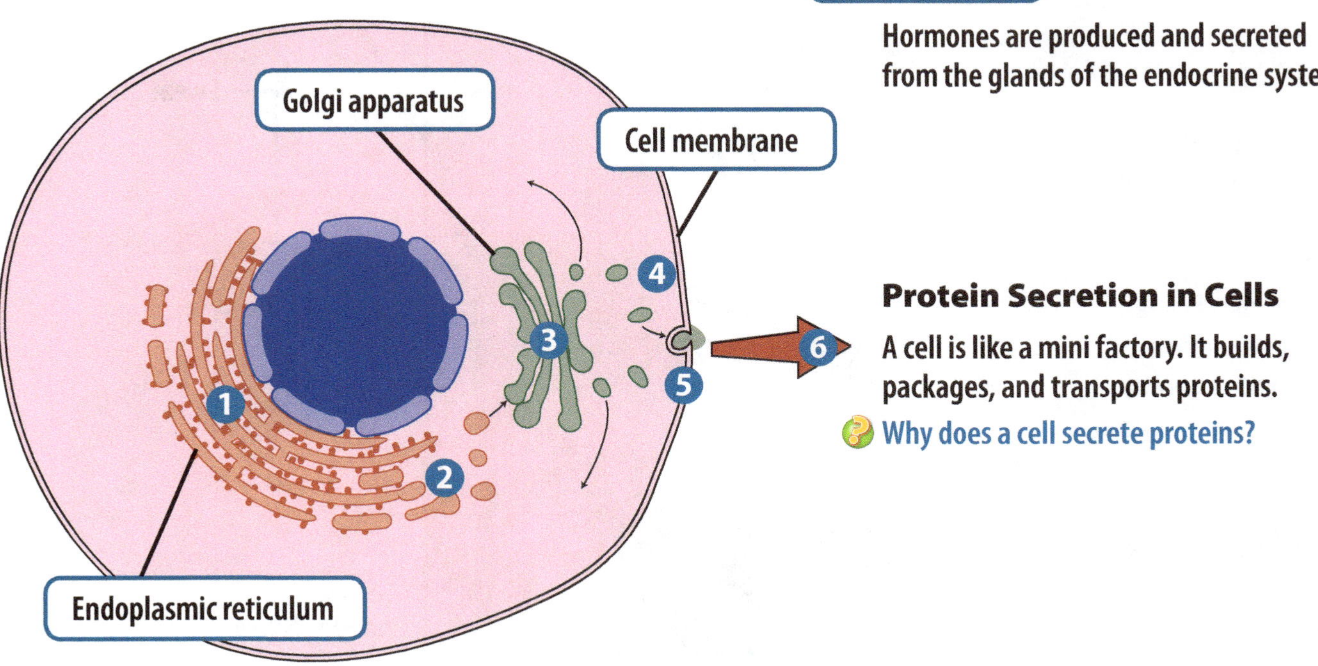

Protein Secretion in Cells

A cell is like a mini factory. It builds, packages, and transports proteins.

Why does a cell secrete proteins?

Maintaining Homeostasis Explain

How can you keep a comfortable temperature inside your house? Temperature is only one factor in a "balanced state." But homeostasis involves more than temperature. Just like central heating and cooling systems, your cells maintain a state of homeostasis, or a stable internal environment. Hormones help the cells do this. Why is it important for your body to maintain a state of homeostasis? How does your body respond to internal and external changes?

Most people have a body temperature near 37°C (98.6°F). A part of the brain contains temperature sensors. As soon as your skin temperature rises above 37°C (98.6°F), you start to sweat. When the water from your sweat evaporates, you feel cooler. You maintain your internal temperature because hormones allow your blood vessels to contract or widen. When you are warm, hormones send signals that cause blood vessels to widen, increasing blood flow. This brings heat to the surface of the skin. The brain also sends out signals if your skin temperature drops. You stop sweating and you might start shivering. When was the last time you remember shivering? Did you shiver for a long time? While shivering, you may also develop goose bumps. This happens when tiny muscles at the base of your hairs cause your hairs to stand up straight. How do goose bumps help your body? Besides when you are cold, when else might you get goose bumps?

The skin responds to changes in temperature to maintain a state of homeostasis.

How does the skin help the body respond to changes in temperature?

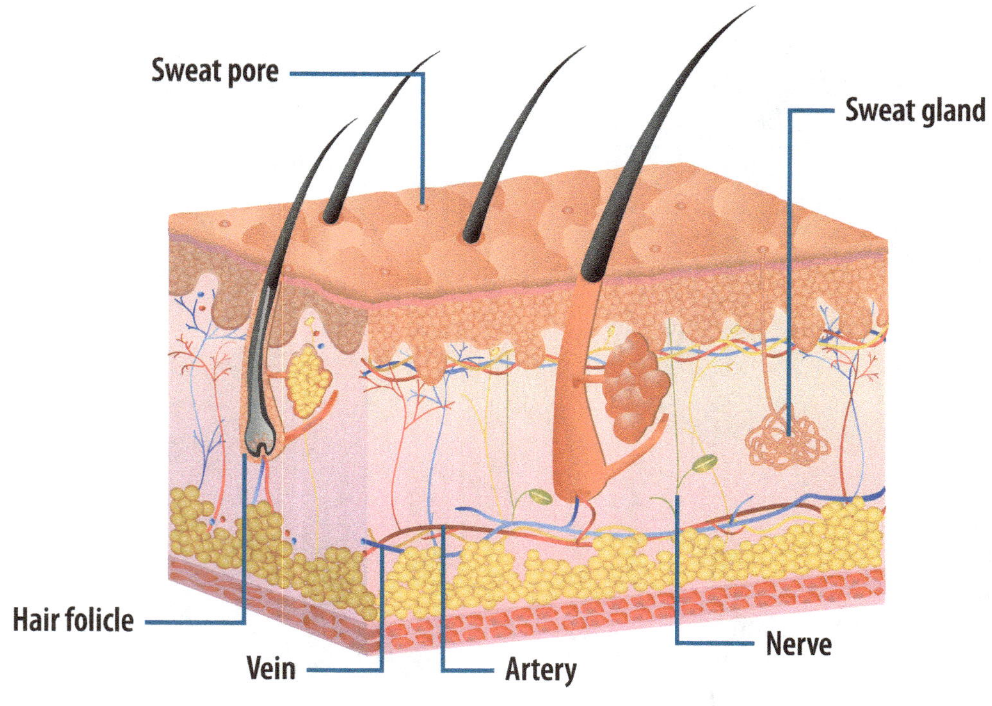

Homeostasis refers to more than body temperature. How would you describe your breathing right now? How would your breathing change if you were running? When you breathe, you take in oxygen that is transported to all the cells of your body. The cells use the oxygen to help make energy. There must be a balance between how much oxygen you take in and how much carbon dioxide you breathe out. As you run, why do your heart rate and breathing increase? How do your heart rate and breathing respond when you stop running? How do these changes maintain homeostasis?

Nerve signals and hormones interact to respond to danger. Our cells produce many stress hormones to prepare the body to deal with a threat. They raise blood sugar and increase blood flow and respiration rates so the body can run fast, fight harder, and think faster. Once the threat is over, nerve signals help restore normal function. How do these changes maintain homeostasis? How do you think your body would feel after overcoming a dangerous or scary situation? What might happen in the body if these situations occur too frequently?

How can you help to maintain your body's homeostasis when it is cold outside? When it is hot?

Explore-a-Lab

Structured Inquiry

How well do different body coverings help cold-climate animals' bodies maintain homeostasis?

Animals have different techniques to maintain homeostasis. With a partner, go to one of the insulation stations. Suspended in ice water are a bag with bubble wrap in it, a second bag with solid shortening in it, and a third bag with nothing in it. Test how the different insulation options compare. Place one hand into the bubble wrap bag and your other hand into the bag with no insulation for one minute. Have your partner keep time. Remove your hands from the bags after one minute. Record your observations on a sheet of paper. Wait two more minutes to let your hands readjust to room temperature. Compare the bag with the shortening and the bag with no insulation for one minute. Write down your observations. Switch roles with your partner. How do the different types of insulation compare? How does this relate to how cold-climate animals maintain homeostasis?

The Immune Response Explain

The Bible tells us that "a cheerful heart is good medicine" (**Proverbs 17:22**). Not only has God given us the ability to have joy and laughter in our lives, He has provided us with an *immune system* to protect our bodies against infection. How do you think a cheerful heart helps the immune system work better? What might happen to your body if your immune system did not function properly?

The immune system produces special cells that attack foreign substances. White blood cells, or **leukocytes**, are parts of the blood that defend the body against disease. They engulf, or "swallow," foreign particles. When the body is injured, leukocytes increase in numbers and move to the injury. The immune system releases chemicals that increase body temperature. This can be experienced as a fever or as a temperature increase around a wound. An increase in temperature helps because many disease-causing organisms live within a narrow temperature range.

Leukocytes learn to tell the difference between your body tissues and foreign substances. These cells "remember" the foreign substances so they can respond faster if you are exposed to them again. This protection is called **acquired immunity**.

While immunity happens naturally, you can also gain immunity through vaccinations. Vaccines are made from a weak, dead, or partial version of a disease. When you receive a vaccination, the cells of your immune system learn to fight the disease without you having to get sick. If you are exposed to that disease in the future, your body will recognize it and know how to respond.

Allergies

The immune system does not always work like it should. *Allergies* involve an immune response to a substance that the cells of most people treat as harmless. Common allergies include reactions to dust, pollen, food products, or bee stings. These substances cause immune cells to release chemicals that produce an **allergic reaction**. Allergic reactions vary between people. They include swelling, rashes, hives, sneezing, and itching. What are some allergic reactions that might be fatal?

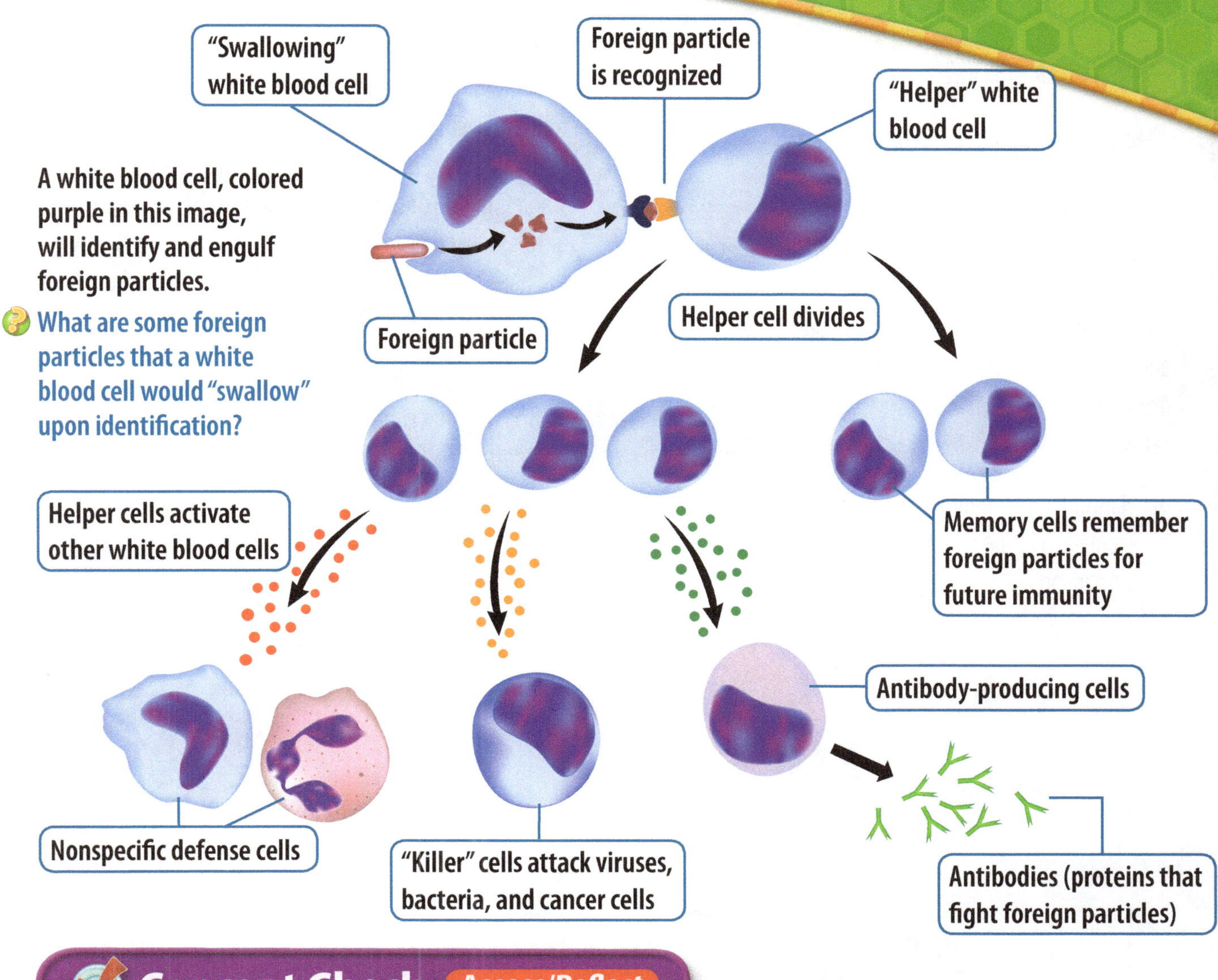

A white blood cell, colored purple in this image, will identify and engulf foreign particles.

What are some foreign particles that a white blood cell would "swallow" upon identification?

Concept Check Assess/Reflect

Summary: How do cells function? Cells regulate processes in the body through the cell cycle, which consists of interphase, mitosis, and cytokinesis. The proteins created regulate cell functions. In addition, glands produce hormones that regulate body functions and maintain homeostasis. Finally, the body protects itself from foreign invaders using the cells of the immune system.

1. Describe two ways that the cell cycle is regulated.

2. What is the purpose of leukocytes?

3. What are some things that hormones control?

4. Suppose you got a paper cut on your finger. Describe some of the ways your cells would respond to the damage.

5. How does your body react in a potentially dangerous situation?

DNA Nanorobots

Imagine a time when being diagnosed with cancer would simply mean a trip to a technician for a shot of nanorobots. Then, as you went about your daily life, the nanorobots would scurry throughout your body, unseen and unfelt, and kill off all the cancer cells. This may sound like science fiction, but it is nearly science fact, thanks to the research of scientists like Harvard researcher Shawn Douglas.

Nanorobots, or nanobots, are tiny devices designed by scientists that are on a scale of about 100 nanometers. In comparison, a human hair has a diameter of about 100,000 nanometers!

A DNA nanobot is created out of DNA components called nucleotides. When the specific nucleotides are assembled in the right way, they pair up with one another. Shawn Douglas's team went one step further and used "DNA origami." Origami is the Japanese art of paper folding. Douglas and his team developed a way to fold the DNA nucleotides into more complex shapes. The folds are held together with nucleotide strands called staples. In one design, the nanobot could be "unlocked" only by leukemia cells (a specific type of cancer). When the nanobot opens, it releases a molecule designed to kill the leukemia cell. No healthy cells were harmed.

In addition, scientists are also working on nanobot technology that could warn people of an impending heart attack. A nanosensor injected into a person's bloodstream would regularly monitor the blood. If the nanosensor determined that a heart attack was imminent, it would send a special ringtone to the person's cell phone, warning them to see the doctor immediately.

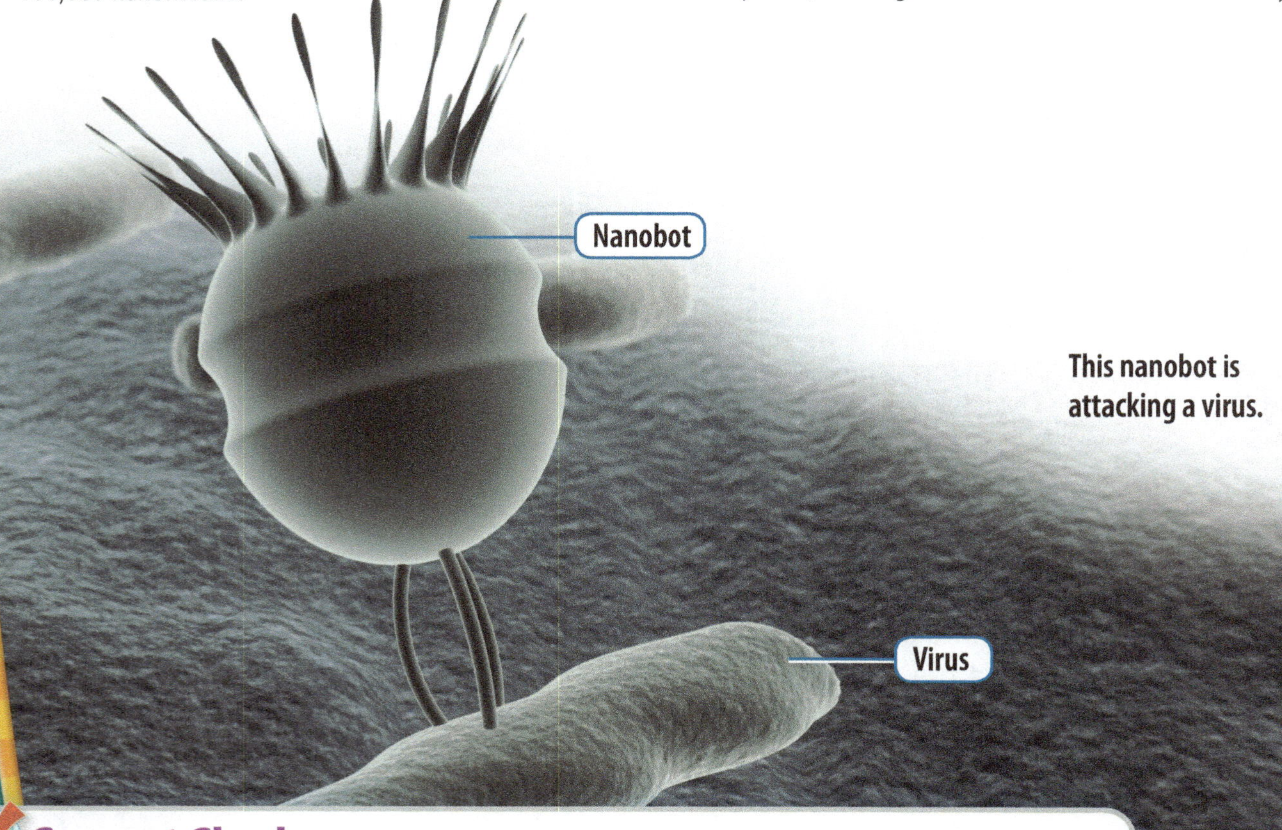

This nanobot is attacking a virus.

✔ Concept Check

1. How can DNA nanobot technology help save lives?

2. How did scientists use origami to create nanobots?

Immunologist

Immunologists are medical doctors who specialize in the study and treatment of diseases that involve the immune system. They may also specialize in the study of disease immunity. Diseases, such as allergies, asthma, and pneumonia, are of particular interest because they affect people's natural immunity. In fact, immunologists are sometimes referred to as allergists.

People who are interested in becoming immunologists must first complete medical school. Once they have a medical degree, three years of additional training in either internal medicine or pediatrics is required. Pediatrics is the branch of medicine that relates to children and illnesses specific to children. Then, two more years of specialized training, called a fellowship, in allergies or immunology is required. And finally, an examination for certification is a necessary final step.

Many immunologists focus their careers on clinical work. This means they see patients and help individuals with their immunological illnesses and allergies. They may work in children's hospitals, community hospitals, or have a private medical practice.

Other immunologists may prefer to work in scientific research. These scientists work in laboratories to study cell interactions and conduct tests. Research helps doctors understand the immune system better and find new treatments and medicines for immunological illnesses.

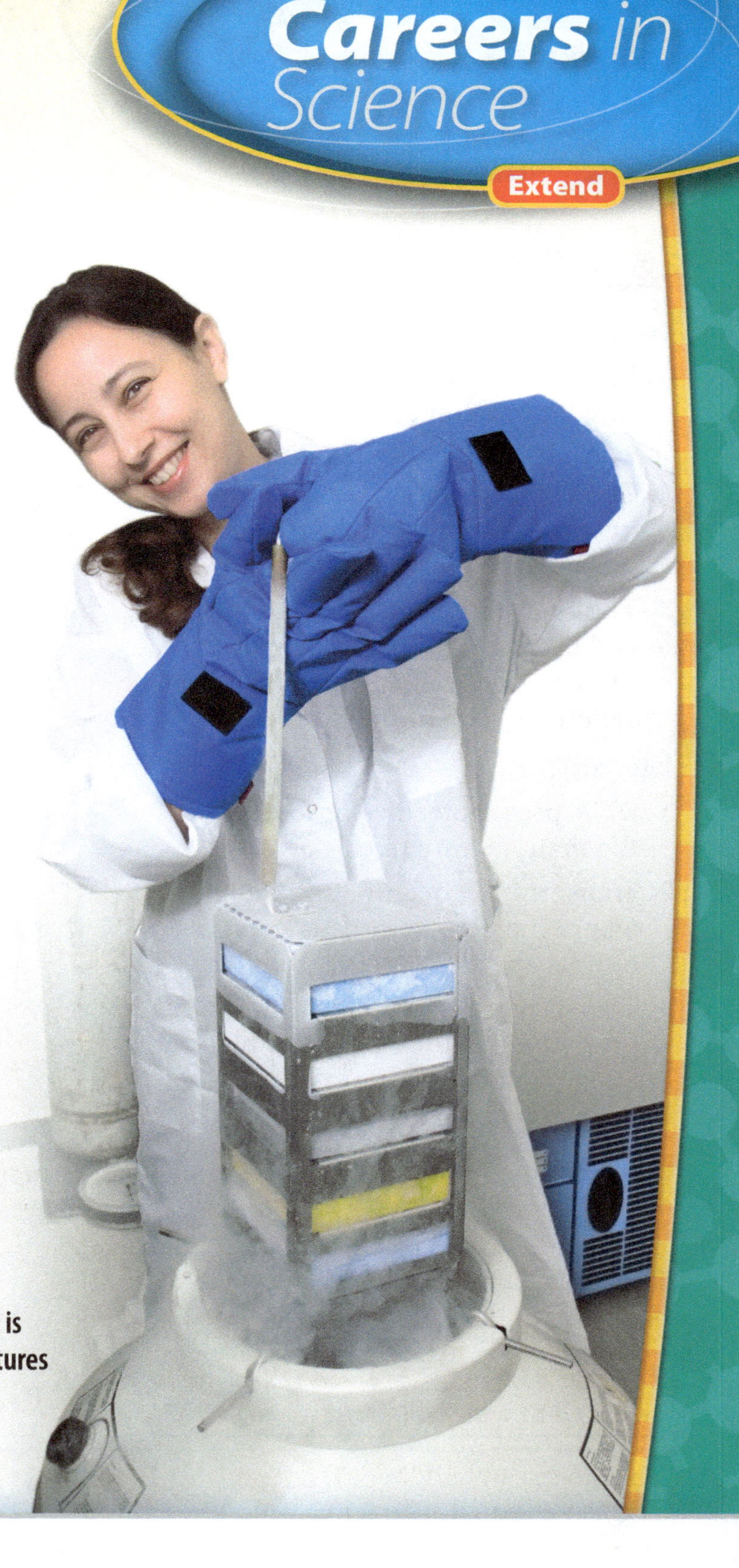

This immunologist is freezing tissue cultures in liquid nitrogen.

Concept Check

1. Why might immunologists need training in pediatric medicine?
2. How might a fellow classmate who has a nut allergy benefit from an immunologist?

Study Guide

Lesson 1

1. The work of many scientists paved the way for Watson and Crick to construct a 3-D model of the structure of DNA.

2. The DNA molecule takes the form of a twisted ladder with its base steps inside the sugar-phosphate sides.

3. The four bases—adenine and thymine and cytosine and guanine—in DNA can be arranged in many ways to store genetic information.

Lesson 2

1. DNA uses molecules of RNA to direct the manufacture of cell proteins.

2. Transcription occurs in the nucleus, where the genetic code is copied to build a specific protein.

3. Translation occurs in cytoplasm, where protein molecules are assembled.

4. A ribosome makes a protein. The protein is packaged for transport in the Golgi apparatus.

Lesson 3

1. The cell cycle is a set of steps a cell goes through that ends in cell division.

2. Proteins are made in cells and can be secreted out of cells.

3. Hormones are special proteins that are released into the blood. They carry messages to cells telling the cells what to do.

4. Cells use nerve signals and hormones to maintain homeostasis, or a stable environment.

5. The immune system protects cells and the body against infection or invaders by producing specialized cells that attack foreign substances.

Show What You Know

Visualize It Complete the following concept map in your *Science Journal*.

Situation		Outcome
The basic unit of hereditary information is a(n) **(1)** ___________.	→	Hair color is controlled by a(n) **(2)** ___________.
The four **(3)** ___________ in DNA can be arranged in many different ways.	→	The order of the base pairs in DNA makes up the "letters" of the **(4)** ___________ code.
The process of building **(5)** ___________ begins with information in a gene.	→	A gene that codes for a protein is **(6)** ___________ as an RNA molecule.
During **(7)** ___________, the genetic information is read on the RNA.	→	**(8)** ___________ are assembled.
One **(9)** ___________ path moves proteins out of the cell in response to signals.	→	Many protein molecules help cells **(10)** ___________ body processes.

Explain how each pair of terms is related.

11. gene—protein
12. nucleotide—double helix
13. RNA—translation
14. hormone—secretion
15. leukocyte—acquired immunity

Multiple Choice
Choose the best answer.

16. Who developed the model of DNA?
    A. James D. Watson and Frances Crick
    B. Alfred Hershey and Martha Chase
    C. Oswald Avery
    D. Friedrich Miescher

17. Which cell structure reads the RNA codons (three bases at a time) during protein synthesis?
    A. endoplasmic reticulum
    B. ribosomes
    C. Golgi apparatus
    D. cell membrane

18. Which base pairs with adenine in a DNA molecule?
    A. cytosine
    B. guanine
    C. thymine
    D. uracil

19. When are chromosomes duplicated?
    A. anaphase
    B. cytokinesis
    C. telophase
    D. interphase

20. How do hormones work?
    A. They control homeostasis with electrical signals.
    B. They monitor and regulate the events in the cell cycle.
    C. They bind with DNA in target cells and direct protein synthesis.
    D. They produce special cells that defend the body against disease.

Check Point
Answer the following questions.

21. Explain the difference between DNA, chromosomes, and genes.

22. **Compare** transcription and translation.

23. Explain what causes an allergic reaction.

24. **Sequence** and describe the phases of mitosis.

25. How do proteins respond to help heal a wound?

26. How might you **conduct an investigation** to show that a body adjusts to maintain homeostasis?

27. A person's leukocyte count is high. **Make a hypothesis** as to why this may be so.

2

The Human Body

Unit Overview

The human body is made up of different types of cells, tissues, organs, and body systems. These structures help you stay alive and healthy. Each system plays an important role, and they work together. As you better understand how your body works, you will learn how to care for it and protect it.

Think about what you do when you jump up and down, walk, sleep, or anything else. Different parts of your body have certain jobs to carry out so that you are able to do all these things and much more.

The functions of the human body depend on all of its body systems.

Some people are experts on different parts of the body or specific problems associated with the body. These experts might be orthopedists, cardiologists, or endocrinologists. The help of these professionals is necessary if one of the body's systems is not working right. To help your body systems continue to function properly, professionals may prescribe certain medicines and drugs.

Living things are organized in special ways. To understand organisms, we need to understand how our bodies work and the systems that keep them healthy.

Your teacher may assign an Open Inquiry lab and a Lifestyle Challenge activity. Use your *Science Journal* to record your work.

How Your Body Is Organized

At any given moment, your brain is directing thousands of processes, including circulating blood, breathing, and perhaps solving a complex problem. God created your body as a dependent and connected unit. Each part relies on other parts as you grow and develop.

Scripture Spotlight

God created your body as a temple in His likeness. Knowing how your body functions can help you care for your body, which honors God. The Bible tells us many ways to respect our bodies. You will read the following passages in this chapter.

Genesis 32:24–32 (p. 120)
Matthew 17:14–18 (p. 138)
Matthew 22:37 (p. 142)
Colossians 3:2 (p. 144)
Psalm 34:15 (p. 153)

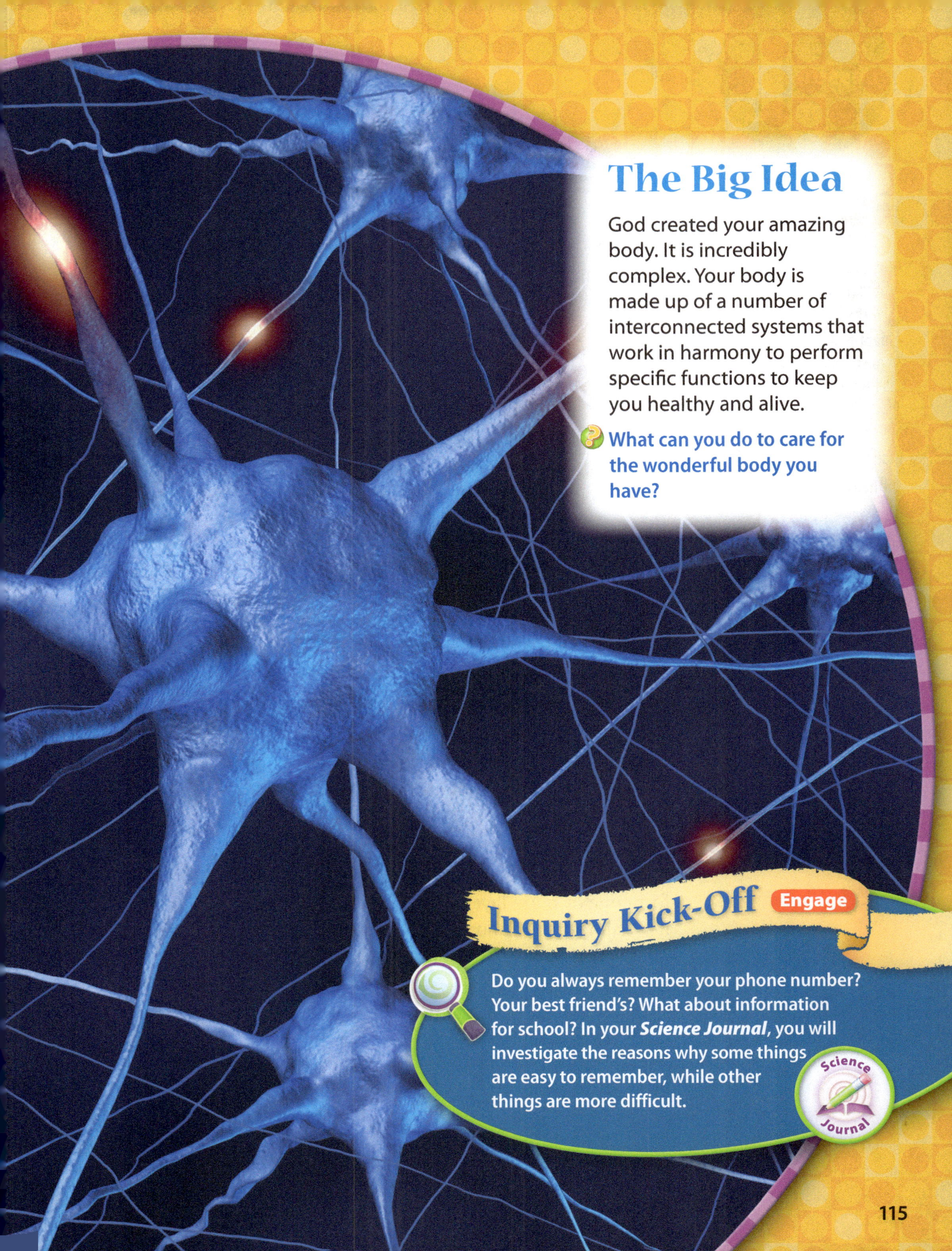

The Big Idea

God created your amazing body. It is incredibly complex. Your body is made up of a number of interconnected systems that work in harmony to perform specific functions to keep you healthy and alive.

What can you do to care for the wonderful body you have?

Inquiry Kick-Off Engage

Do you always remember your phone number? Your best friend's? What about information for school? In your *Science Journal*, you will investigate the reasons why some things are easy to remember, while other things are more difficult.

Science Journal

Objectives

- Distinguish between different types of tissues.
- Identify where each type of body tissue is found.

Vocabulary

tissue

epithelial tissue

connective tissue

muscle tissue

nervous tissue

What Are Tissues?

You know that cells are the basic unit of life, and that living things are organized in specific ways. How is your body organized? How are cells in your right hand connected to those in your left hand? What is the relationship between your eyes and your stomach? Why is this important?

Tissues of the Body (Explain)

Groups of similar cells working together to perform specific jobs are **tissues**. Your body has four types of tissue: epithelial, connective, muscle, and nervous.

Epithelial Tissue

Epithelial tissue protects all internal and external body surfaces. What do you think is inside your body that some tissues might need to be protected from?

Epithelial tissue has several functions, including:

- It acts as a shield to protect other tissues.
- It controls movement of substances between the internal and external environments.
- It absorbs nutrients from the digestion of food.
- It excretes waste products from the body.
- Its cilia removes dust particles and other substances from air that enters the air passages.

Epithelial tissue is made of tightly packed cells that form one or more layers. It regenerates or repairs itself almost continuously, allowing injuries to heal relatively quickly. How does your body react when you get hurt?

Different groups of epithelial tissue have different shapes and different purposes. Where are epithelial tissues found in your body? Use the chart to identify some places in your body where epithelial tissue can be found. What do you think would happen to your body if some of your epithelial tissue disappeared?

Faith Connection

Each cell is designed with a shape to match its function. How might your talents and interests be a clue to God's plan for you?

Types of Epithelial Tissue		
Characteristics	**Location**	**Visual**
flat and irregularly shaped cells	covers skin, lungs, heart, blood vessels	
cube-shaped cells	found in kidneys, middle ear, brain	
column-shaped cells	found in small intestine and some other body parts	

Epithelial cells fit tightly together to provide protection.

What do you think would most likely happen if epithelial cells were not tightly packed together?

Tissues in a Chicken Wing

Why does a chicken wing need more than one type of tissue?

SAFETY: Protect against germs when using raw chicken. Wear gloves during the lab. Wash your hands with hot water and soap after the lab and wipe down your lab bench with disinfecting wipes.

Materials
- raw chicken wing
- disposable gloves
- hand lens
- paper towel roll
- scissors
- tweezers
- 10 round toothpicks
- disinfecting wipes

Procedure

1. Put on disposable gloves. Place the chicken wing on several layers of dry paper towels. **Observe** and examine the skin covering the chicken wing.

2. Cut the skin with the scissors. Carefully remove all or most of the skin. When you peel the skin, you need to remove some of the connective tissue. **Record** what each tissue looks like.

3. Locate the fatty tissue that lies between the skin and muscle. **Observe** that the fatty tissue is found in clumps. **Record** what you see.

4. Straighten the chicken wing and hold it horizontally. Pull open the wing. **Observe** the bundles of muscle surrounding the bones. **Observe** any blood vessels. **Record** what you see.

5. Use the hand lens to find a tendon that connects muscles to bones. Use tweezers to free the tendon from the bone and to pull the tendon. **Observe** and **record**.

6. Find a ligament. Use the scissors to cut the ligament and **observe** how the bones fit into each other.

7. Snap a bone and use the toothpick to extract some of the marrow. Use a magnifying glass to **observe** this tissue. **Record** what you see.

Analyze Results

Describe the tissues that make up the chicken wing and label their locations in the drawing in your *Science Journal*.

Create Explanations

1. Why does a chicken wing need more than one type of tissue?

2. What tissues that make up the chicken wing were you unable to see? Why?

3. Which tissue do you think moves the chicken wing? What role do the tendons have in making the body move?

Connective Tissue

Look closely at the skin on your arm. Use your thumb and forefinger to stretch the skin. Why do the cells not spread apart? What keeps the cells together? **Connective tissue** connects one part of the body with another and separates one group of cells from another. It is the most abundant tissue found in the body. Connective tissue is composed of a few specialized cells embedded in a thick material that contains protein fibers. How does the structure of the connective tissue aid its purpose? What are some connective tissues in your hand?

Types of Connective Tissue		
Type	**Description**	**Visual**
Loose Tissue	small fibers found under the skin, where fat is stored	
Dense Tissue	large fibers for strength, found along tendons, ligaments, and the neck and back	
Supportive Tissue	strong tissue that supports, found in bone and cartilage	
Fluid Tissue	moves dissolved materials throughout the body, helps your immune system function properly, and thickens to prevent excessive bleeding during injuries, found in blood and lymph	

Scripture Spotlight

There was a certain muscle tissue the Israelites would not eat. Read **Genesis 32:24–32** to find out which one and why.

Muscle Tissue

What steps go into you standing up from a chair? How about shooting hoops? Think about how living things move. Specialized cells with the ability to contract and produce movement make up **muscle tissue**. The cells of this type of tissue are called fibers.

Think about your body. Different types of muscle tissue are found throughout it. Muscle tissue only contracts. For this reason, most muscles work in pairs. As one muscle contracts, the paired muscle lengthens, or stretches. Look at your body and try to find an example of a muscle pair. In what ways can your muscles respond? Name a muscle in your body that you cannot control.

Muscle tissue

How do the cells of muscle tissue differ from epithelial cells? Why?

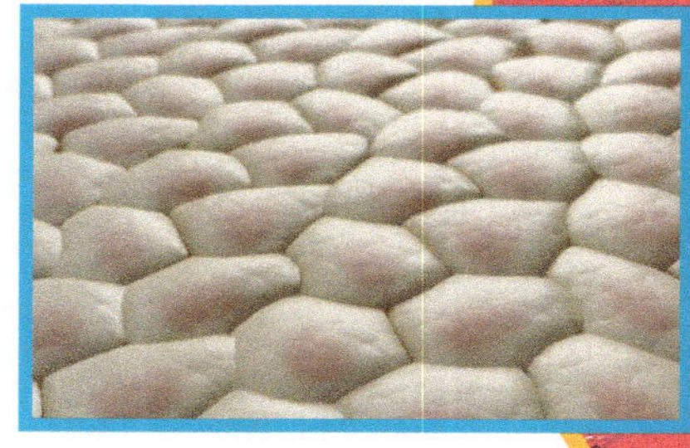

Types of Muscle Tissue			
Type	**Description**	**Where Found**	
Skeletal Muscle	most abundant muscle tissues in body; function in pairs; causes voluntary movement; involved in breathing; enclosed in connective tissue	attached to bones of skeleton; facial muscles attached to skin	
Smooth Muscle	made up of thin, elongated muscle cells; controls slow, involuntary movements	digestive tract, bladder, uterus, walls of blood vessels	
Cardiac Muscle	causes the heart's rhythmical beating, circulating blood throughout the body; involuntary movement	wall of heart	

 **How do the cells of each kind of tissue differ?**

Obtain a set of slides of tissue samples from your teacher. Look at the slides under the microscope. Carefully bring each slide into focus. Draw pictures and write down your observations for each slide. Review the characteristics of the different kinds of tissue. Using this information and your observations, identify the tissues in the slide.

Nervous Tissue

How does your brain tell your toe to wiggle? Why do you pull your hand away from a hot plate? **Nervous tissue** transmits messages throughout the body. The nerve cells, located in the brain, spinal cord, and nerves, sense stimuli and send impulses to different parts of the body in response. Nervous tissue coordinates and controls many of your body's activities. Nervous tissue stimulates muscle contraction, provides information, and enables emotions, memory, and reasoning. What happens if nervous tissue is torn or severed?

How do you move? What forms your bones? Your skin? Why does your heart beat? How do all your bones stay together? The answers to all of these questions can be found in four important tissue systems in your body. By studying how each tissue system works, the mysteries of the human body are revealed.

Check for Understanding

Review what you have learned about the four types of tissue found in the human body. How are they alike and different?

 **Concept Check** Assess/Reflect

Summary: What are tissues? Cells work together to perform specific jobs as tissues. There are four main types of tissues: epithelial tissue, connective tissue, muscle tissue, and nervous tissue. Each set of tissues has specific functions within the body's organs. Epithelial tissue protects all internal and external body surfaces. Connective tissue joins and supports other types of tissue. Muscle tissue produces movement. Nervous tissue sends and receives messages throughout the body.

1. What type of muscle tissue is found in your arms? How do you know this?
2. Where is loose connective tissue found in the body?
3. Which type of tissue heals the quickest? Why?
4. How are skeletal, smooth, and cardiac muscle tissue similar and different?

What Are Your Body Systems?

Essential Question

Suppose that the body systems did not work together. What would happen if the muscles and the circulation did not work together? What evidence do you see that the organs in your body work together? The fact that the tiniest atom can combine with other atoms to form molecules and eventually make an organism that functions as well as you do, shows that a master plan was in place. What does this organization show you about God?

Organs and Organ Systems Explain

The human body is organized from the simplest to the most complex. *Cells* are the basic units of life. A group of similar cells working together to perform a specific job is a *tissue*. A group of tissues that work together to perform a certain task is an **organ**. And a group of organs working together to perform a series of related functions is an **organ system**. Which organ systems work together to help you understand the information about levels of organization presented on this page?

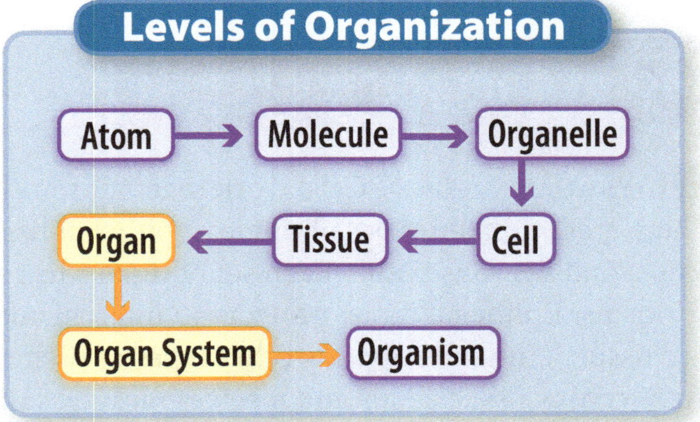

The diagram shows levels of organization in the human body.

What is the advantage of organs working together in a system?

Organ Systems

Your body has several organ systems. Each depends on the other systems to function properly. Using the image below, what other systems do you think the respiratory system relies on to function? What would happen if one of those systems became diseased? Your body has several organ systems. Each depends on the other systems because they are interrelated. You need all your organ systems to keep your body working properly.

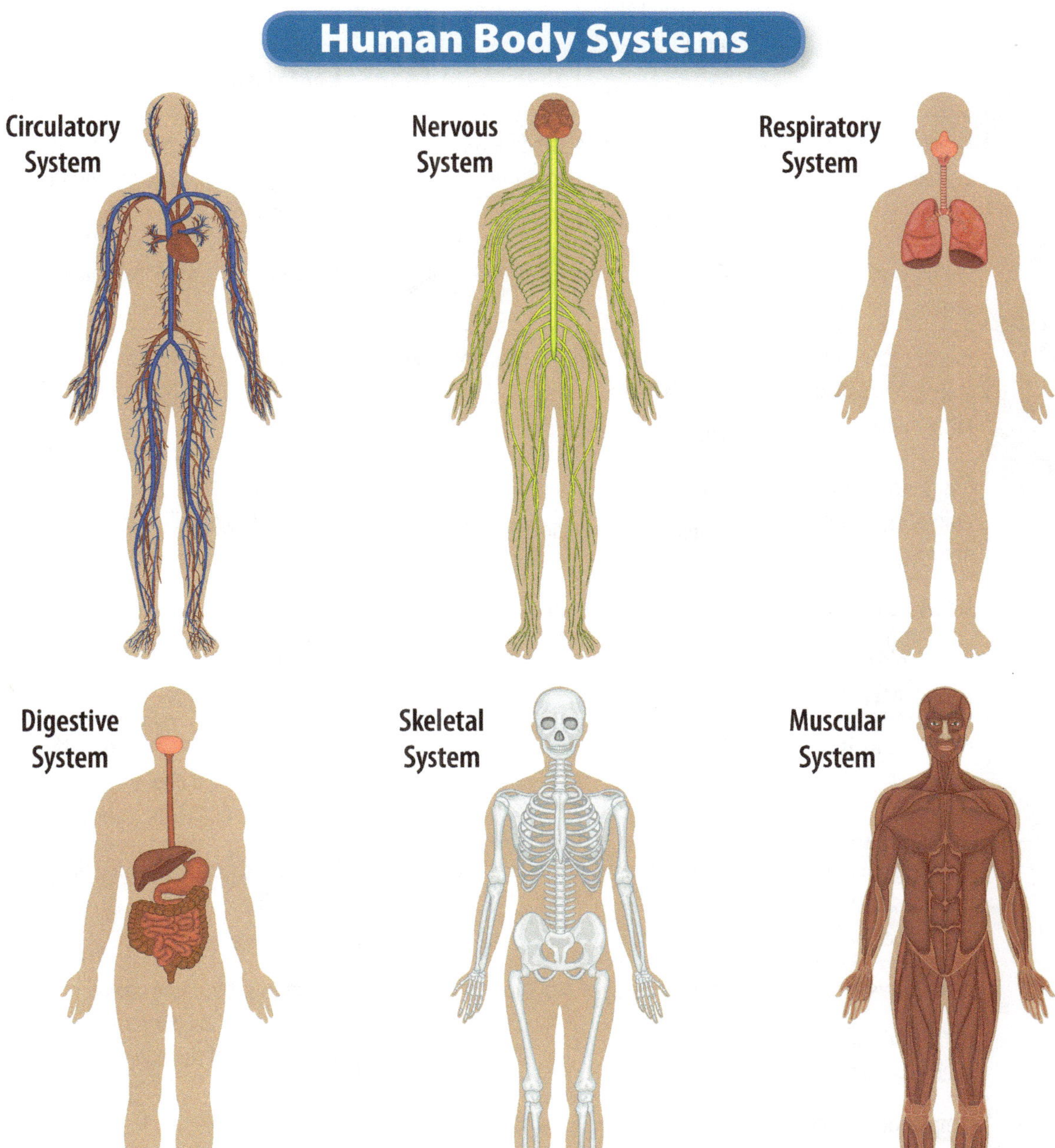

Study the chart of major organ systems. Work with a partner. Rank each organ system in order of importance. Make a three-column chart that names the body system, ranking, and the reason for your ranking. Compare your ranking with your classmates. Talk about the criteria you used.

Organ System	Major Parts	Function	Interesting Facts
Skeletal	bones, cartilage, joints, ligaments	• Move, support, and protect the body. • Cells are produced inside bones.	The smallest bone is found in the ear.
Muscular	muscles, tendons	• Work with bones to help the body move. • Muscles protect some body organs.	There are 650 muscles in the human body.
Circulatory	heart, blood, blood vessels	• Circulate nutrients and blood throughout the body. • Collect waste products.	All the blood vessels in your body are long enough to go around the equator, twice.
Lymphatic	lymph vessels, tissues, organs	• Collect fluid from tissue and return it to the blood.	The body contains more than 500 lymph nodes.
Digestive	mouth, esophagus, stomach, small and large intestines	• Break down food into nutrients and absorb them. • Eliminate wastes.	An empty stomach is about the size of two closed fists.
Respiratory	lungs, nose, trachea, epiglottis, bronchi, alveoli	• Take in oxygen. • Eliminate carbon dioxide and water.	Human lungs have approximately 2400 km (1500 miles) of airways.
Excretory	kidneys, ureters, urethra	• Remove waste from blood.	The bladder is as big as the brain.
Nervous	brain, spinal cord, nerves, nerve endings	• Control and coordinate body activities.	The brain consumes $\frac{1}{4}$ of the oxygen you breathe.
Integumentary	skin, hair, nails, glands	• Protect the body.	In the time it took you to read this far, you have lost about 40,000 cells, about 4 kg (9 lb) a year.
Endocrine	endocrine glands, pancreas, testes, ovary, liver	• Regulate body functions through chemicals.	The thyroid gland is shaped like a butterfly.
Reproductive	male: penis, testes female: uterus, ovaries, vagina	• Provide a way of producing offspring.	Sperm uses the same receptors that you have in your nose to hunt for the scent of the egg.
Immune	lymph nodes, bone marrow, spleen, thymus	• Defend the body against attacks by infectious organisms.	The number one way to boost the immune system is to reduce stress.

Keeping Safe from Infections

How does skin keep infectious organisms out?

SAFETY: Do not eat apples at any time during or after the activity. Do not remove goggles until after clean-up. Dispose of all bags properly.

Materials
- 5 fresh apples and 1 rotting apple
- rubbing alcohol
- soap and water
- zip-top plastic sandwich bags
- disposable gloves
- labels and permanent marker
- paper towels
- goggles
- coarse sandpaper

Procedure

1. Put on disposable gloves and goggles. Label the bags 1–5 and with your last name. Put a fresh apple in bag 1. **Record** its appearance in the chart.

2. Take one of the fresh apples and rub it with a piece of sandpaper that is 10 cm by 10 cm (4 in. by 4 in.)

3. Rub the rotting apple over the four apples unbagged. Put one of these apples in bag 2. Put the apple rubbed with sandpaper in bag 3.

4. Wash one of the remaining fresh apples with soap and water. Dry it, and put it in bag 4.

5. Rub rubbing alcohol on the remaining fresh apple. Let it dry. Put it in bag 5.

6. Be sure all the bags are sealed. Set the bags aside in a section of the classroom where they will not be disturbed for two days. On day 3, **observe** the apples but do not remove them from the bags. **Record** what you see. On day 7 look at the apples again and **record** your observations.

7. **Compare** the apples using the 1(freshest) to 5 (most decay) rating scale of stages of decomposition found in your *Science Journal*.

Analyze Results

Record the appearance of the apples on day 1, day 3, and day 7. Analyze the differences you **observe**.

Create Explanations

1. How does skin keep infectious organisms out?

2. Why is your immune system important?

3. Why should you wash your hands often?

4. Look at the apple in bag 3. How is apple skin similar to your skin?

The Endocrine System Explain

The **endocrine system** helps control body functions through the secretion of hormones. Because of these hormones, the endocrine system controls the rate at which you grow, the growth and development of your reproductive organs, and other body functions. How does the endocrine system work with your nervous system to help regulate all your body functions?

Endocrine glands produce **hormones**, which help regulate other cells and organs. Hormones go directly into the bloodstream and travel throughout the body. Different endocrine glands produce hormones that act on specific tissues or organs. What hormone below helps your body deal with stress?

Organs of the Endocrine System			
Gland	**Hormone**	**Affects**	**Function**
Pituitary	growth hormone	all body cells	Control bone growth, the release of hormones from other glands, and kidney functions.
Thyroid	thyroxine	all body cells	Control the release of energy in the body.
Adrenals	adrenalin	kidneys, liver	Increase heart and breathing rates, raise blood pressure and amount of sugar in blood; help the body deal with stress.
Pancreas	insulin	liver, muscle	Regulate the use of sugar; let the liver store sugar.
Ovaries (female)	estrogen	female sex organs	Control the female reproductive organs and characteristics.
Testes (male)	testosterone	male sex organs	Control the male reproductive organs and characteristics.

Some glands play a greater role in the human body than others. The **pituitary gland** is considered a master controlling gland in the endocrine system. What do you think it controls?

The pituitary gland is attached to the hypothalamus and produces hormones only when they are needed. The hypothalamus is a tiny section of the brain no larger than a pea. If the hypothalamus senses a decrease in the hormone adrenalin, for example, it sends a chemical message to the pituitary gland. The pituitary gland responds by releasing a hormone that increases the activity of the adrenal gland. What other hormones does the pituitary gland regulate?

The Endocrine System

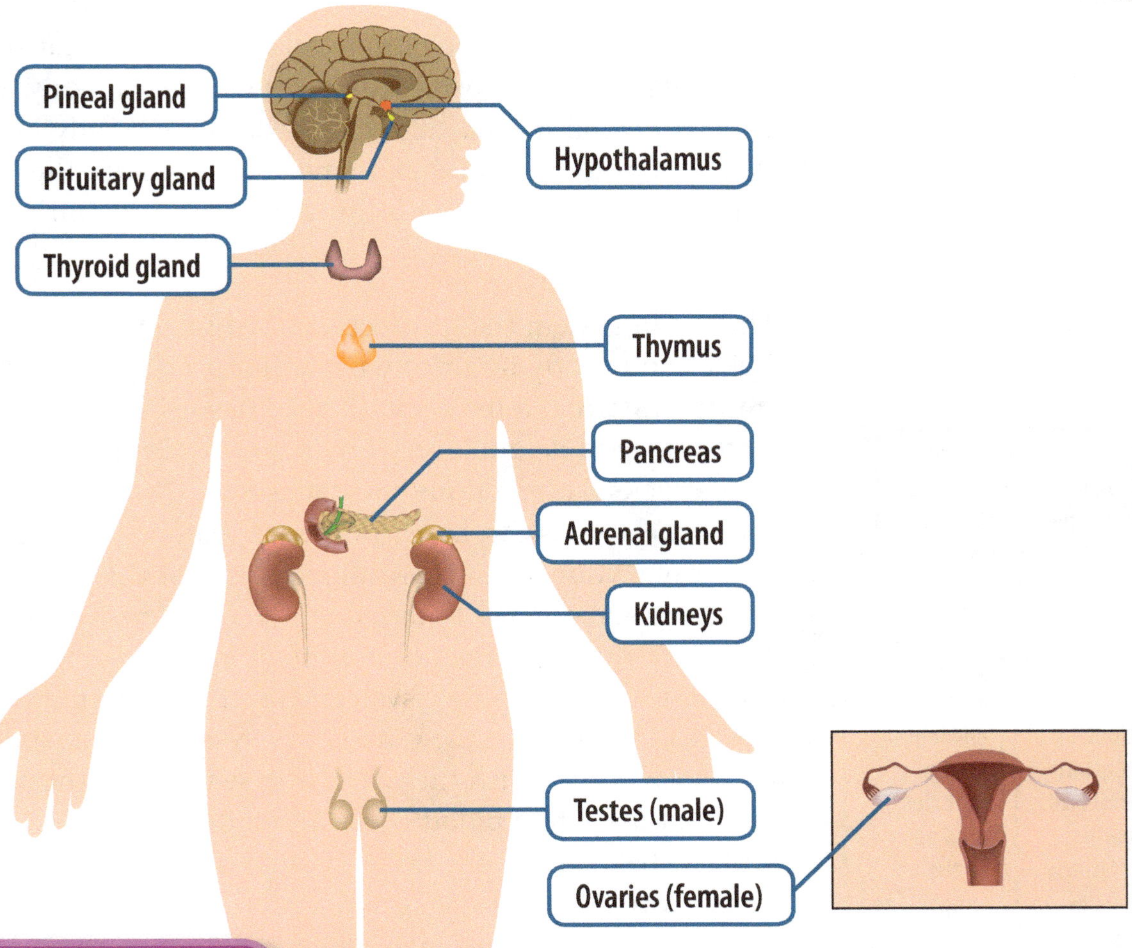

Lesson Activity

Work in small groups to research the parts of the endocrine system (the glands, hormones, and body parts they affect) and learn more about this important body system. Find a picture of the major organs of the human body. Enlarge and laminate the diagram. Use the diagram to trace the paths that hormones take through the human body from their starting point to the target organs. On index cards, record the outcomes of or responses to each hormone's action in different parts of the body.

How does your endocrine system communicate with the rest of your body? What body system does it work with to achieve this goal?

The Immune and Lymphatic Systems Explain

How does your body fight off illness? The immune and lymphatic systems are two closely related systems that share several organs and defensive functions within the body. What does your body defend itself against? How does your body know when its defenses need to be on heightened alert? What other systems or body parts might help with this defensive function?

The Immune System

The organ system that protects against infectious organisms and gives your body immunity is the immune system. The **immune system** is a network of cells, tissues, and organs that work together to defend every organ and tissue in your body against attacks by infectious substances.

The same organs can be part of more than one body system. For example, your skin is considered a part of the immune system. It is actually the body's first line of defense against disease. It is also the primary organ of the integumentary system, and acts as part of the excretory system as well. The immune system also works closely with the circulatory system to transport its defensive cells, and with the lymphatic system to produce and store immature white blood cells. All white blood cells, known as **leukocytes**, start in the bone marrow as stem cells. There are two basic kinds of leukocytes: lymphocytes and phlagocytes.

The lymphnoid organs, including lymph nodes, the spleen, and the thymus also work with the immune system. The thymus is an organ located between the breastbone and the heart. It is responsible for storing the immature white blood cells and preparing them to become specialized cells called *T cells*. Why do you think these cells are called T cells? The T cells have a unique receptor that helps your body fight off infection. Other lymphocytes, called *B cells,* mature and remain in the bone marrow until activated. These B cells circulate in the blood and make the antibodies that recognize pathogens. They may be stored in other lymphatic organs such as the spleen. The spleen filters the blood for foreign cells and recycles old or damaged red blood cells. Why is bone marrow so important to the immune system? If infectious organisms do get into your body, the cells and organs of your immune system detect them and try to get rid of them before they can reproduce.

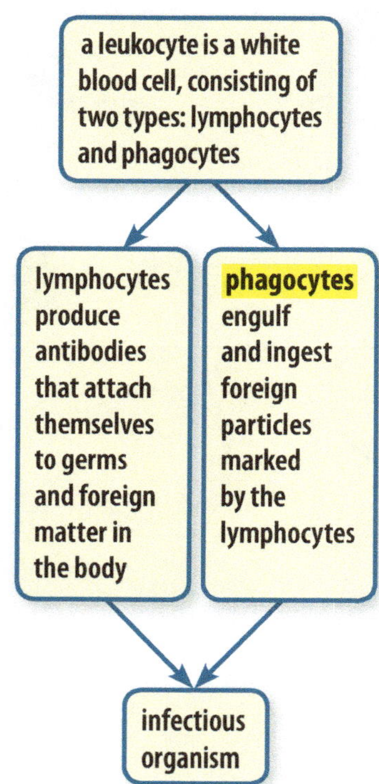

The Lymphatic System

How much of your body is made up of water—25%, 50%, or more? The body is actually more than half water! Where is all of that water? In your bones, in your blood? While bones and blood both contain water, most of the water that makes up your body is in your *lymph* fluid. **Lymph** is a clear, fluid tissue that surrounds all body cells and saturates them with water and nutrients. Lymph flows through the **lymphatic system**, which is a network of tiny tubes and organs found throughout the body that collects fluid from tissues and returns it to the blood. The major lymphatic organs include the lymph nodes, the tonsils, the spleen, bone marrow, and the thymus gland.

Other major organs, such as the heart, lungs, intestines, and even the skin, also contain lymphatic tissue. The **lymph nodes** are tiny organs shaped like beans that are found in the neck, armpits, and groin. These nodes filter the lymph and trap foreign materials. How does the lymphatic system play a vital role in the body's immune defenses?

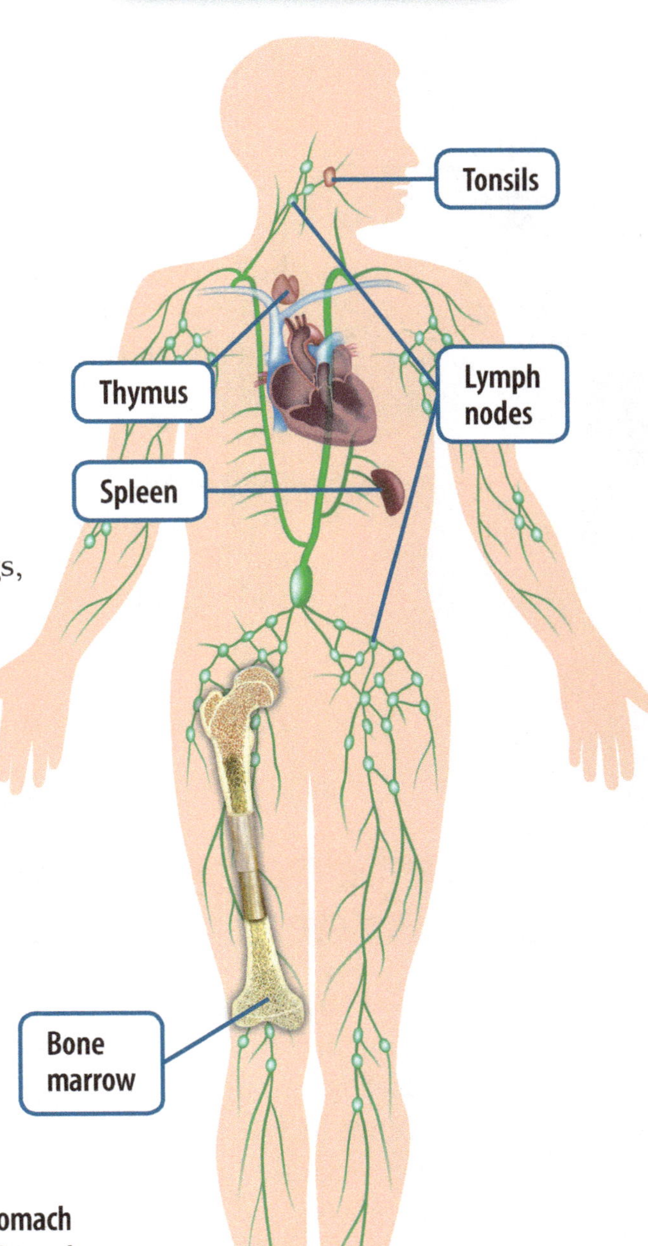

The spleen is an organ above the stomach and under the ribs on the left side. It produces lymphocytes and stores healthy red blood cells.

What happens if you rupture your spleen and it has to be removed? What will be the effect on your immune system's abilities?

Recall that specialized blood cells mature and develop in the lymphatic system. These **lymphocytes** are the white blood cells that fight bacteria and viruses that are harmful to the body by releasing antibodies or destroying foreign material. They also help prevent abnormal cell growth. They come in three varieties—the B cells and T cells, and natural killer cells. The killer cells engulf and digest foreign matter. These lymphocytes are stored in lymph nodes and circulate in the blood and lymph fluid.

The lymphatic system, which releases these specific cells that fight disease, has another important function. To prevent the build-up of excess fluid in the tissues of the body, small tubes called *lymphatic capillaries* extend into the tissues to absorb fluids and return them to the circulatory system. The lymph also carries fats from the small intestine. These capillaries merge into larger vessels that contain one-way valves that prevent the lymph fluid from flowing backwards. It flows into ducts in the upper part of the body and back into the blood. If the lymphatic system does not function properly, how would that affect both the circulatory and immune systems?

A healthy body works like a well-oiled machine. The endocrine, lymphatic, and immune systems work together and use chemicals and special cells to keep you healthy.

 Concept Check Assess/Reflect

Summary: What are your body systems? There are twelve body systems that work together to keep your body functioning properly. The endocrine system produces hormones that regulate your growth as well as the development of your reproductive organs. The body is protected from infectious organisms by the immune and lymphatic systems. These systems include lymphoid organs, tonsils, the spleen, bone marrow, and the thymus gland. These organs produce and store specialized cells that fight infection.

1. If your lymph nodes are swollen, what might this show?
2. Which glands cause the body to have the greatest response?
3. Why is it important for the feedback control system to work properly?
4. What might happen to the body if the lymphatic system does not function properly?

What Is the Integumentary System?

Think about the organs in your body. What do all the organs in your body have in common? What organs can your body function without? What do you think is the most important organ in your body? Why? What other organs offer that organ support so it can function properly?

The Integumentary System **Explain**

The integumentary system is one of many organ systems that make up your body. The **integumentary system** is a group of organs that form a protective outer layer. This system is made up of the skin, hair, fingernails, and sweat glands. What does each organ in this outer layer of the human body do? How do you think the integumentary system of a bird is different than yours?

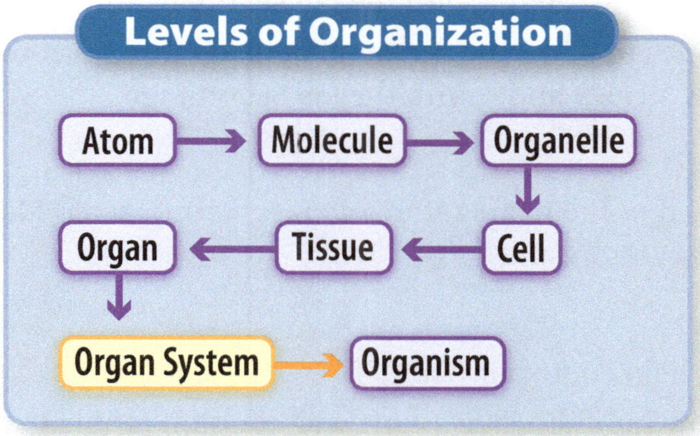

Objectives

- Describe the function of the integumentary system.
- Describe the structure of the skin.
- Describe ways to protect your skin.
- Explain the importance of skin care.

Vocabulary

integumentary system

epidermis

melanin

dermis

follicle

sebaceous gland

sweat gland

Levels of Organization

Atom → Molecule → Organelle

Organ ← Tissue ← Cell

Organelle → Cell

Organ → Organ System

Organ System → Organism

Look again at the diagram of levels of organization in the body.

Is there any level of this organization that could be left out? Explain.

Lower Your Body Temperature

How does sweating help cool your body?

Procedure

1. Begin by reading the temperature of each thermometer. **Record** the starting temperature.

2. Gently tear paper towels into strips about 2–4 cm (0.8–1.6 in.) wide and 10–15 cm (4–6 in.) long.

3. Soak one strip of paper towel in water and wrap it around the bulb of a thermometer. Affix with a small rubber band.

4. Hold the bulb of each thermometer 10–20 cm (4–8 in.) in front of a fan set on low for five minutes. **Compare** and **use numbers** for the temperature of each thermometer. Repeat Steps 3 and 4 two more times and record your average results.

5. Wait about 10 minutes to allow the thermometers to return to room temperature. Repeat Steps 3 and 4 with the fan set on medium.

6. Wait about 10 minutes to allow the thermometers to return to room temperature. Repeat Steps 3 and 4 with the fan set on high.

Materials
- paper towels
- water
- small rubber band
- two thermometers
- small fan

Analyze Results

Describe what happened to the temperature of each thermometer. Make a graph to **communicate** your results.

Create Explanations

1. How does sweating help cool your body?

2. How do high and low humidity affect how well the skin is able to cool the body?

3. How do animals that do not have sweat glands stay cool?

The Skin

Your skin, the primary organ of the integumentary system, is the largest organ in your body. It has several functions:

- It prevents water loss.
- It protects the body from harmful substances.
- It guards the body against injury, extreme temperatures, and damaging sunlight.
- It helps to regulate your temperature.
- It helps get rid of waste.

The skin is made up of nervous, muscle, connective, and epithelial tissues. How do you think your skin uses these tissues? Depending on its location, your skin may be thin or thick. Where on your body is skin thick? Where is it thin? What do you think would happen if it was the same thickness everywhere on your body?

Layers of Skin

Skin cells are epithelial cells that are found in layers. In the illustration, find the epidermis and the dermis. What similarities do you see between these two layers? What differences do you see? What do you think the fat layer does?

The **epidermis** is the skin's thin outer layer. It consists of five layers of cells. Keratin is made from a special protein and makes up the outer layer of the epidermis. In fact, keratin is the same protein that makes up hair, skin, and nails.

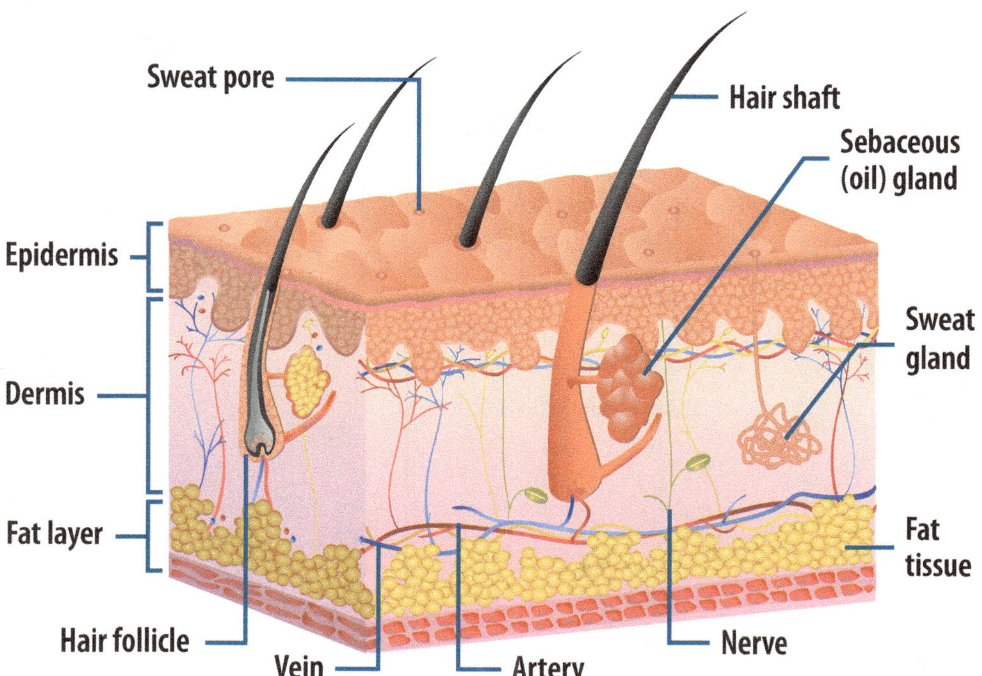

The epidermis consists of epithelial tissue, while the dermis consists of connective tissue.

How does epithelial tissue support the function of the epidermis?

Keratin is also found in the exoskeleton of insects, the talons of owls, and the baleen of whales. The lowest layer of the epidermis contains cells with **melanin**, a skin pigment. Differences in skin color result from variations in the pigment these cells produce. How does melanin protect the body?

The epidermis is anchored in place by the **dermis**, the skin's thick inner layer. The dermis is held together by *collagen*.

Nails protect the sensitive tips of your fingers and your toes from injury. They are composed of keratin. *Hair* is found on skin over almost your entire body, except a few places. What are those places? Hair **follicles** determine whether your hair is curly, straight, or wavy. Hair follicles are small folds of epidermis that push into the dermis. *Blood vessels* at the base of the follicle provide the hair roots with nutrients. When cells divide near the base, new hair is made. How does this explain hair growth?

Skin Layers				
Skin Layer	**Description**	**Purpose**	**Part of Skin Layer**	**Function**
Epidermis	thin outer layer containing mostly dead epithelial cells	• prevents most bacteria and viruses from entering the body • protects internal body parts from injury • filters out ultraviolet radiation from sunlight	Keratin	makes dead cells waterproof, producing a dry layer between living cells and the outside environment
			Melanin	a pigment that gives the skin its color
Dermis	thick inner layer of loose and dense connective tissue	• equipped with nerve endings to sense pain, touch, pressure, and temperature • contains the blood vessels that supply the skin with nutrients • uses hair follicles that produce hair, which sense stimuli • contains sebaceous glands and sweat glands	Collagen	a protein that provides the skin with its strength and elasticity

Other Parts of the Skin

Look back at the skin diagram earlier in the lesson. What are other features embedded in the dermis? A *gland* is a group of cells that makes special chemicals for your body. **Sebaceous glands** found around the bottom of hair follicles help prevent your hair and skin from drying out. They produce an oily material that reaches the skin's surface where hair comes out from skin.

Sweat glands connect to openings on the skin's surface. They rid the body of excess water and certain wastes. What is another function of sweat? What happens when sweat and oil block follicles?

How does oil protect your skin? Why is this important?

Oil glands are found throughout your body, except on your palms and the soles of your feet. Rub a cotton ball with rubbing alcohol over the crease in your arm. Put two or three drops of water onto that arm. What happens? Then put two or three drops of water on your other arm. What happens? Compare the results.

Your body loses thousands of skin cells every day because they are cut off from their supply of nutrients and oxygen. Your body sheds the dead skin cells constantly when you scratch or rub against a piece of clothing. These cells are replaced by the new skin cells underneath. How does you body benefit by losing all these cells?

Math in Science

You lose about 40,000 skin cells every hour. How many skin cells do you lose in one day, one month, and one year? How many skin cells do you lose in 70 years?

Caring for Your Skin Explain

Sun exposure is important. It provides the body with vitamin D. However, you only need 30 minutes of Sun exposure on skin without sunscreen once a week to receive what you need. How much time do you spend in the Sun each day? How do you limit Sun exposure?

Protection from the Sun

When your skin is exposed to sunlight, extra melanin is made, which makes your skin darker. This helps protect you from getting burned by the ultraviolet radiation (UV) in sunlight. Sunburn destroys cells of the epidermis. Other problems associated with too much Sun exposure include skin cancer, such as *melanoma*.

To avoid skin cancer and other skin problems, take safety precautions to guard against the Sun's damaging rays:

- Avoid Sun exposure when the Sun is strongest, between 10:00 in the morning and 4:00 in the afternoon. Stay in the shade.
- Wear protective clothing such as long pants, a long-sleeved shirt, and a hat with a wide brim.
- Wear sunglasses with 100% UV protection.
- If you are going to be out in the Sun for very long, wear sunscreen to protect your skin from damaging UV rays. Reapply sunscreen often.

Explore-a-Lab

Structured Inquiry

How does photoreactive paper mimic melanin in the skin?

Cut four pieces of thin clear plastic into 12-cm (5-in.) squares. Use a permanent marker to label three panes with different SPF treatments. One pane should be reserved for no SPF treatment. Smear a few drops of the appropriate sunscreen on the corresponding plastic pane with a cotton swab. Draw the blinds in your classroom. Working quickly, remove two pieces of photoreactive paper from the container and place on a cookie sheet. Place two panes per paper, sunscreen side up. Place the cookie sheet in the full sun for five minutes. Bring the cookie sheet indoors, write the SPF level of each plastic pane on the photoreactive paper in the proper location. Wash the photoreactive paper in a shallow dish of water to "fix" the images. How might a color change from your natural skin color be a sign of skin damage?

Hygiene

How does the loss of skin cells described earlier make regular bathing and washing important? Good hygiene is essential to maintain health. The skin is your outer layer of defense against bacteria, viruses, and other foreign cells. Bathing regularly with soap and water removes dead skin and bacteria that can cause body odor. Using deodorants and drinking plenty of water also help to control body odor.

A common condition that can be easily managed with proper bathing is athlete's foot. This condition is caused by a fungus that grows best in warm, moist environments. The fungus causes odor and itching, especially around the toes. This condition can be avoided by wearing socks, bathing, and keeping your feet dry. A severe condition may require a visit to the doctor to obtain a prescription to kill the fungus.

In addition to keeping your skin clean and healthy, it is also important to keep your hair clean with regular shampooing and brushing. These simple measures remove dead cells, dirt, and oils that accumulate on the hair. A common condition that can be easily managed is dandruff. This condition forms when many skin cells drop off your scalp at one time. Regular brushing and shampooing can prevent dandruff.

Bathing regularly also controls acne. Acne is a common skin disorder that starts during adolescence and usually appears on the face, back, and upper chest. It forms when oil, dirt, and bacteria get trapped in the skin. The bacteria cause an infection that forms a pimple. Mild cases can be treated with topical over-the-counter treatments, but a severe case may require a visit to the doctor. It is important to care for acne, because severe cases can cause scarring and pitting of the skin.

Skin First Aid

You have learned that the skin is the largest organ of the human body. It consists of several layers that replace cells rapidly. This unique property provides the body with near-constant protection from the outside environment. However, skin can be damaged. Most small bruises, blisters, or cuts to the skin will heal themselves, but some problems, such as puncture wounds and deep cuts and scrapes, need treatment called first aid.

Burns can be more serious. A first-degree burn only damages the epidermis and leaves the skin red and swollen. Second-degree burns damage the epidermis and the top dermis and cause red and blistered skin. Third-degree burns are the most severe, destroying the entire skin layer. How do doctors treat severe burns? Does a first-degree burn require a bandage? What are other ways that you care for your skin when it is injured?

Concept Check — Assess/Reflect

Summary: What is the integumentary system? The integumentary system is an example of an organ system. It prevents water loss and protects the body from the outside environment. It consists of skin, hair, nails, and glands. Skin consists of several layers and features, including oil glands, fat tissue, hair follicles, and nerves. While your skin can repair itself most of the time, it also needs protection and sometimes needs first aid. You can properly care for the integumentary system with daily cleaning and maintenance.

1. How do oil glands affect your skin and hair?
2. Compare and contrast sweat and sebaceous glands.
3. Describe three steps to take for good hygiene.
4. Describe the structure of skin and include an illustration with labels.

Essential Question

What Is the Nervous System?

If you were to compare the body to a computer system, what parts represent the nervous system? How would these parts function similar to a computer? How would they function differently? Your nervous system gathers information from your environment, processes that information, and acts on that information. All of this happens faster than the time it took for you to read this line.

The Nervous System Explain

The **neuron** or nerve cell is the basic functional unit of the nervous system. Neurons consist of the cell body, dendrites, and axons. The **cell body** contains the nucleus. **Dendrites** are fibers that carry impulses from other neurons toward the cell body. **Axons** are fibers that carry impulses away from the cell body to other neurons. These nerve cells transfer stimuli to other cells.

Neurons come in two varieties. *Sensory neurons* are nerve cells that convert external stimuli from the organism's environment into internal electrical impulses. These neurons are located on muscles, joints, and skin, and in all organs that communicate temperature, pain, and pressure. *Motor neurons* communicate with muscle tissue. This interaction causes the muscles to contract and relax, resulting in mobility. These two types of neurons work together. The sensory neuron responds to a stimulus in the environment and activates the motor neuron. The motor neuron activates muscles for reflexive, involuntary motion. Why do you think God wanted humans and other organisms to have such an amazing system to sense our surroundings?

Objectives

- Describe how nerve impulses travel.
- Explain differences between the central and peripheral nervous systems.
- Identify the parts of the brain and their functions.
- Describe the types of thinking the brain performs.

Vocabulary

neuron

cell body

dendrite

axon

synapse

central nervous system

peripheral nervous system

cerebrum

cerebellum

brain stem

inductive reasoning

deductive reasoning

How Nerve Impulses Travel

Nerve impulses travel through the nervous system through the fibers of neurons. When a cell receives a stimulus, it sends an impulse to a nearby neuron's dendrite. The impulse moves through the dendrite to the next neuron. Neurons, however, are not connected. There is a small space that separates them. The place where an impulse crosses from one neuron's axon to the dendrites of another is a **synapse**.

When the impulse reaches the end of the axon, the axon produces chemicals, called *neurotransmitters,* that enable it to cross the synapse. The neurotransmitters produce a new electrical impulse in the dendrites of the next neuron. Information travels in this way from one neuron to another until it gets to the central nervous system. There, the information is processed and a response impulse is sent over another set of neurons. What do you think neurons do when you touch a hot pan? How do you react? How long does this process take?

Follow the path an impulse takes in a neuron.

❓ **How could a problem with your neurotransmitters affect your nervous system?**

Lesson Activity

Make a model of a neuron. Count out 65 beads. Select different colored beads to represent the dendrites, cell bodies, and axons. Use one bead to act as the synaptic terminal. The neuron will have seven dendrites. Determine how many beads you need to represent each part of a neuron. Assemble the beads on a string or thin, flexible wire to illustrate the structure of a neuron. Draw a picture of your model and label the parts. How did you decide on the best number of beads to represent each structure in the neuron?

❓ **How does this structure relate to how neurons communicate with each other?**

Brain Dissection

How does the structure of a sheep's brain compare to that of a human brain?

SAFETY: Follow all dissection safety precautions. Wear gloves and goggles. Wash your hands and work station after dissections. Report broken tools and injuries to your teacher immediately.

Materials
- preserved sheep brain
- long dissection pins
- masking tape
- dissecting scissors
- plastic knife
- dissection tray
- hand lens
- goggles
- plastic gloves

Procedure

1. Before obtaining the sheep brain, **predict** what you think the brain will look like. Then, obtain the brain and rinse it with tap water to remove most of the preservatives.

2. **Measure** the length and width of the brain. **Record** the measurements.

3. Find the mass of the brain and **record** the data.

4. Identify and **observe** the back and front of the brain.

5. **Observe** the cerebrum, cerebellum, and brain stem. Use the dissection pins and masking tape to make flags with the names of the structures. Stick the flags into the brain to indicate the correct locations of the structures. Draw the brain and label the parts on your drawing. Get approval from your teacher before moving on to the next step.

6. Slice the brain along the centerline, starting at the cerebrum. Separate the two halves of brain and lay them with the insides facing up.

7. Identify and **observe** the grey and white matter in the brain and the following structures: corpus callosum, left hemisphere, right hemisphere, and where the pituitary gland meets the brain stem. Use the dissection pins and masking tape to make flags with the names of the structures. Stick the flags into the brain to indicate the correct locations of the structures. Draw and label the structures inside the brain.

8. Properly dispose of the sheep brain, clean your area, and wash your hands with soap and water. Remove your goggles and return them.

Analyze Results

Describe the structure of the brain using your dissection sketches.

Create Explanations

1. How does the structure of a sheep's brain compare to that of a human brain?

2. Explain the location of the cerebellum, cerebrum, brain stem, and corpus callosum in the brain.

The Central Nervous System Explain

The nervous system is divided into two parts. The **central nervous system** consists of the brain and the spinal cord. The **peripheral nervous system** consists of a network of nerves that extend outside the central nervous system. How is this like a power grid? Study the diagrams and learn about the parts of the central nervous system.

The Brain

Your brain helps you to do everything. It is the control center of your body. The brain has three main parts:

- **cerebrum** (forebrain): largest part of the brain that controls thinking and voluntary movements of the skeletal muscles.
- **cerebellum** (hindbrain): controls muscle coordination, balance, and muscle tone.
- **brain stem**: carries information between the spinal cord and the brain and controls breathing and heartbeat rates, swallowing, and coughing.

Your skull offers protection for your brain. How can you protect your brain even more?

The cerebrum is divided into two hemispheres, which are in turn divided into four lobes that are involved in sensory perception, memories, thinking, personality, language, and motor functions. The two halves of the cerebrum are connected by a tough band of nerve fibers called the corpus callosum, which joins the two hemispheres and allows communication between them.

Human Brain Anatomy

The spinal cord is the main pathway for connecting the peripheral nervous system to the brain. Messages about pain, movement, temperature, touch, and vibration are sent through the spinal cord.

- The spinal cord contains millions of nerve fibers that send signals to the limbs, organs, and throughout the body.
- The spinal column protects the spinal cord.

The spinal cord is not able to heal itself because some of its cells are so complex and specialized that they cannot easily be regenerated. If a person has a spinal cord injury, what other body systems are affected? Why?

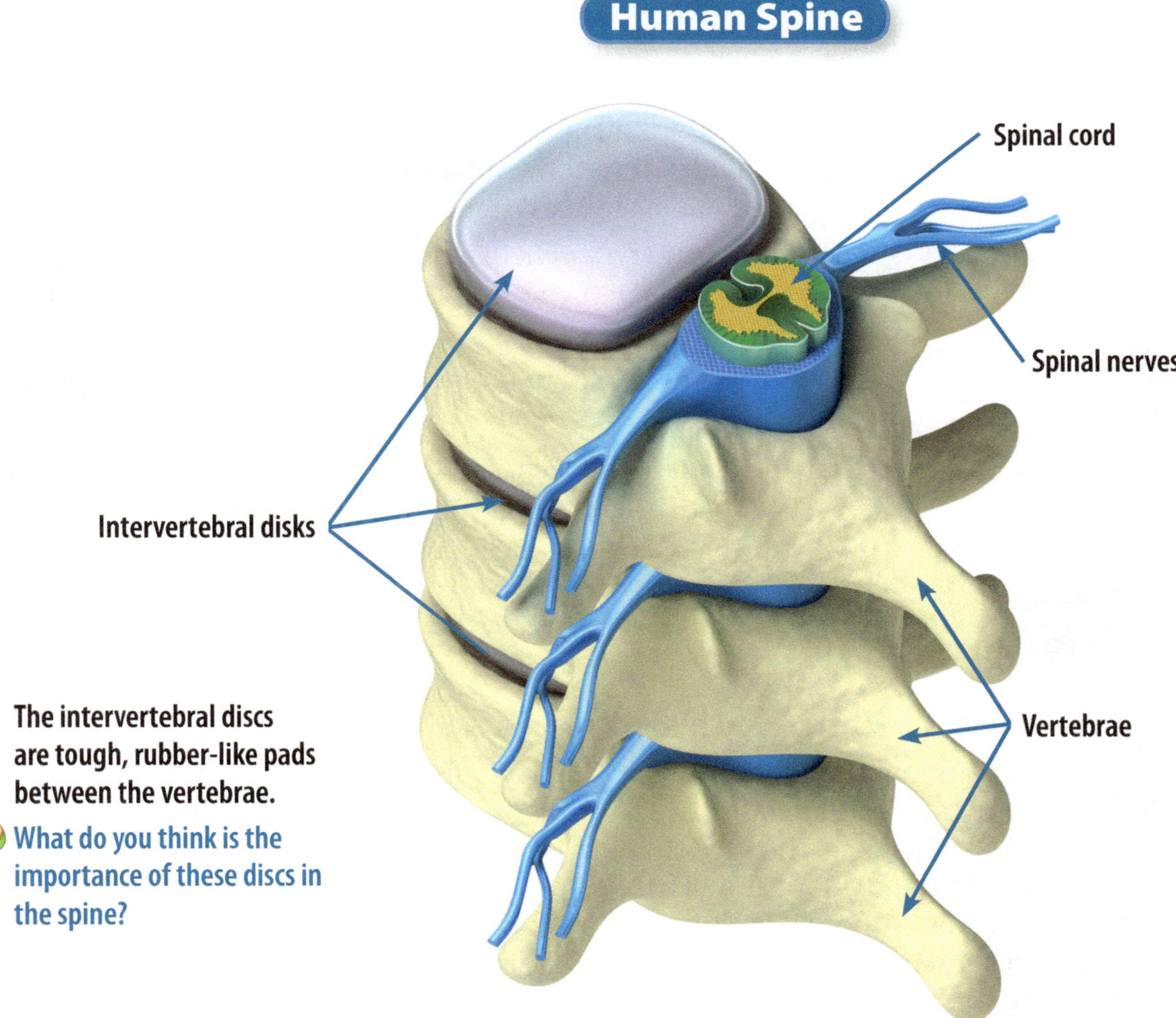

The intervertebral discs are tough, rubber-like pads between the vertebrae.

What do you think is the importance of these discs in the spine?

The Peripheral Nervous System Explain

The network of nerves that links the body to the brain and the spinal cord is called the peripheral nervous system. It consists of the somatic nervous system and the autonomic nervous system.

Somatic Nervous System

The *somatic nervous system* is referred to as the voluntary nervous system because it deals with your senses and sends impulses to and from your skeletal muscles, most of which you control. This system also controls involuntary motions, called reflexes. What are examples of reflex actions? How do reflex actions compare to voluntary actions?

Autonomic Nervous System

The *autonomic nervous system* controls most of the body processes you never have to think about, like digestion, breathing, and heartbeat. What are some other functions that your body does without your thinking about it? The autonomic nervous system is further divided into the sympathetic nervous system and the parasympathetic nervous system.

- The *sympathetic system* helps the body prepare for stress by providing extra energy to the skeletal muscles so the body can handle the stress.
- When the stressful situation is over, the *parasympathetic system* returns the body to normal.

Can you think of a scenario when your body would shift from the parasympathetic system to the sympathetic system? How would your body change during the episode?

Explore-a-Lab

Structured Inquiry

Do you control how often you blink?

Working in pairs, one student should hold a strip of transparent plastic wrap in front of his or her face. The second student should stand about 1 m (3 ft) from the partner. Without warning, the second student should take one of the ten cotton balls and toss it at the partner's face. Do not substitute materials in this exercise. Write down the partner's reaction. Repeat the process with the remaining cotton balls. Record your results. Switch roles and repeat. How would your response change if the balls came from different directions? Have two people throw cotton balls from different directions. Record your results.

The Nervous System

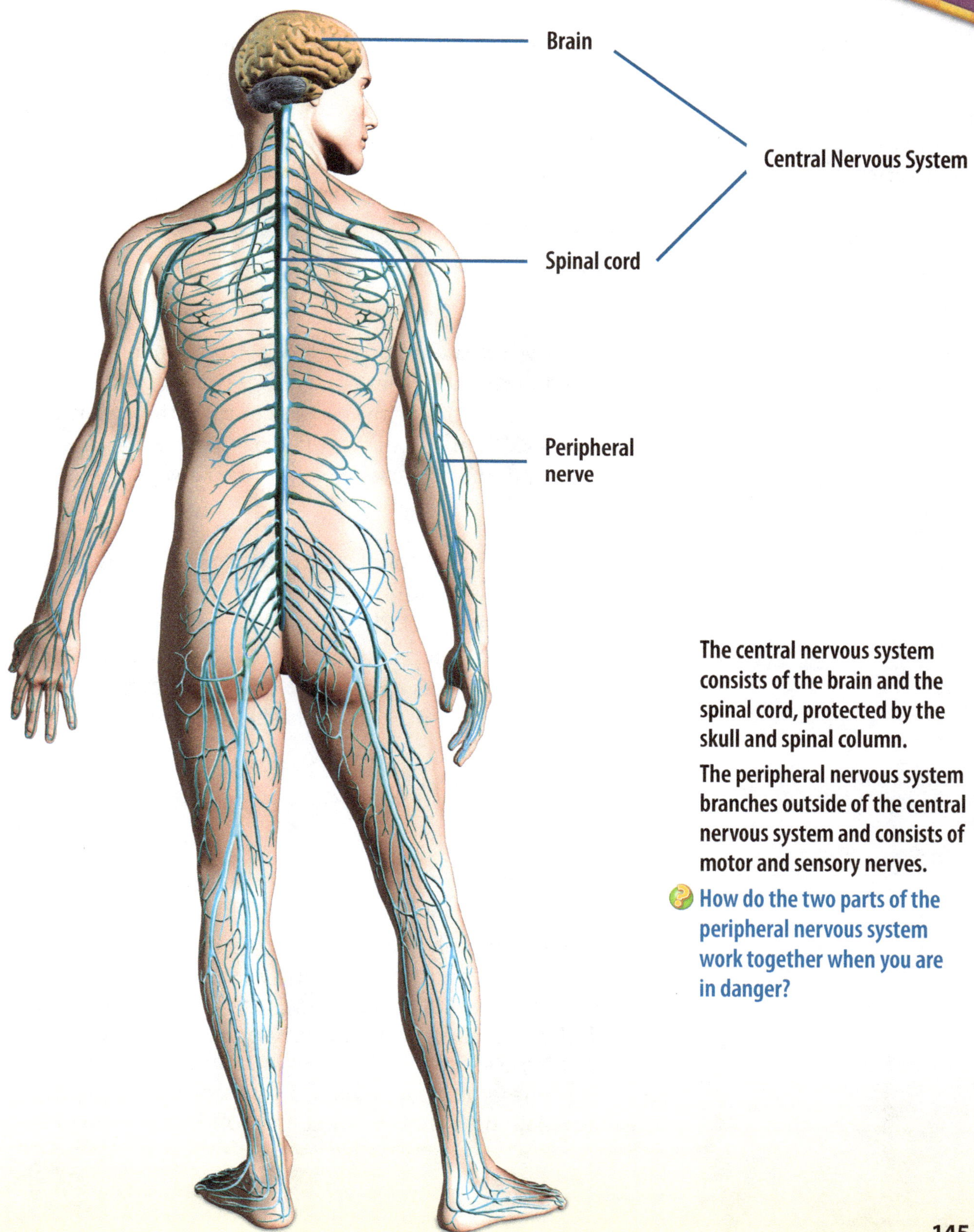

The central nervous system consists of the brain and the spinal cord, protected by the skull and spinal column.

The peripheral nervous system branches outside of the central nervous system and consists of motor and sensory nerves.

How do the two parts of the peripheral nervous system work together when you are in danger?

Check for Understanding

Explain how nerve signals are sent and messages are received throughout the body.

Thinking `Explain`

Every day, your brain handles thousands of ideas and bits of information. Thinking falls into many categories, including perception, memory, imagination, logical thinking, and spiritual thinking. Each type of thinking is interlinked with the others. Your mind works as a whole. You cannot do one type of thinking without relying on another. For example, it is difficult to solve a problem (logic) without remembering (memory) what you have learned in the past. What kinds of thinking are you using right now? What kinds of thinking do you use when playing soccer or a similar game?

Perception

You use your five senses to perceive your environment. Interpreting the information from your senses is a type of thinking called perception. Perception is the process of gathering information through our senses. What senses do you use to perceive things around you? What if one of those senses is taken away?

Memory

Memory is the ability to store learned information that you can retrieve for future use. You hold an idea in your mind while you develop it, elaborate it, clarify it, and use it. For example, while performing a multistep math problem, you remember math procedures and specific facts to help you solve the problem. Which items are easy to remember? Which items are harder?

Explore-a-Lab

Guided Inquiry

 **How does your brain perceive moving images viewed through a vertical slit?**

Using a knife, carefully cut a slit in the cap of a mailing tube 2.5 cm (1 in.) long and 3 mm (0.1 in.) wide. Replace the cap. Hold the tube so the slit is vertical. Close one eye and look through the open end of the tube. Keeping your body still, sweep the tube in an arc while looking through the slit. Try again at a different speed. What do you see? How does this activity explain how moving pictures work?

Imagination

Your imagination allows you to form images of things that are not present. This means you are able to think about things in ways that are different from the way they are. Imagination involves creative thinking. People use imagination in almost every aspect of life. How do scientists use their imagination?

Logical Thinking

You use logical thinking each time you solve a math problem, do a science experiment, or play a game that involves strategy. Logical thinking is the process in which you use reasoning to reach a conclusion. It may involve inductive reasoning or deductive reasoning. **Inductive reasoning** is reasoning from a specific case or cases to make a general rule. **Deductive reasoning** is the process that begins with general statements to develop a specific conclusion.

Suppose you are riding your bike and when you press the hand brakes, the bike does not stop completely. You reach the conclusion that something is wrong, and so you decide your bike needs to be serviced. Is this inductive or deductive reasoning? Now suppose someone tells you, "Snakes are reptiles." You know that all reptiles have scales. So you conclude that snakes have scales. This is deductive reasoning. What are some areas of everyday life in which logical thinking might help you?

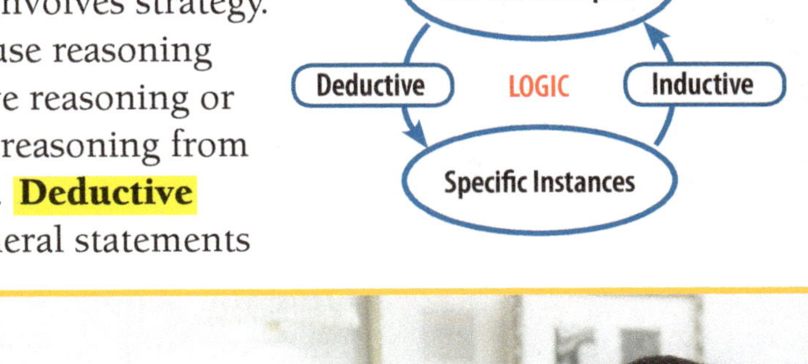

Our actions reflect various kinds of thinking.

🔎 Which kind of thinking is demonstrated here?

Math in Science

Test your thinking skills by solving this problem without using paper or a calculator. Your school has $4800 to buy new computers. If computers cost $600 each, how many can the school buy? What kind of thinking did you use to solve this problem?

Spiritual Thinking

You have already learned about the nervous system and about several aspects of thinking. God created us so that we can connect with Him. Because God created us in His image to be in a relationship with Him, He gave us the ability to think in another way—spiritually.

Have you ever done something you knew was wrong, but you did it anyway? Perhaps your conscience bothered you afterward. The Holy Spirit speaking through your conscience can help you know right from wrong. Listening and responding to your conscience is part of spiritual thinking.

Another part of spiritual thinking is spending time in fellowship with God. In prayer you can invite God to examine your mind (**Psalm 26:2**) and to transform it to be like Him (**Romans 12:2**). Through the Bible, the Holy Spirit can teach you about God and His gift of salvation, show you how to make decisions that will bring you happiness, and help you learn to trust God completely. The more time you spend with God, the more you will come to enjoy it and the better you will become at thinking spiritually.

The Bible explains that things of God have to be understood spiritually instead of from a merely human perspective (**1 Corinthians 2:14–16**). Wanting to know and obey God will help you understand spiritual things more than any other kind of thinking or reasoning. Like other forms of thinking, spiritual thinking improves with practice. In the Bible God has revealed Himself to us. The Bible provides a perspective that makes it possible for us to use all kinds of thinking to better understand the world in which we live.

Check for Understanding

Revisit the discussion of how your brain handles information. Describe the different types of thinking and give an example of how you use each type throughout the day.

Concept Check   Assess/Reflect

Summary: What is the nervous system? Your nervous system gathers information from your environment, processes that information, and acts on that information in some way. It has two parts, the central nervous system and the peripheral nervous system. Different kinds of thinking are perception, memory, imagination, logical thinking, and spiritual thinking.

1. What might happen if the spinal column becomes damaged?

2. What is the function of the cerebrum and the cerebellum?

3. What happens when an impulse reaches the synapse?

4. Which kinds of thinking do you think are unique to humans? Why?

? Essential Question

What Are Sense Organs and Senses?

What is your favorite sense? Why? Is this sense your favorite because you use it the most or are most dependent on it? What would happen to your other senses if your ears were damaged and no longer worked correctly? What if something happened to your eyes? How many senses can you think of in your body? Try to come up with as many as possible and then prioritize your list. Why do you think God gave you your senses?

Sense Organs and Senses Explain

You are aware of what is going on around you and inside your body because of receptors. A **receptor** is a specialized cell that receives information from its surroundings and provides it to the brain. Receptors let you see huge trees, hear delightful melodies, taste delicious flavors, smell flower scents, or feel soft textures. These receptors also keep you safe from harm, because they make it possible for you to see, hear, taste, smell, or feel things that might be dangerous. These kinds of receptors are part of your sense organs.

The sense organs include the eyes, ears, mouth and tongue, nose, and skin. What body systems work together with these sense organs to make it possible to respond to your surroundings?

Objectives

- Identify the major sense organs of the body.
- Explain the structure and function of each sense organ.
- Identify ways to protect your sense organs.

Vocabulary

receptor

iris

pupil

cornea

lens

retina

optic nerve

cochlea

olfactory

Service animals are trained to assist people, who have disabilities, such as sensory disorders, and help them lead fuller lives.

How can the loss of one of your senses affect your perception of things?

Eyes and Sight

Humans are able to look at objects with both eyes. They can see an object's height, width, depth, and color. Your eyes can tell you if an object is moving, or how far away it is from you. How would the world seem different if your eyes were not positioned at the front of your head?

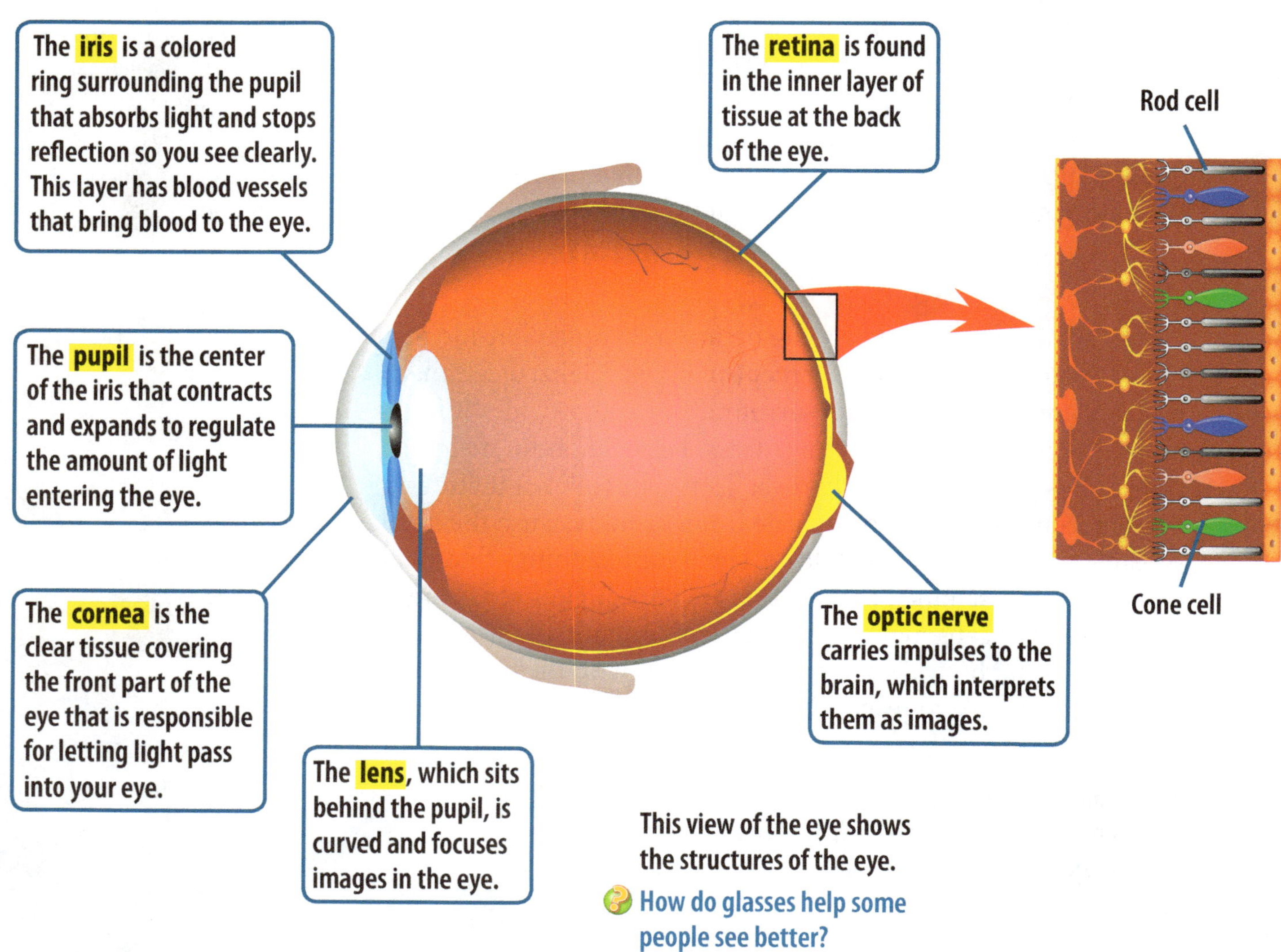

This view of the eye shows the structures of the eye.

How do glasses help some people see better?

The retina has rod- and cone-shaped light receptors. The rods allow you to see black and white and shades of gray. The cones allow you to distinguish colors. What would you see if the cones of the eye did not work?

There are no light receptors in the area where the optic nerve leaves the eye on its way to the brain. Since there are no receptors, there is no vision here. This area is the blind spot.

The Shape of a Lens and Vision

How does the shape of a lens affect the size of the image?

Materials
- metal washer
- petroleum jelly
- 2 small straight-sided plastic cups
- clear microscope slide
- eyedropper
- water
- newspaper

Procedure

1. Place the cups upside down about 2–3 cm (1 in.) apart on a page of newspaper. Place the clear slide over the cups so that about 1 cm (less than 0.5 in.) of the slide overlaps each cup. Refer to the illustration in the **Student Journal** for the set-up.

2. Spread some petroleum jelly on the washer bottom and place the washer, with the jelly side down, on the center of the slide so it is between the cups.

3. **Observe** the newsprint through the hole in the center of the washer. **Record** what you see.

4. Add three drops of water to the hole in the washer's center. Spread the water around inside the hole. Draw a cross-section diagram of the "lens." **Observe** the newsprint through the hole in the center of the washer. **Record** what you see.

5. Add three more drops of water to the hole in the washer's center. Make sure the water is higher than the washer. Draw a cross-section diagram of the "lens." **Observe** the newsprint through the hole in the center of the washer. **Record** what you see.

Analyze Results

Compare the size of the print before and after adding water drops to the slide. Show the different shapes of the lenses in your drawings.

Create Explanations

1. How does the shape of a lens affect the size of the image?

2. How would a lens that curves outward in the middle help a person who has trouble seeing things that are near?

3. What does the water act like when it changes the direction of light?

4. How might this information be helpful for someone who has trouble seeing?

When people who have perfect vision look at an object, the light rays enter the eye and focus on the retina. Many people do not have perfect vision, however. They may have trouble seeing far or near. If a person is farsighted, the shape of the eye directs light rays toward a point behind the retina. So the person sees a blurry image of objects that are nearby. If a person is nearsighted, the shape of the eye directs light rays toward a point that is in front of the retina. So the person sees a blurry image of objects that are far away.

What do you see when an object moves into your blind spot?

On a light-colored piece of paper, make a small dot on the left side. About 15–20 cm (6–8 in.) away, make another dot. Hold the paper at arm's length and close your left eye. Look at the dot with your right eye, and then slowly move the paper toward you. What happens? How can you see the whole picture, if there are parts of your eye that cannot see?

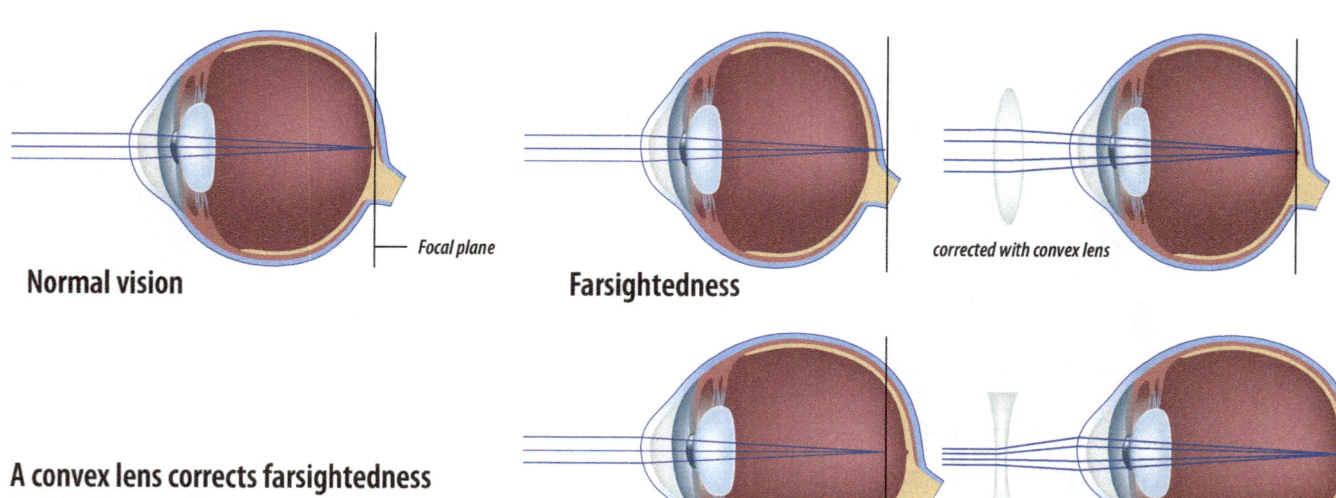

A convex lens corrects farsightedness by bending light rays. A concave lens corrects nearsightedness by spreading the light rays so they focus on the retina.

Which lens is most like a magnifying glass lens, concave or convex? Why?

Ears and Hearing

Your ears are the sense organs that are responsible for your hearing. They are also important in helping you maintain your balance. When the inner ear membrane begins to move, the liquid in the cochlea also moves. This stimulates the hairs of the receptor cells. These cells send signals to the brain, which interprets them as sounds. Are there some sounds the human ear cannot detect?

The semicircular canals are also in the inner ear. These organs have fluid and help you maintain your balance by responding to head movement.

People can damage their hearing by continually listening to sounds that are too loud. Music volumes, especially when listening through earphones, should be controlled so hearing loss does not occur. Hearing loss can also be caused by disease or injury to the cochlea. How do hearing aids help people with damaged hearing?

Follow the path that sound waves travel to produce sound.

Scripture Spotlight

What does **Psalm 34:15** say about the Lord's eyes and ears?

Sound waves are created when objects vibrate. Humans can hear sounds of between 20 and 20,000 vibrations a second.

How do you hear?

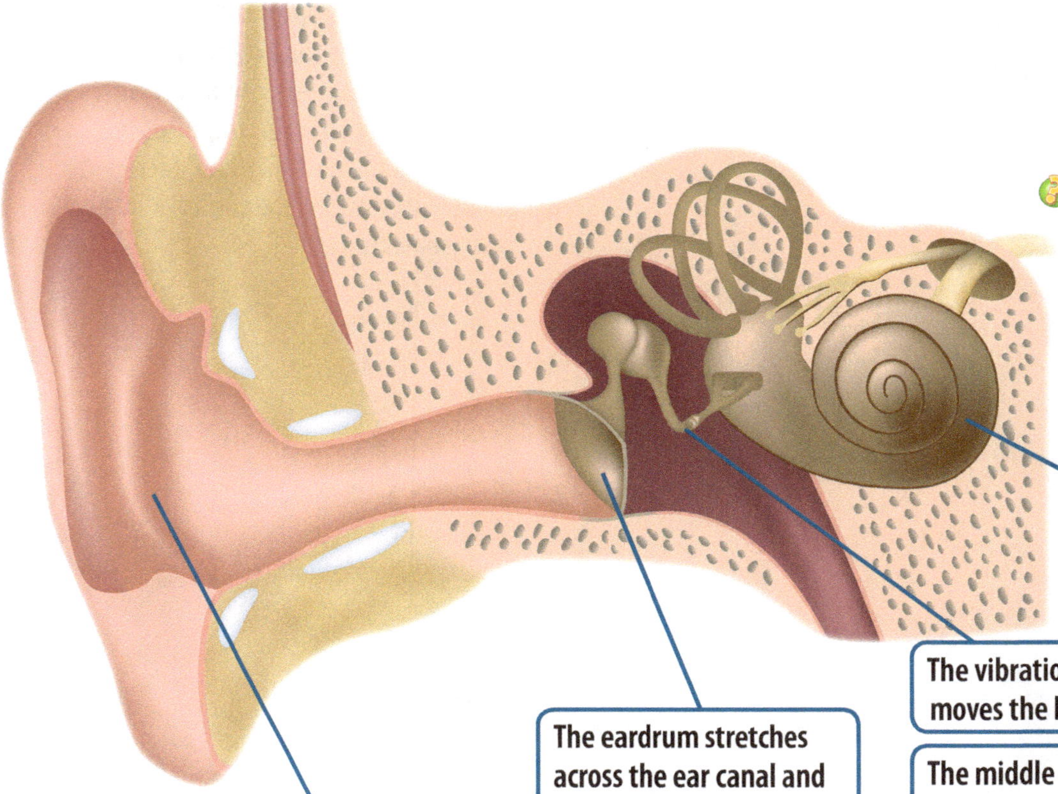

The **cochlea** is the bony, coiled tube that is filled with fluid and lined with hair cells.

The vibration of the ear drum moves the bones of the inner ear.

The middle ear contains three tiny bones called the hammer, the anvil, and the stirrup. These tiny bones amplify, or increase, the effect of sound waves.

The eardrum stretches across the ear canal and vibrates when sound waves hit it. The eardrum separates the outer and middle ear.

The outer ear is the part you see and where sound travels down to the ear canal.

Your mouth has about 10,000 taste buds. Every taste bud has receptor cells that produce one or a combination of four main taste sensations: sweet, sour, bitter, and salty. Taste relies on the chemical reactions that take place in saliva. When the taste buds are stimulated, nerve impulses are sent to the brain. Do you and your classmates all have the same taste sensations when you eat the same food? How do you know?

Your sense of taste may keep you safe from eating unsafe foods. If you put something rotten in your mouth you would find the taste disgusting. Your sense of taste can also help you keep your body in balance. When you eat foods with sugar and salt, your body's need for minerals and carbohydrates is satisfied. When you eat sour foods such as oranges and lemons, your body's need for some necessary vitamins is met. You need to be careful, however, that you eat the right amount of salt, sugar, and carbohydrates as part of a balanced diet.

Sticking out of each receptor cell is a taste bud that identifies the food chemicals in saliva. The binding of these chemicals with the taste receptors generates impulses in nearby nerve fibers. Most likely, all taste cells can interpret sweet, sour, salty, and bitter taste sensations.

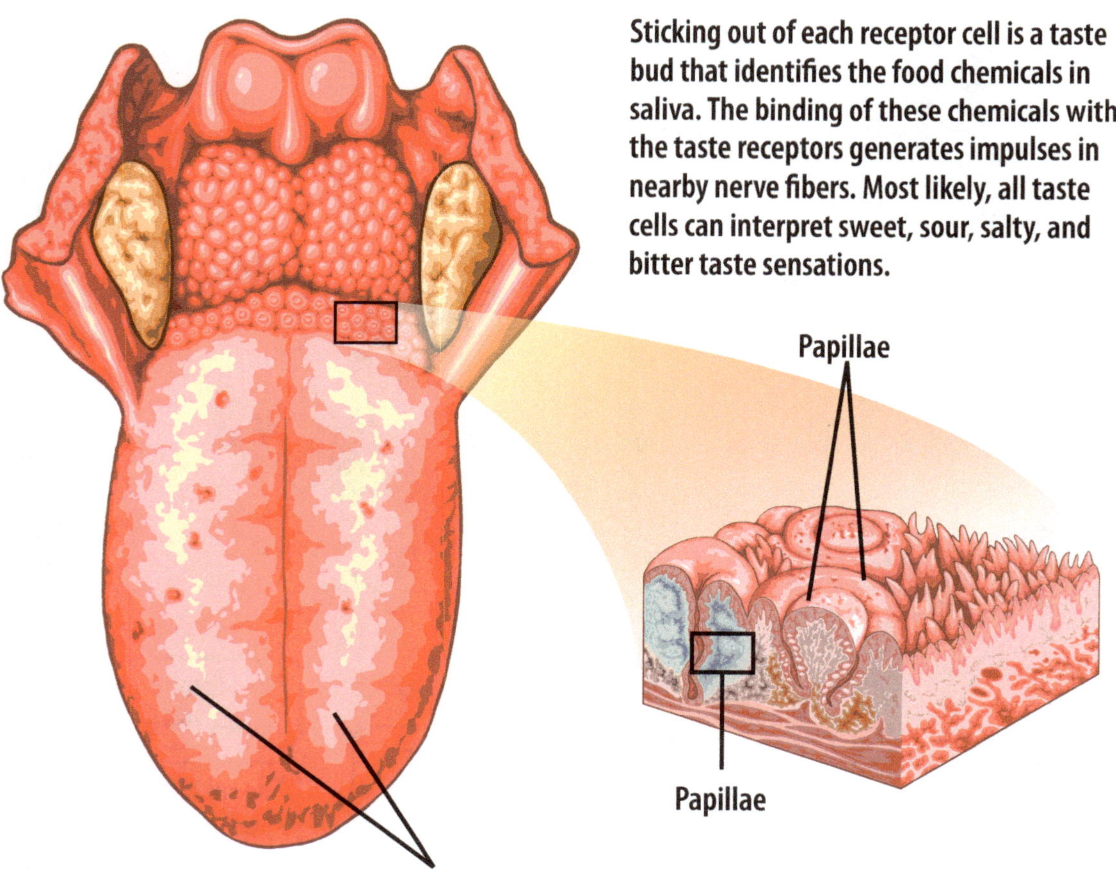

The taste buds are mainly found around the tiny bumps on your tongue called papillae.

 **How accurate is your sense of taste?**

Pair up with a classmate. One student will hand five or six different foods to a blindfolded classmate, one at a time. The blindfolded student should describe the food in terms of the four taste sensations—sweet, sour, bitter, and salty. After describing the food, the student should name the food. The non-blindfolded student should take notes. The students should switch roles and repeat with different foods. How accurate were your descriptions of the food? What senses did you use to identify the food?

Nose and Smell

For you to enjoy the taste of most food you need more than your tongue and your sense of taste. You also need your nose and your sense of smell. Think about what happens when you have a cold and your nose is stuffed up. You lose your sense of smell for a little while. While your tongue still can identify the main tastes, what you eat will not taste the same. Why is that?

Review what you learned about your sense organs. Explain how the five senses allow you to perceive the outside environment.

Skin and Touch

Your sense of touch comes from receptors in your skin. In addition to sensing touch, these receptors sense pressure, pain, heat, and cold. Receptors for touch, pressure, and temperature are not spread evenly over the body. Touch is most sensitive in the fingertips, palms, and lips. Suppose you have a piece of hair on the inside of your mouth. Why are you able to feel it so easily?

Pain receptors are found in almost every part of your body. Each receptor helps protect you from injury by sending warnings of danger to your brain.

Protecting Your Sense Organs (Explain)

You should take care to protect your sense organs.

Safety tips for protecting your eyes

- Use proper lighting when reading, writing, or using the computer.
- Wear eye protection when doing something that could cause harm to your eyes.
- Wear sunglasses that block ultraviolet radiation from the Sun.
- Walk, do not run, when carrying sharp objects.
- Never look directly at the Sun.
- Point spray products away from your face before spraying.
- Get your eyes checked by an eye-care professional.

Safety tips for protecting your ears

- Clean the outside of your ears with soap and water and wipe the inside with a soft tissue.
- Avoid cleaning your ears by poking anything into them.
- Lower the volume on your listening device.
- Get your hearing tested.

Safety tips for protecting your mouth and tongue

- Wear mouth guards in sports where injuries to the mouth may occur.
- Wear a seat belt to reduce injuries during a car accident.
- Never walk or run with objects in your mouth.

Safety tips for protecting your nose

- Don't smoke.
- Blow your nose gently.
- Do not put anything in your nose.

Safety tips for protecting your skin

- Apply sunscreen with an SPF of at least 15 to your skin and reapply every two hours.
- Cover up with clothing to protect exposed skin when in the sunshine.
- Wear a hat when in the Sun.
- Avoid tanning beds and sunlamps.
- Check your skin regularly for any changes in the size, texture, or color of a mole or sore.

The human body experiences five senses—sight, smell, touch, taste, and sound. You see with your eyes, hear with your ears, taste with your mouth, smell with your nose, and touch with your skin. Each sense allows you to interpret stimuli from the outside environment so you know what is happening in your surroundings.

 **Concept Check** Assess/Reflect

Summary: What are sense organs and senses? Sense organs are the eyes, ears, mouth and tongue, nose, and skin. These sense organs provide you with the five senses: sight, hearing, taste, smell, and touch. Receptors in each sense organ allow you to use your senses and protect yourself. There are many ways you can keep your sense organs protected as well.

1. How are the sensory receptors in the retina different?

2. What is the function of the cochlea?

3. How can having a cold affect your sense of taste?

4. What problems might you have if you could not feel pain?

Get to Know
Robert Hooke

English scientist Robert Hooke (1635–1703) is best known for discovering and naming the cell. He was the first scientist to write a book about his microscopic observations to share with other scientists and students. In addition to this key discovery, Hooke also made improvements to meteorological instruments.

Like many children of his day, Hooke was sickly and not expected to reach adulthood. Because of his poor health, his parents did not educate him. He learned by observing the plants, animals, farms, rocks, cliffs, sea, and beaches around him. He was fascinated by mechanical toys and clocks. He made many things from wood, such as a working clock and even a model ship with working guns. Hooke never did one thing at a time; he was happiest when his mind was jumping from one idea to another.

Hooke was also skilled at architecture. After the Great Fire of London, he helped rebuild the city. The fire was a tragic event, destroying about 80% of the city. Hooke worked on redesigning many of London's landmarks and helped to make the city a safer and cleaner place to live.

Watching living things through the microscope was one of Hooke's favorite things to do. He used his technical abilities to improve the design of the compound microscope by controlling the height and angle of the microscope. He changed the lighting system, which allowed him to see objects in much greater detail.

Hooke began to make detailed illustrations of his observations under the microscope. These magnificent drawings appear in his book *Micrographia,* which is famous for its scientific drawings and observations. The book has a wide range of topics, from the building of a microscope, to the color spectrum, to the anatomy of insects. Published in 1665, *Micrographia* became an instant best seller. Hooke's work started the spark of cell theory and set scientists to work in making discoveries by looking through microscopes.

Concept Check

1. Robert Hooke liked to observe living things under the microscope. How did he see his specimens in greater detail?
2. How did Hooke's work influence other scientists?

Scanning Electron Microscope

Scanning electron microscopes (SEMs) help scientists and researchers study a greater variety of specimens than ever before. The technology of the microscope provides more detailed information about the specimen. Using an SEM, scientists are able to see a 3-D view of the specimen, details of its chemical make-up, and its internal structure. Since their development in the early 1950s, SEMs have helped create new areas of study in scientific communities, opening a window into the unseen world of micro space.

The SEM uses electrons instead of light to form an image. A beam of electrons is produced at the top of the microscope by an electron gun. This electron beam moves in a straight path through the microscope. The beam moves through electromagnetic fields and lenses, which focus the beam down toward the specimen. When the beam touches the specimen, electrons and X-rays are scattered from the sample. Detectors in the SEM collect these X-rays and electrons, and make them into a signal that is sent to a screen similar to a television screen. This signal produces the final image, which is observed by the researcher.

The specimen is held in the sample chamber of the SEM. The specimen must be kept extremely still for the microscope to produce clear images, so the chamber must be very sturdy and insulated from vibration. In fact, SEMs are so sensitive to vibrations that they are often set up on the ground floor of a building. The sample chambers of an SEM do more than keep a specimen still. They also move the specimen, placing it at different angles and moving it so that researchers do not have to remount the specimen to take different images.

SEMs can magnify objects up to 300,000 times. Standard optical microscopes usually have a magnification power of a few hundred times. However, SEMs are not always the perfect solution for scientific research. Even the most affordable unit could cost tens of thousands of dollars. SEMs also take up a lot of room and are expensive to operate.

Concept Check

1. Standard microscopes use light to form the specimen's image. What does a scanning electron microscope use to form the image?
2. What are some of the challenges scientists face when using a SEM?

Study Guide

Lesson 1

1. Groups of similar cells that work together are tissues.

2. Epithelial tissue includes the skin, lungs, kidneys, small intestine, and other body parts. Connective tissue and muscle tissue are found throughout the body. Nervous tissue is found in the brain, spinal cord, and nerves.

Lesson 2

1. The human body is organized in levels from simpler parts to form systems. Organs are made of groups of tissues and organ systems are made of groups of organs that each work together to perform specific functions.

2. The endocrine system helps control body functions through the secretion of hormones.

3. The immune system defends the body against attacks by infectious substances.

4. The lymphatic system collects fluid from tissues and returns it to the blood.

Lesson 3

1. The integumentary system is made up of the skin, hair, nails, and glands.

2. While the skin can repair itself, it needs protection from solar radiation, daily maintenance, and at times, first aid.

Lesson 4

1. Nerve impulses cross synapses, sending signals through the central and peripheral nervous systems.

2. The brain consists of the cerebrum, the cerebellum, and the brain stem.

3. Humans use different kinds of thinking, including perception, imagination, logical thinking, memory, and spiritual thinking.

Lesson 5

1. The major sense organs are the eyes, ears, mouth and tongue, nose, and skin.

2. The major senses are sight, hearing, taste, smell, and touch.

Show What You Know

Visualize It Complete the concept map in your *Science Journal*.

Endocrine System

1. ___________________ ___________________

Lymphatic System

2. ___________________ ___________________ ___________________

Body Systems

Nervous System

5. ___________________ ___________________

Immune System

3. ___________________ ___________________

Integumentary System

4. ___________________ ___________________

Vocabulary Check
Complete each sentence.

6. Tissue that consists of cells that contract and produce movement is __________ tissue.
7. A group of tissues is a(n) __________, while a group of organs is a(n) __________.
8. The body system that controls body functions through the secretion of hormones is the __________ system.
9. The part of the neuron that carries impulses from other neurons toward the cell body is the __________, while the part of the neuron that carries impulses away from the cell body to other neurons is the __________.
10. The part of the brain that controls muscle coordination, balance, and muscle tone is the __________.
11. A specialized cell that receives information from its surroundings and provides the brain with that information is a(n) __________.
12. The colored ring that gives the eye its color is the __________.

Multiple Choice
Choose the best answer.

13. Which type of tissue is composed of cells specialized for communication?
    A. epithelial   C. muscle
    B. connective   D. nervous
14. Which of the following protects the others?
    A. skin   C. muscle
    B. cartilage   D. bone
15. Which of the following focuses images in the eye?
    A. retina
    B. lens
    C. pupil
    D. optic nerve
16. Which part of the skin removes extra heat from the body?
    A. sweat glands
    B. oil glands
    C. keratin
    D. melanin
17. Which of the following glands controls the endocrine system?
    A. thyroid gland
    B. adrenal gland
    C. pituitary gland
    D. oil glands

Check Point
Answer the following questions.

18. **Compare** the endocrine and nervous systems. How are they alike? How are they different?
19. Explain how the functions of the immune system, circulatory system, and lymphatic system are interrelated.
20. **Observe** the diagram. What must happen for an impulse to move to the next neuron's dendrite?
21. What senses do you use to communicate with God?

Systems Work Together in Your Body

This rock climber is using many bodily systems to complete a strenuous task. His skeletal system provides the support that the muscle system uses to propel him upward. The respiratory and circulatory systems work together to supply blood and oxygen to his muscles so they can continue this activity. Each individual system has a specific function to sustain life.

Scripture Spotlight

You can see God's thoughtful design in the way your body systems work together. As you learn about your body systems you can appreciate God's care when he designed humans.

You will read the following passages in this chapter.

Hebrews 4:12 (p. 168)   Mark 16:15 (p. 184)   Psalm 104:29 (p. 190)   Psalm 150:6 (p. 190)

Deuteronomy 6:5 (p. 182)   Ezekiel 36:26–27 (p. 185)   Psalm 146:4 (p. 190)

Mark 12:30 (p. 182)   Genesis 2:7 (p. 190)   Psalm 146:2 (p. 190)

The Big Idea

Living things are highly organized and have structures and systems with specific functions. God organized the structures and systems to work together to give life.

Do you think an organism could live if its cells were disorganized and the body systems did not work together? Explain.

Inquiry Kick-Off **Engage**

Respiration is necessary to provide energy to the cells of living things. How long can you go without breathing? In your *Science Journal*, you will explore the respiration rates of three different animals and then compare them with your own respiration rate.

Essential Question

What Do Bones Do?

Carefully move your hand and arm to reach out and turn the page of this book. What bones are moving to help you make this movement? How many bones are involved? What moved the bones? What controlled and coordinated the movement of your hand so that you didn't rip the page when you turned it? Your bones are part of an elaborate system of tissue and structures that work with other body systems to make it possible to move in nearly an infinite number of ways.

The Skeleton (Explain)

Look closely at your hand. How many bones make it up? Count the bones and find out. When you were born, you had more bones than you do now. When you are fully grown, you will have less than you do now. Why do you think this is so? Your bones, along with the tissues that connect them, form the **skeletal system**. This system helps you in many ways. Your leg and arm bones help move you. Your spine supports your body and keeps it upright. Your ribs and skull protect your internal organs. Your blood cells produce and carry nutrients and minerals to all parts of your body.

How are the children's skeletal and muscular systems working together as they rollerblade and ride bikes?

Living Bone

What layers make up bone?

Procedure

1. Identify what color of clay you will use to represent each layer of living bone.

2. Write the name of each layer on a slip of paper. Attach the slips of paper to toothpicks with masking tape.

3. Roll out each layer of clay.

4. Assemble the layers based on your knowledge of the structure of living bone.

5. Use fishing line to cut the model bone crossways.

6. Use the fishing line to cut one half of the model bone lengthwise.

7. Make labels using toothpicks and masking tape. Place the labels into the appropriate clay layers.

8. Draw a cross section and a lengthwise section of the bone model. Use a ruler to **measure** sections that you can depict on the diagram.

Materials
- molding clay (blue, white, yellow, red)
- rolling pin
- monofilament fishing line (10 lb test)
- toothpicks
- masking tape
- metric ruler
- paper
- pencil

Analyze Results

Observe the model. **Compare** the results with the diagram in the book. Record similarities and differences between your model and the diagram.

Create Explanations

1. What layers make up bone?

2. How do the layers of bone work together to allow bones to heal?

3. What do you think a model bone would look like for a bird or other animal?

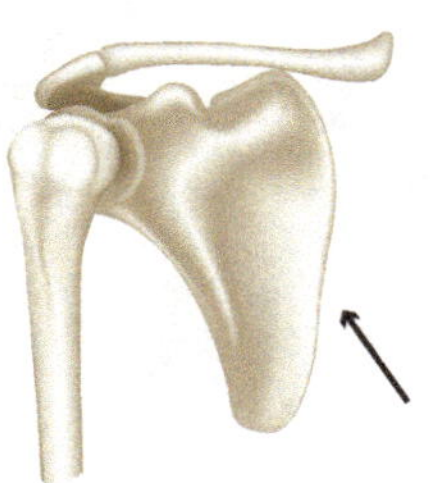

irregular bone

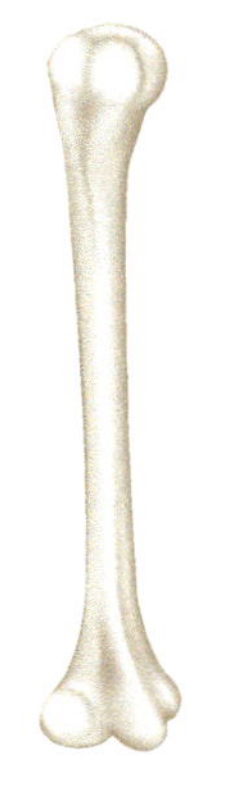

flat bone

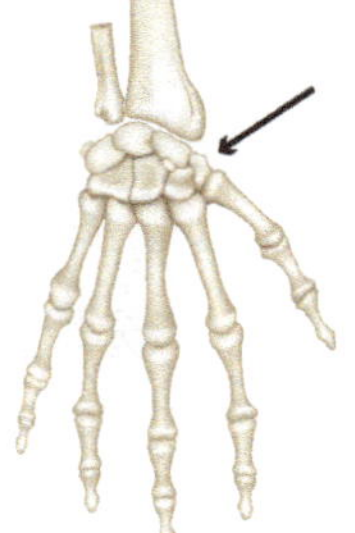

long bone

short bone

Structure and Formation of Bones

Bones have different shapes, and each shape helps bones perform their specific function. Your arms and legs have long bones. Flat bones make up your skull and rib cage. Your wrists and ankles are short bones. What do you think the most likely purpose is for each of the three types of bones? How do you think these features show the Creator's design?

Your bones may vary in shape, but they are all made up of cells called **osteocytes**. These cells are responsible for growth and repair. They constantly remove old bone and create new bone.

God created bones with a lightweight, efficient design. Engineers have not developed a way to mimic this material.

Bones have four layers.

- outer covering This contains blood vessels that carry nutrients and waste to and from the bone cells. There are also nerve cells in this outer layer. Can you describe a time you might experience the fact that there are nerves covering your bones?

- *compact bone* This hard thick layer is what you see when you look at a skeleton.

- spongy bone A soft layer filled with cavities that gives flexibility to the bone and lighter for ease of movement.

- **marrow** Hollow space or cavity deep in side bone where the red, white blood cells, and platelets are produced.

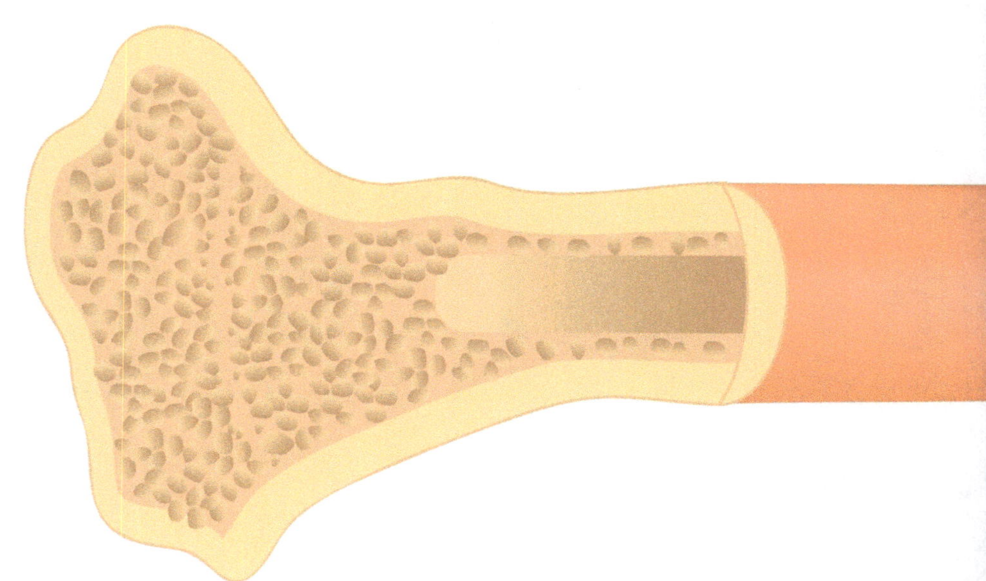

If blood cells are produced in the marrow, why does the soft outer layer provide blood to the bone? Think about the layers of the bone. Which layer do you think provides the most support to the skeleton? Describe what would happen if these same materials were arranged in a different way.

Cartilage

Cartilage is connective tissue that forms part of your skeleton. It is found at the ends of bones and helps cushion them and reduce friction. Sometimes, the cartilage between bones can wear away. What do you think it would feel like if you had less cartilage between your bones?

Bend your ears and the end of your nose. How does it feel? These body parts are also made up of cartilage. It gives them shape and support. God designed the ears and nose to be made of cartilage. What would it be like if the nose was made of thin bone instead? What if your ears and nose were made only of skin?

Spongy bone is filled with many tiny holes.
How do you think these holes relate to the lightweight structure of bones?

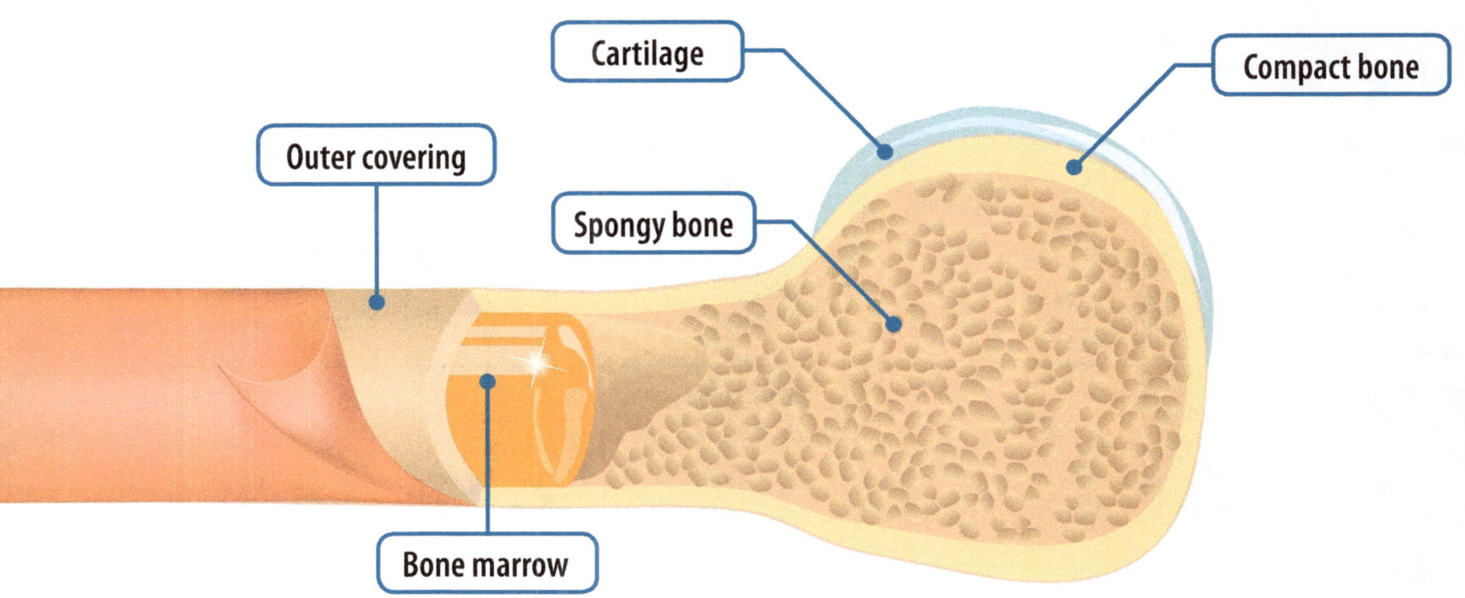

Scripture Spotlight

Does the Bible mention joints? Yes! Look up **Hebrews 4:12** to see what it says. **What other part of the skeletal system is mentioned in the same verse? What do you think the verse means?**

Joints

Look at your hand again. How many places can you find where bones meet? What can you feel as you hold your ankle while you wiggle your foot? You need to be able to move your body, but your bones are hard, so they can bend very little without breaking.

Movement happens where bones meet, at the **joint**. There are two classes of joints: fixed and movable. All bones of the skull are fixed, except for the lower jaw. Fixed joints do not move. Movable joints allow varied ranges of motion. How many movable joints can you identify in your body? Do they all move in the same direction? How do different joints move? **Ligaments**, strong bands of connective tissue, hold the bones in the movable joints together. Ligaments stretch a little so the bones move but do not separate. Overstretching a ligament causes a sprain. Have you ever had a sprain? How does it feel? What activities or actions might increase the risk of sprains?

Four Basic Types of Movable Joints			
Ball and Socket	moves in many directions	examples: hip, shoulder	
Hinge	moves back and forth	examples: knee, elbow, knuckle	
Pivot	rolls or rotates	example: top of the neck	
Gliding	slides	examples: bones in ankle and wrist	

How does a joint work?

Use a pencil, modeling clay, rubber bands, and string to build an elbow joint and a shoulder joint. After you finish your model, work in groups to compare models. What does each item used in the model represent? Describe what worked and what did not work for all group members' models.

Caring for Bones Explain

What is your favorite possession? How do you take care of it? Like a car or a bicycle, the body needs to be cared for to function properly. Recall **1 Corinthians 10:31**: "So whether you eat or drink, or whatever you do, do all to the glory of God." Your body is a temple of the Holy Spirit. Bones must be cared for in order to function properly. Proper diet and exercise are the best ways to ensure the health of your bones.

Front view of right knee

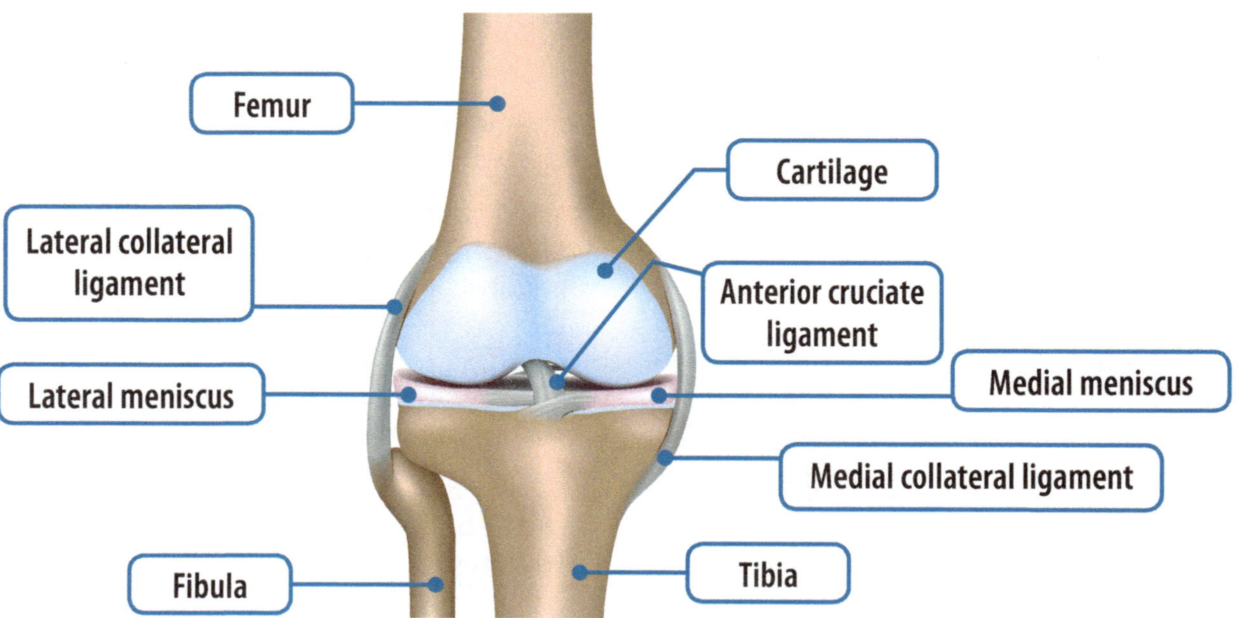

Eat a balanced diet of foods that have a minimum amount of processing. This diet supplies adequate amounts of calcium and phosphorus, two minerals necessary to maintain strong, healthy bones. Phosphorus is found in almost every type of food. What types of food are rich in phosphorus? What are some healthy sources of calcium?

Bones and muscles are strengthened by exercise. As a muscle becomes stronger, the bone to which it is attached becomes stronger, too. What activities and exercises will increase the strength of your bones and muscles?

Sometimes bones are injured or become diseased. The chart below outlines a few problems bones could face, but it also tells you how to minimize the risk so you can stay healthy.

Osteoporosis is a common bone disorder, especially among the elderly. Sometimes, if extra calcium is needed somewhere else in the body, bone cells may dissolve minerals out of the bone around them. If too many minerals dissolve out of the bone, it is weakened. Osteoporosis is also common in women. Eating food rich in calcium while you are young ensures rich calcium deposits in your bones. This is one of the best protections against developing osteoporosis when you are older. Pregnant women also take supplements to ensure that the growing fetus does not strip the calcium from the mother's bones. How might one treat osteoporosis?

Skeletal System Injuries and Disorders		
Type	**Description**	**Minimize the Risk**
Fractures	broken bones	exercise to increase bone strength; decrease high-risk activities
Sprains	overstretched ligament	warm up before activities; avoid high-impact activities that could stretch joints
Osteoporosis	loss of bone tissue	eat foods high in calcium; get more vitamin D; exercise regularly
Scoliosis	abnormal curve of the spine	advanced cases need a doctor's care, a brace, or an operation
Arthritis	inflammation of the joints	exercise; eat a healthy diet; keep a healthy weight

 **How does a lack of calcium affect a chicken leg bone?**

Get three cooked, clean, dry chicken leg bones. Observe the bones' texture, hardness, and flexibility. Record your observations on a sheet of paper. Measure and record the mass of each bone. Place the three bones into three different jars. In one jar, cover the bone with water. In another jar, cover the bone with distilled white vinegar. Do not put anything with the bone in the third jar. Screw lids on all the jars. After five days, remove the chicken bones from the jars. Rinse the bone that was in vinegar with water. Dry each bone with a paper towel. Observe and compare the bones' texture, hardness, and flexibility. Measure and record the mass of each bone. What differences do you notice? What do you think happened to the bone in the vinegar? What might cause your bones to not have enough calcium in them, or to lose calcium?

Take a look at yourself in a long mirror. Imagine what the bones look like in your skeletal system. Imagine how the bones fit together in this system to provide your body the support it needs to remain upright and move. Now, imagine how the skeleton protects the body's internal organs. The skeletal system is the impressive internal architecture of the body.

 **Concept Check** Assess/Reflect

Summary: What do bones do? The skeletal system is made up of the bones in your body—long, short, and flat. Bones make coordinated movements at joints. There are four types of joints; ball and socket, hinge, pivot, and gliding. It is important to eat healthy foods, exercise, and stay fit in order to maintain healthy bones.

1. How does good nutrition help the skeleton?
2. List two different protective bones and the organs that they protect.
3. What might be the cause of pain in the knees.
4. Why is it important for your bones that you have good nutrition?
5. How is your skeleton different from that of a grasshopper? Explain how your skeleton helps you.

What Do Muscles Do?

Objectives

- Describe the structure of the muscular system.
- Explain how to care for the muscular system.
- Explain how muscles and bones work together.
- Identify some injuries and diseases of the muscles.

Vocabulary

muscular system

tendon

Think about runners in a professional race. What factors are held constant so the competition is fair? Men and women do not compete against one another because the design of their bodies is different. For example, a man's pelvis is designed to allow for powerful muscles to develop. The hip sockets also are closer together to allow for more efficient motion. Meanwhile, a woman's pelvis is broader. This design allows her to to give birth to a baby. Would it be fair to allow men and women to compete against each other in a running competition? What is the impact of how muscles develop and attach to the pelvis that can affect the competition? This simple example illustrates a difference between the design of men's and women's bodies. How do these differences show evidence of God's design?

Muscles Explain

Your body has more than 600 muscles that form the **muscular system**. This system works with the skeletal system to help you move. Together, they are called the musculoskeletal system. You are able to control some muscles, such as those in your arm. They are called voluntary muscles. Other muscles, such as those in your heart and stomach, you are unable to control. They are called involuntary muscles. Can you think of other muscles that you are able to control? What other muscles are in your body that you cannot control? Do you think there are any involuntary muscles that you can control at times? If so, what are they?

Muscles move your body.

Skeletal Muscle

- Moves muscle with voluntary motion
- Bundled long cells that form fibers
- Bundled fibers that make muscle
- Appears striated or striped

Smooth Muscle

- Moves muscle with involuntary motion
- Made of small cells that are connected directly to one another
- Lacks striated appearance
- Makes up the esophagus, stomach, intestines, and blood vessels

Cardiac Muscle

- Moves muscle with involuntary motion
- Found only in the heart
- Appears striated

Tendons

Wiggle your fingers as if you are typing. How would you describe the movement you see in the top of your hands? What you see are your tendons at work. **Tendons** are tough bands of tissue that fasten your muscles to your bones. Based on your experience dissecting the chicken wing in a previous chapter, how are tendons essential for movement? One of the more well-known tendons is the Achilles tendon. It fastens the calf muscles to the heel bone. Why was this tendon named for Achilles? There are many tendons in your hands and your feet. What is the benefit of having so many tendons in these parts of your body?

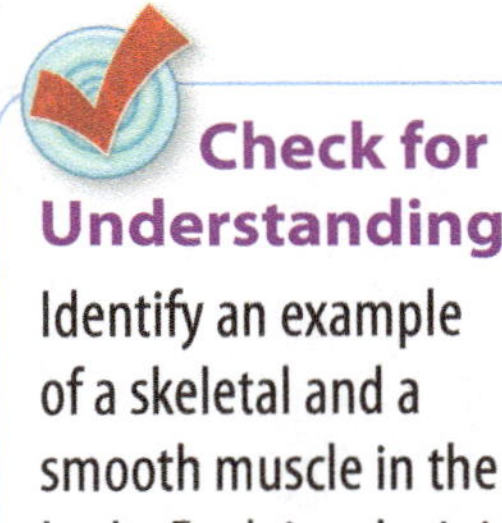

Check for Understanding

Identify an example of a skeletal and a smooth muscle in the body. Explain why it is categorized that way.

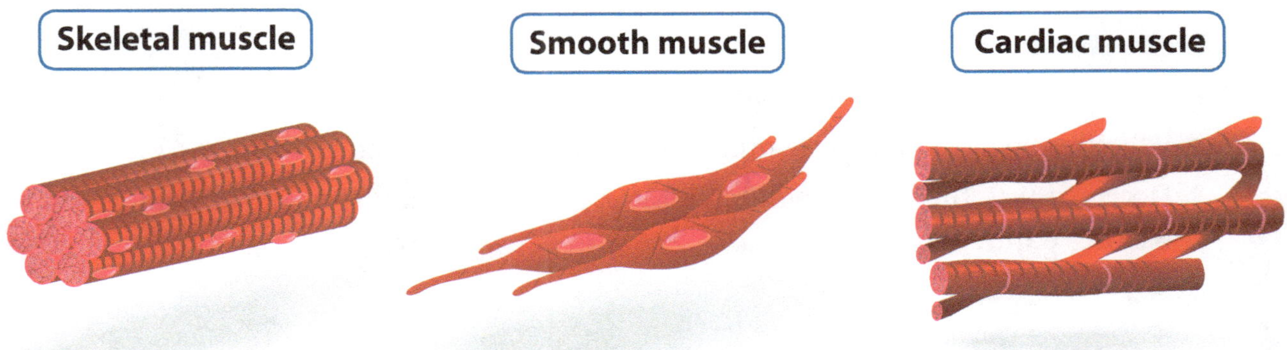

Tired Muscles

How do muscles react to repeated activity?

Procedure

Materials
- book
- stopwatch
- graph paper
- colored pencils

1. You are going to do three activities to see how your hand muscles tire. Use your non-writing hand for all activities. You will work with a partner. Gather and **record data** for each activity using the chart in your *Science Journal*.

2. Holding your palm upward, lift your index finger up and down as fast as you can. Count each time you lift your finger. Every 20 seconds, your partner will **record** the last number you counted. Continue for 120 seconds. Then, switch roles with your partner.

3. For the second activity, open and close your hand as fast as you can. Count each time you complete one cycle. Every 20 seconds, your partner will record the last number you counted. Continue for 120 seconds. Then, switch roles with your partner.

4. For the third activity, find out how long you can hold a book in your open hand with your arm extended straight out from you.

5. Switch roles with your partner. Repeat Steps 2–4 to get three sets of data for you and your partner.

Analyze Results

Make a graph using colored pencils to **display the data**. How did your data change from the beginning to the end of each activity? Explain.

Create Explanations

1. How do muscles react to repeated activity?

2. Which bones assisted your muscles in each of the movements?

3. Do you think the results would be different if you had used your writing hand? Explain.

Study the diagram below. The muscles of your body are on the right side of the diagram, and the bones are shown on the left side. The muscular system and the skeletal system cannot function without each other. The smaller image shows a back view.

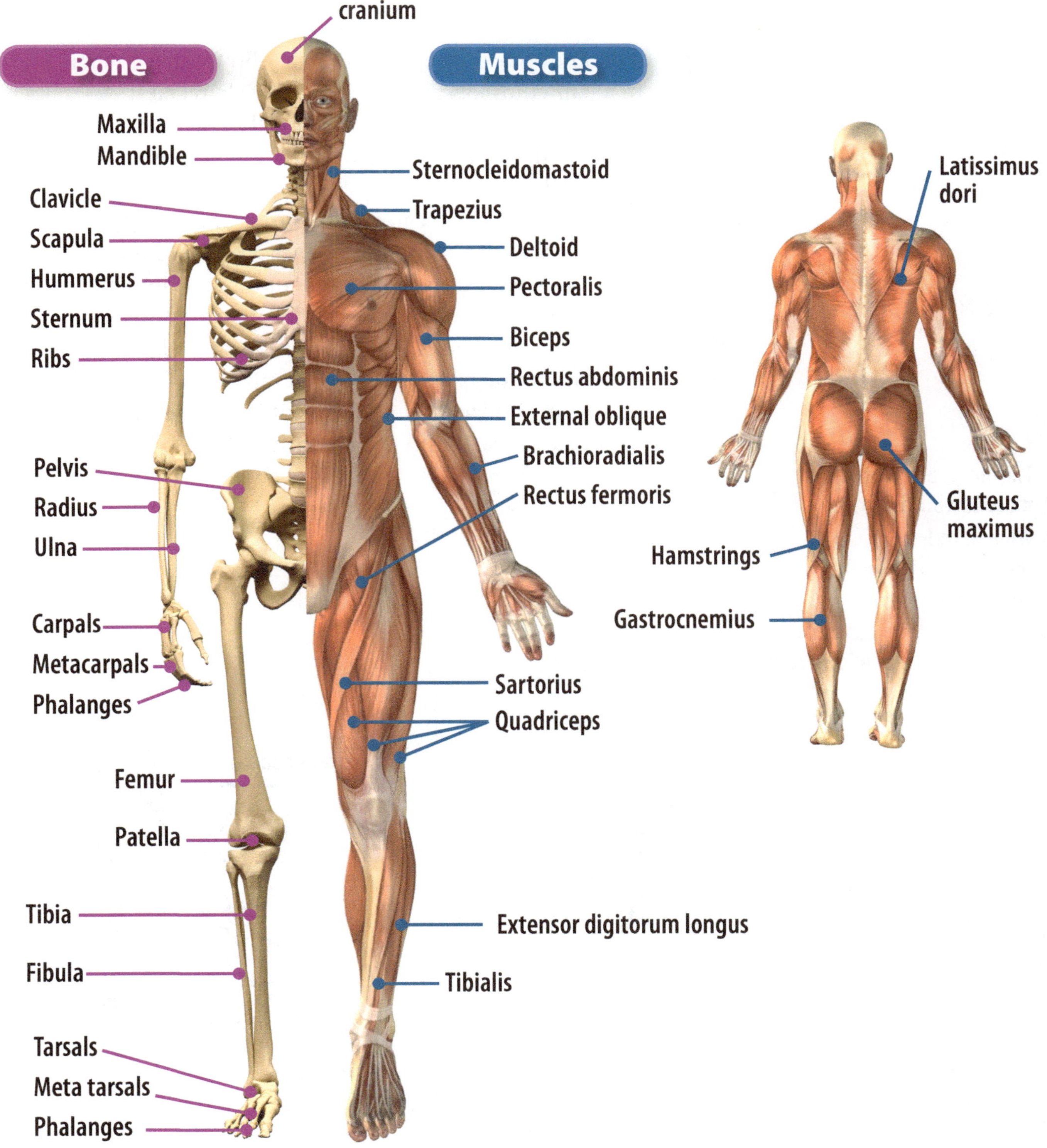

Muscles and Bones (Explain)

In a world where machines do most of our hard work, why do you think that muscles need to get tired for good health? Your muscles and bones need each other to support and move your body. Muscles move body parts when they contract, or shorten. This contracting of muscle cells creates a pulling action. Because muscles can only pull and not push, they work in pairs. Bones move when one muscle in a pair contracts and the other muscle relaxes.

One muscle in a pair is a flexor, or a muscle that bends a joint. The other muscle in the pair is an extensor, or a muscle that straightens a joint. Think about the biceps and triceps to understand how this works. The biceps is located in the front of the upper arm, and the triceps is at the back of the upper arm. Bend one arm at the elbow and place your opposite hand around your upper arm. What do you feel happen as you move your arm? When your biceps contracts, your arm bends at the elbow. As you relax your biceps, your triceps contracts and your arm straightens. How is the movement in the arm similar to the movement in the leg? What would happen if your muscles did not work in pairs?

The biceps and triceps muscles work as a pair.

What must happen with the two muscles for the arm to straighten?

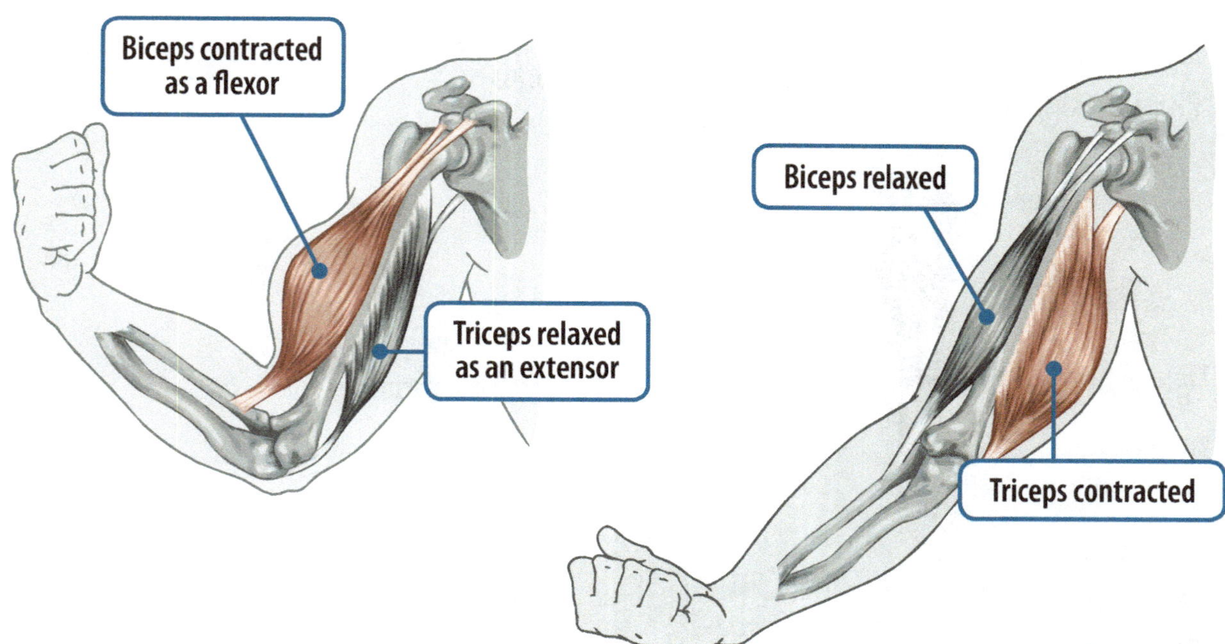

Caring for Muscles

Just like your bones, it is important to maintain healthy muscles. Proper diet and exercise are the best ways to ensure the health of your muscles. A proper, balanced diet supplies the nutrition necessary to keep muscles healthy. In particular, carbohydrates provide the energy needed for muscles to do their work. What are healthy sources of carbohydrates?

What are some exercises you can do to strengthen different muscle groups in the body? As a muscle becomes stronger, the bone to which it is attached becomes stronger, too. Without exercise, muscles become weak and tire more easily. What activities can you do to increase the strength of your muscles?

Muscular System Injuries and Disorders		
Type	Description	Reduce the Risk
Strain	overstretched muscle	warm up before activities; lift and bend properly
Tendonitis	inflammation of a tendon	use proper posture; stretch before physical activities
Muscular dystrophy	a genetic disorder that gradually weakens muscles in all parts of the body	there is no cure; treatment includes physical therapy, bracing joints, taking medication, and using devices such as motorized wheelchairs

What are common injuries a runner might experience? Explain.

Take a look at yourself in a long mirror. Imagine all of the muscles that make up your muscular system. Imagine how the skeletal muscles attach to the bones to provide your body the ability to move. Now, imagine how the smooth muscles and cardiac muscles work without you having to think about it. The musculoskeletal system does not work alone. Later in this chapter you will learn about the circulatory and respiratory systems.

 How can I model the movement of my hand?

You will need a piece of thin cardboard, a pen, scissors, string, and tape. Trace an outline of your hand with the fingers spread out on to the cardboard and cut it out. Cut five pieces of string to be about the length of your handprint. Use a small piece of tape to attach one string to the tip of the thumb on your cardboard hand. Use another small piece of tape to attach the string to the joint of your cardboard thumb.

Use a third piece of tape to attach the string at the joint connecting the palm and thumb. Do this for the remaining fingers, placing tape at each joint. Tape the five string ends to the palm of the hand. Pull on the strings at different points to make the fingers move. What part of your fingers do the strings model? Does your hand work in a way similar to the model? Explain.

 Concept Check   Assess/Reflect

Summary: What do muscles do? The muscular system is made up of the muscles in your body—smooth, skeletal, and cardiac. This system works in concert with the skeletal system to move the body. The health and strength of your muscles are dependent upon how well you take care of yourself. It is important to eat healthy foods, exercise, and stay fit in order to maintain healthy muscles.

1. Explain why muscles work in pairs.
2. How does running keep the body strong and healthy?
3. Suppose you have a fractured humerus and your arm is in a cast. Explain what would happen to your biceps during that time. Why?
4. If muscles work in pairs, what muscles work to open and close the eye?
5. How might a rock climber and runner train differently to maximize performance?

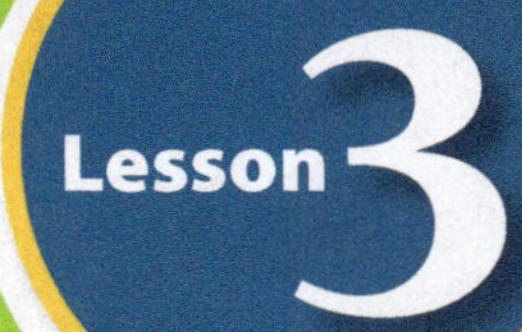

Essential Question
How Does Circulation Work?

If you sit down to a bowl of cereal for breakfast, do you ever wonder where the food came from? How were the ingredients brought together to create the cereal? How did the cereal make it into your home? All of these questions are tied to a vast transportation system that brings raw materials from the farm to where they are manufactured in a factory and then delivered to a store. The transportation system moves materials in an orderly way. The pattern of roads, bridges, and tunnels is similar to the network that moves nutrients, blood, and oxygen throughout your body. God designed a most amazing transportation system inside our bodies—the circulatory system. It transports all the materials needed for us to live all day, every day.

Circulation Explain

The busy transportation system inside your body is moving 7500 liters (2000 gallons) of materials each day to keep you alive! The group of organs that transports materials from one place in your body to another is the **circulatory system**. It consists of blood, the heart, and a network of blood vessels.

Objectives

- Describe the structure of the circulatory system.
- Describe the make-up of blood.
- Describe the path blood follows as it moves through the body.
- Explain how to care for the circulatory system.
- Identify diseases of the circulatory system.

Vocabulary

circulatory system
red blood cell
platelets
hemoglobin
atrium
ventricle
capillaries

As these runners race to the finish line, their bodies work to adjust for the energy they are exerting.

How would the circulatory system be responding as these runners race to the finish line?

179

Blood

What is the purpose of blood? How does it transport materials to the different organs in the body? Blood consists of solid and liquid parts, including plasma, red blood cells, white blood cells, and platelets. The liquid part of blood is *plasma*. It consists of about 90% water and 10% nutrients and other materials. The solid part of blood includes red blood cells, white blood cells, and platelets. **Red blood cells**, or *erythrocytes* are large microscopic cells without nuclei that carry oxygen from the lungs to body cells and carry carbon dioxide from body cells to the lungs. The white blood cells, also called *leukocytes,* help protect you from infection and disease. They are larger than red blood cells, but fewer in number and are also found in your spleen, liver, and lymph glands. Some white cells called *lymphocytes* are your immune system first responders. They seek out bacteria, viruses, and fungi in your body. **Platelets** are small colorless bodies that release coagulating chemicals to form clots to stop the flow of blood.

Blood transports nutrients and oxygen to your body's cells. It carries carbon dioxide and other waste products away from the body's cells. In addition, blood helps maintain a proper temperature for your body and fights disease. Why is the color of red blood cells significant? Red blood cells contain an iron-rich molecule called **hemoglobin**. This gas transporting molecule gives blood its color and makes up 95% of a red cell. It attaches to oxygen molecules. Where does the exchange of oxygen and carbon dioxide occur?

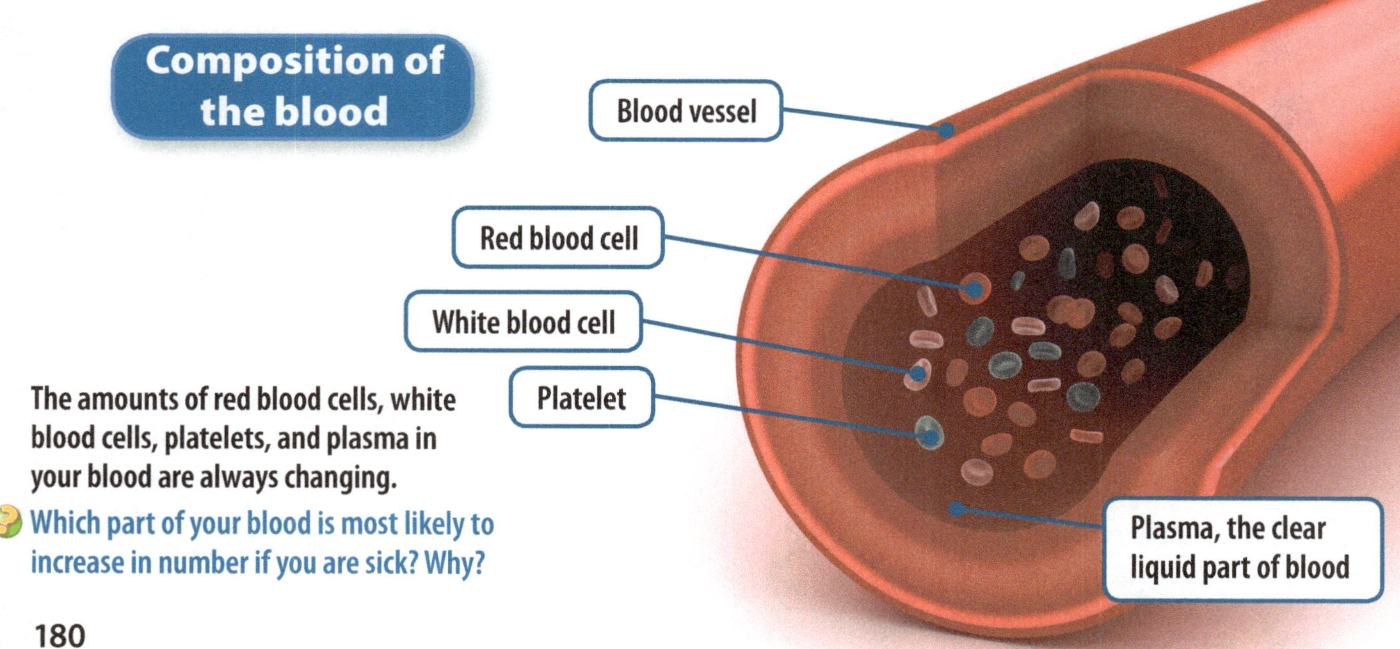

The amounts of red blood cells, white blood cells, platelets, and plasma in your blood are always changing.

Which part of your blood is most likely to increase in number if you are sick? Why?

Measuring Heartbeats

How does heart rate change with exertion?

Procedure

1. Wipe the earpieces of a stethoscope with an alcohol swab. Place the earpieces in your ears.

2. Place the bell of the stethoscope on your partner's back under the left scapula, and listen for the heart sounds.

3. Listen for the loudest and strongest sound of the heartbeat. This sound signals the contraction of the ventricle and is called systole. Now listen for the second softer sound. This softer sound signals the relaxation of the ventricles and is called diastole.

4. Once you can recognize the two sounds of the heart, use the stop watch to identify the time it takes between the soft sound and the louder sound. Tell your partner the time and have your partner record the data.

5. Reset the stopwatch, count the number of heartbeats for 15 seconds.

6. Estimate the average heart rate at rest for one minute (the time between the soft and loud sounds of the heartbeat). Your partner will **record the data**.

7. Complete ten jumping jacks. Repeat Steps 6–7.

8. Complete 30 jumping jacks. Repeat Steps 6–7.

9. Switch roles and repeat the exercise.

Materials
- stethoscope
- stopwatch
- pencil
- paper

Analyze Results

Graph your data to show how heart rate changes. Then use different colors to graph the rate of change for each person. Use different symbols to differentiate the data for each activity. **Compare** your data and your partner's data. **Use numbers** to explain.

Create Explanations

1. How does heart rate change with exertion?

2. What is the benefit of a change in heart rate with exercise?

3. How can exercise strengthen your heart?

4. How can monitoring heart rate provide information on heart health?

What do **Deuteronomy 6:5** and **Mark 12:30** say about the heart? Do you think they are talking about the cardiac muscle? Similar to the way we sometimes use the word *heart* today, the Bible describes it as the center of our thoughts, feelings, affections, and morals, as well as the dwelling place of Christ in us.

The Heart `Explain`

God designed the heart to pump blood to all parts of the body. It is made up of cardiac muscle, nerve tissue, and connective tissue. Place your hand over your heart. Do you feel the thumping of your heart? Your heart is actually located between your lungs and is protected by the sternum. Make a fist. That is about the size of your heart. What is the width of your heart? What is its circumference? Your heart beats about 100,000 times a day! It can pump approximately 26.5 liters (7 gallons) of blood every minute. As you grow, your body needs more oxygen. At 12 years of age, your heart beats between 70 and 110 times per minute. As you age, your heart rate slows to correspond to changes in your growth rate.

The heart not only pumps blood through the body, but also pumps blood back to the lungs, where it will be replenished with oxygen. Recall that your heart is divided into two sides separated by a wall of muscle called the septum. The heart contains four chambers. As you may recall, each upper chamber is an **atrium**. Each lower chamber is a **ventricle**.

Why does your heart make the distinct lub-dub sound? The sound relates to the closing of the heart valves. As the heart contracts, it pushes blood from one chamber to the next. A one-way valve closes behind each chamber. Why do you think it is important that these valves be one-way?

The pumping action of the heart creates your blood pressure which is measured in two ways; *systolic* and *diastolic*. Systolic is the pressure in your arteries from when your heart muscle contracts in a heart beat. Diastolic is the pressure in your arteries between heartbeats when the heart muscle rests between beats and refills with blood. Both systolic and diastolic pressure are measured to monitor heart and blood flow. A reading of less than 120 mm Hg systolic pressure and less than 80 mm Hg are considered healthy. How does blood circulate through the heart? Follow along with the diagram on the next page. Each of the following steps corresponds to the number in the diagram.

1. *Veins* carry oxygen-poor blood from the body back to the heart. The *superior vena cava* brings blood from the upper body and the *inferior vena cava* from the lower body. The blood is deposited into the right atrium.

2. When the heart contracts, the blood is pushed from the right atrium through the tricuspid valve, into the right ventricle. This and other valves in the circulatory system open by the force of pumping blood in one direction and close preventing blood flow from backing up. The right ventricle also contracts and pushes blood through the pulmonary valve into the pulmonary artery. An artery is a blood vessel that carries blood away from the heart.

3. The blood travels from the pulmonary artery to the lungs. In the lungs, the blood releases carbon dioxide and picks up oxygen.

4. Blood returns to the heart through the *pulmonary veins*. The blood can only pass one way through the mitral valve that closes each time blood is forced through and is deposited in the left atrium, which contracts and pushes blood into the left ventricle.

5. From the left ventricle, the blood is pumped through the aortic valve, which closes once the blood is pumped through into the aorta, the largest artery.

6. The aorta branches to other arteries, and blood flows to all parts of the body. The blood circulates through the body depositing oxygen, before returning as oxygen-poor blood back to the heart by the superior and inferior vena cava. Then the cycle begins again.

The pathway of blood flow through the heart

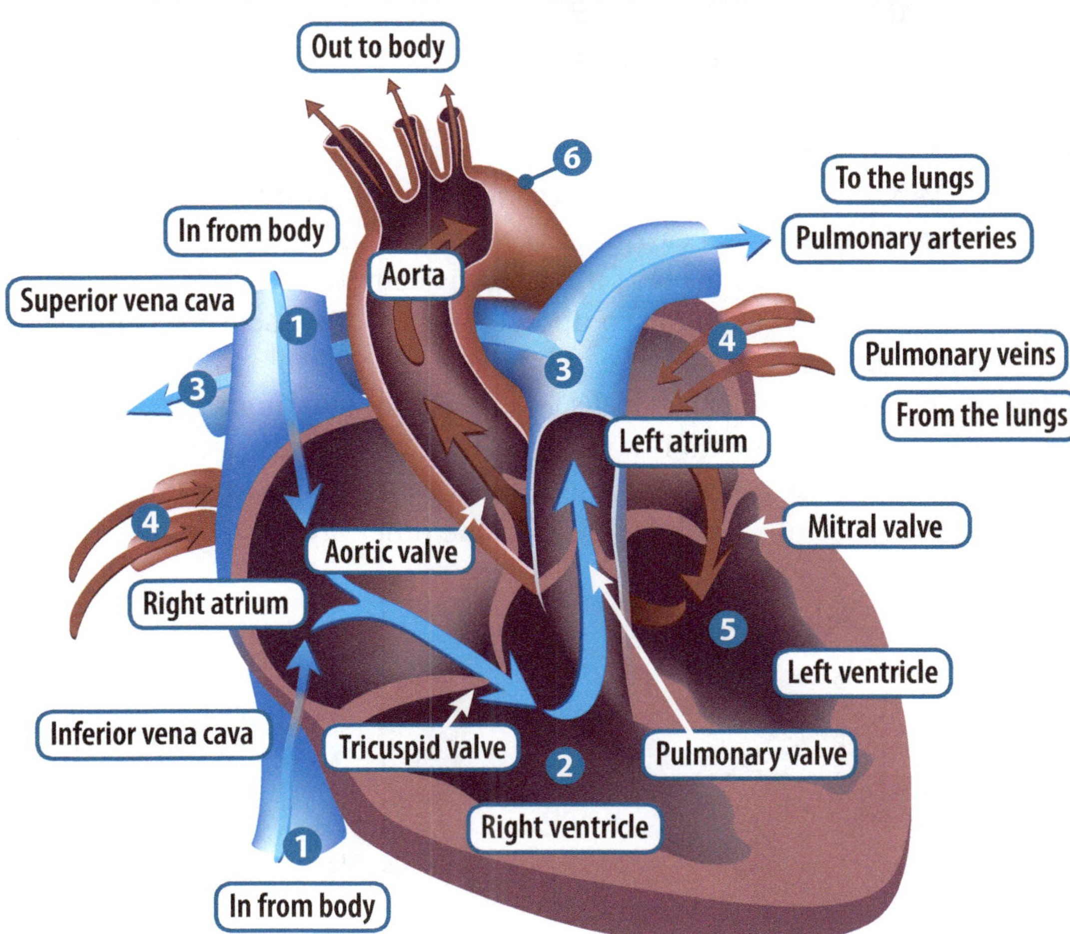

Check out your *Science Journal* for an exploration of the heart's structures.

Extend

How can you observe your heartbeat on the outside of your body?

Why is it important to monitor your heart rate? When are some times when your heart rate should be monitored? Taking your pulse is an easy way to monitor your heart rate. You can make a pulse rate detector, using a flat thumb tack, large wooden match, and stop watch. Place a thumb tack into the base of the match. Rest your arm on a table, palm facing upward.

Place the match and thumb tack on the pulse point of your wrist with the match pointing up. Monitor the match. Count how many times it moves in 15 seconds. Multiply by four to get your pulse rate (heartbeats per minute). Repeat the experiment after completing 30 jumping jacks.

SAFETY: Always use care when handling sharp objects.

Scripture Spotlight

In **Mark 16:15**, Jesus commands his disciples to go into all the world preaching the gospel. Think about Christians being like blood. How could the flow of blood through your body be like Christians taking the gospel to all the world?

Check for Understanding

How does the structure of veins and arteries relate to their function? Explain.

Blood Vessels `Explain`

Recall that blood vessels transport blood throughout the body. These blood vessels vary in size and structure. Your system of blood vessels, lined up end to end, is more than 96,500 km (60,000 mi) long! How many times around Earth is that distance? Take a guess and then research to find out.

There are three types of blood vessels:

- Arteries
- Veins
- Capillaries

What type of blood vessel carries oxygen-poor blood? What type of blood vessel carries oxygen-rich blood?

The outermost layer of the artery is composed of connective tissue. The next layer is composed of smooth muscle cells and elastic tissue. The inner layer is in contact with blood. It is composed of *endothelial cells* and an elastic membrane. Blood flows through the hollow cavity at the center of the artery. What is the purpose of smooth muscle in the artery? What is the significance of endothelial cells in the inner layer?

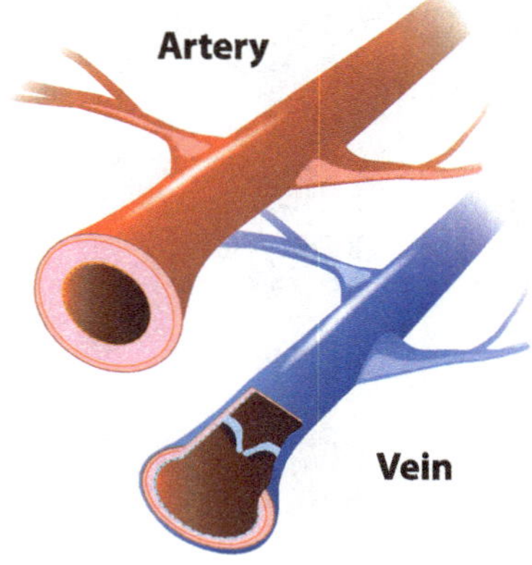

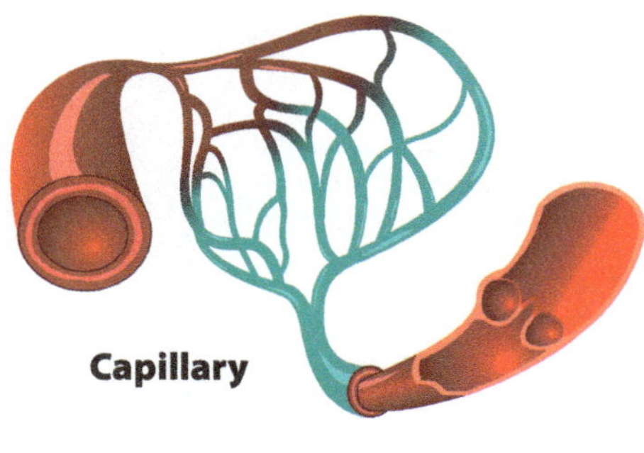

184

The outermost layer of the vein is composed of connective tissue. It covers a band of smooth muscle. This layer is thinner than in arteries. Why might the smooth muscle layer be thicker in arteries rather than veins? Like arteries, the interior of veins is lined with endothelial cells. It also contains valves. What might be the function of valves in the system of veins?

Arteries branch into smaller and smaller blood vessels until they become capillaries. **Capillaries** are the smallest blood vessels. In fact, they are so narrow that red blood cells pass through them single file. Why is it important that arteries branch into smaller and smaller vessels?

Scripture Spotlight

Read about God's heart surgery in **Ezekiel 36:26–27**.

Lesson Activity

Use a map of the United States showing major river systems. Locate the Mississippi River. Where does the river end? Where does the river begin? How many tributary rivers can you count that flow into the Mississippi River? Select one of the larger tributary rivers. How many smaller rivers flow into the tributary river? Compare the movement of water in river systems to the movement of blood throughout your body.

How does your circulatory system compare to Earth's river systems?

How does eating a nutritious salad keep your heart healthy?

Caring for the Circulatory System

What are some things you do every day to make sure you stay healthy? Which of your ideas relate directly to the health of your circulatory system? The best way to care for your circulatory system is to eat a balanced diet and get regular exercise. A vegetarian diet can help with maintaining heart health. When choosing healthy foods to eat, avoid excess salt and fatty foods. What are some common fatty foods? What is the danger of fatty foods to the circulatory system? Exercising vigorously for 20 minutes at least three times a week will help your heart stay healthy. How does exercise help your heart and blood vessels? Avoiding alcohol, tobacco, and other drugs will further protect your body. What damage can these substances do to your heart and circulatory system?

The chart below lists three common circulatory system diseases. What are some actions that may reduce the risk of all three of these circulatory system diseases?

	Circulatory System Diseases	
Type	**Description**	**Reduce the Risk**
Hypertension	high blood pressure or high systolic and diastolic pressure; increases risk of heart attack and stroke	eat healthy foods low in salt and fat; keep a healthy weight; exercise; do not smoke; controlling anger can lower systolic and diastolic blood pressure
Atherosclerosis	buildup of fat, cholesterol, and other substances on inner lining of arteries; reduces blood flow	eat a balanced diet; exercise; do not smoke; advanced disease may require surgical treatments and diet change
Coronary artery disease	results from plaque buildup in arteries; usually caused by atherosclerosis; can cause heart pain and heart attack	eat healthy foods low in fat; exercise; keep a healthy weight; do not smoke; maintain low blood pressure; control anger

The heart pumps 24 hours a day, 7 days a week, 52 weeks every year for your entire life. It pumps when you are awake, active, and sleeping. This vital service moves oxygen- and nutrient-rich blood to all of the cells in your body. It swiftly carries oxygen- and nutrient-poor blood back to your heart where wastes are exchanged for oxygen. The cycle continues again and again allowing you to grow, thrive, and enjoy the world around you.

Concept Check Assess/Reflect

Summary: How does circulation work? The circulatory system consisting of the heart, arteries, veins, and capillaries moves blood containing oxygen and other nutrients to cells throughout the body in a series of arteries that become smaller and smaller. Blood is composed of red blood cells, platelets, and hemoglobin. The oxygen-poor blood returns to the heart through a series of veins. The blood is pumped through the four chambers of the heart composed of two atriums and two ventricles to the lungs, where the wastes are removed in return for oxygen. The heart then pumps the oxygen-rich blood back through the body to circulate again.

1. Describe the structure of the circulatory system.

2. How are arteries and veins different?

3. Describe three main functions of blood.

4. Explain one disease of the circulatory system and how to reduce the risk of getting the disease.

How Does Respiration Work?

A lake is a wonderful place to spend a hot summer day. What activities can you do in the water? Many organisms can live in the lake, such as fish. How do fish breathe? Fish do not have lungs; rather, they have gills to exchange carbon dioxide for oxygen that is dissolved in the water. Why can you not breathe underwater? How does your anatomy differ from that of the fish?

Respiration Explain

How do you obtain oxygen? Take a deep breath. What happened to your body as you took the breath? Everyday your lungs inhale and exhale about 20,000 liters of air! That's more than 10,000 2 L soda bottles full of air. The **respiratory system** enables you to breathe in and exhale this air. It takes in oxygen and eliminates carbon dioxide. The circulatory system has a very close relationship with the respiratory system. How does the circulatory system use the oxygen you breathe in?

The respiratory system is broken into the conducting zone and the respiratory zone. The conducting zone provides the passageway for air to flow from the outside environment to the lungs. It consists of the nose, pharynx, larynx, trachea, bronchi, and bronchioles. The conducting zone also is called the upper respiratory tract. The respiratory zone is where oxygen and carbon dioxide are exchanged. It consists of bronchioles, alveolar ducts, and alveoli, found deep inside the lungs. The respiratory zone also is called the lower respiratory tract. Have you ever had a runny nose or a chest cold? Both of these are symptoms of infections of these parts of your respiratory tract.

The respiratory tract is a commonly infected system.

Objectives

- Describe the structure of the respiratory system.
- Describe the path of oxygen as it moves through the respiratory system.
- Explain how to care for the respiratory system.
- Identify diseases of the respiratory system.

Vocabulary

respiratory system

pharynx

alveoli

Measuring Lung Capacity

How much air can your lungs exhale?

Procedure

1. Fill a plastic graduated cylinder completely with water.

2. Fill the dishpan with 3–5 cm (1–2 in.) of water.

3. Cover the top of the graduated cylinder with the palm of your hand and tip it upside down. Be careful that no water leaks out.

4. Place the graduated cylinder in the dishpan.

5. Place the tubing in the opening of the graduated cylinder. Steady the cylinder so that it does not tip.

6. Take a normal breath and blow a steady stream of air into the tubing.

7. Look at the measurements on the graduated cylinder. **Record** the amount of air that fills the graduated cylinder.

8. Repeat Steps 2–7 two more times. **Record the data**. Find the average.

9. Rest for three minutes. Then, repeat Steps 2–7 after taking three deep breaths. **Record the data**. Find the average.

10. **Record** the averages of your classmates for normal breaths and deep breaths.

Materials
- dishpan (10–15 cm deep)
- flexible plastic tubing
- 1000-ml graduated cylinder
- water
- ruler
- permanent marker

Analyze Results

Graph your breathing and the class data to show how much air was breathed out in both tests. Determine how the amount of air breathed out changed from one test to the other. **Use numbers** to explain.

Create Explanations

1. How much air can your lungs exhale?

2. How did your data compare to that of your classmates? Why might there be a difference?

3. Suppose you ran in place for several minutes. Do you think your data would be the same? Why or why not?

4. Why might a trumpet player have greater lung capacity than someone who plays the piano?

Breathing

Take a deep breath through your nose. Then, take another deep breath through your mouth. How does it feel when you breathe through your nose compared to breathing through your mouth? Why might it be beneficial to breathe through your mouth rather than your nose, and vice versa?

How do you breathe? Look at your chest as you breathe in. How does your chest move as you breathe out? What makes your chest cavity change shape? How do hiccups disrupt normal breathing patterns? One structure required for breathing is a thick muscle called the *diaphragm*. It is located below the chest cavity. When the diaphragm contracts, your chest cavity enlarges. This act allows you to take in, or inhale, air. This is the process of breathing in. When the diaphragm relaxes, your chest cavity diminishes. This act allows you to release air, or exhale. This is the process of breathing out.

The act of breathing brings oxygen into the body and eliminates carbon dioxide and other wastes from the body. Oxygen-poor blood is circulated past the lungs, where oxygen is exchanged for the waste carbon dioxide. The heart circulates the oxygen-rich blood through the body. Each day you breathe about 20,000 times. How many breaths do you take in one year? A lifetime?

Is the motion of the diaphragm voluntary or involuntary? Explain.

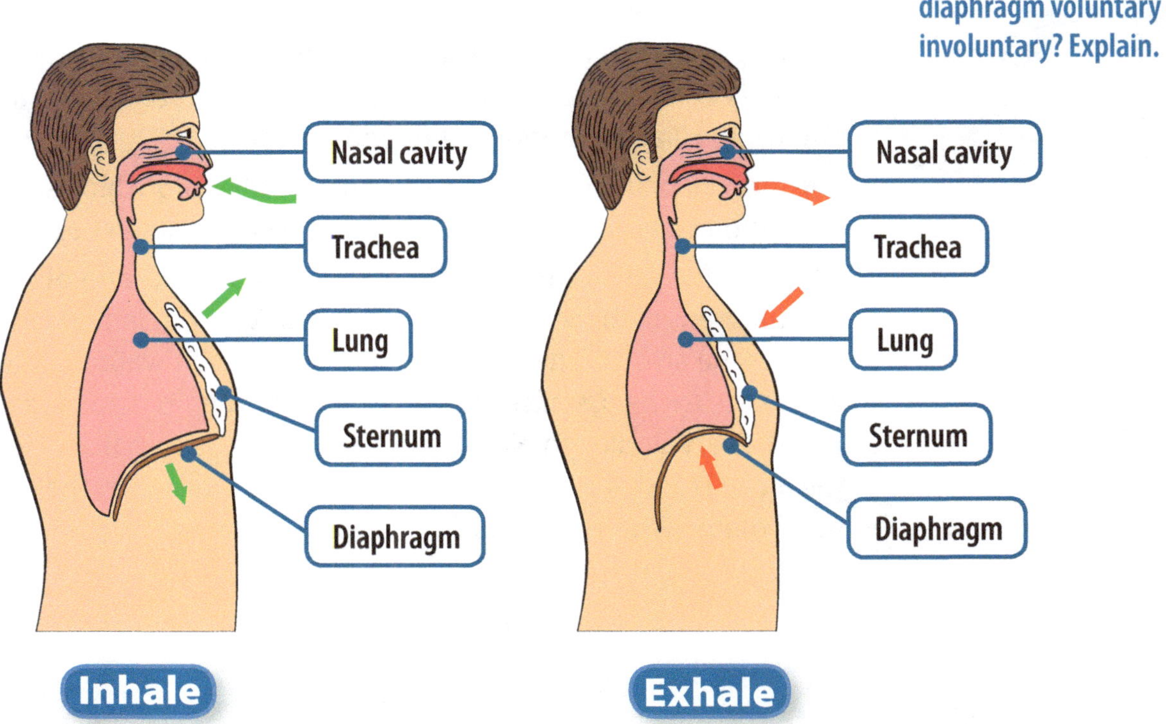

Air Pathway

Like blood in the circulatory system, air must follow a certain path through the respiratory system. Study the diagram while following the steps described below.

1. Air enters your body through your nostrils, the two openings in your nose. From there, air flows to the nasal cavities, which are lined with mucous tissue, cilia, and blood vessels. The mucous tissue warms, moistens, and cleans the air. Cilia are tiny hairs in the nasal cavities. They act as a filter to keep dirt and dust from getting to the lungs.

2. The nasal cavities connect to the pharynx, located at the back of the mouth.

3. The **pharynx** is connected to the trachea, also known as the windpipe. It is also lined with cilia to trap dirt and foreign particles from entering the lungs. Why do you think the name windpipe is an appropriate alternate name for the trachea?

4. The epiglottis is a flap of tissue that moves over the trachea when you swallow. Why is the epiglottis necessary to survive? Have you ever heard the saying "it must have gone down the wrong pipe"? How does this saying relate to the trachea and epiglottis?

5. The upper part of the trachea is called the larynx, or voice box. Two ligaments, your vocal cords, are stretched across the larynx. Air passes trough the space between the vocal cords producing sounds. This is your voice!

6. The end of the trachea divides into two bronchi, or air tubes. Each bronchi connects to a lung. Like the trachea, the bronchi also are lined with cilia to continue the filtering process.

7. The bronchi branch off into smaller tubes called bronchioles. Some bronchioles are no wider than the diameter of a human hair.

8. At the end of each bronchiole lie small air sacs called **alveoli**. There are hundreds of millions of alveoli in the lungs. Each inhalation inflates the alveoli. Tiny capillaries line the alveolar walls. This is where the exchange of carbon dioxide for oxygen is made in the blood. Every single red blood cell in your bloodstream flows through the alveoli for this gas exchange. Oxygen is picked up by the hemoglobin where it is held for transport to the waiting cells of the body.

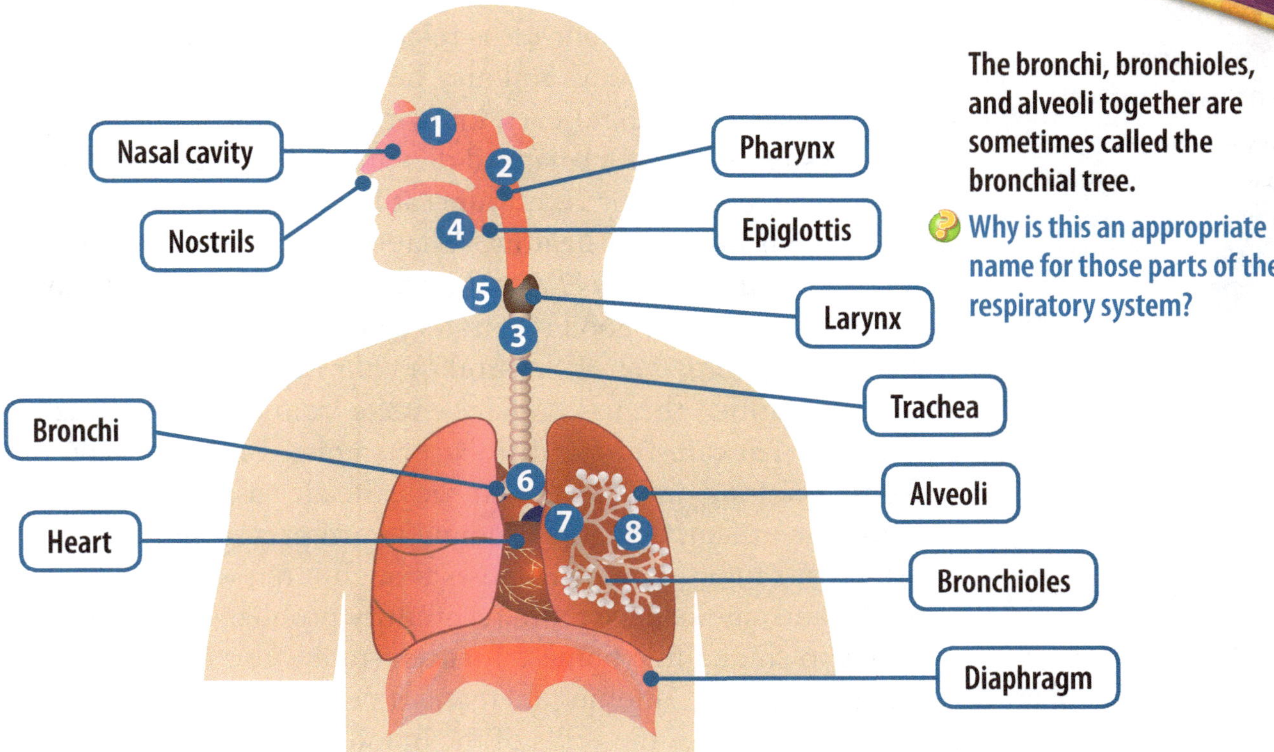

The bronchi, bronchioles, and alveoli together are sometimes called the bronchial tree.

 Why is this an appropriate name for those parts of the respiratory system?

Exchange of Gases

When you breathe, oxygen moves from the air into capillaries that surround the alveoli. The hemoglobin in red blood cells absorb the oxygen and transport it throughout the body. Carbon dioxide will move from the red blood cells into the alveoli. The carbon dioxide is exhaled as a waste product. This process is known as gas exchange. God created the circulatory system and respiratory system to work without conscious effort on our part. Imagine what life would be like if we had to remember to breathe and pump our hearts!

Explore-a-Lab

Structured Inquiry

How long can you hold your breath?

Take a deep breath and time how long you can hold it. Record this time on a sheet of paper. Rest for a couple of minutes. Take ten deep breaths and hold your breath. Record the time on the paper. Rest for another two minutes.

Take ten deep breaths and hold your breath while jogging in place. Record how long you held your breath while jogging. Compare the data you collected. Explain the reasons for differences in the data.

Caring for the Respiratory System **Explain**

What are some things you do every day to make sure you stay healthy? Which of your ideas relate directly to the health of your respiratory system? You may be promoting a healthy respiratory system without even knowing it! The best way to care for this system is to eat a balanced diet and get regular exercise. Exercising vigorously for 20 minutes at least three times a week will help your lungs stay healthy. What other activities can you do to keep your lungs healthy? Avoiding alcohol, tobacco, and other drugs is important for good respiratory health, as well.

Breathing polluted air is harmful to your health. What substances pollute the air? Perhaps smoke from a cigarette was a pollutant that came to mind. Tobacco smoke contains many poisons. Smoking damages the air sacs, the air passages, and the cilia that line them. Being in a room where a person smokes can damage your lungs. Research shows that children who breathe second-hand smoke have more respiratory problems then children who live in smoke-free homes. Tobacco smoke also damages the circulatory system. Smoking increases a person's blood pressure. The smoke damages the walls of the blood vessels and causes the blood to thicken. This results in an increase in blood pressure, which strains the heart muscle.

Lesson Activity

In your group, determine how you will use the supplies provided to build a model of the respiratory system. Make a sketch of your proposed ideas on large drawing paper. Label each part of the system on your sketch. Have your teacher approve your ideas. Then, use the supplies to build the 3-D model. Be sure to label important parts and functions of the respiratory system on your model. Be prepared to share and explain your model to the rest of the class. Your teacher may even display the models in the classroom or school library!

What are some limits of a model of the respiratory system?

The respiratory system is vulnerable to chronic disorders.
Some of these disorders are genetic, while others are caused
by pathogens. Some of the disorders result from following an
unhealthy lifestyle. All of the disorders can be managed with
proper diet, healthy living, and, when necessary, medication
and other treatments.

Respiratory System Diseases		
Type	**Description**	**Reduce the Risk**
Chronic obstructive pulmonary disease (COPD)	consists of a number of diseases; difficulty breathing; chest tightness; shortness of breath; chronic—does not go away	eat a balanced diet; drink plenty of fluids; wash hands often; get enough sleep; do not smoke; manage stress and anger
Asthma	chronic lung disease; inflames airways; causes wheezing and shortness of breath	avoid triggers such as infections, allergens, cigarette smoke, air pollution, and strong odors; take medication as prescribed by your health professional
Pneumonia	infection of lung tissue caused by bacteria or a virus; coughing, fever, and shortness of breath	wash hands frequently; clean your living area; do not smoke; be sure vaccinations are up to date

Every day you draw life-sustaining breath. As you inhale, the
air follows a circuitous path through your respiratory system,
where it is warmed, moistened, and cleaned. The oxygen in the
air is exchanged for carbon dioxide, a waste product from your
cells. The carbon dioxide is released with each exhale. The process
continues again and again, allowing you to grow, thrive, and enjoy
the world around you.

Concept Check   Assess/Reflect

Summary: How does respiration work? The respiratory system takes in
oxygen and releases carbon dioxide. It consists of many structures that clean
and filter the air to prevent illness and ensure the respiratory system brings
oxygen into the body. Blood receives oxygen in exchange for carbon dioxide
in tiny air sacs in the lungs.

1. Why should breathing take place through the nose and not the mouth?
2. Your mother tells you not to talk with your mouth full of food. Explain why this is good advice.
3. Describe how air is cleaned in the respiratory system.
4. Why would smoking be unhealthy for the respiratory system. Explain why this is so.

Get to Know
Leonard Bailey

Dr. Leonard Bailey became the first doctor to offer a solution for babies born with bad hearts. The left side of some newborns' hearts did not work correctly. In 1984, he transplanted the heart of a baboon into an infant known as Baby Fae. She lived less than a month after the surgery.

The following year, Dr. Bailey successfully replaced a baby's sick heart with a healthy human one. That baby, called Moses, lived to be an adult. Because human hearts were not easily available, few transplants on infants had been done. Now, more than 6000 babies around the world have had their hearts replaced. More than three-fourths of these babies have lived at least 10 years.

Dr. Bailey grew interested in newborns with bad hearts while he was still in medical school. He attended Loma Linda University and returned to work there. Until he became willing to try transplants, nothing could be done to help those babies.

To test his ideas about transplanting human hearts, Dr. Bailey used baby lambs, pigs, and goats. The lambs and pigs were the donor animals. Because the goats were hardier, they received the transplanted organs. The hospital had a farm where the animals were raised and could be watched after surgery. The transplants were successful, and the goats responded well to their new hearts.

After the successful transplant for baby Moses, whose real name was Eddie, people began to consider donating the organs of human infants and children to help others who might need organ transplants. Because of Dr. Bailey and others on his team, many more babies had a chance to live.

Called to Serve

Leonard Bailey showed his deep regard for life in becoming a doctor. He cared for the tiniest humans God made, giving them a chance to live.

Concept Check

1. Why were so few heart transplants done on babies?
2. What steps did Dr. Bailey take in preparing to transplant human hearts?

Prosthetic Heart Valves

If the heart's valves are not working properly, blood can collect in the heart or be unable to enter it. Since the 1960s, however, doctors have been replacing valves with prosthetic ones. A *prosthetic* is an artificial device that repairs or replaces a missing body part.

Doctors worldwide replace more than 300,000 valves annually. No perfect devices exist, although scientists continue creating new designs. More than 80 different kinds of devices have been developed. The two basic types are mechanical and bioprosthetic. A bioprosthetic valve comes from an animal. Both types of prosthetic valves mimic the action of a healthy heart valve.

Mechanical heart valves (MHVs) may solve some heart problems, but they do have some risks. One problem is the design of MHVs. The flaps that open and close the heart's atrium and ventricle are different sizes. The MHV flaps have been the same size. Scientists observed that this design interrupts the smooth flow of blood. They are working to redesign the flaps to mimic human heart valves more closely. Patients who receive an MHV need to take medicine to thin the blood.

To avoid the need for blood thinners, doctors developed bioprosthetic heart valves (BHVs). They may come from the patient's own tissue, or from pigs or cows. Valves from a pig can be used after being sterilized. Valves made from cow tissue must be constructed from the sac around the cow's heart. BHVs made from the patient's body also have to be formed. No option is perfect. BHVs suffer from structural deterioration, leaks, valve blockage, and infections.

Prosthetic heart valves have been used for more than 50 years. During that time, doctors have improved the designs and technology. During your lifetime, even more advances will be made.

During heart valve surgery, the patient's heart is stopped. A machine is used to keep blood circulating through the body.

Concept Check

1. Describe a problem that has been discovered in mechanical heart valves.
2. Why were bioprosthetic heart valves developed?

Study Guide

Lesson 1

1. The skeletal system is made up of bones that support and protect the body.

2. Diseases and disorders like arthritis can affect the skeletal system.

3. A healthy diet and exercise keep the skeletal system strong. Phosphorus and calcium are needed in the diet to maintain healthy bones.

Lesson 2

1. The muscular system is made up of muscles that work together with the skeletal system to make the body move.

2. Muscles that move bones work in pairs. One muscle contracts, while the other muscle relaxes.

3. Diseases and disorders like tendonitis can affect the muscular system.

Lesson 3

1. The circulatory system is made up of blood, the heart, and blood vessels that transport materials from one place in your body to another.

2. Blood is the connective tissue that transports nutrients and oxygen to your body's cells. It is made up of plasma, red blood cells, white blood cells, and platelets.

3. The heart pumps oxygen-rich blood, which is carried throughout the body in arteries. The oxygen-poor blood is returned to the heart by veins.

4. Healthy eating and activity are helpful practices for maintaining a healthy circulatory system.

Lesson 4

1. The respiratory system enables you to take in oxygen and eliminates carbon dioxide and water when you breathe.

2. The respiratory system consists of the nose, pharynx, larynx, trachea, bronchi, lungs, bronchioles, and alveoli.

3. Exercise and a balanced diet are two important factors to keep your circulatory and respiratory systems healthy. Avoiding tobacco smoke and other pollutants also helps the respiratory system.

Show What You Know

Visualize It Complete the concept map in your *Science Journal* by listing the function of each system.

Skeletal System
1. ________________
____
____

Muscular System
2. ________________
____
____

Body Systems

Circulatory System
3. ________________
____
____

Respiratory System
4. ________________
____
____

Complete each sentence with the correct term.

5. The air in the respiratory system is cleaned by __________.
6. The spongy and soft living part of the bone is __________.
7. Tissues that hold muscles to bones and keep different muscles together are __________.
8. Each upper chamber of the heart is a(n) __________.
9. Each lower chamber is a(n) __________.
10. Blood vessels are lined with __________ muscle.

Multiple Choice

Choose the best answer.

11. Why is your knee classified as a hinge joint?
    A. It moves in many directions.
    B. It moves only in a back-and-forth direction.
    C. It moves up and down and side to side.
    D. It rotates or turns from side to side.

12. Which of the following can be the result of a calcium deficiency?
    A. heart disease
    B. muscle strain
    C. osteoporosis
    D. respiratory infection

13. Air moves through from the nose to the lungs. Which of the following shows the order of the structures the air follows?
    A. nasal cavities, pharynx, larynx, trachea, bronchi, bronchioles, alveoli
    B. nasal cavities, larynx, pharynx, bronchi, trachea, alveoli
    C. nasal cavities, trachea, pharynx, larynx, bronchi, alveoli
    D. nasal cavities, pharynx, larynx, bronchi, bronchioles, trachea, alveoli

Check Point

Answer the following questions.

14. How do the skeletal and muscular systems work together in the elbow joint?

15. Your body has three different types of muscle tissue. Is it possible for smooth muscle to replace skeletal muscle? Why or why not?

16. How are tendons and ligaments the same and different?

17. What role does the diaphragm have in helping you breathe?

18. **Compare** the structure and function of veins and arteries.

19. **Infer** what would happen if the oxygen content in a room decreased by 20%.

Safety and Care of the Body

Scripture Spotlight

When Jesus was on Earth, a large part of His ministry was healing people. Learning what to do in an emergency and how to help people who are sick or hurt is one way you can be more like Jesus.

You will read the following passages in this chapter.

Colossians 4:14 (p. 201)
Matthew 9:35 (p. 203)
Psalm 91:15 (p. 215)
Luke 10:30–37 (p. 217)
Luke 33:49–51 (p. 217)

Numbers 21:6–9 (p. 218)
2 Kings 4:38–41 (p. 218)
Acts 28:3–6 (p. 218)
Proverbs 25:20 (p. 220)

From the time you were a toddler, adults have told you to be careful when doing certain things. Following safety precautions is one way we take care of our bodies. Sometimes, accidents happen, and we may need help from health care professionals.

The Big Idea

Because our bodies are the temple of the Holy Spirit, we should take the best possible care of them. When illnesses or injuries happen, we should seek to relieve suffering as Christ did during His ministry on Earth.

How can you behave like Jesus when someone is injured?

Inquiry Kick-Off Engage

What are some places in your home where accidents can happen? Where do you think the most common injuries occur? What can you do to reduce the risk of accidents in your home? Investigate these questions in your *Science Journal*.

Objectives

- Identify different types of health professionals.
- Describe the focus of specific health professionals.
- Describe situations requiring professional health services.

Vocabulary

pediatrician

oncologist

psychiatrist

orthodontist

dietitian

Some serious illnesses require a team of health-care professionals to work together to achieve the best treatment for a patient.

What professionals might you find on a team treating a cancer patient?

What Do Health Professionals Do?

Is there a difference between a doctor and a health-care professional? What does a health-care professional do? Would you like to help people be more healthy? What would you do to help people have healthy lives? Most health-care professionals spend a lot of time studying science. What specific area of science do you think would be more important? Why?

Who Are Health Professionals? (Explain)

Health professionals are people who work in a field that is related to health-care. They all treat patients or help people fulfill their health-care needs. Physicians, nurses, psychologists, counselors, dieticians, and dentists are just a few of the heath professionals that are trained to help people and relieve suffering.

Physicians

What type of doctor do you see when you are sick? Is your doctor a specialist or a doctor who treats many different illnesses? A family doctor is also known as a *general practitioner*. These doctors have training in medical care and can give you the care you need for most common illnesses and injuries.

Doctors recommend getting a medical checkup every one to two years, even if nothing seems wrong. Why do you think a regular medical checkup is a good idea? When you need care in a specific area, a family doctor may refer you to a specialist.

A **pediatrician** is a physician who specializes in taking care of children from birth until about 18 years of age. Pediatricians are specially trained to recognize diseases and illnesses that are specific to young and developing children. They also monitor their patients' development by checking height, weight, eyesight, and overall health at regular examinations. Most pediatricians treat illnesses, minor injuries, infectious diseases, and administer vaccinations. What are some illnesses that a pediatrician would treat that other doctors might not treat?

If you need an operation, you would go to a surgeon. *Surgeons* specialize in a specific area of the body. There are several types of *surgeons*. If you needed a hand, leg, or knee operation, you would go to an *orthopedic* surgeon. If you need heart surgery, you would see a *cardiovascular* surgeon. A *neurosurgeon* performs surgery on the brain and nervous system. What other types of specialty surgeons can you name?

An **oncologist** is a physician who studies, diagnoses, and treats cancerous tumors. There are different types of oncologists: medical, radiation, and surgical. Some oncologists help find the best treatment options for patients, while others perform surgeries for the treatment of cancer. Like many other physician specialties, oncologists have additional training in specific types of cancers and cancer research. Also like many physicians, oncologists often work with a team of specialists to manage a patient's care and recovery. What other health professionals might be on an oncology team? Why is it important for physicians to work as a team?

A physician who specializes in assessing and treating mental illnesses is a **psychiatrist**. Like physicians, psychiatrists are medical doctors. They meet with their patients to discuss many different types of problems and to try to help them find solutions. Psychiatrists can prescribe medications to help patients with various mental and emotional problems.

Scripture Spotlight

Read **Colossians 4:14**. Which New Testament writer was a physician?

Focus on Health

For healthy development it is recommended that you see a doctor regularly every one to two years or have a checkup visit annually one at age 6, age 8, age 10, and every year until age 21. Checkups for adults are also recommended. Research to find out how often they should see a doctor.

Seeing a Health Professional

Which health professional should you see for each illness you identified?

Procedure

1. In groups, brainstorm a list of minor illnesses and injuries. Remember to keep the list to minor illnesses, such as a sore throat, and minor injuries, such as a twisted ankle or splinter.

2. Write each illness or injury on a slip of paper. Place the slips of paper in a hat or jar.

3. Have one group member choose an illness or injury from the hat, and **communicate** the symptoms to the group by acting them out. For example, if you choose a cold, you might **communicate** that you have a runny nose and a cough. The rest of the group will ask questions to **gather data** about how the "patient" is feeling.

4. Then, decide if the illness or injury requires the help of a health professional. If so, as a group, decide which health professional would be most appropriate.

5. Take turns being the "patient" until your group has acted out the number of scenarios specified by your teacher.

6. Use a graphic organizer to display your decision-making process in your *Science Journal*.

Materials
- slips of paper
- hat or jar

Analyze Results

In which cases do illnesses or injuries require the help of a health professional?

Create Explanations

1. Which health professional should you see for each illness you identified?

2. If you are unsure about a situation involving an illness or injury, what course of action should you take? Explain.

3. Your doctor's office is closed for the weekend and someone needs medical care. Where and how do you seek medical care for him or her?

Nurses

Nursing is a diverse field with many opportunities. Nurses have different levels of education, training, and certification. Most people are familiar with *registered nurses* (RNs). They commonly assist doctors in an office, a hospital, or other medical care facility. Nurses will carry out a doctor's instructions, called *orders,* such as giving patients medicine, observing the patient's status, and using different types of medical equipment. When you have been to the doctor's office, how did the nurse help you?

Nurse practitioners (NP) can provide the same services as a primary health-care provider. They treat both physical and mental conditions through comprehensive history taking, physical exams, and ordering tests for interpretation.

Certified nurse midwifes (CNM) are trained to provide care to women and their babies. They often work with obstetricians to ensure that all health services are available, especially for high-risk pregnancies.

Certified registered nurse anesthetists (CRNA) are anesthesia professionals who safely administer anesthetics to patients during surgery, pain management, and trauma stabilization.

What does **Matthew 9:35** say that Jesus did when He visited different cities and villages?

Psychologists and Counselors

A psychologist studies human behavior and mental processes. Psychologists may teach, do research, or work in clinical or counseling positions. They use knowledge and research to solve problems and treat mental illnesses. There are many different types of psychologists.

Have you ever gone to see a counselor, like the guidance counselor at your school? A counselor is a professional who provides support and help to people of all ages depending on their needs. Psychologists and counselors help people with their mental and emotional health, as well as life choices. They can also help people with making educational and career choices, or with a family's well-being. They encourage people to discuss their problems like Jesus did. In addition to healing him or her physically, Jesus ministered to people's emotional needs as well. Jesus is even called Counselor in **Isaiah 9:6**. By showing compassion, Jesus helped people overcome feelings of hurt, rejection, betrayal, and guilt. He can still do that today. Psychologists and counselors who use Christian principles today can partner with God to bring healing. Why do you think talking about your problems can help your mental health?

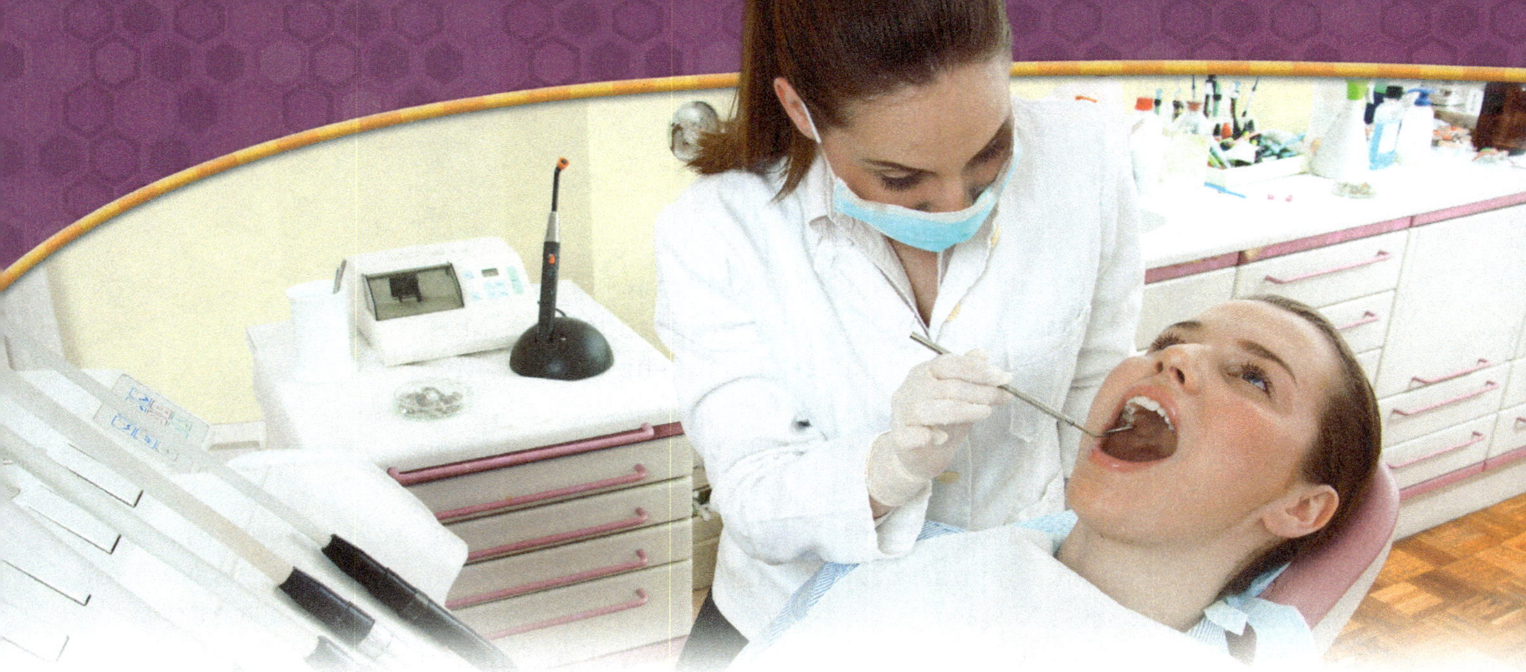

Why do you think one dentist has several dental hygienists working in the same office?

Dentists

When was the last time you went to a dentist? Was your last visit a routine visit or was there a different reason? Dentists help people properly care for their teeth and may perform corrective surgery on gums, extract teeth, and replace missing teeth. Why are regular visits to a dentist important to your health?

Dental hygienists work with dentists to care for your teeth. They may clean your teeth, examine your teeth and gums, and show you how to properly care for your teeth and gums.

Like physicians, some dentists specialize. An **orthodontist** is a dentist who specializes in straightening teeth with the help of braces and other appliances. Straight teeth are important for more than just good looks. Teeth that are out of alignment are not always able to chew food properly. This can lead to jaw and stomach disorders if not corrected. Can you name any other types of specialty dentists?

Dietitians

Do you ever plan and make meals for your family? Do you need to consider a special diet for anyone in your house? What do you try to make sure the meal includes or avoids? A person who plans food and nutrition programs, supervises meal preparation, and manages the serving of meals is a **dietitian**. Most dietitians manage food service programs for hospitals, schools, and other facilities. *Clinical dietitians* may work in a hospital, nursing home, or other care facility to help patients get the best nutrition for their needs. They may work with other health professionals to create diets for people with special dietary needs or who are on certain medications. Why is it important for dieticians to work with other health professionals?

When to Seek a Health Professional Explain

Do you know how to administer first aid? Do you have a first aid kit or a first aid station at home? How do you know when it is time to seek medical aid? With your family, put together a first aid kit and learn some common first aid practices, such as how to clean a minor wound (a cut or a scrape) and apply a bandage.

When should you seek more than just first aid for an injury or illness? For example, if you have a mild fever, you might not need to go to the doctor. You may just need some rest. If your fever is high (temperature above 103°F or 39.4°C), however, and lasts more than 24 hours, then you should seek medical help. Injuries that do not get better or are serious need the attention of a health services professional. If your family physician is not available or if you have a serious health issue, a hospital or medical clinic will have a health specialist to help you at any time of day. If you have a serious injury or illness, some professionals will even come to your location to help. You will learn more about this in the next lesson.

Concept Check Assess/Reflect

Summary: What do health professionals do? Health professionals help people live a healthy lifestyle. Specialists have extra training and treat patients with specific medical concerns. Physicians and nurses often care for patients' physical well-being. Psychologists, psychiatrists, and counselors care for people's emotional and mental needs. Dental professionals care for your teeth. Dieticians work to keep people healthy through their diets.

1. A 12-year-old needs a medical checkup. Which kind of physician should the person go to? Explain.

2. What are some situations that require professional health services? Why?

3. What type of counselor might be most likely to help someone who has an eating disorder, such as anorexia or bulimia? Would the same counselor be able to help someone with a weight-loss problem? Explain.

4. Where can someone go for medical care when the doctor's office is closed? When do you think it is okay to not see a doctor? Explain.

How Can We Stay Safe?

What are some practices at home that could prevent accidents? What rules at school are designed to reduce accidents and injury? What emergency personnel help you?

Emergency Responders `Explain`

Firefighters, law enforcement officers, paramedics, safety engineers, emergency medical technicians (EMTs), and other emergency and rescue workers are trained to respond to emergencies and to help people who may be injured or in danger. How do they help citizens to stay safe?

Firefighters

Some firefighters are volunteers and some are professionals, but all have hundreds of hours of special training in fighting fires, emergency first aid, and other rescue training. They also help in traffic accidents and medical emergencies. They are always ready to take action and rescue people in danger. Firefighters must be able to carry hoses, climb ladders, and enter burning buildings. They follow specific procedures to put out fires and keep the public safe.

Some firefighters specialize in battling fires in forests or areas that suffer droughts or lack of rain.

What makes the work of firefighters particularly dangerous? Why is proper training critical for firefighters?

Objectives

- Identify types of emergency responders.
- Describe ways of avoiding falls.
- Identify ways of preventing suffocation.
- Explain ways of preventing injuries caused by fire.
- Identify common household poisons.
- Describe ways of preventing poisoning.
- Describe ways of being safe around water.
- Explain how to keep safe around guns.
- Explain practices that can prevent accidents and injuries when using roadways.
- Explain how to keep safe when participating in outdoor activities.

Vocabulary

emergency medical technician (EMT)

paramedic

safety engineers

suffocation

poison

Accidents Happen

Where do most accidents happen at school?

Procedure

1. Draw a detailed map of your school on the graph paper. Label all the rooms and areas of the school.

2. With your partner make a list of the 10 most recent injury causing accidents that have happened at school.

3. On your map, mark the location of each accident.

4. Gather and record information about each accident. Include the date, the location, a description (type of injury), the cause, and the level of seriousness (minor or serious) of the accident.

5. Use the chart in your *Science Journal* to record your data.

6. Use your data to propose possible ways to correct any hazard that may have contributed to each accident.

Materials
• graph paper

Analyze Results

Draw a conclusion about how each accident could have been prevented.

Create Explanations

1. Where do most accidents happen at school?

2. Why do you think so many accidents happen where they do?

3. What are two common injuries that result from the accidents that happen at school?

4. Look over the list of accidents and their causes. What is the most common cause of the accidents you have investigated?

Law Enforcement Officers

How would you describe what a police officer does? What does an officer at a school do? What are the duties of the officers who work at airports?

In addition to arresting people who break laws, police officers, highway patrol officers, federal agents, and other law enforcement officers also help to educate citizens. Many police officers participate in public outreach and public safety programs. What might an officer be able to teach your family to help you stay safer? Officers may lead or participate in youth education programs against drugs and violence. Law enforcement officers may work with communities to set up programs, such as a neighborhood watch.

When you call 911, police are often the first to respond to an emergency situation, whether it is a safety emergency, a medical emergency, or a fire. Emergency personnel are often busy helping people. That is why it is important to only dial 911 when you have an emergency. Calling for other reasons is against the law.

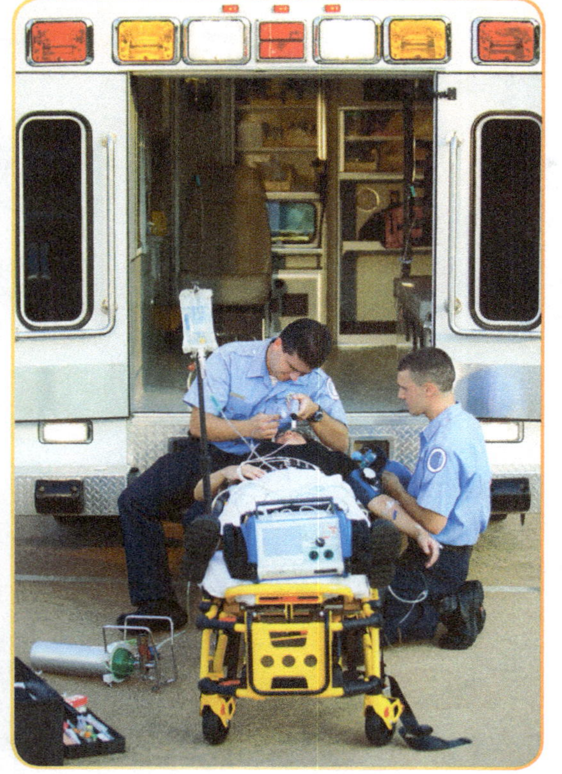

EMTs and Paramedics

Have you ever seen an ambulance on its way to an emergency? Who do you think was inside? **Emergency medical technicians (EMTs)** are health-care professionals trained to treat and transport emergency victims. EMTs are able to administer cardiopulmonary resuscitation (CPR) and oxygen and provide other basic life support to victims. They may work in hospital emergency departments and fire departments. They also provide care in ambulances. Like firefighters and law enforcement officers, EMTs and paramedics must have many months of specialized training, and then, they must take exams to qualify for a career in their field.

Paramedics are trained to work in emergency situations. They may ride in an ambulance, or they may be part of a response team with the fire department.

Safety Engineers

Another group of workers concerned with protecting people are safety engineers. **Safety engineers** help to keep people free from danger, risk, or injury in the workplace. Their job includes developing safety programs and eliminating unsafe conditions in plants, mines, construction sites, and other public places.

Safety engineers also work on the designs of buildings and other structures to make sure they are safe. For example, an engineer needs to think about how much weight might be on a bridge at a given time. Most bridges are designed to hold more weight than they would have at one time, and they have multiple other safety features.

Keeping Yourself Safe Explain

You can learn a lot from emergency responders about how to keep yourself safe. For example, firefighters often teach other people about fire safety. One way to keep yourself safe is to always listen to firefighters, police officers, and other community helpers. Another way is to learn and follow rules and guidelines for safety in different situations. What safety rules do you follow every day?

Fire Safety

Fire drills at school, home, and church prepare you for what to do during a real fire. Here are some guidelines to follow.

Fire Prevention Tips
• Do not play with matches or lighters.
• Do not overload electrical outlets.
• Turn handles of pans toward the center of stoves.
• Use caution around open flames so not to ignite loose clothing.
• Check smoke detector batteries regularly.
• Know your escape route at home, school, and church.

In a Building Fire
• Check closed doors for heat before opening them. If a door is hot, do not open it! If it is safe to do so, exit through a window. If you cannot escape through the doors or windows, hang something white outside the window. Firefighters will see it and know where you are.
• If a door is cool, open it slowly. If it is clear, leave quickly. If the path is blocked, shut the door and look for another path.
• Crawl low under smoke as you exit.
• Close doors behind you as you leave.
• Call 911 from the nearest phone. Do not go back into the building.
• If your clothes catch on fire, you should stop, drop, and roll. Then, get immediate first aid for any burns.

Prevent Falls

- Remove tripping hazards from walkways.
- Use nonslip strips or mats in tubs and showers.
- Use a sturdy ladder or stepstool instead of a chair or table.
- Use handrails on the stairs.
- Clean up spills immediately.
- Walk in buildings and round the pool.

Prevent Choking

- Keep small objects and plastic bags away from small children.
- Chew your food well before swallowing.
- Do not put plastic bags over your head.
- Stay out of old freezers and refrigerators. Remove the doors to avoid entrapment.
- Do not crawl under an object that may fall on you.

Prevent Falls

Have you ever tripped on a rug or some other obstacle? Were you injured? Could the accident have been prevented? Tripping over something may seem like a minor issue, but falls send thousands of people to the emergency room each year, and sometimes people die from their injuries. Many injuries caused by falls can be prevented with some simple precautions.

Falls can happen anywhere and to anyone. What age group do you think has a higher risk of being injured by a fall? Why? What could be done to prevent accidents for that age group?

Prevent Choking

Many accidents are the result of suffocation. **Suffocation** happens when air is prevented from entering the lungs. Suffocation can happen because of accidental poisoning, from smoke, or from choking. Choking is a common cause of suffocation in small children. In older people, choking is usually caused by food stuck in the air passage. You can take steps to prevent suffocation.

Lesson Activity

 **How can being aware of community events help people prevent accidents and live safely?**

Think about your community. What accidents or emergency situations have happened in your community in the past five to seven days? Checking the Internet, listening to local radio stations, and watching local news reports are all ways to stay informed about what is happening in your community. Review newspaper articles from the past five to seven days to find out what local public safety or emergency situations have occurred. Read the articles, prepare a presentation, and share the information with your classmates.

Prevent Poisoning

A **poison** is any substance that causes injury, illness, or death when taken into the body. Your home likely has more poisons in it than you think. Cleaning supplies, paints, bug sprays, weed killers, and gasoline are just a few poisons that are likely in your home. Even medication that is usually helpful can be poisonous if not used properly. There are safety precautions you can take to prevent poisoning.

Water Safety

Swimming, boating, fishing, and other water sports can be great exercise and provide hours of recreation. However, you must know how to be safe around the water. Follow these common-sense precautions when you are around water.

Drowning is one of the highest accident rates among children.

Why are accidents involving water such a hazard?

Prevent Poisoning

- Keep poisonous materials locked and away from children.
- Dispose of unused poisons properly.
- Do not store poisons in unmarked containers.
- Do not taste unknown substances.
- Avoid breathing toxic fumes.
- Avoid poisonous plants and animals.
- Teach children to ask an adult before eating something.

Water Precautions

- Do not swim alone.
- Wear a life jacket in a boat.
- Look before you leap—do not dive.
- Be aware of strong currents and avoid them.
- Do not swim in a thunderstorm.
- Do not try to rescue a drowning person on your own.

What kinds of poisons are kept in your home?

Survey your home with an adult for poisons. Make a chart listing what the product is and what the poisonous ingredients are. If the container lists what to do if the poison is eaten, record that as well. Finally, on your chart, indicate if the poison is properly stored.

How many poisonous products did you find? Was it more or less than you expected? What warnings were on the package to indicate the product was a poison? Did you find any hazardous products that you did not know were hazardous? Which ones? Write a brief explanation of what you would do if someone was poisoned in your home.

Gun Safety

Many people have guns in their homes for protection, hunting, or collecting. Gunshot wounds can cause serious injuries or death. Guns can be kept and handled safely as long as a few simple rules are followed.

- Never take a gun to school.
- Never play with a gun like it is a toy.
- Never point a gun at a person, even if it is not loaded.
- Never pretend or threaten to shoot someone.
- Store unloaded guns in a locked cabinet or box.
- Keep the safety on until ready to shoot.
- Only touch the trigger when you are ready to shoot.
- If you find a gun, do not touch it and report it to a trusted adult.

Roadway Safety

Because so many people use different types of vehicles on the road, there is a high potential for accidents. Often, using common sense, being aware of your surroundings, and being courteous to others on the roadway can prevent accidents. The following chart lists some other precautions you can take on the roads.

Pedestrian	Bicycle	Motor Vehicle
• Use sidewalks and paths when possible. • Walk or run in the opposite direction as traffic. • Obey traffic signs. • Cross streets only at crosswalks. • Wear bright or reflective clothing. • Avoid narrow or winding roads. • Anticipate what the drivers of motor vehicles may do. • Never run out into the street.	• Wear a helmet. • Make sure any bicycle you ride is working properly and is adjusted to suit you. • Do not ride double unless you have a tandem bike. • Use bike lanes when possible. • Ride with the flow of traffic. • Obey traffic signs. • Use hand signals when making a turn. • Wear bright or reflective clothing. • Avoid narrow or winding roads. • Yield the right-of-way to pedestrians.	• Wear a seat belt. • Keep body parts inside the vehicle. • Do not distract the driver or interfere with his or her view. • Lock the doors especially when small children are passengers. • Do not operate a motor vehicle without a license.

Outdoor Safety

Outdoor activities can provide hours of fun and exercise as well as appreciation for God's Creation. It is important to know how to be safe when you are outdoors, whether you are enjoying a day at the park or camping with your family. Always check the weather before you go outdoors, and wear appropriate clothing. How do you think sunglasses can be used as safety equipment?

Be Prepared for the Outdoors

- Never go hiking, skiing, or participate in other outdoor activities alone.
- Always tell someone where you are going and when you expect to return.
- Use maps and compasses when hiking, backpacking, or cross-country skiing.
- Wear proper shoes, clothing, and gear for the activity you are doing.
- Take first aid supplies, matches, and a flashlight.
- Build campfires only in designated areas. Keep them small. Be sure they are completely extinguished and cold to the touch before leaving.
- When hiking or skiing, stay on marked trails.
- Take plenty of water with you.
- Wear sunscreen.

Concept Check Assess/Reflect

Summary: How can we stay safe? Emergency responders, such as firefighters, law enforcement officers, EMTs, paramedics, and safety engineers, work to keep citizens safe. They also teach us how to be safe. We can keep ourselves safe by following rules and guidelines about how to be safe in a fire, near water, around guns, on the road, and outdoors. We can also learn about preventing dangers such as falls, choking, suffocation, and poisoning.

1. How do emergency responders help keep people safe?

2. What are some safety practices that you use every day?

3. Why should you tell someone what time you expect to return from a hike in the woods?

4. What do you think is the best way to avoid an accident?

Essential Question

What Do You Do in an Emergency?

What is the difference between an emergency and a problem? Emergencies are situations that are outside your control. You often have to call on the assistance of professionals to help during an emergency. What emergencies have you seen? During an emergency, a person calls 911 for assistance. What have you learned from emergencies you have encountered? What can you do to prepare for an emergency event.

Emergencies (Explain)

An **emergency** is a situation that requires an immediate response. Can you name some emergency situations?

In the case of a medical emergency, immediate help is needed to prevent further injury and even save lives. It is important to stay calm in emergencies. Why?

Some physical injuries or illnesses require immediate medical care. These types of injuries or illnesses include severe symptoms that may occur without warning.

Causes for Immediate Medical Attention

- broken bones or back or neck injury
- chest pain or severe stomach pain
- uncontrolled bleeding
- heatstroke or vomiting blood
- difficulty breathing or shortness of breath
- loss of consciousness
- very high fever
- sudden dizziness, severe headache, or change in vision
- severe or constant vomiting
- major burns
- poisoning

If possible, stay with the victim while calling 911. Remember to use 911 only for emergencies.

 Why should 911 only be used for emergency situations?

Calling 911

In most areas of the United States and Canada, the emergency number is 911. Calling 911 immediately connects you to a trained operator who will get you the help you need. If you are not sure the situation is an emergency, it is best to call 911. Once you explain the situation to the 911 operator, he or she can assess whether emergency help is needed.

During an Emergency

Look around you. Make sure there are no dangers in the area. Do not move an injured person unless he or she is in danger. → If you feel you are in danger, go where it is safe and call 911.

↓

Call 911 or ask someone close by to call 911. Listen to the operator's questions and speak slowly and clearly. Describe the emergency. State your location and phone number. Follow any directions the 911 operator gives. Stay on the phone until the operator tells you that you may hang up.

↓

Perform life-saving first aid

Lesson Activity

❓ Are you prepared for a medical emergency?

Knowing what to do during a medical emergency can be the difference between life and death. Make a list of the steps you should follow during a medical emergency. Remember to include 911 and the numbers of family members. Take some time to discuss and share the list with your family members. Make sure you post the list near a telephone where everyone in your family can see it. Make copies for everyone to carry in their wallets, purses, or backpacks. It will be important to review and update the list every so often.

❓ Why would you want to avoid hanging up too soon when you call 911?

Design to Protect

How can you protect an egg in a fall?

Procedure

1. Work with a group to **research** and come up with a design that will protect an egg from the impact of a fall.

2. **Analyze** the materials that have been made available. Decide on some additional items to include in your design.

3. Build your device for holding the egg. **Compare** your design to those of your classmates. How do they differ? In what ways are they alike?

4. With the help of your teacher, measure out heights of 1 m (3 ft), 3 m (10 ft), and 5 m (15 ft). Begin at the first height and drop your constructed device. **Record** the results each time. **Observe** and **compare** to see which design protects the egg best overall.

Materials
- raw egg
- several plastic straws
- masking tape
- newspaper
- items of your choice
- metric tape measure

Analyze Results

What items were more helpful in a protective design? What items were the least effective? Which additional items do you think would be good to protect people against impacts during falls?

Create Explanations

1. How can you protect an egg in a fall?

2. What do you think are the three most important items to have to guard against impact? Explain your choices.

3. Do you think it is possible to design a device to protect against any impact or fall? Explain.

4. How could the design you used to protect an egg be used for people? Give some examples.

Giving First Aid Explain

You may find yourself in a situation where you need to help someone, whether it is an emergency or a common household injury. Before you help someone else, you need to make sure that you can also keep yourself safe. Talk to and listen to the victim to find out how you can help. Be aware of the things and people around you. Call 911 or tell someone to call—do not assume that others will help. Wear protective gear, if possible. If you do not have protective gear or a first aid kit, how could you improvise to protect yourself or to help someone?

The following sections describe some basic first aid you could provide to a victim. However, you should always follow the instructions given by a 911 operator or other emergency responder.

Scripture Spotlight

Read about two examples of first aid in **Luke 10:30–37** and **Luke 33:49–51**. Which is like something you could do to help someone in need?

Concussions

A **concussion** is a serious brain injury that temporarily changes the way the brain works. Refer to the list at the right for signs of a concussion. There may be bleeding inside the skull or the brain may be bruised. Serious concussions may lead to death. If you are with someone who may have suffered a concussion, call 911. Have the victim sit and rest, but do not let him or her go to sleep. Most concussions are caused by a blow to the head by an object or from a fall. They can also be caused if the head is shaken, such as in a car accident. How does a helmet work to prevent concussions? When should you wear a helmet?

Symptoms of a Concussion

- unclear thinking or inability to concentrate or remember new information
- headache
- blurry vision
- nausea and dizziness
- easily upset or angered, sad, nervous, or emotional
- change in sleeping habits

Stopped Breathing or Heartbeat

Cardiopulmonary resuscitation (CPR) is an emergency procedure that involves chest compressions and rescue breathing to keep oxygen and blood flowing in the body. CPR can also be used in the case of drowning. If you are not certified to perform CPR, the 911 operator will give you instructions on what to do. Call 911 and follow the operator's instructions.

It is best to get formal training before performing any type of CPR. Check with your teacher or local health organizations for a list of CPR classes. Your actions might save a life.

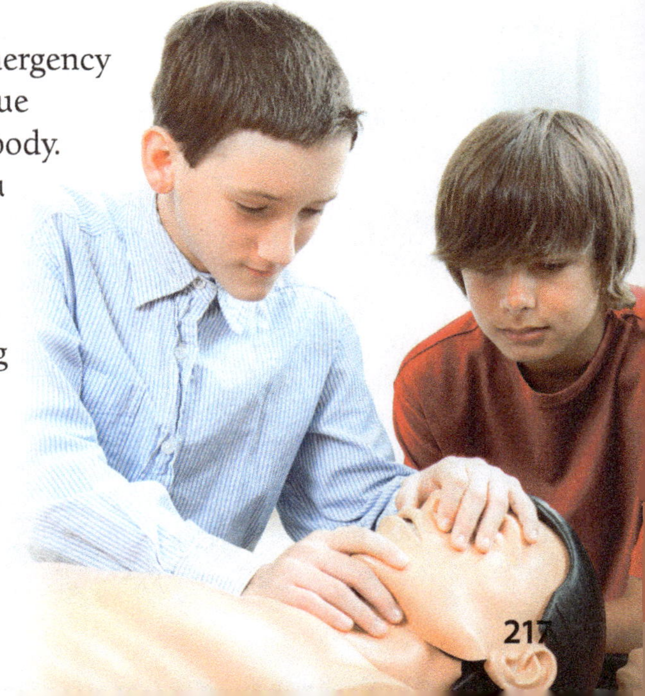

Knowing how to use CPR can help you save someone's life.

Why is it important to be properly trained in CPR before using it?

Check out these stories of miraculous cures! Read **Numbers 21:6–9**, **2 Kings 4:38–41**, and **Acts 28:3–6**.

Oftentimes, snakes, like this timber rattlesnake, are encountered on hiking trails.

How can people protect themselves from animal bites of any kind while hiking?

Bleeding

Large bleeding wounds and deep puncture injuries can be life threatening if a victim loses too much blood. You can try to help a bleeding victim by following the steps outlined in the box. What would you do if protective gloves were not available?

Animal Bites, Scratches, and Stings

Always ask permission before you touch someone's pet. Give the animal space, move slowly, and never pull on or hit an animal.

What if you encounter an animal in the wild? Do not approach the animal, and do not try to pet it. Wild animals are not pets. The animal may be sick, injured, or scared, and might bite you. When an animal bites, the skin is broken, producing minor or serious wounds. Some animals carry rabies, a deadly viral disease of the nervous system. If you are bitten by an animal with rabies, then you must be treated immediately to prevent infection.

Some animals, such as some snakes, spiders, and scorpions, are venomous. They have a type of poison that can be injected into you through their teeth (fangs) or stingers. Some venom can cause a person to become sick or stop breathing. What should you do if you are bitten by a venomous animal?

1. Call 911 or ask someone else to do it.
2. Wear protective gloves and remove any dirt that is easily seen. Do not attempt to clean the wound thoroughly.
3. Cover the wound with a sterile bandage or cloth and press firmly. Hold the bandage in place for several minutes.
4. Bind the wound with a bandage and tape. If the bleeding continues and seeps through the bandage, do not remove the bandage. Add additional bandages.
5. Elevate the wound higher than the heart. Continue applying pressure.
6. Wash your hands as soon as possible.

Caring for Animal Bites and Scratches

- Wash a minor bite or scratch with soap and water.
- Apply antibiotic cream and a bandage.
- Deep puncture wounds should be treated by a doctor.
- Seek medical attention for snake bites.

Caring for Insect Stings

- Scrape out the stinger if necessary.
- Wash the area with soap and water.
- Apply a cold compress.
- Apply an antiseptic cream.
- Seek medical help for any reactions.

Heat Exhaustion and Heatstroke

When your body gets too hot, you may experience **heat exhaustion** or **heatstroke**. If you experience heat exhaustion, you should immediately stop any activity, sit in a cool place, and drink cool water or a sports drink. If you have heat exhaustion, then you are also at risk of heatstroke. The chart to the right details the difference between heat exhaustion and heatstroke.

If you are with a person who is experiencing heatstroke, call 911. Help the victim cool off with a cold bath; place ice packs or cold wet towels on the head and neck and under the arms; or place a cold damp sheet on the person and blow air on him or her with a fan. How can you prevent heatstroke?

Hypothermia

Hypothermia occurs when your body loses heat faster than it produces it, causing your body temperature to drop. If your body temperature drops below 35°C (95°F), your heart, nervous system, and other organs may not work correctly. Some other signs of hypothermia are outlined in the box.

In some cases, like when someone falls into frigid water, hypothermia can occur as quickly as 15 minutes. A victim of hypothermia needs his or her body to be warmed back to normal body temperature 37°C (98.6°F). Removing any wet clothing and covering the person with several blankets can raise the body temperature back to normal. What can you do to prevent hypothermia?

Symptoms	
Heat Exhaustion	**Heatstroke**
• headache	• body temperature above 40°C (104°F)
• nausea	• lack of sweating
• faintness	• nausea and vomiting
• dizziness	• red skin
• heavy sweating	• rapid breathing and heart rate
• skin that feels cool and moist	• headache
• goose bumps in the heat	• confusion
• fatigue	• unconsciousness
• weak, rapid pulse	• muscle cramps or weakness
• muscle cramps	

Signs of Hypothermia
• shivering
• clumsiness
• slurred speech
• confusion
• poor decision making
• sleepiness
• weak pulse

Emergency responders are treating this girl for hypothermia and other injuries.

Why do you think the blanket is silver?

Frostbite

If someone has hypothermia, he or she may also have frostbite. **Frostbite**, which is caused by extreme cold weather conditions, is damage to the skin and its underlying tissues. Ears, nose, toes, and fingers are most likely to experience frostbite. Symptoms of frostbite are a feeling of "pins and needles" followed by numbness. Severe frostbite may cause blisters. As the frostbitten area warms, it may become red and painful.

Look at the chart to find out the proper way to treat frostbite.

Do	Do Not
Immediately take the victim to a warmer place. Take him or her to the hospital or call 911.	Rub or massage the frostbitten body parts or wet them with snow.
Remove wet clothing.	Apply direct heat, such as a heating pad or air from a hair dryer, to thaw the body.
Wrap affected areas up so they do not become frozen again. Put dressings between frostbitten fingers and toes.	Thaw frostbitten parts if they might get frostbitten again. Refreezing could make the damage worse.
Warm frostbitten areas gradually with warm water (40–42°C or 104–107.6°F). Wrap or cover other areas in a warm blanket.	Walk on frostbitten feet or toes. This further damages the tissue.
Give the victim warm drinks, such as soup and herbal tea.	Give the victim alcohol. It decreases blood circulation and makes recovery more difficult.

Simple measures can help prevent frostbite.

 **How does dressing between frostbitten fingers help a victim?**

 ## Math in Science

Normal body temperature is about 37°C (98.6°F). If your body temperature falls below 35°C (95°F), then hypothermia can occur.

Solve this problem: Before he went out, Ian's body temperature was 37.2°C (98.9°F). He spent 4 hours outside in −25°C (−13°F) temperature. At the end of 4 hours, his body temperature had dropped by 3.1°C (5.6°F). Was Ian experiencing hypothermia? How do you know?

Fractures, Sprains, and Strains

Slips, falls, and tripping are the most common kinds of accidents. A serious fall could result in an injury, such as a fracture. A **fracture** is a break in a bone. The symptoms of a fracture include a limb or joint looking out of place, swelling, bruising, bleeding, severe pain, numbness and tingling, and trouble moving the limb. Once treated, fractures may take several weeks or even months to heal. The box below lists some first aid you can offer for a broken bone before arriving at the hospital.

Sometimes when people walk, run, or jump, they may twist an ankle. This might result in a sprain or a strain. A *sprain* is an injury to a ligament. A *strain* is an injury to a muscle or tendon. Sprains and strains cause pain and swelling. Usually, these injuries may be treated by rest, ice, compression, and elevation. If the sprain or strain is very painful and does not seem to get better, a health-care professional may need to be consulted. How do you think you can prevent these types of injuries?

Caring for a Fracture

- Immobilize the injured area.
- Apply ice.
- If the bone has punctured the skin, do not push the bone back in place.
- Control bleeding.
- Seek medical help.

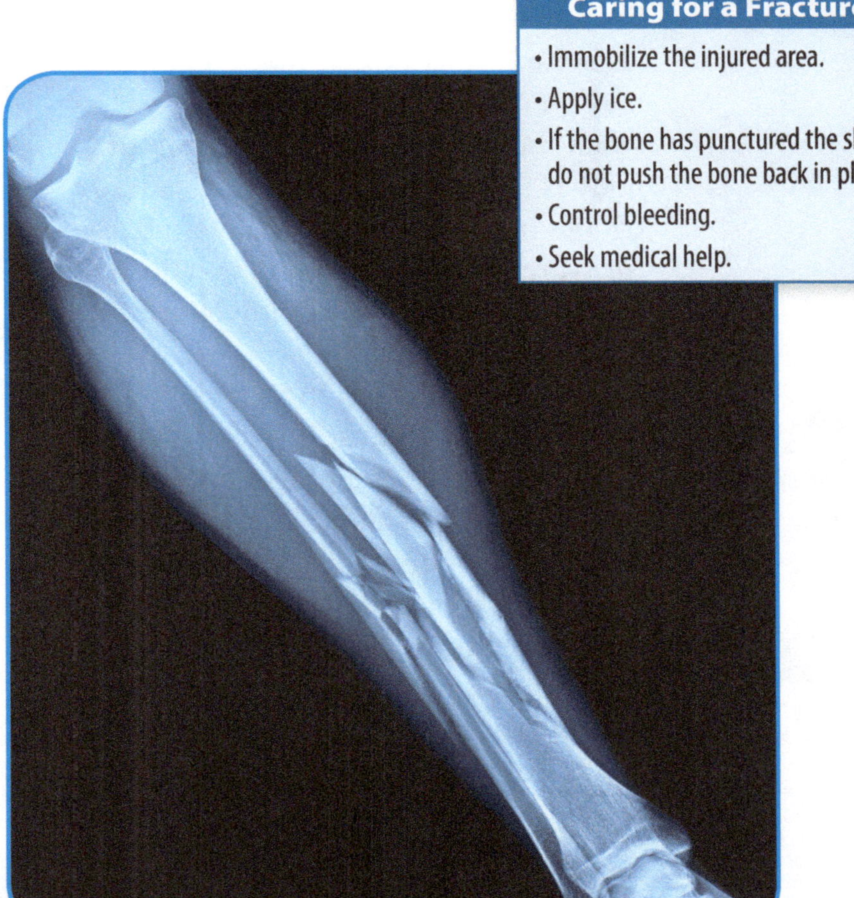

This X-ray shows a serious bone fracture.

What common household items could you use to make a splint or a sling until you could get medical help?

Check out your *Science Journal* for a Structured Inquiry that explores how you make choices in an emergency.

Extend

Burns

Burns can be caused by exposure to the Sun, heat, chemicals, and electricity. All burns require first aid treatment. The chart covers some basic first aid you could offer to a burn victim.

A first-degree burn affects the outer layer of skin and can be treated at home. Second-degree burns involve damage to the outer layer of skin and the second layer of skin. Third-degree burns are serious and affect the deep layers of skin as well as fat, muscle, nerves, and bones. Call 911 in cases of serious burns or go to a medical facility. Why do you think you should not remove burned clothing? Why would elevating a burn help relieve pain?

Caring for Burns

- Run cool water over first-degree burns.
- Treat second-degree burns with water and wrap in a sterile dry dressing.
- Seek medical treatment for second-degree and third-degree burns.
- Do not remove burned clothing in cases of third-degree burns.
- Cover third-degree burns with a cool, moist cloth, and elevate the burned area.

Shock

Sometimes a sick or injured person may go into shock. **Shock** occurs when the flow of blood throughout the body is reduced. Shock may be caused by loss of blood or other body fluids (*dehydration*), as well as trauma, heart attack, allergic reactions, infections, heatstroke, and burns.

A person experiencing shock may be restless, have nausea, weakness, confusion, dizziness, bluish lips, pale and cool skin, and irregular breathing. Place the person on his or her back with the legs raised about 30 cm (12 in.). Never move a person with a head, leg, neck, or spine injury. See the accompanying box for other ways you can help a person who might be experiencing shock.

1. Call 911 or ask someone else to do it.
2. Give first aid for wounds or injuries.
3. Loosen tight clothing and keep the person warm.
4. Offer comfort.
5. Never give the person anything to eat or drink.

Choking

When an object becomes lodged in the throat or windpipe, blocking the flow of air, the person is choking. A person who is choking can die from lack of oxygen within a few minutes if his or her airway is not cleared. If the person cannot cough, cannot speak and/or grabs his or her throat with the hands, the person is suffocating. For choking emergencies, the American Red Cross recommends you respond as listed in the box on this page. Why would you want to administer abdominal thrusts to a choking victim?

Poison

More than 80 deaths occur each day due to accidental poisoning. Most poisonings happen at home, so it is essential to know what to do. Call the poison control center or 911 and follow these listed first aid procedures. Why would you not want a person with something in his or her eyes to rub them?

All of this information should now make you more aware of and better prepared to deal with most emergency situations you might have to face.

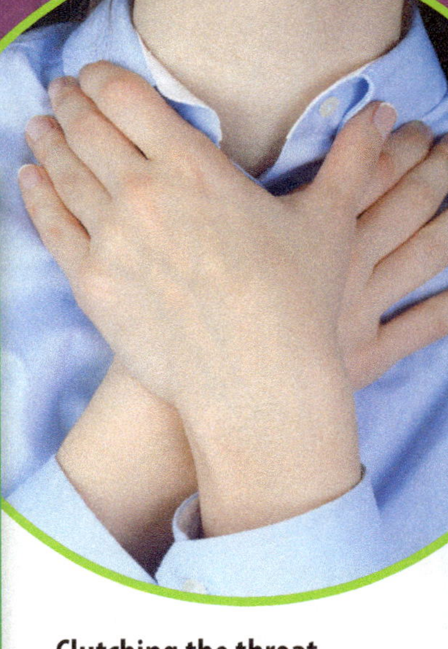

Response to Choking

- Ask the person if he or she is choking.
- If a person is choking, give five forceful blows with the heel of your hand between his or her shoulder blades.
- Use the Heimlich maneuver: five abdominal thrusts done by making a fist with one hand and grasping it with the other and pulling inward and upward under the person's rib cage, while standing behind the person.

Clutching the throat with both hands is the international sign for choking. You should then ask the person if he or she needs help.

If the person cannot respond to your question, what should you do?

Response to Poisoning

- If the poison was inhaled, move the person to fresh air.
- If the poison is on the skin or clothing, remove the affected clothing and rinse the skin with lukewarm water for about 20 min.
- If the poison gets in the eyes, flush the eyes with lukewarm water for 20 min. Have the person blink. Do not rub the eyes.

✔ Concept Check — Assess/Reflect

Summary: What do you do in an emergency? An emergency is a situation that requires an immediate response. You may need to call 911, and perform CPR or abdominal thrusts. You may need to provide first aid to victims of bites, heatstroke, hypothermia, frostbite, fractures, burns, shock, and poisoning.

1. What safety actions should you follow before giving first aid?

2. What might happen if you give a shock victim something to eat or drink?

3. How are hypothermia and frostbite similar? How are they different?

Restoring Nerve Connections

When cells in the body are damaged, they can be repaired or replaced through our bodies' natural healing processes. However, cells in the central nervous system, such as those in the spinal cord, are not able to renew or repair themselves. Once damaged, a nerve cell will remain damaged. If damaged seriously enough, a nerve cell will die. People with damage to the spinal cord may lose the function of their legs or arms (paralysis), or even the ability to breathe on their own.

Doctors have had success in "rerouting" nerve processes and restoring nerve functions. The damaged nerves are not repaired, but doctors can use other nerves nearby to go around the damaged nerves. It is like taking a detour around a damaged part of a highway by using side streets.

Recently, doctors have discovered a way to encourage new nerve cell growth. When damaged, nerve cells quickly create scar tissue to protect against further injury. This scar tissue prevents cell repair and new nerve cell growth. It fills in the spaces of the damaged cells, preventing the cells from growing back together. Using special enzymes that break down scar tissue, doctors can encourage new nerve cell growth.

A team of doctors in Texas has discovered that the body releases a calcium-rich mechanism to heal damaged axons, the connecting portions of a nerve cell. However, this calcium also seals the damaged or severed ends of an axon. By removing the calcium, doctors have discovered that the axons remain unsealed and therefore can be repaired.

Concept Check

1. Why is damage to nerve cells so dangerous?
2. What ways have doctors discovered to repair nerve cell damage?

Emergency Room Nurse

Nursing is an important part of health care. Nurses may work in schools, nursing homes or retirement facilities, doctor's offices, or hospitals.

Nurses usually complete three years of full-time education. Once they have finished their courses, they are eligible to take a written exam to become a registered nurse. Nurses must complete this exam at a certain grade level to be allowed to practice nursing.

Once a nurse has completed these requirements, he or she may wish to specialize. Emergency room nursing is also sometimes called trauma nursing. This high-stress, fast-paced environment is challenging and fulfilling. Nurses work as part of a highly trained medical team to save lives. They provide comfort and compassion to people in pain, and administer medications and fluids to help stabilize people who are seriously hurt. They may also assist as part of a skilled medical team in emergency surgeries. In some cases, nurses in hospital emergency rooms have the authority to order tests and X-rays for patients who have a clear need. This speeds up the care and decreases the waiting time in emergency rooms.

To be an emergency room nurse, a registered nurse must take additional training. In most cases, nurses must take a 9- to 10-week, full-time training course. The courses consist of over 150 hours of classroom study and lecture and over 150 hours of practical experience and laboratory work in a real emergency room, with supervision.

This career offers a new challenge every day in the demanding field of emergency medicine.

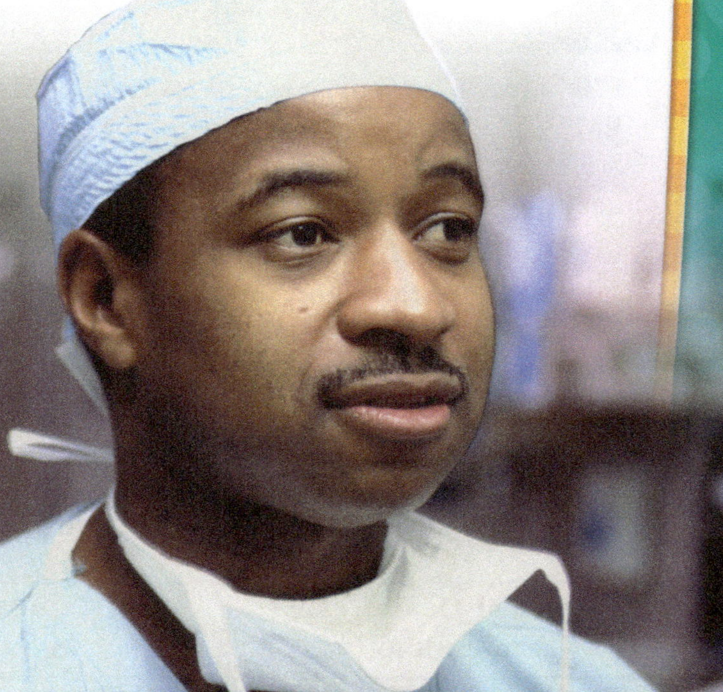

✅ Concept Check

1. How much education is required to become an emergency room nurse? How is this different from a registered nurse?
2. What duties do emergency room nurses have?

225

Study Guide

Lesson 1

1. Health professionals are specially trained to care for patients of different ages and to meet their different needs.

2. Physicians and surgeons care for the body and mind, while dentists focus on a patient's teeth and gums.

3. Psychiatrists, psychologists, and counselors help patients deal with mental issues.

4. Complementary care professionals, such as dietitians, psychiatrists, and dentists, provide specialized treatments for specific areas of concern.

Lesson 2

1. Firefighters, law enforcement officers, EMTs, paramedics, and safety engineers are some professionals who focus on keeping people safe.

2. Learning rules and guidelines for fire safety, water safety, gun safety, roadway safety, and outdoor safety can help you prevent accidents and emergencies.

Lesson 3

1. In an emergency, call 911 and give first aid to a victim.

2. Life-threatening conditions require professional treatment rather than first aid.

3. Situations that require first aid include concussions, stopped breathing or heartbeat, bleeding, bites, heat exhaustion, heat stroke, hypothermia, frostbite, fractures, sprains, strains, burns, shock, choking, and poisoning.

4. Slips, falls, and tripping are the most common kinds of accidents. A fracture is a break in the bone, and a concussion is a serious head injury.

5. Burns, bites, and poisonings need to be treated immediately to protect against infection and possible death.

Show What You Know

Visualize It Complete the graphic organizer. List at least four ideas in each category.

Types of Health-care Professionals	Careers in Safety	Emergency Care for a Bleeding Victim
1. _______________	5. _______________	9. _______________
2. _______________	6. _______________	10. _______________
3. _______________	7. _______________	11. _______________
4. _______________	8. _______________	12. _______________

Vocabulary Check

Fill in the blank with the correct vocabulary word.

13. Planning food and nutrition programs and supervising meal preparation are some responsibilities of a(n) __________.

14. A dental specialist who straightens teeth is a(n) __________.

15. When air is prevented from entering the lungs, __________ occurs.

16. When your body loses heat faster than it produces it, you have __________.

17. A serious brain injury that temporarily changes the way the brain works is a(n) __________.

18. The role of a(n) __________ is to help keep people free from danger, risk, or injury in the workplace.

19. Substances that cause injury, illness, or death when taken into the body are called __________.

Multiple Choice

Choose the best answer.

20. Which of the following best describes a psychiatrist?
    A. a professional who provides personal, social, and educational counseling
    B. a professional who assesses, diagnoses, and treats mental disorders
    C. a physician who treats mental and emotional illnesses
    D. a physician who treats children's illnesses

21. Which of the following should you do to help a person in shock?
    A. Give the person something to eat and drink.
    B. Call the poison center.
    C. Loosen tight clothing and keep the person warm.
    D. Wash the injury with soap and water.

22. Which of the following is true about third-degree burns?
    A. affect the deep layers of skin and nerves
    B. can be treated at home
    C. damage only the top layers of skin
    D. involve only the outer layer of skin

23. Which of the following should you do during a fire?
    A. Crawl low under smoke as you exit.
    B. Return to the house to check that pets and everyone else is out.
    C. Call 911 from the kitchen.
    D. Leave all the doors open.

Check Point

Answer the following questions.

24. How do emergency responders help people?

25. A friend trips and falls on a concrete sidewalk. You notice that his nose and face are bleeding, and he tells you he is dizzy. What should you **infer** about his condition, and what should you do?

26. List three examples of life-threatening emergencies.

27. What makes some accidents more serious than others?

3

Earth and Space Science

This photo shows the moon above Earth's blue, hazy atmosphere. In this unit, you will take a much closer look at our planet, what it is made of, and how it has changed.

Unit Overview

Describe the view of Earth you see in this photo. Is it smooth or rough? As inhabitants of Earth, how would you describe our planet's terrain?

The Earth is composed of many different minerals and rocks. What minerals or rocks can you name? In Chapter 7, you will learn to differentiate different minerals and rocks. In addition, you will learn how these materials form and are used. Can you identify any minerals or rocks in your classroom? If so, how are they used?

Minerals and rocks form Earth's surface. These materials make up the thin outer shell of the planet. In Chapter 8, you will learn about Earth's structure. In particular, you will see how Earth's outer layer moves, producing earthquakes and volcanoes. Have you ever experienced an earthquake?

If so, what did it feel like? Why do volcanoes and earthquakes only occur in certain parts of the world?

The surface of Earth is constantly changing. Some of these events are rapid, while others happen more slowly over hundreds and even thousands of years. In Chapter 9, you will learn about the processes that affect the planet's surface and the beneficial outcomes.

Hidden within rocks, fossils provide clues about the animals and plants that lived on Earth many years ago. In Chapter 10, you will learn about Earth's rock layers, the fossils found in them, and how they are consistent with the story of the biblical Flood. Have you ever seen a fossil? How would you describe a fossil? What are some interesting ways scientists use fossils to understand Earth history?

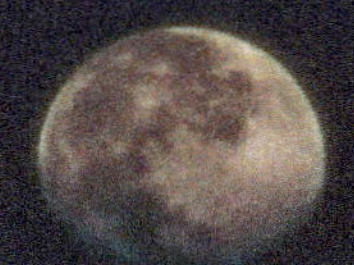

Your teacher may assign an Open Inquiry lab and a Lifestyle Challenge activity. Use your *Science Journal* to record your work.

Minerals and Rocks

Scripture Spotlight

God created minerals, rocks, and other Earth materials for us to use. You can strengthen your faith by studying about rocks and minerals that God created, how they form and change, and their value.

You will read the following passages in this chapter.

Exodus 28:9–12 (p. 237)
Exodus 17:5–6 (p. 245)
Deuteronomy 8:15 (p. 245)
Matthew 7:24–25 (p. 246)
Genesis 2:11–12 (p. 248)
Matthew 16:18 (p. 251)

The Big Idea

When God created Earth, He included a variety of minerals and rocks. People value rocks because they are useful and beautiful.

How do you use minerals and rocks?

Tourists visit caves to see beautiful rock formations, such as the limestone stalactites and stalagmites in this cave, found in Carlsbad Caverns in New Mexico.

Inquiry Kick-Off **Engage**

What processes did God design to form different rocks? You may not be able to see the processes, but you can model them. In your *Science Journal*, you will model and investigate how one kind of rock is formed.

What Are Minerals?

Objectives

- Identify characteristics used to identify minerals.
- Describe common properties of minerals.
- Identify common minerals using their properties.

Vocabulary

luster

streak

cleavage

fracture

Think about the last time you walked on dirt. What color was it? If you look closely at a handful of dirt, what can you see? You have learned that Earth's crust is made of minerals and rocks. What makes a mineral a mineral? In this lesson, you will learn about the characteristics of minerals and the properties of minerals that help to distinguish one mineral from another.

Minerals Explain

Recall that a mineral is a solid, inorganic, natural material with a certain structure and composition. Minerals are the same throughout their composition. What does all of this mean?

Characteristics of a Mineral

- Minerals are solid.
- Minerals are inorganic. This means that they were never alive or part of anything that was alive. Graphite is made of carbon but was never alive.
- Minerals form in nature.
- Minerals have a specific chemical make-up, or composition. For example, the formula for halite (sodium chloride, or table salt) is $NaCl$. So, one molecule of halite is always made of one atom of sodium (Na) and one atom of chlorine (Cl).
- Minerals have a definite structure. A mineral's structure depends on the atoms that make up the mineral and how the atoms combine. Diamond and graphite are both made only of carbon (C). However, the carbon atoms in diamond are arranged in a different structure than the carbon atoms in graphite.

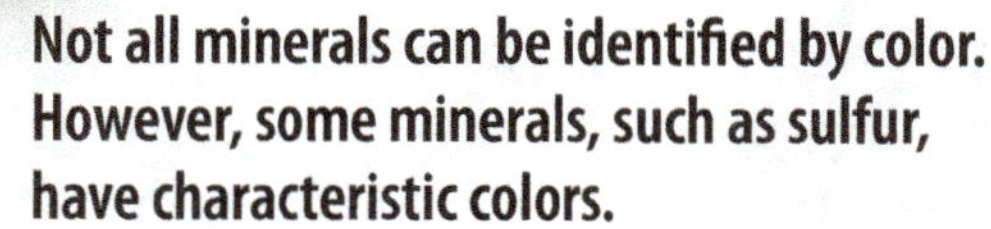

Not all minerals can be identified by color. However, some minerals, such as sulfur, have characteristic colors.

What other properties of sulfur can you see in this picture?

Identifying Minerals
How are minerals identified?

Procedure

1. You will **observe** a different mineral at stations that you rotate through. At each station, perform Steps 1–7. **Record** your observations in the data table in your **Science Journal**. **Record** the color and luster of the mineral.

2. Use a penny, a nail, a glass plate, and the hardness scale found later in this lesson to determine the hardness of the sample. Make a small scratch on the mineral to test its hardness. **Record** the hardnesss value.

3. Use the streak plate to determine the streak color of the mineral. **Record** this data. If the mineral does not leave a streak, write *None*.

4. Carefully squeeze one or two drops of white vinegar onto the mineral. **Observe** and **record** what happens. Use water and paper towels to completely wipe the white vinegar off of the mineral.

5. Test the mineral with a magnet. **Record** your observation.

6. **Record** any other observation of the mineral.

7. Use the mineral identification key in your **Science Journal** to determine the name of the mineral and **record** the name in the table.

8. Repeat Steps 1–7 with at least four other minerals.

Materials
- safety goggles
- streak plate
- penny
- glass plate
- nail
- magnet
- dropper bottle of white vinegar
- various unknown minerals
- paper towels
- Mohs Hardness Scale (found later in this lesson)

Analyze Results

Look at the data you have collected for each mineral and explain how the various properties helped you to identify the sample.

Create Explanations

1. How are minerals identified?

2. Which property was most helpful in identifying each mineral? Why?

3. Why was it necessary to test several properties before identifying the unknown minerals?

Properties of Minerals Explain

What are the properties of a mineral? Each type of mineral has certain characteristics or properties. Different properties are the result of the chemical composition and physical structure of each mineral.

Color is likely the first property that you might observe. Some minerals are always the same color, such as gold. Can all minerals be identified by their color?

Luster is a property that describes the way a mineral reflects light. A mineral with a metallic luster is shiny, like a polished metal. Silver and pyrite have metallic luster. Nonmetallic minerals do not have a shiny luster. The surface of talc resembles the surface of a pearl, so talc's luster is described as pearly. Quartz has a glassy luster. Because one form of the mineral hematite looks like an earthen clay pot, its luster is described as earthy.

Streak is the color of a mineral in its powdered form. A streak test is done by dragging a mineral across an unglazed, porcelain tile called a streak plate. This causes the part of the mineral being dragged to break down into a powder. Do all minerals make a streak? Why do scientists do streak tests?

Explore-a-Lab

Structured Inquiry

Use a streak plate to find out the color and streak properties of a sample of sulfur. Make a scratch across the streak plate with the sulfur and observe the color of the streak made by the sulfur on the plate.

How would you describe the color and streak properties of sulfur?

Color is only one property used to classify minerals.

Why can you not rely on color alone to identify a mineral?

Hardness

Another mineral property is hardness. Recall that hardness is the resistance of a mineral to being scratched. A mineralogist named Friedrich Mohs developed a scale that ranks the hardness of certain minerals. Minerals and objects with higher numbers on the scale will scratch minerals and objects with lower numbers on the scale. Suppose you have a mineral sample and when you test for hardness you find it scratches a penny but not a knife blade. What do you think is the mineral's hardness?

Mohs Hardness Scale		
Index Mineral	**Scale**	**Common Object**
Diamond	10	
Corundrum	9	Steel file (6.5)
Topaz	8	
Quartz	7	Glass (5.5)
Orthoclase	6	Knife blade (5.1)
Apatite	5	Nail (4.5)
Fluorite	4	
Calcite	3	Penny (3.5)
Gypsum	2	Fingernail (2.5)
Talc	1	

Diamond (10) is the hardest mineral, and talc (1) is the softest.

Which minerals can be scratched with your fingernail?

Cleavage and Fracture

Atoms are arranged differently in different minerals. These arrangements cause minerals to break in different ways. Minerals that break along smooth planes have **cleavage**. Graphite is a mineral that has cleavage in only one direction. This causes it to cleave, or break, into thin sheets. Minerals that break along rough, uneven surfaces have **fracture**. Quartz is one mineral that fractures.

Other Properties

In addition to color, luster, streak, hardness, and cleavage or fracture, many minerals have unique properties. The mineral magnetite, for example, is magnetic. Halite tastes salty. Gold does not rust, tarnish, or corrode like other metals. The mineral calcite has several unique properties. It fizzes when it contacts an acid. Calcite also has a property called double refraction, which causes light shining through it to split into two rays. See this for yourself. Place a sample of calcite over the words on this page and describe what you see. Another type of calcite glows bright red under ultraviolet light.

Which image shows cleavage and which shows fracture? Explain how you know.

Check for Understanding

Think about how you identified minerals in the *Structured Inquiry*. What other tests could you conduct to help identify the minerals you examined? Explain.

Common Minerals Explain

Earth is made up of more than 3500 different minerals. However, only a few minerals are commonly found. Think about the minerals you have studied in this lesson. They are made up of some of the most common elements on Earth. For example, silicon and oxygen are the two most abundant elements on Earth. So, it makes sense that quartz, which is made up of silicon and oxygen (SiO_2), is the most abundant mineral on Earth. The list below identifies other common minerals that make up Earth's crust.

Some Common Minerals and Their Properties				
Mineral	**Common Color(s)**	**Luster**	**Hardness**	**Other Properties/Characteristics**
Calcite	Colorless, white, or light pink	Glassy	3.5	Fizzes with vinegar or hydrochloric acid
Fluorite	Colorless, green, or purple	Glassy	4	Cleavage, can be mistaken for quartz, some are fluorescent
Galena	Lead-gray	Metallic	2.5	Lead-gray streak, cubic cleavage, feels very heavy
Graphite	Dark gray to black	Metallic	1–2	Can be scratched with fingernail, feels greasy, dark gray streak
Gypsum	White, colorless, or gray	Glassy or dull	2	Weak cleavage, very soft
Halite	Colorless or white	Glassy	2.5	Cubic cleavage, dissolves in warm tap water
Hematite	Silvery-gray, reddish-brown, or black	Metallic, dull, or earthy	5–6	Dark reddish-brown streak, can be metallic or non-metallic
Hornblende	Dark green or black	Glassy	5–6	Greenish-gray streak; cleavage
Magnetite	Black	Metallic	5.5–6.5	Black streak; magnetic; fractures
Mica	Brownish-black or white	Glassy	2.5–3	Perfect cleavage in one direction
Feldspar	Pink or less commonly gray	Glassy	6	Cleavage in two directions
Pyrite	Brassy yellow	Metallic	6–6.5	Greenish-black streak, often called "fool's gold"
Quartz	Colorless, white, gray, purple, orange, or pink	Glassy	7	Fracture
Talc	Pale green, white, or gray	Pearly	1	Pale green to white streak; feels silky

236

Have you ever found a rock and wondered what it was made of? Some minerals are called *rock-forming* minerals because they make up most of the rocks on Earth. Do you know of any rocks that contain any of these minerals? Which of these minerals do you think might be the most common minerals in rocks? Why?

Lesson Activity

 What rocks contain common minerals?

Your teacher will assign your group one of the common minerals from the chart on the previous page. Use the Internet and other resources to learn what types of rocks contain the mineral. Create a poster, displaying the name and a picture of the common mineral, the name and a picture of the rock, and where the rock is commonly found. Share your information with the rest of the class. Did any groups list the same rock as your group? Which rocks contain multiple common minerals? Which rocks have very few common minerals?

Although many people use the terms *mineral* and *rock* interchangeably, they are not the same. You have learned that a mineral is a solid, natural, inorganic material. A mineral also has a specific chemical make-up and definite physical structure. Keep the definition of a mineral in mind as you study rocks in the next lesson.

Concept Check   Assess/Reflect

Summary: What are minerals? Minerals are natural, inorganic solids with a definite structure, as well as a specific chemical composition. Minerals have specific properties that allow scientists to identify them. These properties include color, luster, streak, hardness, and cleavage or fracture. Of the thousands of minerals on Earth, only a few are common. Some of these common minerals are the ones that make up most of the rocks in Earth's crust.

1. The Mohs scale measures __________.

2. The two main categories of luster are __________ and __________.

3. Define each of these mineral properties: *luster, streak, cleavage,* and *fracture.*

4. Frozen water, or ice, is a mineral. Liquid water is not a mineral. Why is ice considered a mineral but water is not?

What Are Rocks and the Rock Cycle?

Objectives

- Describe how igneous rocks form.
- Distinguish between intrusive and extrusive rocks.
- Describe how sedimentary rocks form.
- Differentiate between clastic, chemical, and organic sedimentary rocks.
- Describe how metamorphic rocks form.
- Distinguish between foliated and non-foliated rocks.
- Describe the rock cycle.

Vocabulary

igneous rock

texture

sedimentary rock

clastic sedimentary rock

chemical sedimentary rock

organic sedimentary rock

metamorphic rock

rock cycle

Rocks are materials that are made up of two or more minerals. Granite is a rock made up mostly of quartz, feldspar, mica, and hornblende. Explain different ways that rocks are formed. What type of rock do you think is most common where you live?

Igneous Rocks Explain

Igneous rocks form when molten rock cools and hardens. Igneous rocks are identified by their composition and texture. **Texture** refers to the size of the mineral grains or the way in which they are arranged in any type of rock. Words used to describe the size of the grains include *fine* (small) and *coarse* (larger). Can you name one example of an igneous rock? How would you describe its texture? How does this rock look different from a mineral?

238

Intrusive Rocks

Igneous rock formed from the cooling of molten rock below Earth's surface is called *intrusive igneous rock*. These rocks have large mineral crystals that can be seen with your unaided eye. The slow rate of cooling allows the mineral crystals to grow fairly large. Igneous rocks that have large crystals have coarse-grained texture. Granite, diorite, and gabbro are intrusive rocks.

Extrusive Rocks

Igneous rock formed from lava, molten rock that reaches the Earth's surface and cools, is called *extrusive igenous rock*. Extrusive igneous rocks have small mineral crystals. These rocks cool and harden quickly, leaving little time for crystals to grow. Extrusive igneous rocks have fine-grained texture. Rhyolite, andesite, and basalt are extrusive rocks. When lava cools and hardens quickly, and there is no time for crystals to grow, the texture of an igneous rock is glassy. Obsidian and pumice are glassy igneous rocks.

Explore-a-Lab

Structured Inquiry

Place three to four drops of a solution of Epsom salt and water onto two glass slides. Let one sit overnight and record what you observe. Hold the second slide over a lit candle until all of the liquid evaporates. Again, record what you observe. Compare the size of the crystals you observed on both slides.

Which salt crystals represent crystals in an intrusive igneous rock, and which ones represent crystals in an extrusive igneous rock? Explain.

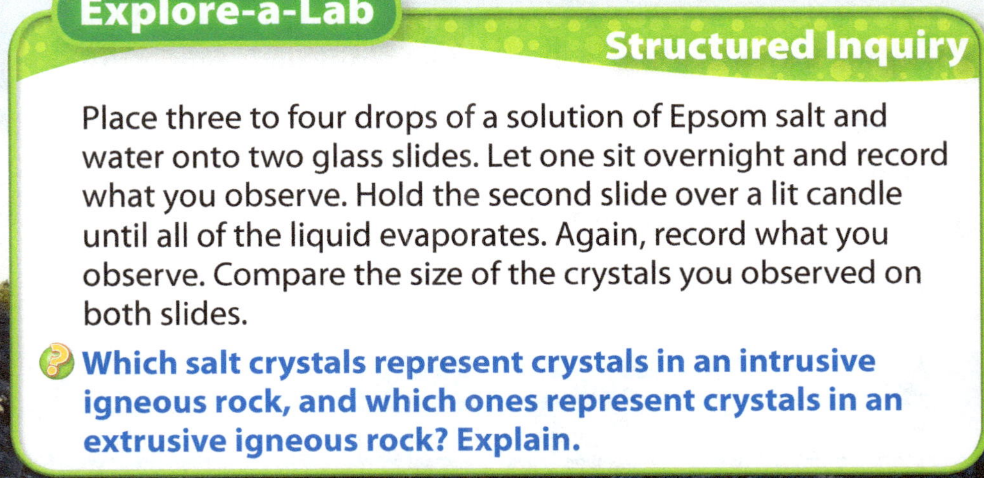

Identifying Igneous Rocks

How are igneous rocks identified?

Procedure

1. Begin at one station. **Observe** the texture and color of the igneous rock. **Record** your observations in the table in your *Science Journal*.

2. At your teacher's direction, move to another station and repeat Step 1. Continue this process until you have **observed** each available rock.

3. Use your observations and the identification table below to determine the name of each rock. **Record** the name in the data table in your *Science Journal*.

Materials
- igneous rock samples
- hand lens

Analyze Results

Compare your descriptions to the table at the bottom of this page. What difficulties did you encounter when determining texture and composition?

Create Explanations

1. How are igneous rocks identified?

2. Was rock texture easy to identify? Explain.

3. Which characteristic was most helpful in identifying the rocks? Explain.

Igneous Rock Identification Table				
Texture	**Composition**			
	Felsic (light colors: white, tan, pink, light gray)	**Intermediate** (gray or combined light and dark colors)	**Mafic** (dark colors)	**Ultramafic** (very dark colors, green)
Very large crystals	Granite pegmatite	Diorite pegmatite	Gabbro pegmatite	
Large crystals	Granite	Diorite	Gabbro	Dunite
Fine-grained, crystals too small to see	Rhyolite	Andesite	Basalt	
Visible crystals of two different sizes	Rhyolite	Andesite	Basalt	
Glassy	Obsidian	Obsidian	Basaltic glass	
Bubbly appearance	Pumice	Pumice	Scoria	
Volcanic fragments	Volcanic tuff	Volcanic tuff	Volcanic tuff	

This sandstone formation appears to have stripes. The red color is caused by iron contained in the sediment that makes up the rock.

Sedimentary Rocks Explain

Sedimentary rocks form when pieces of existing rocks, called *sediment,* are compacted and cemented together. What do you think holds the rocks together?

Most sedimentary rocks form in water, but some form on land. Why do you think most sedimentary rocks form in water? There are three main types of sedimentary rocks: clastic rocks, chemical rocks, and organic rocks.

Type of Sedimentary Rock	Type of Sediment	Examples
Clastic sedimentary rock	• *clasts* range in size from clay to boulder • named by size and shape of clast	• mudstone • siltstone • sandstone • breccia • conglomerate
Chemical sedimentary rock	• form when water evaporates or when water is supersaturated with dissolved ions	• limestone • gypsum • halite • travertine • dolomite
Organic sedimentary rock	• form from compacted and cemented organic material	• coquina • coal • bioclastic limestone

conglomerate

gypsum

coquina

coal

How can you tell the difference between different sedimentary rocks?

241

Math in Science

The clasts in a conglomerate are rounded and usually range from less than 0.004 mm to 256 mm (0.0002–10 in.) in size. Sand ranges from 0.125–1mm (0.005–0.04 in.), while silt is less than 0.004 mm (0.0002 in.). Clasts that make up pebbles are 2 to 64 mm (0.08–3 in.). Clasts that make up cobblestone are 64 to 256 mm (3–10 in.). Use a metric ruler and a compass to draw the clasts of the largest cobblestone and pebble to scale. How do the clasts in a conglomerate compare to those in a sandstone and shale?

Metamorphic Rocks `Explain`

Metamorphic rocks form when existing rocks are changed by high heat, high pressure, and/or very hot fluids. Metamorphic rocks can form when magma comes into contact with existing rocks. This type of metamorphism is called *contact metamorphism*. Metamorphic rocks can also form where Earth's tectonic plates meet. This type of metamorphism takes place over very large areas, or regions, and is called *regional metamorphism*. Metamorphic rocks can be divided into two groups based on their texture: foliated and non-foliated.

Foliated Rocks

Foliated metamorphic rocks form when high pressure squeezes existing rocks from opposite sides. What does this do to the minerals in the rocks? In extreme cases, the minerals line up and form alternating dark- and light-colored bands. Slate, schist, and gneiss are examples of foliated rocks. Most foliated rocks form in areas of regional metamorphism.

Non-foliated Rocks

Non-foliated metamorphic rocks usually form when high heat and/or hot fluids come into contact with existing rocks. Non-foliated rocks are not layered or banded. Instead, non-foliated rocks usually appear to have one color and interlocking crystals. Marble and quartzite are two non-foliated rocks. Most non-foliated rocks form as the result of contact metamorphism.

schist

hornblende

The texture of metamorphic rocks can be either foliated or non-foliated.

Which of these rocks is the foliated rock and which is the non-foliated rock?

Create a model rock using different chocolate candies to represent sand grains. Place the chocolate candies into a small, thick, wide-mouth jar. Use your fingers or a large wooden dowel covered in plastic wrap to press down on the chocolate pieces. Which type of rock did you model? Place the jar on a hot plate and turn the hot plate on low. After the chocolate pieces have softened, use the dowel to press down on the chocolate pieces. Now which type of rock did you model? After the jar cools, use a plastic knife to cut your model rock. Remove your rock and record your observations.

What types of rocks did you form? Which rock-forming processes did you model?

Check for Understanding

Now that you have studied the different types of rocks, look at your igneous rock classifications from the *Structured Inquiry*. Would you change any of your classifications? Why or why not?

Parent Rocks

The rocks that existed before metamorphism occurred are called *parent rocks*. Parent rocks can be sedimentary, igneous, or metamorphic rocks. The photos below show some common metamorphic rocks and their parent rocks.

Heat and extreme pressure can change a parent rock into a metamorphic rock.

What differences do you see in the metamorphic rocks? How do they relate to the differences you see in the parent rocks?

Some Common Rocks Explain

Recall that even though more than 3500 minerals have been identified on Earth, only about 12 are common. Likewise, there are about 500 different rocks on Earth, but only 20 are common. They are listed and described in the table below.

Common Rocks and Their Characteristics			
Rock Type and Texture		**Examples**	**Characteristics**
Igneous	Intrusive Coarse-grained crystals	Granite	Coarse-grained; light-colored; made of interlocking crystals of feldspar, quartz, and mica
		Diorite	Coarse-grained, gray and white in color, contains mostly interlocking crystals of feldspar and hornblende
		Gabbro	Coarse-grained, dark-colored, made of interlocking crystals of feldspar and hornblende
	Extrusive Fine-grained crystals	Basalt	Fine-grained, gray to black in color, made of interlocking crystals of feldspar and hornblende
		Pumice	Resembles a sponge, floats on water, made of volcanic glass and contains microscopic bubbles
		Obsidian	Volcanic glass is quartz that has been colored by ash. It often contains flecks of white or red and black swirls.
Sedimentary	Clastic	Conglomerate	Coarse-grained, contains rounded pebbles in a fine cement
		Sandstone	Can be coarse-grained or fine-grained, mostly quartz, feels gritty, minerals are often rounded and look like they are glued together
		Shale	Finely grained, made mostly of quartz and feldspar clay, splits into layers
	Chemical crystalline	Limestone	Can be fine-grained or coarse-grained, contains mostly calcite, can contain fossils
		Rock salt	Fine-grained; can be clear, white, or pink; contains halite; tastes salty
		Chert	Fine-grained; made of silica; usually white or gray; can contain small fossils
	Organic clastic	Bituminous coal	Can be black or brown; may contain plant fossils; can be burned
		Coquina	Fine-grained or coarse-grained, made mostly of shells and other organic matter
Metamorphic	Foliated	Slate	Fine-grained, has a dull surface, usually gray or black; parent rock is shale
		Schist	Medium- to coarse-grained, can be split into sheets; contains mica that gives it a sparkly appearance; parent rock is usually slate
		Gneiss	Medium- to coarse-grained; alternating bands of light and dark minerals; parent rock is often granite or schist
	Non-foliated	Marble	Fine- to coarse-grained; usually made of calcite grains that interlock; can be white, grey, or pink; parent rock is usually limestone
		Quartzite	Medium- to coarse-grained; made mostly of quartz grains that are fused together; can be white, grey, pink/purple; parent rock is usually quartz sandstone
		Anthracite (coal)	Fine-grained; very hard and shiny; parent rock is bituminous coal

How Rocks Change

Rocks may seem permanent, but they are always changing. These changes may happen quickly or over a long period of time. Some of the changes take place at or near Earth's surface. Other changes take place deep beneath Earth's surface. The natural processes that change rocks make up the **rock cycle**. Refer to the drawing below as you read about these processes.

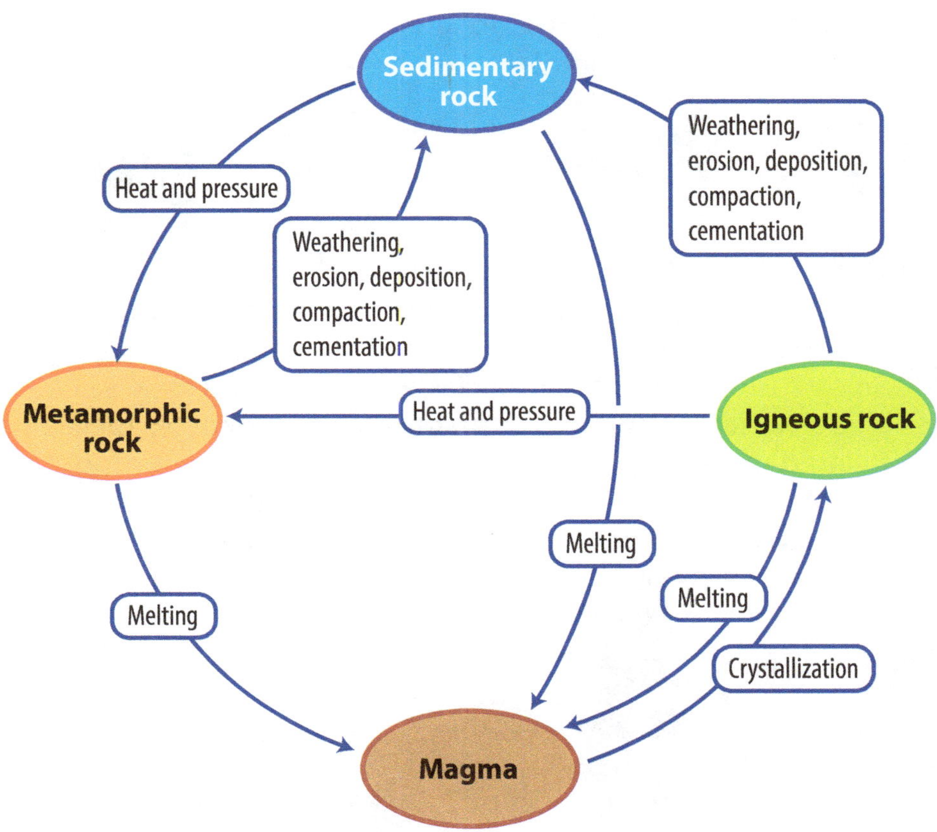

Scripture Spotlight

What do **Exodus 17:5–6** and **Deuteronomy 8:15** say about a rock?

The rock cycle includes any natural process that changes a rock.

Describe how a sedimentary rock might become an igneous rock.

Like most natural cycles, the rock cycle has no beginning or end. For this discussion, let's start with the processes that take place at or very close to the Earth's surface. *Weathering* is any process that breaks rocks into smaller pieces. Recall that these pieces of Earth materials are called sediment. *Erosion* is the carrying of sediment from one place to another. Water, wind, and ice are common agents of erosion. When these agents slow down or stop, they drop the sediment they are carrying. *Deposition* is the dropping or laying down of loose Earth materials. Once sediment is deposited, it is slowly *compacted*, or squeezed, and *cemented*, or bound together. These processes form sedimentary rocks.

Rock cycle processes that take place deep within Earth result in the formation of either igneous rocks or metamorphic rocks. Melting rock deep within Earth produces magma. As magma slowly cools, minerals *crystallize*. These minerals make up intrusive igneous rocks. If the magma rises to the surface, it becomes lava. Lava crystallizes and forms extrusive igneous rocks.

If rocks are changed by heat, pressure, and/or hot fluids, but they do not melt, then metamorphic rocks form. High temperature and hot fluids usually produce non-foliated rocks. High pressure usually produces foliated rocks.

Explore-a-Lab

Guided Inquiry

 **How can you model different types of rock?**

Use the materials provided by your teacher and model one of the following types of rocks: an intrusive igneous rock, an extrusive igneous rock, a clastic sedimentary rock, a chemical sedimentary rock, an organic sedimentary rock, a foliated metamorphic rock, and a non-foliated metamorphic rock. Exchange models with another student, but do not tell each other which type of rock you modeled. Identify the type of rock represented by your classmate's model.

Concept Check Assess/Reflect

Summary: What are rocks and the rock cycle? Rocks are natural materials that contain two or more minerals. Molten rock that cools and hardens is igneous rock. Cooling below Earth's surface forms intrusive igneous rock, while cooling at the surface forms extrusive igneous rock. Rock that is cemented together is sedimentary rock. The three types of sedimentary rock are clastic, chemical, and organic. Rocks changed by intense heat, pressure and/or hot fluids are metamorphic rock. The two types of metamorphic rock are foliated and non-foliated. All rocks form and change through a series of natural processes called the rock cycle, which has no beginning and no end.

1. What are the three main types of rocks?
2. What type of rock results from weathering, erosion, and deposition?
3. Compare and contrast three types of sedimentary rocks.
4. Can sandstone become shale? Explain your answer.

What Minerals and Rocks Do We Use?

Objectives

- Explain what an ore is.
- Identify common minerals.
- Identify common uses of rocks.
- Describe how minerals and rocks are obtained.
- Describe the effects of mining.

Vocabulary

ore
malleable
ductile
gem
quarry
conservation

Look around your home or school. What objects are made from minerals or rocks? Many minerals and rocks are valuable because they are used in inventions that make life easier. Consider iron as an example. How many ways did you use iron today?

Many of these minerals and rocks come from ores. An **ore** is any solid Earth material that can be mined at a profit. These resources are nonrenewable. Which resources can you think of that are nonrenewable? What are some common ores, and how are they used in everyday life?

Metallic Minerals `Explain`

Why are many of the ores used in everyday products metallic minerals? For example, gold and silver are two rare but important minerals commonly used to make items of beauty, coins and electronic devices, such as cell phones, televisions, and computers. These two minerals exhibit *tenacity*. Tenacity describes a mineral's ability to resist crushing, bending, breaking, or tearing.

Silver is **malleable**, or able to be flattened by pounding with a hammer. It is used to make spoons, knives, and forks, and to purify water. Gold is **ductile**, meaning that it can be stretched into a wire. Platinum is another ductile metal used in wiring and electrical contacts. All true metals are ductile. Gold is also used to treat and diagnose some rare illnesses. And, because gold does not react with most other substances on Earth, it is used in dental crowns, fillings, and even some types of braces.

Gold is used in computer hardware, such as this memory chip. Because gold is a high-quality metal, it allows information to be transmitted very quickly.

Scripture Spotlight

What does **Genesis 2:11** say was in the land of Havilah? What else was there according to verse **12**?

Aluminum is obtained from the rock bauxite, an aluminum ore.

How does recycling help aluminum production?

Bauxite ore

Aluminum cans

In addition to silver and gold, other metallic minerals have a variety of uses because of their properties. Read about some of these metals in the table below. Can you think of any other uses of these metals? Which tenacity properties do they exhibit?

Metallic Mineral	Properties	Uses
Titanium	• silver color • lightweight • as strong as steel • does not corrode easily in saltwater or interact chemically with the human body	• used in sporting equipment, such as golf clubs, sports helmets, bicycle frames • used in parts for boats, aircraft such as helicopters, and spacecraft • used in surgical instruments, artificial joints, pins used for setting broken bones, dental implants
Iron	• gray color • can have a magnetic field around it • can withstand high temperatures • good conductor of heat and electricity	• primarily used to make steel and stainless steel • used in manufacturing automobiles, tools, buildings, machine parts • powdered for use in magnets, paints, plastics, printer ink, fertilizer, cosmetics
Copper	• reddish-gold color • good conductor of heat and electricity • tarnishes to a green color when exposed to oxygen	• pulled to make wires used in electrical switches, cables, and electromagnets • used in plumbing pipes and roofing material • used to make brass and bronze, two common metal alloys • used in ceramic glazes, musical instruments, lightning rods, electronics, and coins
Aluminum	• silver color • lightweight • good conductor of heat and electricity	• pulled to make wires • pounded into sheets to make beverage cans, foil, ladders, tennis racquets • used to manufacture car, truck, boat, plane, and train parts • used to make bicycles, street light poles, ship masts, baseball bats, cooking utensils, and coins

Nail File or Emery Board?

What type of material makes the best nail file?

Procedure

1. Put on your safety goggles. Place one spoonful of each type of abrasive sediment in separate paper trays. Spread out the abrasives evenly in the trays.

2. Use a magnifying lens to examine each type of abrasive sediment. Describe the characteristics of each abrasive. **Record** your observations.

3. **Predict** which abrasive you think will be the best at reducing the mass of a craft stick.

4. Write your initials on one side of a cardboard strip. Use the paintbrush to coat the other side of the cardboard strip with glue. Lay the cardboard, glue side down, in the tray so that it becomes coated with one of the abrasives. Carefully turn it back over and set it to dry on the sheet of newspaper.

5. Repeat Step 4 for the other abrasive samples.

6. Allow the cardboard strips to dry overnight.

7. Label each of the three craft sticks with one of the abrasives used to make the abrasive stick. Use the balance to **measure** the mass of each craft stick. **Record** the mass.

8. Test the abrasiveness of each stick by rubbing the matching craft stick 10 times across the length of the abrasive stick. **Record** the mass.

9. Repeat Step 8 with the other two abrasive sticks and craft sticks.

Materials
- cardboard strips
- white glue
- fine quartz sand
- coarse quartz sand
- crushed pumice
- small paper trays
- plastic spoons
- craft sticks
- paintbrushes
- newspaper sheets
- magnifying lens
- balance
- safety goggles

Analyze Results

Compare the effect each abrasive had on your craft sticks.

Create Explanations

1. What type of material makes the best nail file? Explain.

2. Compare the abrasive materials. Include color, shape, and hardness.

3. Explain why many nail files have a coarser abrasive on one side and a finer abrasive on the other side.

Nonmetallic Minerals Explain

Quartz is a nonmetallic mineral with many uses. Quartz crystals are sometimes used as gems. A **gem** is a mineral that is valued for its beauty. Gems are used extensively in the jewelry industry. Amethyst, rubies, emeralds, and sapphires are just some of the gems used. Many of these dazzling gems are found on the crowns and scepters of royalty. Amber is an organic gem with an interesting past. What unique features does amber sometimes include? What are some other gems you can name?

Diamond is another nonmetallic mineral that is probably best known as a precious gem used in making rings and other jewelry. However, because it is the hardest known mineral on Earth, diamond has other important uses. Crushed diamond is often used as an abrasive. Drill bits used to penetrate Earth's rocky crust are coated with diamonds. What other materials do you think diamonds could cut through?

The mineral fluorite is used to make hydrofluoric (HF) acid. This acid is corrosive and is used to etch and polish glass. It is used to clean brick and stone buildings. Many cleaning products contain HF acid because it removes rust and polishes tarnished metals. It is also used to make plastics, pesticides, opaque glasses, and gasoline. Why would this acid be dangerous? What safety precautions should someone take when using HF acid?

The mineral feldspar is used to make dinnerware and other types of ceramic items. It is also used in flooring tiles, soaps, cement, concrete, fertilizers, and paper. Feldspar commonly weathers to produce clay minerals. Clay minerals can be used to make water pipes, bricks, and many other materials used in construction. What kinds of water pipes are made of clay? Where else have you seen clay used on houses?

Drilling through Earth's hard crust would be much more difficult if drill bits such as this were not being used.

Why would a diamond drill bit be needed to drill through rocks, such as gneiss and granite?

Rocks and Sediment

Minerals and rocks are used to make many of the products people use in daily life.

Rock Type	Industrial Use
Granite	buildings, kitchen and bathroom countertops, and tombstones
Gabbro and basalt	crushed for concrete
Pumice	soaps and emery boards
Schist and gneiss	building stone
Slate	floor tiles and roof shingles
Quartzite	building materials, construction filler, and street material
Marble	carved into sculptures and other works of art
Conglomerate, sandstone, and shale	bricks, decorative stone, and cement
Coal	generating electricity
Limestone	construction and animal feed
Rock salt	cooking, de-ice roads and sidewalks
Sand and gravel	concrete and road materials

This building was constructed with sandstone bricks and has a slate tile roof.

What feature of slate makes it ideal to use in this way?

Scripture Spotlight

How many rocks are mentioned in **Matthew 16:18**? If you said only one, look again!

Called to Serve

Matthew 5:13 calls Christians the salt of the Earth. After reading about the uses of rock salt, how many different ways can you think of for Christians to serve as salt?

Math in Science

According to the Mineral Information Institute, every person in the United States needs about 17,260 kg (38,052 lbs) of minerals and other resources every year to maintain his or her standard of living. Use this value to compute how many kilograms of resources you have used so far in your life. About how many kilograms of resources will you likely use in a lifetime?

Obtaining Earth Materials Explain

Some rocks and minerals form at or very close to Earth's surface. Most, however, form deep within Earth. Rocks and sediment at the surface are often quarried. Rocks and minerals that form deep within the crust are obtained by other mining methods.

A **quarry** is a surface mine from which large slabs of rock or large volumes of sediment are removed. Many quarries use special equipment to cut and remove large slabs of rock. Granite, marble, sandstone, and slate are some types of rocks mined in the United States. Other quarries use bulldozers and dump trucks to mine sand and gravel.

Some valuable Earth materials are obtained by open-pit methods. An open-pit mine is used to recover rocks or minerals that are close to Earth's surface. This type of mining damages the land. Laws now require that mined lands be *reclaimed*, or restored to their original condition. Some types of reclamation make the land better than it was before it was mined. What improvements could be made to an area during mine reclamation?

Underground mines are used when minerals or rocks lie deep within Earth's crust. Some underground mines resemble huge mazes that include many rooms. Other underground mines are vertical shafts that are drilled to reach the ore. What are some benefits and consequences of mining?

Earth's Resources

There are two types of natural resources. Nonrenewable resources are those that are used faster than they can be replaced. Minerals, rocks, sediment, land, and soil are all nonrenewable resources. Energy resources, such as oil, natural gas, coal, and uranium, are also nonrenewable.

Renewable resources are those that can be replaced in a relatively short period. Resources that will likely never run out, such as wind and solar energy, are also renewable resources. What are some resources that you use regularly?

Check out your *Science Journal* to explore the processes of mining.

Extend

Land and Soil

Do you think land and soil are renewable resources? Why or why not? Land not only provides mineral resources, but it also provides the space required for living, transportation, growing crops, and many other uses. Land can also be used for grazing livestock. Soil is used to grow crops and other plants. Unfortunately, unwise farming methods can make land and soil unusable for many years by stripping the soil of its nutrients.

Minerals, Rocks, and Sediment

Do you think the minerals and rocks you have studied are renewable resources? Why? Even though it may seem that there is an endless supply of sediment, rocks, and minerals, most of these Earth materials are nonrenewable resources. Most minerals, rocks, and sediment take a long time to form. Some exist only in very small amounts. Iron, for example, is an abundant natural resource. However, iron is nonrenewable because it takes a very long time to form. Gold is a nonrenewable resource for two reasons: it takes gold a very long time to form and it exists in very small amounts in Earth's crust. Why do you think gold exists in such small amounts?

Uranium

Uranium is a nonrenewable natural resource. Some forms of this mineral naturally break down, or decay. As it decays, enormous amounts of energy are released in the form of nuclear energy. This energy can be used to generate electricity. Unfortunately, harmful wastes form when uranium decays. Storing these wastes safely is a constant concern. Before uranium ores are mined, do they produce heat and waste? If so, what happens to the waste? Do you think the dangers of the waste are acceptable because of the energy generated? Why or why not?

How has the land shown here been reclaimed?

Conserving Earth's Resources Explain

God provides us with resources that make our lives easier. Whether a resource is renewable or nonrenewable, it should be used wisely. The wise use of a natural resource is called **conservation**.

Think about all of the resources you use on a typical day. Do you think you use them wisely? How can you help conserve Earth's natural resources? How can using resources wisely help you honor the Creator?

Lesson Activity

Separate into two groups. One group will be damming a local river and the second group will be opposed to the dam. Use the Internet to discover positive and negative consequences to damming a river. Gather your notes and develop your argument. Each group should select a spokesperson to represent their side. After each group has had a chance to speak, take turns debating the topic.

 Concept Check Assess/Reflect

Summary: What minerals and rocks do we use? Many of the minerals and rocks on Earth have value and are used as they are found or ground for use in construction of everyday products. Rocks and minerals are natural resources used by humans. There are renewable and nonrenewable resources. Wise use of resources is called conservation.

1. Name five different mineral resources and at least one use for each.
2. Marble is made primarily of the mineral calcite. How does this make marble useful to an artist who makes statues from this rock?
3. Identify five renewable and five nonrenewable resources.
4. The mineral halite is a valuable nonmetallic mineral that often forms perfect cubic crystals. Yet it is not a gem. Why?
5. What is an ore? Give an example.

Extend

Get to Know

President Theodore Roosevelt

Theodore Roosevelt was president of the United States from 1901 to 1909. During that time, he helped to protect many natural and historic areas.

Even before he became president, Roosevelt was a conservationist. As a young man, he spent a few years travelling and living in the American west. He even owned a ranch in western North Dakota. Roosevelt had asthma as a child and grew up skinny, weak, and pale. During his years in North Dakota, his steady activities of riding, roping, hunting, and hiking helped him develop into a healthy and strong young man. Roosevelt came to believe that outdoor work and play was essential for the health of everyone. To make sure everyone would have access to nature, he became determined to preserve lands of beauty and wonder.

In 1906, Congress passed the Antiquities Act. This allowed the president to set aside lands of historic value. President Roosevelt immediately named four national monuments in the desert Southwest. In 1908, he named a large part of the Grand Canyon as a national monument. The Antiquities Act made sure that many beautiful places, historic settings, and natural resources in the United States would be protected.

Called to Serve

Theodore Roosevelt wrote, "There can be nothing in the world more beautiful than the Yosemite, the groves of giant sequoias and redwoods, the Canyon of the Colorado, the Canyon of the Yellowstone, the Three Tetons; and our people should see to it that they are preserved for their children and their children's children forever, with their majestic beauty all unmarred."

Concept Check

1. What arguments do you think Roosevelt used to win the debate over saving lands of historic value?
2. Do you think the Antiquities Act should only be used by the president? Explain.

Tidal Power

Almost everything we do today requires some form of energy. Listening to the radio, using your computer, or cooking dinner all use energy. Have you ever thought about where that energy comes from? What would happen if all of a sudden your source of energy was no longer available? What could you use instead? Enter alternative energy resources. These resources use the energy of the Sun, wind, or water, rather than a fossil fuel. Tidal energy is one form of alternative energy that has a lot of potential and a lot of controversy surrounding its use. Similar to the hydroelectric power generated by water flowing across a dam, this energy is generated by the natural movement of the changing tides.

Along nearly every ocean shore, the tide changes twice a day. Along a wide stretch of sand, you might not feel the power of the water as it slowly rises toward shore. But, what if the water were channeled into a narrow area where it could flow in and out much faster? Scientists have developed several types of technology that use this fast-moving water. In each case, the flowing water turns a set of turbines which, in turn, causes an electric generator to produce electricity. One type of technology is called a tidal barrage or tidal barrier. It blocks flowing water along the ocean shore or at a river's mouth. Water fills the area behind the barrage gates.

When the water level is high enough, the gates are opened, the water rushes in or out, and the turbines turn. Another technology is a tidal fence. A tidal fence works like giant turnstiles, like the ones you might go through when entering an amusement park. Again, moving water currents turn the turnstiles. The faster the water flows, the faster the turnstiles turn, and the more electricity is generated.

Unfortunately, this energy resource comes with some problems. Opponents argue that building these structures upsets natural habitats and can cause the animals and plant life in the area to die.

Trade-offs will need to be made on both sides as our search for reliable alternative energy resources continues. What arguments would you make for and against these turbines?

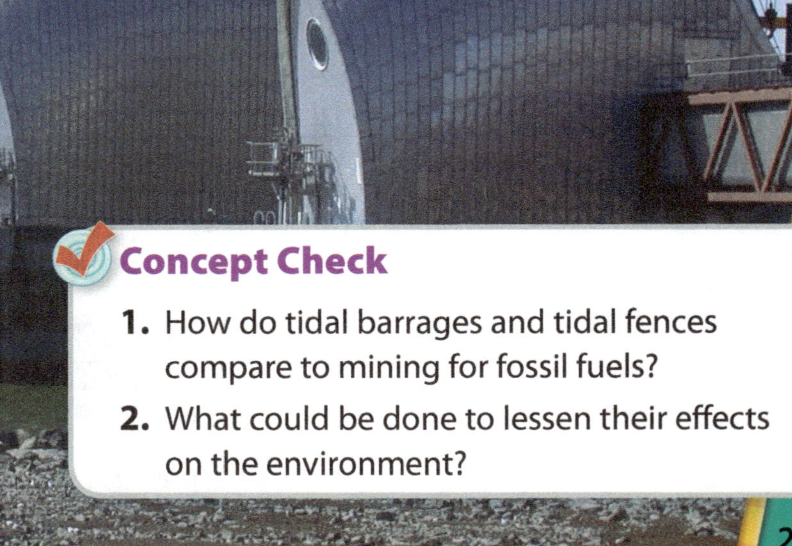

Concept Check

1. How do tidal barrages and tidal fences compare to mining for fossil fuels?
2. What could be done to lessen their effects on the environment?

Study Guide

Lesson 1

1. A mineral is a natural, inorganic solid with a definite structure and a specific chemical composition.

2. Color, luster, hardness, cleavage, fracture, and streak are some common properties that are used to identify minerals.

3. Quartz, feldspar, olivine, pyroxene, amphibole, mica, calcite, and halite are the most common minerals on Earth.

Lesson 2

1. Rocks are made of two or more minerals.

2. Igneous rocks form from magma. Intrusive igneous rocks cool slowly beneath Earth's surface and have large crystals. Extrusive igneous rocks cool quickly on top of Earth's surface and have small crystals.

3. Sedimentary rocks form when pieces of Earth materials are compacted and cemented. They can be clastic, chemical, or organic.

4. Metamorphic rocks form when heat, pressure, and/or hot fluids change existing rocks. They can be foliated or non-foliated.

5. The rock cycle describes the various processes through which rocks form.

Lesson 3

1. An ore is any solid Earth material that can be mined at a profit.

2. Metallic mineral ores are used in making items of beauty, coins, electronics, and construction. Metallic minerals also have medical applications.

3. Nonmetallic mineral ores are used in construction, abrasives, and glass.

4. Ores and other Earth materials are mined at the surface or deep within Earth. This can harm the environment unless care is exercised as mining takes place.

5. Renewable resources can be replaced in a relatively short period. Nonrenewable resources are used faster than they can be replaced.

6. Conservation is the wise use of the natural resources.

Show What You Know

Visualize It Complete the following concept map.

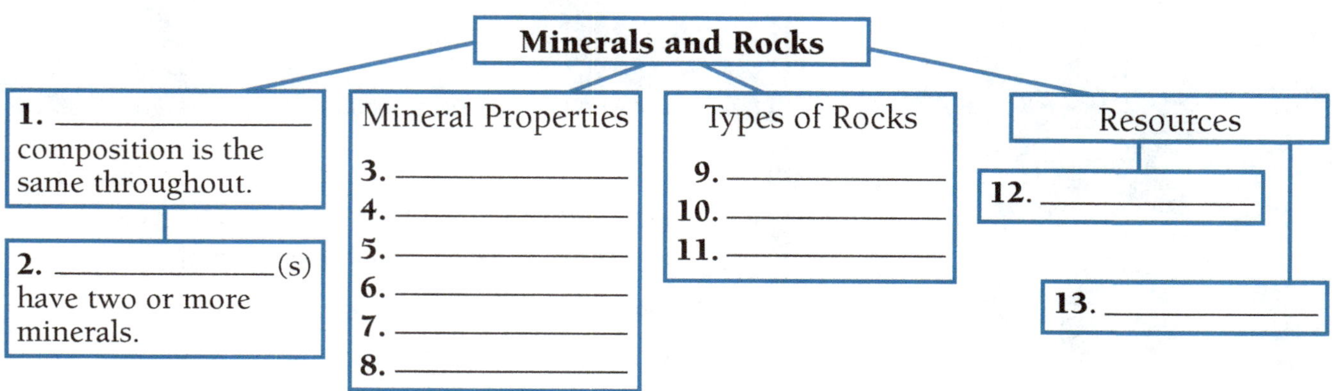

Explain how each pair of terms is related.

14. cleavage—fracture
15. streak—color
16. texture—metamorphic rock
17. ore—quarry
18. coal—nonrenewable

Multiple Choice
Choose the best answer.

19. What is a mineral's luster?
  A. resistance to being scratched
  B. value when it is mined
  C. color on a streak plate
  D. the way it reflects light

20. What rock forms from a lava flow?
  A. foliated metamorphic rock
  B. non-foliated metamorphic rock
  C. clastic sedimentary rock
  D. extrusive igneous rock

21. A rock that is made of large mineral crystals is a(n) __________.
  A. intrusive rock   **C.** foliated rock
  B. clastic rock   **D.** extrusive rock

22. Which processes produce sediment?
  A. weathering and erosion
  B. crystallization and deposition
  C. cementation and melting
  D. changes in heat and pressure

23. Why do nonrenewable resources need to be conserved?
  A. They are expensive to obtain.
  B. They are relatively cheap to use.
  C. They can cause pollution.
  D. They cannot be replaced quickly.

Check Point
Answer the following questions.

24. Why is color not always a useful property to use to identify a mineral?

25. A dark-colored mineral scratches a penny, but not glass, and breaks along smooth planes. **Infer** the mineral's hardness, state whether it has cleavage or fracture, and name the mineral.

26. **Compare** intrusive and extrusive igneous rocks and give an example of each.

27. **Classify** the rock shown here. Explain your answer.

28. **Sequence** the steps that could change granite to sandstone.

29. List three different rocks and explain how each can be used in everyday life.

Chapter 8

Our Dynamic Earth

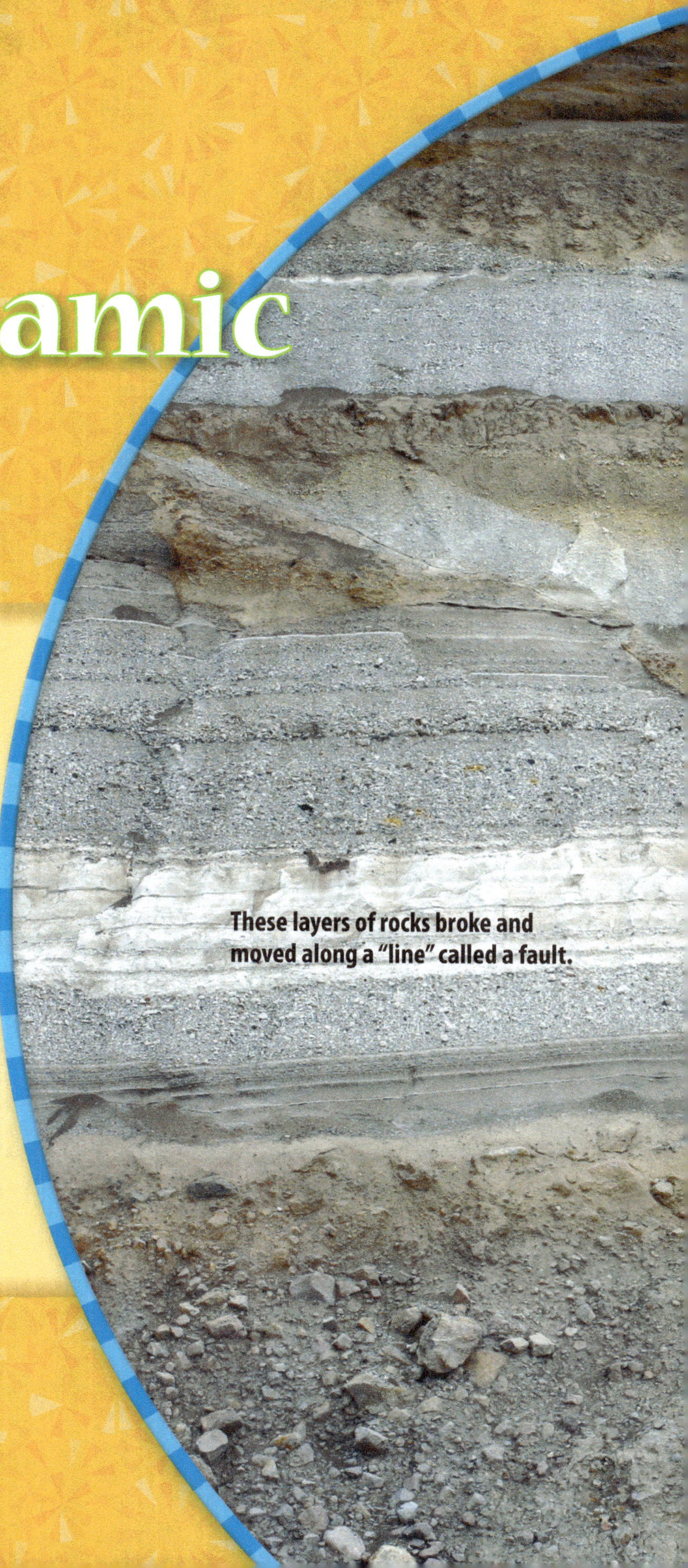

These layers of rocks broke and moved along a "line" called a fault.

Scripture Spotlight

God created a dynamic planet where Earth's surface moves and changes. You can strengthen your faith by studying volcanoes and earthquakes. You will study the following passages in this chapter.

Job 38:4–7 (p. 263)
Psalm 102:25 (p. 265)
Hebrews 11:3 (p. 267)
Nahum 1:3–7 (p. 273)
Matthew 27:50–53 (p. 279)

The Big Idea

Our Earth is always changing. A series of related processes cause these changes and work together to help create our environment and sustain life on Earth. Studying these processes helps us understand how Earth is changing and gives clues to how it may have changed in the past.

What caused these layers of rocks to break and move?

Inquiry Kick-Off

The explosiveness of a volcano depends in part on the thickness of the lava. What variables determine how far a liquid can spread? Examine the thickness of different household liquids to simulate the flow of lava. You will find the procedure in your *Science Journal*.

Objectives

- Describe Earth's crust, mantle, and core.
- Compare continental and oceanic crust.
- Differentiate between Earth's lithosphere and asthenosphere.
- Compare and contrast the inner and outer core.
- Explain why the theory of uniformitarianism requires long ages for Earth's history.

Vocabulary

core

mantle

crust

lithosphere

asthenosphere

uniformitarianism

catastrophes

catastrophism

neocatastrophism

short-age geology

Essential Question

What Is Earth's Structure?

If your first view of Earth were from space, what would it look like? Could you distinguish between different features? What do you think might be below Earth's surface? Is this material the same composition as rocks at Earth's surface? What are some clues that might help you answer these questions?

Earth's Layers `Explain`

Scientists cannot study Earth's interior directly. Instead, they study the material ejected during volcanic eruptions and gather seismic data during earthquakes. They use this information to learn about the composition of Earth's interior. With this data, they discovered evidence that material in Earth's interior is different from rocks at Earth's surface. From this evidence, scientists believe Earth's interior is made of layers.

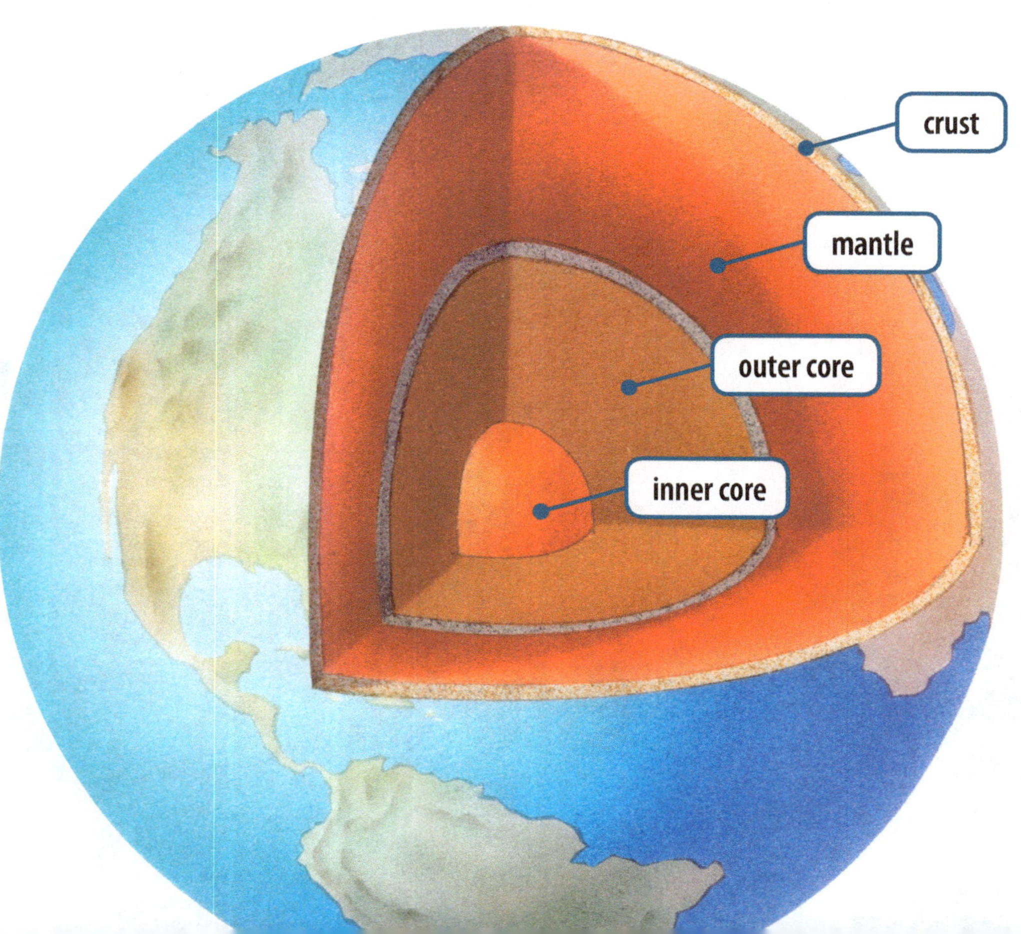

Earth has four main layers— the crust, the mantle, the outer core, and the inner core.

How does the composition of each layer affect its position in Earth?

262

Pick up any rock on the ground. What layer of Earth are you holding? Recall that Earth has four distinct layers. The inner layer is called the **core**. It is composed of a solid inner core and a liquid outer core. The inner core is mostly iron and the outer core is iron and nickel. The next layer is the **mantle**, a very thick layer of mostly solid rock composed mainly of oxygen, silicon, and magnesium. How does the composition of the mantle differ from the core? The surface of Earth is called the **crust**. The crust is composed of the elements oxygen, silicon, and aluminum. Earth's crust, mantle, and core vary in thickness, density, and composition. Study the figure on the previous page and refer to it as you read about Earth's layers.

What is God saying to Job in **Job 38:4–7**?

Earth's Crust

All of the land around you, as well as the land that makes up the ocean floor, is the crust. Compared to the diameter of Earth, is the crust very thick? Take a peach and slice it in half, being careful to cut around the pit. How do the layers in the peach represent the layers of Earth?

Earth's crust is the thinnest layer of the planet. There are two types of crust: continental crust and oceanic crust. Scientists classify the crust based on chemical composition. How can chemistry produce different properties in the crust? What might you measure to determine if the crust is one type or another?

Continental Crust

Continental crust has a density of about 2.7 g/cm³ and is made of mostly granite, sedimentary, and metamorphic rock. In some places, this type of crust is only a few kilometers thick. However, in other places it is much thicker. How thick do you think the crust is under the Himalayan Mountains? Research and find out.

Oceanic Crust

Oceanic crust has a mostly basalt composition, a density of about 3.3 g/cm³, and is much thinner than continental crust. Oceanic crust ranges in thickness from about 5 km (3 miles) to about 10 km (6 miles). When oceanic crust collides with continental crust, this much denser oceanic crust sinks below the continental crust.

Core Sampling

How do core samples of Earth's crust help scientists?

Materials
- dark sand
- light sand
- potting soil
- fine aquarium gravel
- cup
- straw
- water
- spray bottle
- plastic spoons
- white paper
- metric ruler

Procedure

1. Place 1 cm (0.5 in.) of each material in the cup. Mist each layer until damp and compress it.

2. Place a layer of another Earth material that is at least 1 cm (.5 in.) thick on top of the first one. Mist it with water, and compress it.

3. Place a third layer at least 1 cm (.5 in.) thick of Earth material on top of the second. Mist it with water, and compress it.

4. Place a fourth and final layer at least 1 cm (.5 in.) thick of Earth material on top of the third layer. You should now have four layers in the cup with a total thickness of at least 4 cm (2 in.).

5. Use the same process you have been using and add a couple more layers to the cup. Be sure to mist and compress each layer.

6. Use the straw to extract a core sample. Push it straight down through the layers of Earth materials.

7. Place your finger over the top of the straw and draw it out of the cup. Gently lay it on a sheet of white paper.

8. Wad a small piece of paper into a ball so it fits inside the end of the straw. Use the flat end of the skewer to gently push on the wad of paper in the straw and push the core out of the straw onto the paper.

9. Draw, label, and describe the layers in your core and in another group's sample.

Analyze Results

Observe and **compare** the different core samples.

Create Explanations

1. How do core samples of Earth's crust help scientists?

2. Why were there differences between your core sample and another group's core sample?

3. How could core samples of Earth's crust be used in real-world studies?

 **How do minerals differ in density?**

You can calculate the density of minerals present in the crust by using displacement. First, get a sample of quartz and hematite from your teacher. Use the balance to determine the mass of the quartz sample to the nearest 0.1 g. Next, fill a 100-mL graduated cylinder with 50 mL of water. Carefully place the piece of quartz into the water in the graduated cylinder by letting it slide down the side into the water so no water splashes. Subtract the beginning volume (50 mL) from the final volume and determine the volume of the quartz sample. Use the density formula $d = m/v$ to determine the density of the quartz. Repeat this process with the hematite sample. Based on your findings, which mineral is most common to the oceanic crust? Explain your reasoning.

Earth's Mantle Explain

Refer to the image below and locate the mantle. Earth's mantle is its thickest layer. It measures nearly 2900 km (about 1800 miles) from top to bottom. The mantle is made of igneous rocks that are rich in the hard, green, glassy mineral called olivine. The average density of Earth's mantle is about 4.5 g/cm³. The upper 100 km (about 60 miles) of the mantle is rigid. This rigid rock, together with the crust, is called the **lithosphere**. Below the lithosphere lies a partially molten section of the mantle that flows like a soft plastic. This part of the mantle lies 100 to 200 km (60 to 120 miles) below the lithosphere. It is called the **asthenosphere**. You will learn more about the lithosphere and asthenosphere in Lesson 2.

Scripture Spotlight

According to **Psalm 102:25** who laid the foundation of the Earth?

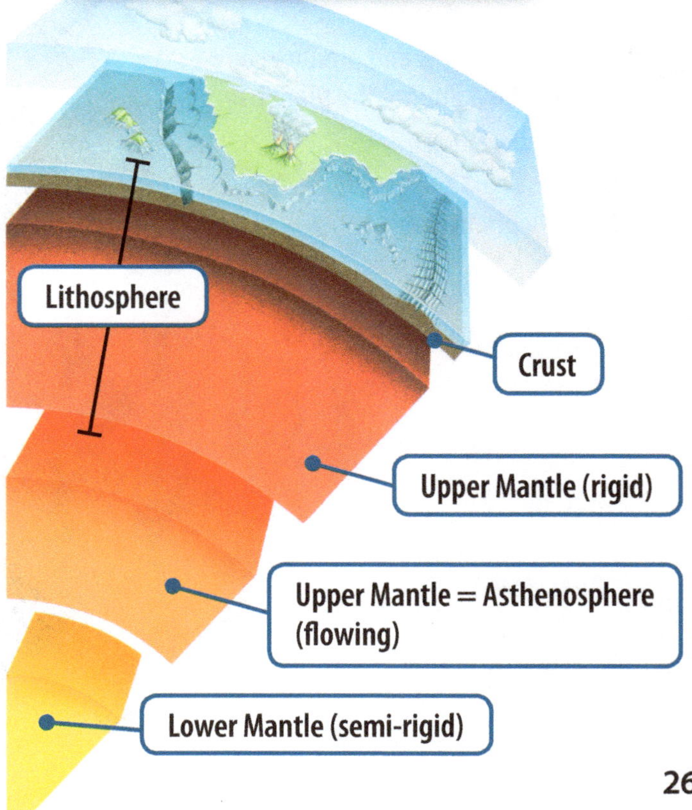

A cross-section of the upper mantle and crust. The crust and very top of the mantle form the lithosphere.

Why does the crust float on top of the mantle?

Earth's Core Explain

Recall that the core is the innermost part of our planet and is made mostly of iron and nickel. The metallic elements make the core very dense. In fact, the density of the core ranges from 10 g/cm^3 to about 13 g/cm^3. How does the density of the core relate to the density of the other layers in Earth? How does the core's density explain its position inside the planet? Why do scientists believe there are two parts to the core?

Outer Core

The outer core is about 2200 km (1375 miles) thick. The intense pressure and heat inside the planet keeps the material in the outer core liquid. As Earth rotates around its imaginary axis, the liquid outer core spins, creating Earth's magnetic field. Does it spin faster, slower, or the same as the other layers?

Inner Core

The inner core is about 1278 km (799 miles) in diameter. Unlike the outer core, this part of the core is solid. The intense pressure at the center of the planet is so great that despite the high heat the material there is a solid.

Earth's core rotates and produces a magnetic field around the planet.

❓ **What do you think life on Earth would be like if the planet did not have a magnetic field?**

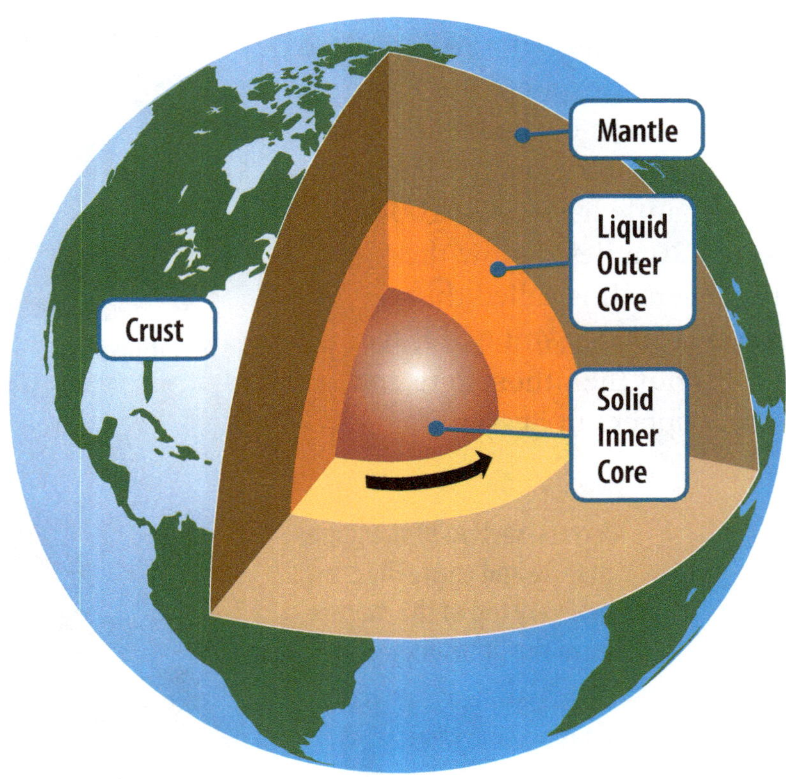

Math in Science

A scale model is one that uses the relative sizes of the parts of an object to create the model. Use your calculator to complete the chart below to find the scale thickness of each of Earth's layers. The crust has already been done for you.

Earth's Layers at a Scale of 1 cm = 1000 km (0.4 in. = 621 miles)		
Layer	Actual Thickness in km (miles)	Scale Thickness in cm (in.)
Crust	35 (22)	0.035 (0.022)
Mantle	2900 (1802)	
Outer core	2200 (1367)	
Inner core	1278 (794)	

Explore-a-Lab

Guided Inquiry

 **How can you make a scale model of Earth?**

Use your values from the **Math in Science** exercise and the materials provided by your teacher to make a model of Earth's structure.

Compare your model with the models of at least two other students. How can you improve your model?

Interpretations of Geology

While scientists agree on the structure of Earth, they do not agree on how that structure came to be. Each scientist's worldview affects his or her interpretation of the data.

All geologists look at the same rocks, mountains, and glaciers. In other words, they see the same data. But because of different worldviews, they reach very different conclusions.

Based on differing worldviews, there are two basic ideas to explain the origin of Earth's matter. Scientists who believe in Earth's natural formation suggest a story similar to this: After the big bang caused the Universe to come into being, Earth began as a huge spinning ball of dust. For millions of years, meteors crashed into Earth, each time making Earth bigger and hotter. As Earth grew hotter, it formed an ocean of lava, or liquid rock. Finally, after meteors stopped crashing into Earth, it cooled and the crust hardened.

Scripture Spotlight

According to **Hebrews 11:3**, where did our world come from?

Contrast this concept with the words of **Genesis 1:1**. "In the beginning God created the heavens and the Earth." This verse describes an absolute beginning when God purposefully created our Earth. Why can scientists not know for certain how Earth was formed?

Uniformitarianism

James Hutton (1726–1797), who is usually identified as the first modern geologist, introduced an idea called ==uniformitarianism==, which was made popular by another scientist named Charles Lyell (1797–1875).

Uniformitarianism included these basic ideas:

1. The laws of nature have always been the same.
2. Natural processes have always happened at the same rate.
3. We should explain the data using processes known to happen now.

Assuming that geological processes have always occurred at the same rate, these scientists suggested that Earth must be much older than previously thought in order for there to be enough time to explain the rock layers, mountains, and other geologic features of Earth. They suggested that instead of Earth being several thousand years old, it must be millions of years old. What evidence do you think these scientists used to support this idea?

Neocatastrophism

A different theory suggests that natural processes are sometimes interrupted by ==catastrophes==, or sudden, short-lived, violent events like floods, volcanoes, and earthquakes. The idea that these events can produce geologic change very quickly is called ==catastrophism==.

Modern geologists still accept the first and third uniformitarianism assumptions listed above, but there has been a shift away from the idea that natural processes have always happened at the same rate. Modern catastrophes have clearly demonstrated how quickly geologic changes can happen. ==Neocatastrophism== retains some of the original assumptions of uniformitarianism while accepting the evidence of catastrophe. While recognizing periodic catastrophes, geologists continue to accept long ages in between.

Short-age Geology

Short-age geology suggests that much of the geologic record formed in thousands, not millions of years. It begins with a biblical worldview that accepts the reality of God's intervention at specific times in history. Short-age geologists recognize evidence in the rocks that would be difficult to explain if Earth had been in existence millions of years. They believe that much of the data can be explained by the Flood described in the Bible.

Uniformitarianism	Neocatastrophism	Short-age Geology
The laws of nature have always been the same.	The laws of nature have always been the same.	The laws of nature have always been the same, however God can intervene in history.
Natural processes have always happened at the same rate.	Catastrophes have interrupted the natural processes.	Catastrophes have interrupted the natural processes.
Whenever possible, we should explain the data using processes known to happen now.	Whenever possible, we should explain the data using processes known to happen now.	Whenever possible, we should explain the data using processes known to happen now.

As you read the rest of the chapters in this unit, you will learn much more about geological processes. Remember that as scientists look at the same data, their interpretations may be different. How will you interpret the evidence?

Concept Check Assess/Reflect

Summary: What is Earth's structure? Earth's interior is layered. The deepest layer, the core, is very dense. It contains a solid inner core and a liquid outer core. The core's rotation produces a magnetic field that protects the planet. The middle layer is the mantle. The asthenosphere is the upper mantle that flows. The outer layer is the crust. It consists of oceanic (basaltic rock) and continental (granitic rock) varieties. The crust and upper mantle form a rigid layer called the lithosphere. Based on the data found in these layers, scientists have different interpretations of the age of Earth.

1. Compare continental crust with oceanic crust.
2. How do the lithosphere and the asthenosphere differ?
3. Explain how Earth has a liquid outer core and a solid inner core.
4. Compare and contrast the assumptions of neocatastrophism and short-age geology.
5. Which do you think would best model Earth's structure—an orange or an apple? Explain your choice.

Essential Question

How Does the Crust Move?

Look closely at a world map. Do any of the continents look as if they might fit together like the pieces of a puzzle? Which continent might fit next to North America? How can Australia fit in? If the landmasses once were together, how did they move apart? What evidences have you observed that suggest Earth's plates are still moving?

Continental Drift Explain

Alfred Wegener used the visual evidence of a world map, as well as several other pieces of evidence, to develop and publish the hypothesis of **continental drift** in 1915. According to this concept, Earth's continents have slowly drifted, by moving through the ocean floor. At one point, all of the continents were joined to form a single landmass. Wegener named this supercontinent Pangaea. By following the outline of the coastline, mirrored by the continental shelf, the continents fit together like pieces of a jigsaw puzzle. Other evidence for Pangaea, found in Africa and South America, includes identical fossils of Lystrosaurus, a barrel-chested dinosaur the size of a pig. In addition, some rock beds on these two continents are similar. Why do you think Wegener used these pieces of evidence to support his theory?

Wegener also used climate evidence to support his idea. Coal beds have been found in Antarctica, yet coal forms in warm, humid climates. Glacial rocks have been found in Africa. What do you think Wegener reasoned about the locations of Antarctica and Africa based on these pieces of evidence?

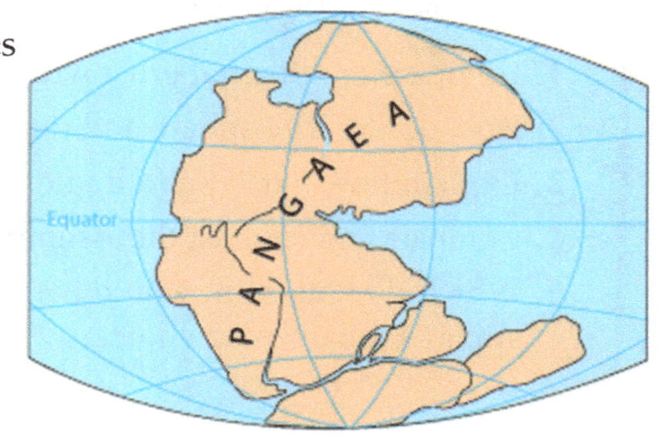

Wegener's proposed model of the supercontinent Pangaea.

Even though Wegener used many different types of evidence to support his hypothesis, he could not explain how the continents had moved. Because of this, his peers did not accept his idea. Can you think of some other reasons Wegener's ideas were not accepted?

Seafloor Spreading

Decades after Wegener presented his hypothesis of continental drift, other scientists found evidence to support his theory. Scientists noticed alternating strong and weak magnetic reflections while mapping the ocean floor. The oceanic crust that formed during a period of normal polar orientation produced a strong magnetic signal, while the oceanic crust that formed during a reversed period produced a weak signal. What is the dominant element in the oceanic crust?

It was hypothesized that crustal boundaries were the central point where rock on the ocean floor was formed. Scientists called these boundaries in the ocean floor *spreading centers*. Lava erupts at the spreading center, producing new oceanic crust. **Seafloor spreading** describes the process of forming new ocean crust. Now, the scientific community had a reasonable mechanism—seafloor spreading—to explain moving continents. Where do you think seafloor spreading occurs in the oceans today?

Evidences of Seafloor Spread	
Evidence	**Explanation**
Magnetic reversal	Changes in the magnetic pattern of the ocean floor indicate reversals in Earth's magnetic field over time.
Depth of sediment	The farther away core samples are taken of the ocean floor from a spreading center, the deeper the sediments are.
Radiometric dating	Radiometric dating of rocks also shows that rock ages increase with distance from a spreading center.

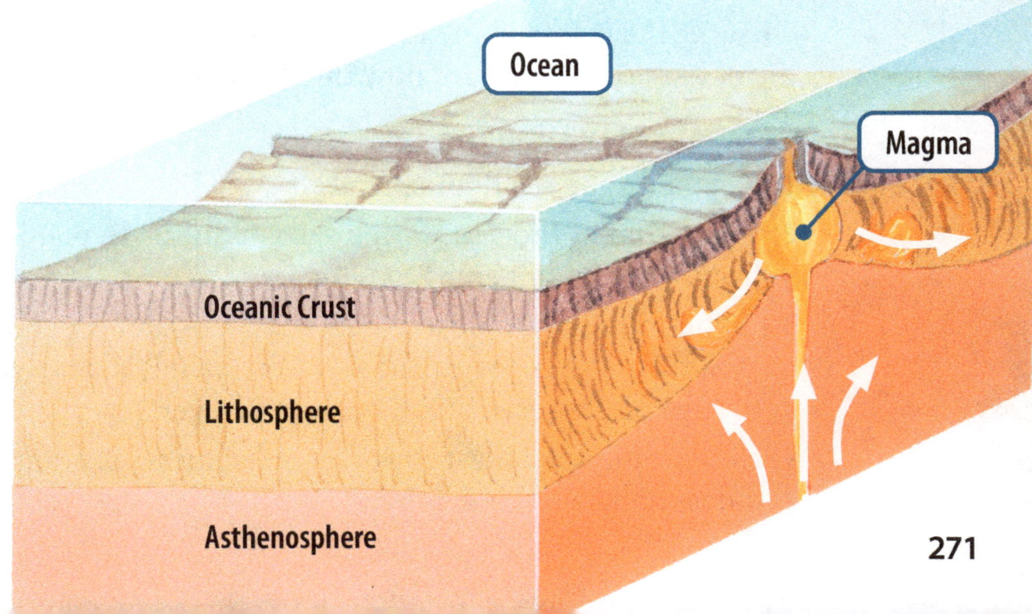

Plates move apart along spreading centers. Most of these centers are on the ocean floor.

What happens to Earth if new crust is forming at a spreading center?

Model Sea Floor Spreading

How does the sea floor spread?

Procedure

1. Fold one sheet of paper so the short ends line
   up. Unfold the paper. **Measure** and draw a
   line 11.5 cm (4.5 in.) long down the crease,
   leaving 5 cm (2.0 in.) from the edge at both
   ends of the line. Cut this line to form Slit B.

2. To the left of Slit B, **measure** 3 cm from the
   short edge of the paper, draw another 11.5 cm
   (4.5 in.) vertical line. Cut the line to form Slit A.

3. Repeat Step 2 to the right of Slit B
   to form Slit C.

4. Draw 11 stripes, 2.54 cm (1 in.)
   wide, on the second sheet of paper.
   The stripes must be parallel to the
   short sides of the paper.

5. Color the stripes alternating colors. Cut the paper in
   half to form two strips 28 cm (11 in.) long. Tape the
   short ends, as shown in the diagram, colored sides facing each other.

6. Push the striped paper up at B. Then, push one loose end down through
   A and the other loose end down through C.

Materials
- metric ruler
- 2 sheets of paper
  22 cm (8.5 in.) × 28 cm (11 in.)
- scissors
- colored pencils (2 different colors)
- transparent tape

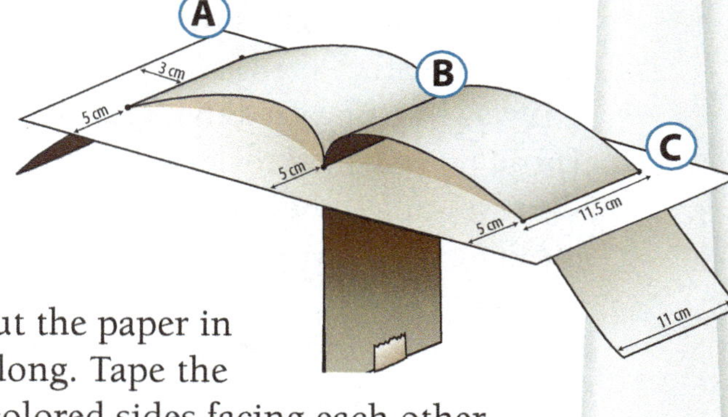

Analyze Results

Observe the colors of the rock that emerges up from inside Earth and onto
the ocean floor. **Compare** how this models the action of seafloor spreading.

Create Explanations

1. How does the sea floor spread?

2. What kind of process is occurring at Slit B? Explain how that differs
   from what is happening at Slits A and C.

3. If you were to sample and date the rocks along a colored strip starting
   at Slit B and moving to Slit A, what change, if any, would you see in
   the age of the rocks? Explain your answer.

4. How would the oldest rocks on the continents compare to the age of the
   oldest rocks on the sea floor?

Plate Tectonics `Explain`

Scientists used the concept of seafloor spreading as the mechanism to explain plate movement. They called their proposed idea the ==theory of plate tectonics==. This theory states that Earth's lithosphere is broken into large slabs of rock. These slabs are called tectonic plates. Plates move around on the partly melted section of the mantle called the asthenosphere.

Study the image of Earth below. Earth's lithosphere is broken into seven major tectonic plates, with each named for the continent or ocean that it holds. The plates move slowly over Earth's surface.

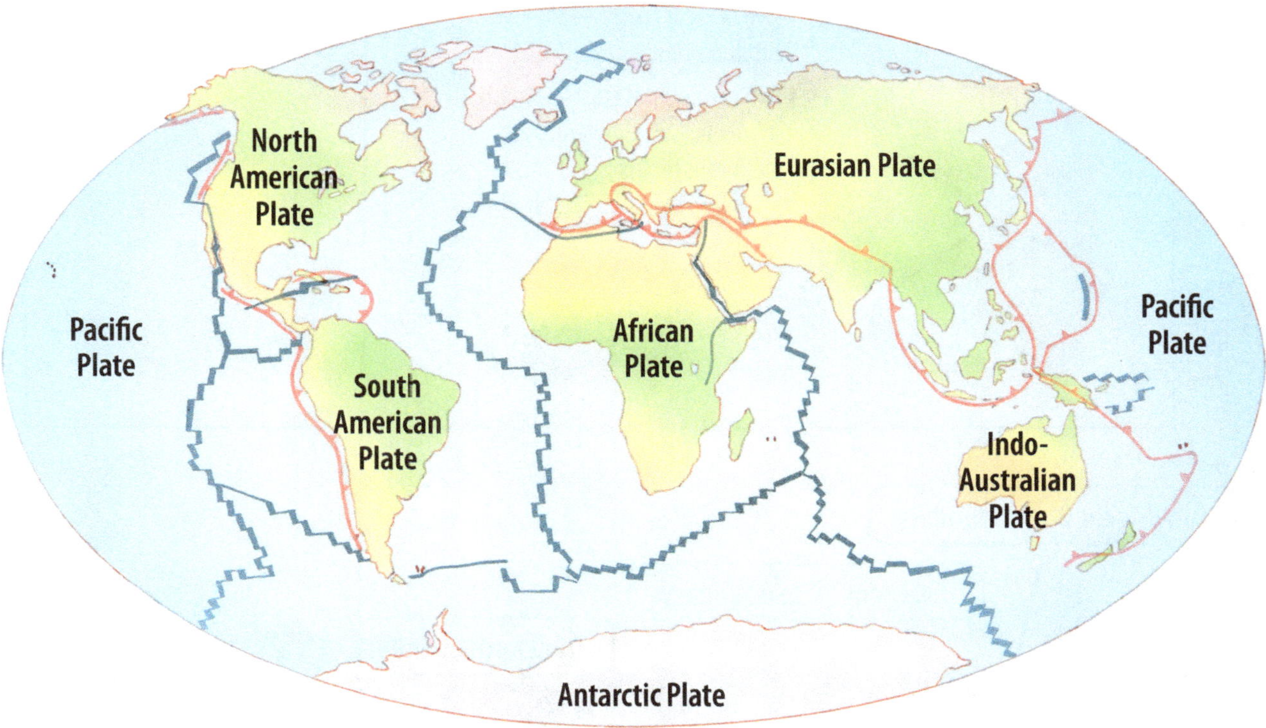

Look at the map of Earth. The blue lines show plate boundaries. The red lines show subduction zones. Locate the seven major tectonic plates.

Which plate do you live on? What countries and oceans are located on the plate where you live?

Types of Plate Boundaries

Tectonic plates meet at locations called plate boundaries. There are three types of plate boundaries: divergent boundaries, convergent boundaries, and transform boundaries.

Divergent Boundary	**Convergent Boundary**	**Transform Boundary**
• Where plates move apart • Found on land and in the ocean • Marked by a spreading center in the ocean • At a spreading center, new floor forms as oceanic crust moves away from the spreading axis • On land, a rift zone forms • Volcanoes form in these regions	• Where plates come together • Found on land and in the ocean • **Subduction** occurs when denser crust is pushed below less dense crust into the mantle, where it melts • The region that outlines where subduction occurs is called the *subduction zone* • Oceanic crust subducts under continental crust • Two oceanic crusts—older crust subducts under newer crust • Two continental crusts—neither subducts; push upward, form mountain chains	• Where plates move horizontally past one another • Found on land and in the ocean • Horizontal movement of rocks grinding past each other produces earthquakes

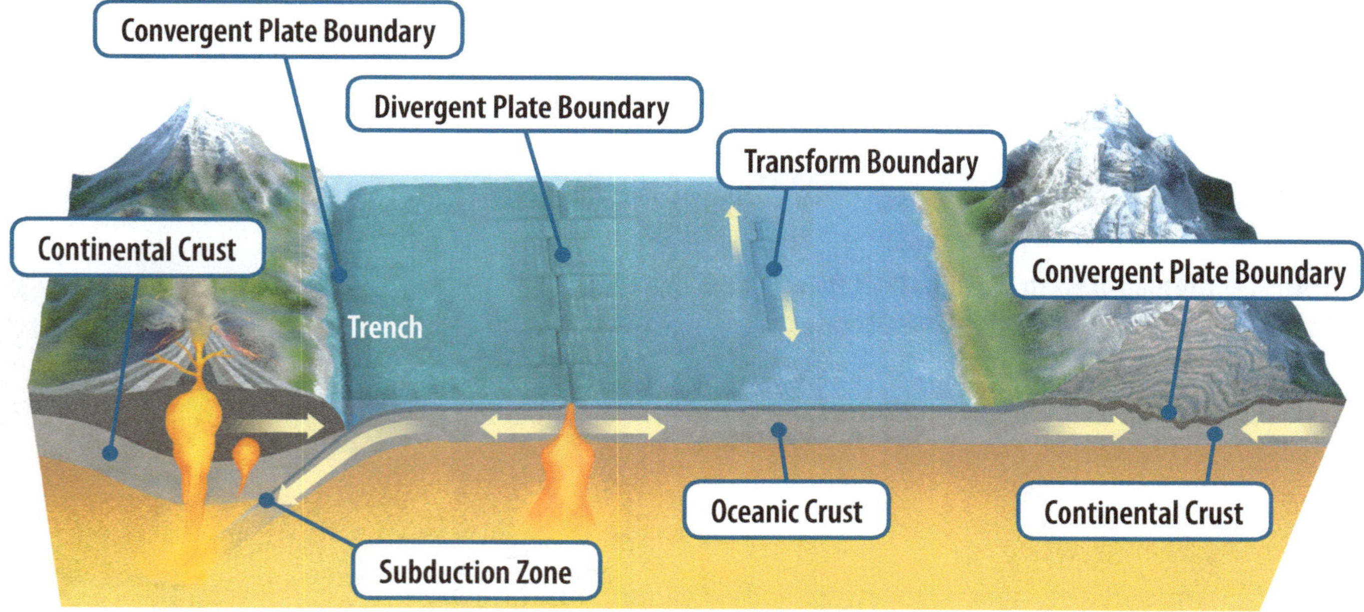

Locate three different plate boundaries in the diagram.

What physical characteristic of the crust would cause one to subduct under another?

Types of Boundaries			
	Divergent	**Convergent**	**Transform**
oceanic crust–oceanic crust	spreading center **Mid-Atlantic Ridge**	subduction zone **Marianas subduction zone**	lateral strike-slip fault **Blanca fracture zone**
oceanic crust–continental crust	does not form	subduction zone **Peru-Chile subduction zone**	lateral strike-slip fault **San Andreas fault**
continental crust–continental crust	rift zone **East African rift zone**	mountain chain **Himalayan Mountains**	lateral strike-slip fault **Haywood fault**

Check out your *Science Journal* to investigate the different types of convergent boundaries.

Extend

What Causes Plate Motion?

Tectonic plates move very slowly over Earth's surface. The slowest plates move at a rate of only about 2.5 cm (1 in.) per year. The fastest plate motion is a little more than 15 cm (6 in.) per year. How far might the slowest and fastest plates have moved since you were born? Do you think people can feel the plates move? Why or why not?

Plates move because of the action of convection currents that are created by the heating and cooling of the magma in the mantle. The circular movement of these currents creates two primary forces that move the plates. One force, *ridge push,* is created as newly formed ocean crust pushes adjoining crust away from the ridge. The second force, *slab pull,* occurs as the magma moves under the plate and locks onto the underside of the plate, pulling it along.

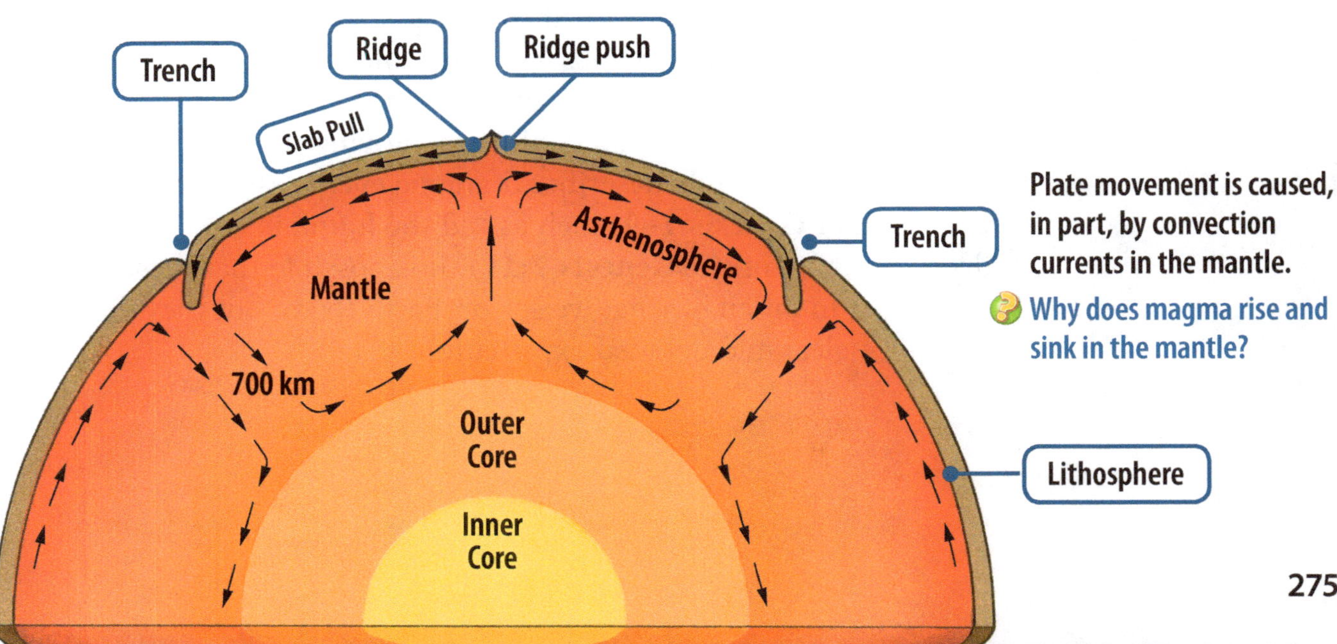

Plate movement is caused, in part, by convection currents in the mantle.

Why does magma rise and sink in the mantle?

Explore-a-Lab

Structured Inquiry

How can you make a model of plate boundaries?

To model plate boundaries, you will use frosting on top of wax paper to create an asthenosphere, graham crackers to represent continental plates, and flat fruit snacks, such as fruit leathers, to represent oceanic plates. Work with a partner, and use these materials to model the three types of convergent boundaries and a transform boundary.

Math in Science

Currently, scientists estimate that seafloor spreading along the Mid-Atlantic Ridge averages about 2.5 centimeters (1 inch) per year. How much new crust will be added to this ocean floor in 1000 years? Explain.

Earth's Crust and the Genesis Flood Explain

Although scientists do not fully understand the causes of plate motion, evidence indicates that dramatic continental movement has occurred in the past. The theory of plate tectonics explains much of what is currently observed in geology.

What could have caused a super continent, like Pangaea, to break up, with the resulting continents moving thousands of miles apart? The Bible describes dramatic events associated with the Flood of **Genesis 7–9**, such as "all the fountains of the great deep" bursting open (**Genesis 7:11**). It is possible that these events led to much of what we see on Earth today, but they do not provide a clear and detailed explanation.

Currently, the continental plates are moving slowly (1–4 cm per year). At that rate, it would have taken millions of years for them to arrive at their present locations, which is exactly what many scientists think happened. Scientists who interpret the early chapters of Genesis literally, however, think the movement must have happened much more quickly. These scientists believe that the violent events during the Flood and geological changes since the Flood might explain how Earth's plates could have moved apart much more quickly than they are moving today.

Although the theory of plate tectonics explains phenomena that we see happening today, it's much harder to know exactly how plate movements might have occurred during events that involved the intervention of God, like the Genesis Flood.

Concept Check Assess/Reflect

Summary: How does the crust move? Pangaea was thought to be a supercontinent made up of smaller continents that broke up and drifted apart. The discovery that the seafloor was spreading supported this theory of continental drift. The theory of plate tectonics, which states the Earth's lithosphere is broken into large slabs of rock was eventually developed as scientists learned that convection currents in the mantle were the force that moved the plates. There are distinct boundaries where these large slabs, or plates, meet. These plates move slowly on top of the asthenosphere, and form divergent, convergent, and transform boundaries. Some scientists think it took millions of years for Pangaea to break apart. Biblical scientists believe the dramatic events described in Genesis could have caused the continents to move faster.

1. Explain the role of a spreading center in plate tectonics.

2. What happens at each type of plate boundary?

3. Explain why oceanic crust always subducts below continental crust.

4. New crust forms along divergent plate boundaries. Yet, Earth is not getting larger. Why?

5. How might the Genesis Flood have affected the breakup of Pangaea?

Objectives

- Describe three types of faults.
- Differentiate among three types of earthquake waves.
- Explain how earthquakes affect Earth's surface.
- Compare and contrast cinder cone, composite, and shield volcanoes.
- Describe the structure of a volcano.
- Explain how volcanoes affect Earth's surface.

Vocabulary

foot wall

hanging wall

compression force

tension force

shear force

P-wave

S-wave

L-wave

caldera

cinder cone volcano

composite volcano

shield volcano

How do the types of faults relate to the types of tectonic plate boundaries?

Essential Question

What Causes Earthquakes and Volcanoes?

Where do earthquakes occur? How often do they occur? Do you live in an earthquake-prone area? One famous earthquake in the United States was the 1906 earthquake in San Francisco. It occurred along a transform fault boundary. The earthquake caused massive destruction throughout the city.

How would you describe a volcanic eruption? Are they always violent events? Mt. Tambora erupted in 1815 along the Java Trench subduction zone in the Pacific Ocean. The eruption released massive amounts of lava and ejected ash high into the upper atmosphere. The ash produced a cooling effect, resulting in a year without a summer.

Earthquakes and Faults Explain

Have you experienced an earthquake? How would you describe it? The Bible describes earthquakes as "great" and "violent" (**Matthew 28:2**, **Acts 16:26**, and **Revelations 6:12**). Earthquakes occur when the rocks in Earth's crust rupture suddenly. The rupture releases a massive amount of energy. This energy travels through the crust, producing violent shaking and movement in Earth's surface. The rupture occurs along large cracks in the crust. The earthquake's *focus* is the location where rock ruptures. Directly above the focus is the earthquake's *epicenter*. The cracks in Earth where earthquakes occur are called faults. Where do you think the most violent shaking occurs?

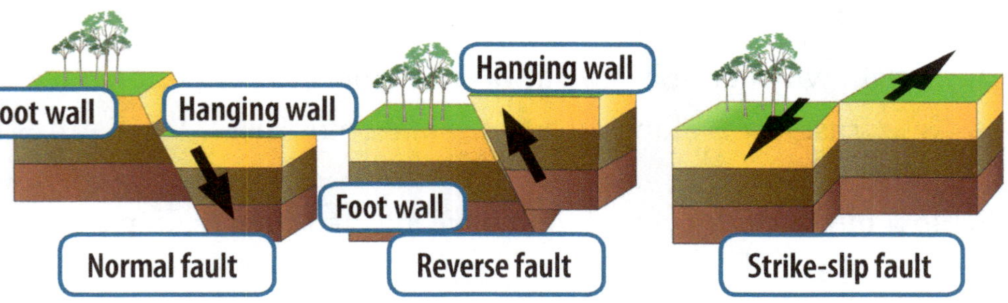

The movement of Earth along a fault can be horizontal or vertical. The horizontal movement produces a strike-slip fault. Vertical movement occurs when one portion of the crust moves upward and the other portion moves downward. Imagine that the fault is a diagonal slant through the crust. The rock below the fault is called the **foot wall**. The rock above the fault is called the **hanging wall**. The hanging wall can be higher or lower than the foot wall, depending on the type of fault. If the hanging wall moves up during the earthquake, it produces a reverse fault. If the hanging wall moves down, it produces a normal fault.

Different forces produce the movement along the three faults. Hold a block of modeling clay in your hands and push it together. What happens to the clay? Just as you can compress clay, this same type of **compression force** produces a reverse fault. Take the same clay block and pull it apart. What happens? This **tension force**, or pulling force, produces a normal fault. Take two blocks of clay and rub them against each other horizontally. What happens this time? A tearing force called **shear force** pulls on the rubbing surfaces of each block of clay. This force creates a strike-slip fault.

Earthquake Waves

Have you ever tried to break a rock? Rocks in the crust experience considerable forces that build up until the rock ruptures. The rupture releases energy that radiates from the epicenter. The energy travels as two main types of waves: body waves, such as **P-waves** and **S-waves**, and surface waves, such as **L-waves**.

Primary Wave (*P*-wave)	• also referred to as compressional waves • travels parallel to the direction of motion • travels through the planet • travels through solids and liquids • the fastest earthquake wave	
Secondary Wave (*S*-wave)	• travels perpendicular to direction of motion • travels through the planet • travels through solids • travels slower than *P*-waves	
Surface Wave (*L*-wave)	• travels across the surface of the planet • causes the crust to move side to side and up and down, like the waves created when you toss a rock into a lake or pond • travels slower than a *P*-wave or *S*-wave	

Finding the Epicenter

How is the epicenter of an earthquake determined?

Procedure

1. **Use the numbers** in the data table and the travel-time graph to determine the distance from each seismic station to the earthquake's epicenter.

Materials
- metric ruler
- drawing compass
- map and travel-time graph

Seismic Station	P-wave Arrival Times (min)	S-wave Arrival Times (min)	Distance from Epicenter (km)
A	1.9	3.9	
B	3.6	7.0	
C	3.3	6.3	

2. From the data table, the *P*-waves took 1.9 minutes to get to Station A. The *S*-waves arrived at the station in 3.9 minutes. Find these two times on the graph. Read the distance from Station A to the earthquake along the *x*-axis. **Record** this distance in the data table.

3. Repeat Step 2 for the other two seismic stations.

4. Use the map scale, ruler, and your compass to draw a circle with a radius of the value you recorded for Station A. Round to the nearest 0.5 cm for each radius. The circle must be centered around Point A on the map.

5. Repeat Step 4 for the distances to the other two seismic stations.

Analyze Results

Study the data displayed on your map. **Communicate** the location of the earthquake epicenter.

Create Explanations

1. How is the epicenter of an earthquake determined?

2. Why did you need data from three stations to locate the earthquake's epicenter?

3. Refer to the map in the previous lesson. What two major tectonic plates are interacting at the location of this earthquake's epicenter?

Measuring and Locating Earthquakes **Explain**

What type of damage is caused by earthquakes? The damage caused by an earthquake depends on different factors. The depth of the focus can impact the strength of the waves and the distance they travel. Typically, an earthquake with a shallow focus produces surface waves that travel farther and cause more damage. The type of rock, soil, or building foundation can also affect the amount of damage. How might damage be different in areas of compact soil and areas with loose sand? Why do deep-focus earthquakes cause less damage than shallow-focus earthquakes?

The *Richter scale* measures the amplitude of the largest seismic waves. The results are presented on a logarithmic scale. Each whole number on the scale represents 10 times the power. A 5.0 earthquake produces 10 times more shaking than a 4.0 earthquake. The *Mercalli scale* measures the power of an earthquake by the observed destruction caused by the shaking.

Scientists use seismographs located around the world to pinpoint earthquake epicenters. A seismograph records the intensity, or magnitude, of an earthquake. Scientists use the arrival time of the P- and S-waves from three stations around the world to triangulate the location of the epicenter.

Both the Richter scale and the Mercalli scale measure the intensity of an earthquake.

Which scale is more scientifically accurate?

Category	Mercalli Scale	Richter Scale (approximate)
I. Not felt	Not felt, detected only by sensitive instruments	1–2
II. Just perceptible	Felt by only a few people, especially on upper floors of tall buildings	3
III. Slight	Felt by people lying down, seated on a hard surface, or in the upper stories of tall buildings	3.5
IV. Perceptible	Felt indoors by many, by few outside; dishes and windows rattle	4
V. Rather strong	Generally felt by everyone; sleeping people may be awakened	4.5
VI. Strong	Trees sway, chandeliers swing, bells ring, some damage from falling objects	5
VII. Very Strong	General alarm; walls and plaster crack	5.5
VIII. Destructive	Felt in moving vehicles; chimneys collapse; poorly constructed buildings seriously damaged	6
IX. Ruinous	Some houses collapse; pipes break	6.5
X. Disastrous	Obvious ground cracks; railroad tracks bent; some landslides on steep hillsides	7
XI. Very disastrous	Few buildings survive; bridges damaged or destroyed; all services interrupted (electrical, water, sewage, railroad); severe landslides	7.5
XII. Catastrophic	Total destruction; objects thrown into the air; river courses and topography altered	8

Effects of Earthquakes (Explain)

What is your biggest concern during an earthquake? Look around you. If an earthquake happened right now, what objects would become hazards? Earthquakes often only last a few seconds, but the damage is enormous. As the ground shakes up and down and back and forth, objects can be hurled across the room and structures can crumble. Which would do more damage to a building, up-and-down or side-to-side movement?

Not only do earthquakes damage human-made structures, they often cause landslides, a movement of surface material down a slope. Liquefaction occurs when water-saturated soil loses strength during the shaking and flows like a liquid. The rapid displacement of crust along the coast can add energy to the ocean.

When an earthquake occurs in the ocean or along the coast, the vertical displacement of land can produce a *tsunami,* a series of waves with an average wavelength of 350 km (217 mi). A tsunami wave travels through the ocean as a shallow-water wave, reaching up to 201 m/s (450 mph). What happens as a tsunami wave begins to interact with the sea floor? Friction slows the wave, causing the wavelength to decrease and the wave height to increase. How does this change the energy in the wave? Why is a tsunami wave a dangerous earthquake aftereffect? Once the initial wave passes, a succession of waves will arrive every 5 to 90 minutes, depending on the period of the particular tsunami wave. The largest wave in the tsunami set is often not the first wave, but is one of the later waves.

Volcanoes

How safe would you feel standing on the flanks of Mt. St. Helen? What about Mt. Vesuvius? Both volcanoes produced spectacular eruptions, but most of the time, volcanoes are relatively quiet. To understand how and why a volcano erupts, you have to understand the anatomy of a volcano. Let's begin the dissection.

A volcano sits over a large *magma chamber* of molten rock, forming a conical structure. The volcano grows with each eruption of molten rock. The base of the volcano is a broad, circular structure. This molten rock is called magma. Do you recall what magma is called when it is expelled at Earth's surface? The magma travels up a long pipe, called a *conduit,* toward the surface. As the conduit nears the surface, it expands, forming the *throat.* The *vent* is the opening in the volcano at Earth's surface. At the summit of a volcano, the vent may form a large, bowl-shaped area called a *crater.* If a crater and the land around it collapse after an eruption, a large depression called a **caldera** forms. If the volcano becomes unstable, smaller conduits, called *branch pipes,* may transport magma to Earth's surface along the side, or flank, of the volcano. The branch pipes feed *parasitic cones* that form along the volcanic flank. If a branch pipe does not reach Earth's surface, the magma forms a *sill* or a *dike.* A sill is a layer of igneous rock that is horizontal to the surrounding bedrock. A dike is a layer of igneous rock that cuts through the surrounding bedrock diagonally.

The cross-section shows the structure of a volcano.

What would happen to the volcano if the chamber ran out of magma?

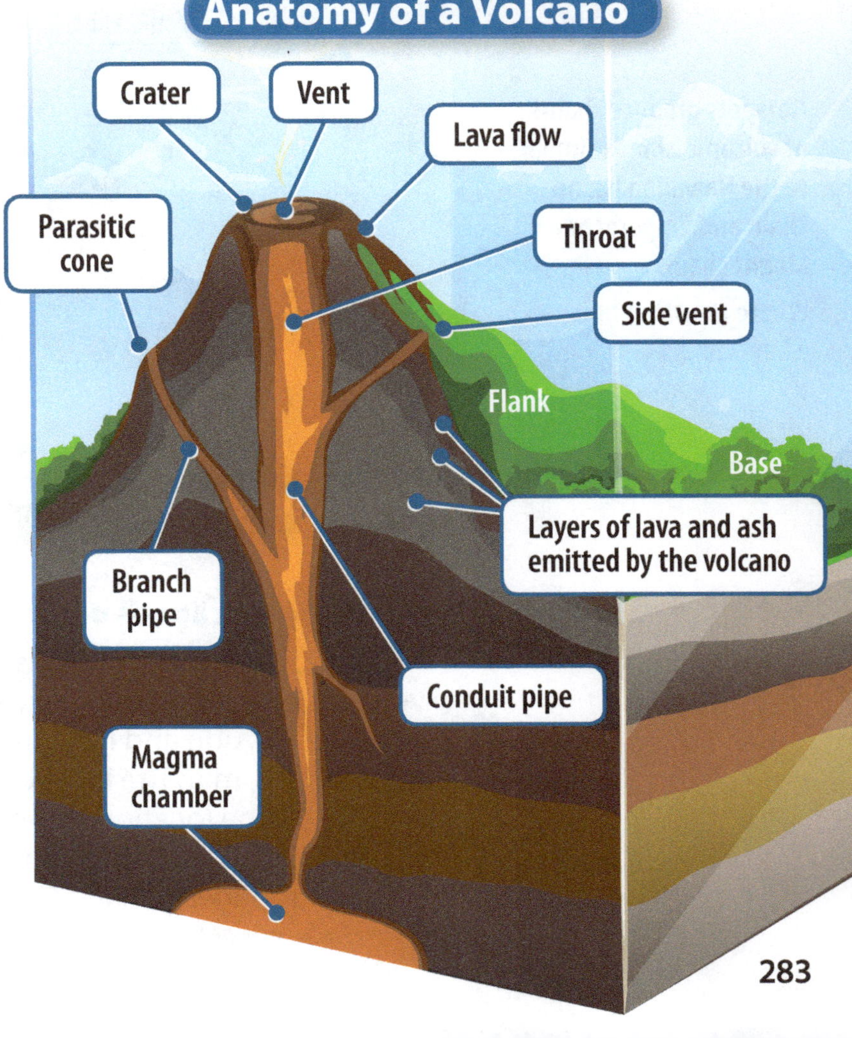

Types of Volcanoes Explain

Volcanoes can be classified according to their composition and their overall shape. There are three main types of volcanoes: **cinder cone**, **composite**, and **shield** volcanoes.

Volcano Type	Description
Cinder Cone	• small, cone-shaped mountain with steep flanks and bowl-shaped crater at summit • erupts volcanic ash and cinders that form the steep flanks of the volcano • are not associated with any specific plate boundary
Composite	• large mountain with steep sloping flanks • erupts thick (silica rich) lava, volcanic ash, and cinders • produces highly explosive eruptions • forms along convergent (subducting) plate boundaries
Shield	• broad mountain with gently sloping flanks • erupts runny lava • produces quiet eruptions • forms at hotspots

Hotspots produce chains of volcanic islands similar to the Hawaiian Island chain and Emperor Sea Mount chain.

Where are other hotspot island chains?

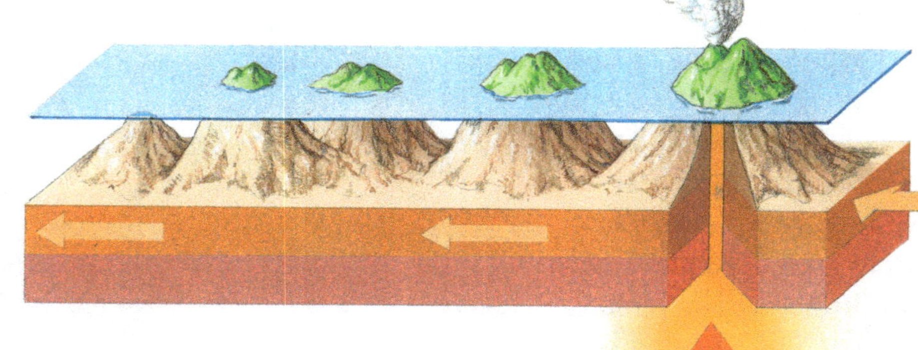

Explore-a-Lab

Structured Inquiry

Where do volcanoes occur?

As a group or individually, use the Internet to identify 10 volcanic eruptions. The events may be recent or historic. Take the latitude and longitude for each location. With this information, map the location of each event on a world map. Mark the locations of the eruptions and the volcanoes in color. Compare the results on the world map to a map of plate boundaries. What is the volcano hazard where you live?

Effects of Volcanoes

What is your biggest concern during a volcanic eruption? Is it the explosion, the lava, the ash, the intense heat? Not all volcanoes erupt the same way. Shield volcanoes are fueled by magma from the upper mantle. The lava that erupts from shield volcanoes has low silica content and low *viscosity*, or resistance to flow. This runny lava is called *pohoehoe lava*. The shield volcano usually has quiet eruptions, but the lava flow moves quickly. Unlike shield volcanoes, composite volcanoes have explosive eruptions. The lava that erupts from these volcanoes has high silica content and high viscosity. This thick, pasty lava that erupts from these volcanos is called *a'a' lava*.

In addition to the lava produced by volcanic eruptions, pyroclastic flows may also be produced. A *pyroclastic flow* is a mixture of deadly gases, like sulfur dioxide and carbon monoxide, and ash that races down the flanks of the volcano. Another danger is a *lahar*, a mixture of flowing water, dirt, and rock. Lahars usually form when a volcano erupts under snow or ice. The heat of the eruption quickly melts the snow and ice, causing torrents of water, mud, and debris. Research different examples of each hazard from volcanoes around the world.

Called to Serve

The Adventist Development and Relief Agency (ADRA) assists people suffering from natural disasters world-wide. Check out the ADRA website at www.adra.org to see what you can do to help bring relief to people all over the world.

Concept Check — Assess/Reflect

Summary: What causes earthquakes and volcanoes? Earthquakes result when rock ruptures. The rupture can produce a normal fault (tension force), reverse fault (compression force), or transform fault (shear force). The energy travels through Earth as P-, S-, and L-waves. Scientists map the arrival of the waves to locate the earthquake's epicenter. They measure the destructive force of the earthquake using the Richter and Mercalli scales. Scientists classify volcanoes as shield, composite, or cinder cone. Volcanoes form over a magma source along plate boundaries and hotspots. They produce numerous hazards based on the flow of the lava.

1. Describe some of the hazards faced by people living at the base of a composite volcano.

2. Explain how the Hawaiian Islands were formed.

3. Draw a normal, reverse, and strike-slip fault. Label the foot wall and hanging wall where appropriate. Draw arrows to show how the rocks move. Describe the forces at each fault.

4. List and describe the different types of earthquake waves.

Get to Know
Ms. Marie Tharp

In 1870, the French author Jules Verne wrote the famous novel *20,000 Leagues Under the Sea*. At the time, no one, not even scientists, had an idea of what lay below the ocean's surface. During the 1940s, that changed. Marie Tharp was an oceanographic cartographer. Her work paved the way for the acceptance of the theory of plate tectonics. An oceanographic cartographer makes maps of the ocean floor. Ms. Tharp's maps were the first comprehensive maps compiled of the ocean floor.

She was once quoted as saying, "I had a blank canvas to fill with extraordinary possibilities, a fascinating jigsaw puzzle to piece together." Little did she know that that was exactly what she was doing. The details that emerged from her maps helped support a new idea. This idea was that Earth's crust is broken into large pieces that fit together like a puzzle.

Her data came from sonar readings taken by ships as they crossed the ocean. As she plotted the data onto her blank paper, an amazing feature was revealed. It was a 64,375-km (40,000-mi) long ridge that circled the globe. Further study showed that the ridge was actually a rift with mountains on either side of it. Ms. Tharp thought it was similar to the East African Rift. She knew that earthquakes and volcanoes in that area showed that the land there was being pulled apart. At first, her colleague and research partner Bruce Heezen doubted her idea. But soon it became clear that she was right. More data showed that earthquakes were common along the ridge on the ocean floor. Marie Tharp published her completed map of the world's oceans in late 1977. Versions of this map are still in wide use today. Marie Tharp's work in science was groundbreaking. She made a major contribution to science long before women were widely accepted in any scientific field.

We can thank Marie Tharp and her detailed maps for the acceptance of the theory of plate tectonics. Her contributions will continue to enhance our knowledge of Earth's process for years to come.

Concept Check

1. What connection is described between the first studies of the ocean floor and the acceptance of plate tectonic theory?

2. How is Ms. Tharp's research inspiring to people who have different thoughts and ideas?

Earth's Liquid Outer Core

Early scientists suspected that Earth was made of several different layers. They thought that each layer had a different composition and temperature. But they had no way to prove this. Even today, we do not have the technology to drill into the different layers to sample them. How do modern scientists know that Earth is made of distinct layers? We can thank the scientists who studied the seismic waves that are released during earthquakes.

When an earthquake occurs, energy is released in the form of seismic waves. Scientists discovered that these waves can be tracked as they travel from their point of origin to a seismographic station some distance away. But even though the seismic waves leave the point of origin at the same time, the different types of waves arrive at different times. This led scientists to think that seismic waves travel through different Earth materials at different speeds. Scientists positioned seismographs around the world to accurately track seismic waves and pinpoint the location of earthquakes.

Researchers paid attention to the way seismic waves travel through Earth. *S-waves* could not pass through liquid and *P-waves* slowed when traveling through liquid. Both waves would refract off of certain boundary layers. These observations helped identify the layering by the rate and way that waves passed through or were refracted.

Research showed that *P*-waves travel through solids and liquids. These waves were detected at every station. But, in some locations, *S*-waves did not arrive. In other places, the waves arrived as if they had curved instead of traveling in a straight path. This led scientists to discover that part of Earth's interior was liquid. It was soon theorized that this liquid portion is the outer core.

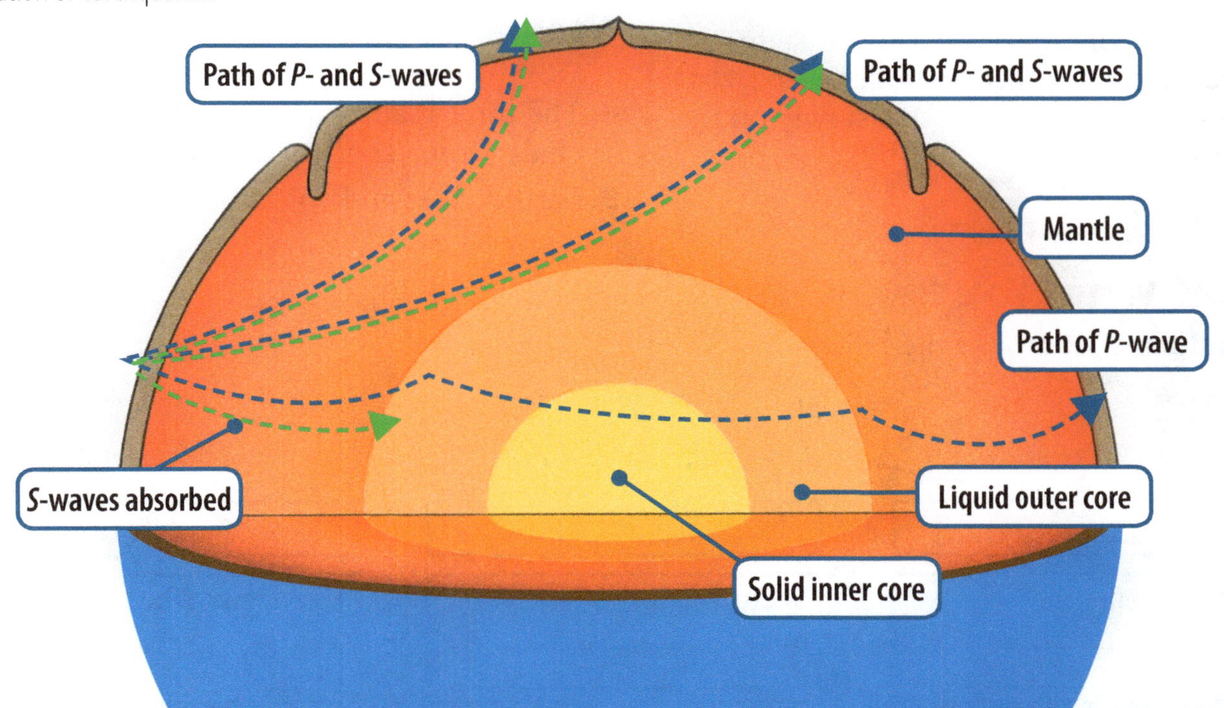

Concept Check

1. Why were scientists interested in the travel of seismic waves?
2. What does the curved path of a *P*-wave or *S*-wave indicate?

Study Guide

Lesson 1

1. Earth has three main layers—the crust, the mantle, and the core.

2. There are two types of crust—continental crust and oceanic crust.

3. Earth's crust and upper mantle make up the lithosphere. This layer floats over the partly melted upper mantle, called the asthenosphere.

4. Earth's metallic core has a solid inner core and a liquid outer core.

Lesson 2

1. Continental drift is the idea that the continents have moved slowly over time.

2. Plate tectonics is a theory that states that Earth's surface is broken into slabs called plates, which move slowly on the asthenosphere.

3. Plate boundaries include divergent, convergent, and transform boundaries.

4. Seafloor spreading provides the mechanism to move plates across the planet's surface, along with the forces of ridge push and slab pull.

Lesson 3

1. Earthquakes occur when rock breaks under extension, compression, or shear force, resulting in normal faults, reverse faults, and transform faults.

2. Scientists use seismic waves (*P*-waves, *S*-waves, and *L*-waves) to identify the epicenter of an earthquake.

3. Volcanoes often form along plate boundaries and hotspots.

4. The three main types of volcanoes are cinder cone, composite, and shield.

5. Earthquakes and volcanoes produce many hazards.

Show What You Know

Visualize It Complete the following concept map.

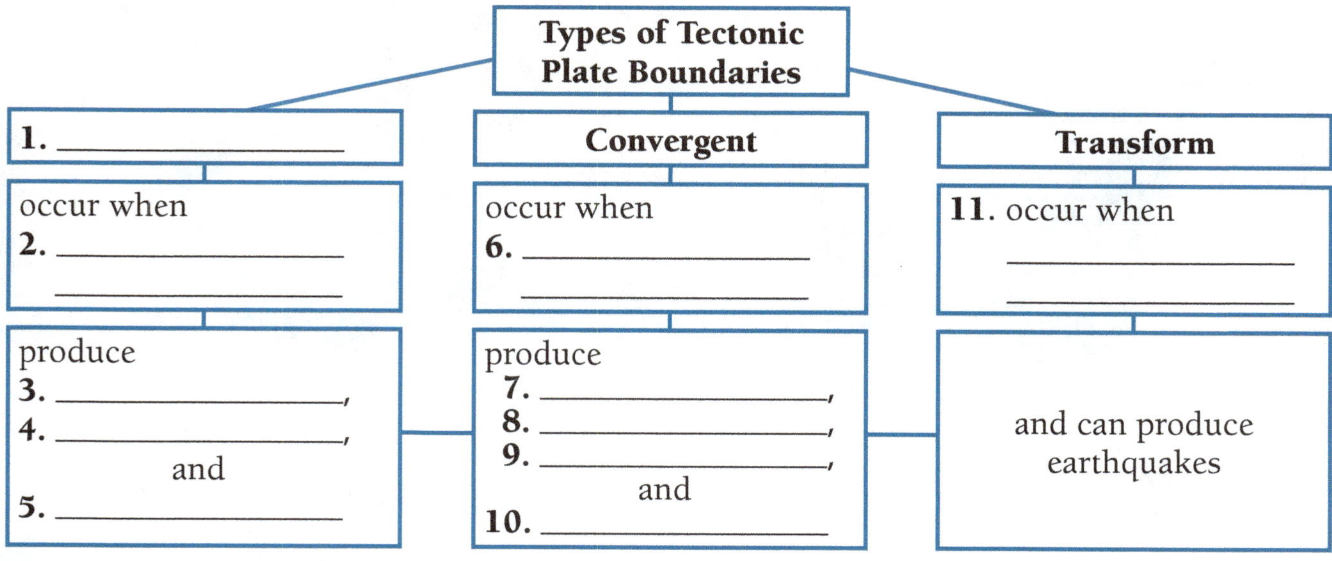

Vocabulary Check

Explain how each pair of terms is related.

12. lithosphere—asthenosphere

13. divergent boundary—convergent boundary

14. caldera—crater

15. earthquakes—faults

16. volcanoes—lahars

Multiple Choice

Choose the best answer.

17. Which is true about continental crust?
  A. thinner than oceanic crust
  B. denser than oceanic crust
  C. made mostly of granite
  D. the thickest layer of Earth

18. Which of these is true of Earth's core?
  A. The outer part is solid.
  B. The inner part is partly melted.
  C. The outer part is the thickest layer of the planet.
  D. The inner part is denser than the outer part.

19. Why was Wegener's idea of continental drift not more widely accepted at the time it was proposed?
  A. He could not explain how continents moved.
  B. He had very little evidence to support his idea.
  C. He was not a well-respected scientist at the time.
  D. He could not explain why rock units were not the same.

20. Along what type of plate boundary did the Appalachian Mountains form?
  A. continent–continent convergent boundary
  B. ocean–ocean convergent boundary
  C. continent–continent divergent boundary
  D. ocean–ocean divergent boundary

Check Point

Answer the following questions using complete sentences.

21. Explain why crust beneath mountains is thicker than crust under a plain.

22. Explain how Adventist scientists use short-age geology to explain Earth's history.

23. Why does neither plate at a continental–continental boundary subduct as the plates move?

24. **Infer** why plate movements today are so slow.

25. The Arabian and African plates are separated by a divergent boundary. This boundary extends north and south down Africa's east side. It is often called a rift valley and has many lakes. **Develop a hypothesis** describing what you think may happen to this area as the plates continue to spread apart.

26. **Compare** cinder cone, composite, and shield volcanoes.

27. Draw and label simple pictures to show how the three types of earthquake waves cause rocks to move.

Earth's Surface Features Change

Scripture Spotlight

As you read about Earth's features and processes in this chapter, keep the biblical story of the Flood in mind. These passages are also related to the topic of this chapter.

Psalm 78:15–16 (p. 295)
Job 14:18–19 (p. 297)
Isaiah 40:15 (p. 299)
2 Chronicles 26:10 (p. 304)

2 Kings 3:16–17 (p. 308)
Isaiah 48:17–19 (p. 310)
Job 38:29–30, 37:10 (p. 311)
Isaiah 40:3 (p. 313)

The Big Idea

Water, glaciers, wind, and gravity are constantly causing the surface of our planet to change. They have left behind clues about the history of Earth.

This is the Grand Canyon. How do you think it formed?

A unique combination of erosional features is visible in the steep-sided Grand Canyon. It is hard to look at the immense gorge and not be curious about the geological forces that shaped this incredible landscape.

Inquiry Kick-Off Engage

The United States boasts some of the most amazing features on the planet. How do these features change over time? In your *Science Journal*, you will study some processes that lead to the transformation of features into amazing landscapes.

Essential Question

What Are Weathering, Erosion, and Deposition?

Earth's surface is constantly being changed by water, glaciers, wind, and gravity. Sometimes the change is slow. As you read about weathering, erosion, and deposition, remember that cataclysmic events can produce changes that are rapid and dramatic. What event in the Bible relates to these processes?

Changes to Earth's Surface `Explain`

You know that Earth's crust and upper mantle are broken into huge slabs called plates. Forces deep within Earth cause these plates to move and change the surface. For example, most mountains form when plates come together. Many volcanoes form when plates collide as well as when plates move apart. All plate motions can cause earthquakes that shake the ground. Why are you not always able to see these movements?

In addition to plate movements, other processes change the planet's surface. These are agents of change that are at or very close to Earth's surface.

Weathering

In Chapter 7, you learned that weathering is any process that breaks rocks into smaller pieces. Now you will find out that there are two types of weathering—mechanical weathering and chemical weathering. Which one do you think happens more often? Why?

Mechanical Weathering

Mechanical weathering, also called physical weathering, is the movement or breakdown of rocks. Mechanical weathering can take many forms.

Type of Mechanical Weathering	Definition
Abrasion	The movement of rocks or sediment past one another causes rocks to break down.
Pressure	Rapid or sudden removal of rocks reduces pressure on the underlying rock, causing it to expand and fracture.
Freezing	Water freezes in the cracks of rocks and expands, pushing out on the rocks and breaking them into smaller pieces.
Temperature	The repeated expansion and contraction created by heating and cooling of rocks weakens rocks.
Salt wedging	As water flows into the cracks of rocks and evaporates, the remaining salt crystals wedge in the cracks, breaking the rocks more.
Animals	Animals that burrow into the ground cause mechanical weathering by breaking rocks and soil into smaller pieces.
Plants	Plant roots grow into cracks in rocks. As the roots become larger, they push outward on the rocks, breaking them into smaller pieces.

Why do you think *mechanical* is a good way to describe this type of weathering? Find a partner and look around the school and grounds for examples of mechanical weathering. Create a list of examples and identify the processes involved. Can humans cause mechanical weathering? If so, how?

Chemical Weathering

Chemical weathering is the breakdown of rocks by chemical means. Chemical weathering can take many forms. Can humans cause chemical weathering? If so, how?

Type of Chemical Weathering	Definition
Oxidation	Oxygen in the air reacts with many different minerals to form oxides—aluminum, copper, magnesium, and calcium, to name a few—in rocks.
Hydrolysis	Ions in water exchange with ions in a mineral.
Solution	Carbon dioxide dissolves in water to form a carbonic acid solution that eats away at many rocks and minerals, especially limestone.
Hydration	Reaction of a substance with water.
Organisms	Chemicals, like acids, are released by plants as they grow or decay.

Morphing Landscapes
How do landscapes change?

Procedure

1. With a partner, cover a board with plastic wrap. Tape it on the backside with duct tape and lay it flat.

2. Cover the entire surface of the plastic-covered board with soil and pat it into place.

3. Position the ring stand so that it is above the soil. Fill the container with water and place on a ring stand shelf. Have your partner hold the filled container in place.

4. Insert a tube into the container of water and have your partner hold it in place. Place the open end in your mouth and apply suction like on a straw to create a stream of running water.

5. Direct the water onto the soil at the end of the board in one spot and allow it to flow for 30 seconds.

6. **Observe** what happens. **Record** the observations in your *Science Journal*.

7. Repeat Steps 1–6, only this time raise the board with a small block of wood to create a slope.

Materials
- large plastic container
- water
- long sheet of plastic wrap (to cover board)
- duct tape
- potting soil
- a ring stand and ring
- a board 40–50 cm (15.7 in.– 19.7 in.) long
- baking pan
- a small rectangular wood block
- 1 m (3 ft) plastic or rubber tubing

Analyze Results

What was the overall difference between the two trials? During which trial did you witness the greatest change? Explain.

Create Explanations

1. How do landscapes change?

2. What does the wood board represent? Provide an example in nature. Explain.

3. How might this experiment represent a local or regional landscape? Explain.

 **How does weathering affect the surface of rocks?**

Obtain four different rocks from your teacher. Label the rocks A, B, C, and D. You will also receive four medicine bottles with droppers. The medicine droppers are labeled W, X, Y, and Z. Select one rock. Place a drop from bottle W onto the rock and observe. Use a piece of paper towel and wipe away the liquid. Record your results. Repeat this step for medicine droppers X, Y, and Z. Repeat this entire process for rocks B, C, and D. Remember to record all of your observations. What type of weathering, if any, occurred? If this happened on a much larger scale, like a hillside, would the landscape change?

Many factors affect weathering. Climate is one of these. Rocks in warm, wet climates weather faster than rocks in cold, dry climates. Why? The table below provides a list of factors that affect weathering. How do temperature and moisture play a role in weathering? Limestone and marble, for example, weather more easily than granite and basalt. The minerals in limestone and marble react more readily with acidic water than the minerals in granite and basalt. What other factors affect weathering? How could the size of a rock or a crack in a rock affect weathering?

Scripture Spotlight

What is described in **Psalm 78:15–16**? Do you think it was like mechanical weathering? Chemical weathering? Both? Neither?

Factors	Effect on Weathering
Climate	Rocks in warm, wet climates—faster Rocks in cold, dry climates—slower
Vegetation	Rocks with many plants—faster Rocks with few plants—slower
Structure	Rocks with cracks and holes—faster Rocks with no cracks and holes—slower
Type of rock	Limestone and marble—faster Granite and basalt—slower
Surface area	Rocks with large surface areas—faster Rocks with small surface areas—slower

Moving water causes much of the erosion that takes place on Earth.

How are these boulders being eroded?

Check out your *Science Journal* for a Structured Inquiry that explores erosion by water. **Extend**

Erosion **Explain**

The materials produced by any type of weathering rarely stay in the same place. **Erosion** is the movement of weathered material. Water, both as a liquid and as ice, along with wind and gravity constantly move weathered materials to new places.

Erosion by Water

Water causes much of the erosion on Earth. Creeks, streams, rivers, and other bodies of water that flow through channels erode the land over which they move. Erosion by moving water depends on several factors. The two most important factors are the speed and amount of moving water. How does the size of a stream or river affect the rate of erosion?

Erosion by water is not always slow. A rapid release of water, can erode large amounts of rock and sediment. When Mount St. Helens erupted in 1980, it spewed large amounts of volcanic material out onto the mountain. Streams flowing over this new sediment rapidly carved a canyon 100 feet deep in less than a year. In a field near Walla Walla, Washington, another canyon more than 100 feet deep was carved through sedimentary rock in just a few days when water from an irrigation canal broke through.

Besides speed and amount of moving water, the shape of a stream or river also affects how it erodes the land. The fastest water moves along the outside of a river bend. Banks will be cut away and steeper in the outer bends, and the channel will be deeper. The slower water moves along the inside of the bend. What other factors would affect erosion caused by rivers?

The way that water flows and its speed can make changes in a stream or river.

What are some well-known examples of water erosion?

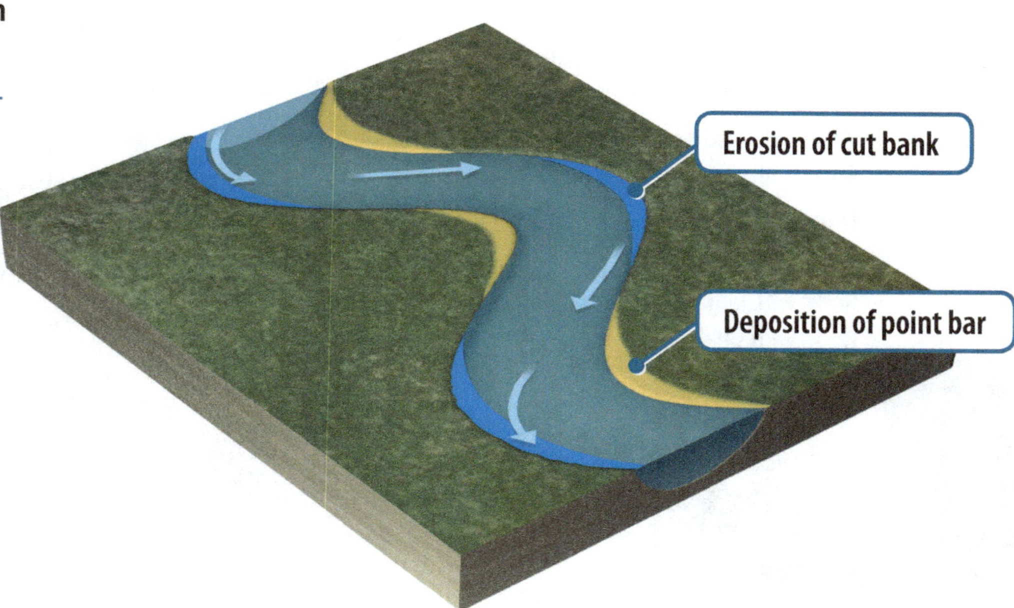

Waves and currents are other forms of moving water that erode the land. As waves break against the shore, they pick up and carry weathered materials back into the lake or ocean. Currents cause erosion as they move within a body of water. Along the Pacific Coast of Washington, Oregon, and Northern California, the waves crash onto the shore, moving tons of rock and sand. The same is true in many other coastal regions. How does this erosion by waves and currents change the shoreline?

Explore-a-Lab

Structured Inquiry

How does a glacier change a landscape?

Work with a partner to construct a three-dimensional mountain or continental landscape from modeling clay on a tray or baking sheet. Your teacher will give your group a model glacier made from a mixture of water and gravel frozen overnight. Remove the paper carton from the model.

Use gloves to scrape the glacier over the model landscape to observe the erosional effects of glaciers moving over a landscape. How will your glacier alter the environment? How closely does the behavior of this model resemble the motion of a real glacier?

Erosion by Glaciers

Glaciers are frozen masses of water that change the land beneath them. **Alpine glaciers** form high in mountain valleys. The sides or bottoms of these glaciers melt and refreeze. As this happens, they pluck rocks from the valleys and carry them to other places.

Continental glaciers form over large areas of continents. Currently, continental glaciers form in Earth's polar regions. These types of glaciers cause erosion as they move under their own weight. Rocks embedded in these glaciers scrape the land over which they move to produce deep grooves in the underlying surface.

These deep scratches were made when a glacier moved over this area.

What clues do the scratches leave about the size of any rocks the glacier may have been pushing?

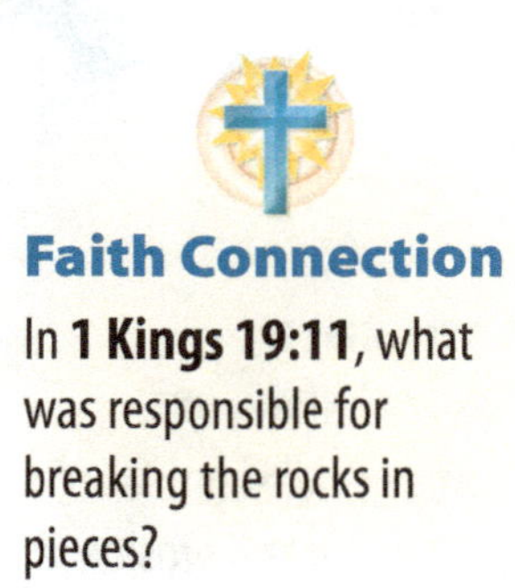

Faith Connection

In **1 Kings 19:11**, what was responsible for breaking the rocks in pieces?

Erosion by Wind

Have you ever been pushed about by a strong wind? What objects have you seen picked up and moved by wind? Did they move a long distance? Do you think the wind could move soil particles long distances?

Wind, or moving air, is another agent of erosion. Like moving water, erosion by wind depends on the speed of the wind. Wind erosion also depends on the time that the wind blows. A gentle wind, for example, will carry small particles only short distances. A stronger wind that lasts for hours, however, will move large volumes of the same particles over the surface and can create dust storms. Erosion by the wind is most effective in arid regions. Do you think wind or water causes more erosion in an arid region? Why?

Erosion by Gravity

Gravity also produces erosion. As with all erosion, several factors affect erosion by gravity:

- landscape slope
- water flow and saturation of soil
- type of rock or sediment

What are some examples of gravitational erosion that you have heard about in the news?

Great Sand Dunes National Park in Colorado boasts the tallest sand dunes in North America.

What force created these sand dunes? What is the source of the sand?

Deposition Explain

When flowing water, ice, and wind slow down or stop, they drop some or all of the materials they are carrying. Likewise, when loose materials reach a lower elevation, they will stop moving—at least for a time. The dropping of loose materials is called **deposition**. Like erosion, deposition is caused by water, ice, wind, and/or gravity.

Deposition by Water

As a stream or river rounds a bend or empties into another body of water, the water slows down. As the water slows, it drops some or all of the materials it is eroding. The first materials deposited from the water are the larger pieces, such as gravel. Gravel is any sediment that is greater than 2 mm (0.8 in.) in diameter. Next, a stream or river drops its sand-sized load. Sand is sediment that ranges from 0.5 mm (0.2 in.) to 2 mm (0.8 in.) in diameter. Silt, which is fine particles of sand and clay, is carried farther by the water and is deposited in very calm water. Where in a river will the most deposition occur?

Water in the form of waves and currents also causes deposition. You will learn more about coastal deposition in Lesson 3.

Deposition by Glaciers

Glaciers, like water, also deposit materials when they slow down, stop, or melt. However, unlike water, many glacier deposits are usually not well sorted. This is because such materials are deposited directly by the ice. Also unlike water, glaciers can carry huge rocks great distances. These rocks, which are called *erratics*, can be as large as small houses!

Deposition by Wind

Like water, wind erodes sediments of different sizes. Unlike water, however, wind usually only erodes sediment that is sand-sized or smaller. Like water, wind deposits the materials it is carrying when it slows down or stops. Also like water, wind carries smaller sediment farther than larger sediment. Where might wind be the most effective method of erosion and deposition?

Scripture Spotlight

What does **Isaiah 40:15** say about a grain of sand?

Check for Understanding

Think back to the *Structured Inquiry*. Where did the eroded material finally end up?

Deposition by Gravity

Like weathered materials eroded by water, ice, and wind, materials moving as the result of gravity are deposited when they reach a lower elevation. In some cases, the loose materials are eroded and deposited as a single mass. In other cases, such as in rockfalls, the materials are eroded and deposited mostly as individual pieces. You will learn more about these processes and the features they produce in Lesson 3.

The saturated soil becomes unstable and slumps under the force of gravity.

 **How are erosion and a mudslide similar?**

Concept Check   Assess/Reflect

Summary: What are weathering, erosion, and deposition? Earth's surface is affected by wind, water, ice, and gravity. Each of these factors picks up and transports sediment and rock during erosion. Erosion can occur during mechanical or chemical weathering. Deposition occurs when these materials stop moving. Both erosion and deposition change Earth's surface.

1. What is the difference between an alpine glacier and a continental glacier?

2. On a family vacation, you visit the Grand Canyon. Why is the Grand Canyon so deep, and what kinds of weathering actions produced its unique features?

3. Contrast the processes of *erosion* and *deposition*.

4. Imagine you are in a canoe floating down the river. You are approaching a curve in the river. Your mom tells you to steer toward the outside of the river. Explain why.

How Is Soil Formed?

Have you ever squished your toes in a muddy garden? What did it feel like? What materials combine to make mud? Is all mud the same? Why or why not?

Soil Formation **Explain**

Over time, materials are deposited and eventually mix with air, water, and decayed organic matter, or **humus**, to form **soil**. Soil forms from the surface downward and also from the bottom up. Soil forms as the result of weathering. Humus and other decaying materials in the top layer of soil continue to decompose and move lower into the soil profile. Rock particles from weathered rock lower down in the soil profile move higher up toward the surface. The two meet and mix to form the topsoil layer. Silty soil has much smaller particles than sandy soil, so it is smooth to the touch and retains water longer, and is fairly fertile. Weathered materials from the top and bottom layers of soil meet and mix to form the topsoil of what is called the A horizon of a soil profile. You will see this on the next page. How is a knowledge of soil helpful?

Many of the same factors that control weathering also affect soil development:

- climate
- vegetation
- type of rocks
- time
- topography
- biotic activity

A thicker layer of soil takes more time to form than a thinner layer of soil. Soils are thinner on steeper slopes than on level land. Soil develops and matures in areas where animals and other organisms burrow into the ground. Soils also develop more easily in places where plant roots are abundant. Why is this so?

Objectives

- Define soil.
- Compare and contrast the soil horizons.
- Describe common properties of soil.
- Describe the differences among types of soil.

Vocabulary

humus

soil

soil horizons

soil profile

bedrock

permeability

In autumn, leaves fall from the trees onto the ground.

What happens to the leaf litter that falls on the ground?

While many factors affect soil formation, most soils form distinct layers called **soil horizons**. These layers make up what is called a **soil profile**. A typical soil profile is shown below. A soil profile ends where the unweathered rock, or **bedrock**, meets the lowermost soil horizon.

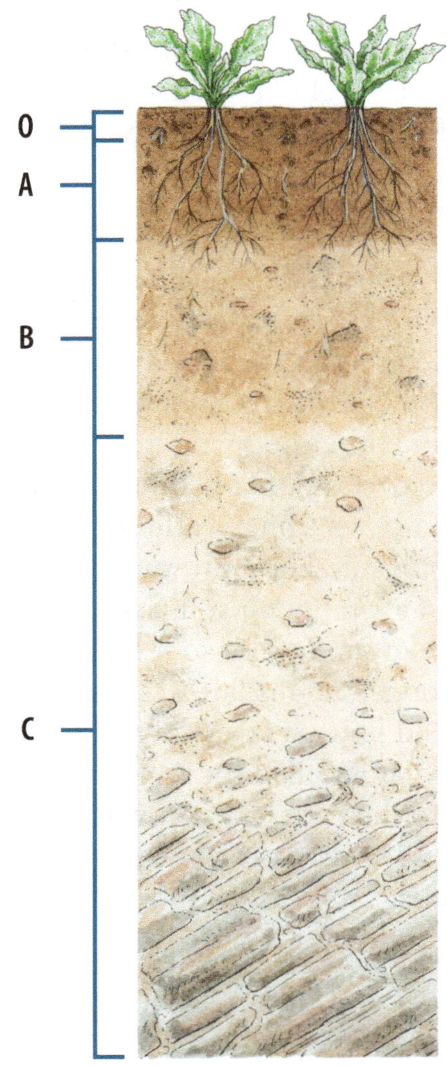

O Horizon

The O horizon includes leaf litter, humus, and animal parts. Although this layer contains the most organic matter, it is often the thinnest horizon in any soil profile. Why do you think this is called the O horizon?

A Horizon

The A horizon is a crumbly mixture of weathered rock and organic material. It is often called topsoil. Water moving down through the A horizon leaches, or removes, certain minerals from this soil layer.

B Horizon

The B horizon is subsoil mostly made up of very fine sediment. Some B horizons are rich in calcium or iron minerals due to the type of underlying rock. These minerals often form hard pans that do not allow good drainage.

C Horizon

The C horizon contains weathered materials as well as some bedrock. The C horizon usually does not contain any organic matter.

What does a soil profile that does not have horizon layers suggest?

Soils form all over the world. Their composition is based on the bedrock in the area. The climate of the region also contributes to how rapidly the soils form. Other factors that help soils form include organisms in the soil, like earthworms. Soil scientists estimate that in dry areas it takes several hundred years for 1.25 cm (less than 0.5 in.) of soil to form from parent material. In warmer regions that have wetter climates, soil forms at a much faster rate.

Comparing Soils

What characteristic of soil has the greatest effect on its permeability?

Procedure

1. **Measure** 50 g (1.8 oz) of each soil sample—a sandy soil with no humus; a clay-rich soil that contains some sand and humus; and a loamy soil mix of sand, clay, and potting mix.

2. Place each sample on a sheet of paper and **observe** under the dissecting microscope or with the hand lens. Draw what the soil particles look like in your *Student Journal*.

3. Attach the funnel onto a ring stand. Place a 500-mL (16-oz) beaker under the funnel.

4. Line the funnel with a paper coffee filter. Place one soil sample in the funnel.

5. **Measure** 100 mL (3 oz) of water in the graduated cylinder. Slowly pour the water into the funnel containing the filter and soil. Repeat this with another 100 mL of water. Be sure not to overfill the funnel.

6. Use the stopwatch to time how long it takes for the water to flow through the soil. Stop the stopwatch when the water stops dripping. **Record** your results.

7. Remove the filter from the funnel. Repeat this with at least three samples of each kind of soil. Construct a graph that displays the results.

Materials
- 3 soil samples
- hand lens or dissecting microscope
- paper coffee filter
- ring stand with funnel
- 2 beakers
- stopwatch
- graduated cylinder
- 1800 mL (61 oz) of tap water

Analyze Results

Make a table in which to **record** your observations for each soil sample. Write a short paragraph summarizing your observations.

Create Explanations

1. What characteristic of soil has the greatest effect on its permeability?

2. How would the sample permeability change if the more permeable sample contains 25% clay or silt? Explain.

3. Which of the soils is better suited for growing crops? Explain your answer.

Properties of Soils `Explain`

Like rocks and minerals, soils have different properties that can be used to describe and identify them.

Scripture Spotlight

2 Chronicles 26 describes a king of Judah who was only 16 years old. He was a successful warrior and a great builder. What does the end of verse **10** say about him?

Color	What colors of soil have you seen? Most soils are some shade of white, gray, brown, yellow, or red. A few soils are nearly black in color. What do you think makes soils such different colors? Can you tell from the color of the soil if it is good soil for growing crops? Why or why not?
Composition	Soil color is closely related to soil composition. Soils that are rich in iron are red or reddish brown in color. Soils rich in the mineral calcite are often white or gray. In general, soils that are rich in organic matter are black in color.
Texture	Soil texture describes the size of the particles in the soil. Sandy soils feel gritty when rubbed between the fingers. Soils rich in clay feel like baking flour. Silt is a mixture of sand and clay.
Permeability	**Permeability** describes the ease with which water will flow through the soil. Sandy soils have rapid drainage. Silty soils have moderate drainage. Clay soils have very slow drainage or no drainage.

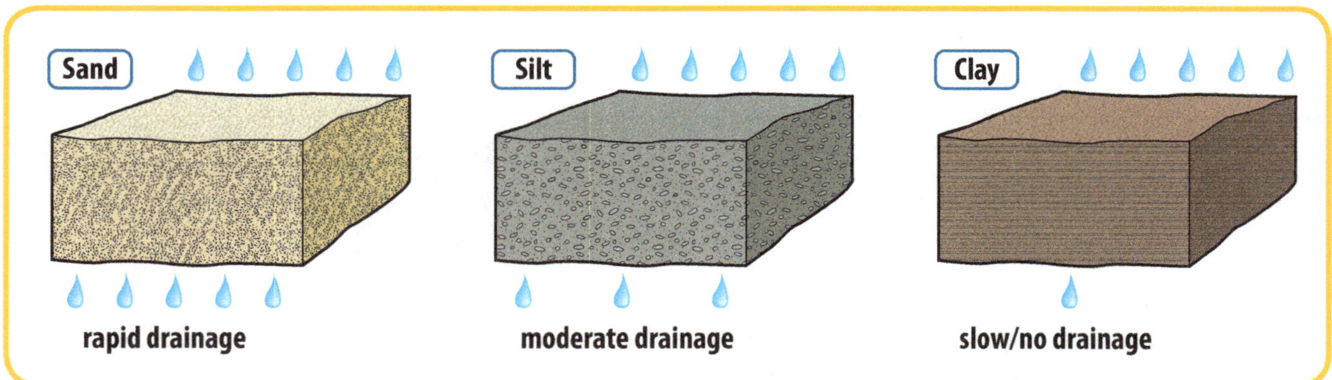

Sediment size affects how water flows through soil.

How does permeability differ for soils composed primarily of clay, sand, or gravel?

What features of the soil determine its ability to hold water?

Obtain four soil samples from your teacher. Label four cups "A," "B," "C," and "D." Measure the mass of a cup. Record the value. Fill the cup with 50 g of a soil sample. Measure the mass of the cup and the sample. Record the value. Place a funnel on a ring stand. Dampen and place a coffee filter in the funnel. Add the sample to the filter. Place a large plastic cup under the funnel. Add 200 mL (8 oz) of water to the sample in the filter. Allow water to flow through the soil and drain into the plastic cup.

Once the water has been added, let the sample sit until no more water drips out. Remove the filter and sample and place it back into its original cup. Measure the mass. Calculate the amount of water the soil holds by subtracting the dry mass from the wet mass. Divide the mass of the water by the original dry mass of the soil (50 g) to find out the number of grams of water retained per gram of soil. Record the value. Repeat this process for three additional samples. Which soil sample held the most water? Describe the color and texture of this soil sample.

Soil Types

Soil scientists have used the properties just discussed and a few more to identify more than 2400 different types of soil on Earth. *Alfisols* are soils that form on forest floors in humid climates. *Andisols* are soils that form from weathered volcanic materials. *Aridosols* are soils that develop in dry regions on Earth. *Gelisols* are cold-climate soils that form on top of permafrost. The permafrost lies within 100 cm (39 in.) of the soil surface. Soils that are rich in organic matter are called *histosols*. Histosols are common in areas such as bogs and swamps that are wet nearly year round. What kind of soil is common around your school? Which type of soil do you think would be best for growing a garden? Why?

Math in Science

Scientists studying some soils in the Ouachita Mountains of Arkansas found that 150–200 mm (6–8 in.) of soil developed in some areas over a period of 30 years. On average, how much soil was formed each year?

The soil type and its properties determine the kinds of plants that grow in an area. Can you remember anyone who has had difficulty in growing a particular plant in their yard or garden? Why do you think they had difficulty getting the plant to grow? The soils in different regions vary greatly. Even soils in the same region can vary. Soil on one side of a valley can be different from the soil on the other side of the valley. Sometimes even soils that may be across the street from one another can be quite different. Why do you think this is so?

Plants range widely in the kind of soil they require for good growth. Some plants, such as cacti, require coarse soil that drains well. Other plants, such as roses, lilacs, and some kinds of maple trees, do best in soils that have high amounts of clay and silt. Sandy soils are preferred by many common plants, including acacia trees, junipers, and pines. Knowing what kind of soil you have and knowing what kinds of soils are required by a particular plant can help to create the right kind of soil in which the plant can thrive. This can lead to happy gardening. How might the farmer who owns the field pictured here improve his soil? What are some of the characteristics of soil that you have just learned about that he might be able to change? What natural ways could he use to alter his soil to grow healthy crops?

Most soils are covered by plants, buildings, and roads. Tilled agricultural fields provide a view of soils in an area.

✓ Concept Check Assess/Reflect

Summary: How is soil formed? Soils form from humus, weathered rocks, and minerals. The weathered remains are mixed with air, water, and organic matter. Soil compositions differ based on climate, vegetation, topography, and rock composition. Soil formation produces distinct layers, but the layers are not always well defined. Scientists classify soils based on the climate and region where they form. Knowing the kind of soil that plants grow well in can yield better crops.

1. Why might a farmer be interested in the soil in his or her region?

2. Why might a farmer choose to plant crops in the valley rather than on the side of a mountain?

3. Cacti live in the desert. Based on soil type, what type of soil is best suited for cacti—a clay-rich soil, silt, or a sandy soil? Explain.

4. The Everglades are located in Florida, where water mostly covers the ground. What type of soil do you think forms in this region? Explain your answer.

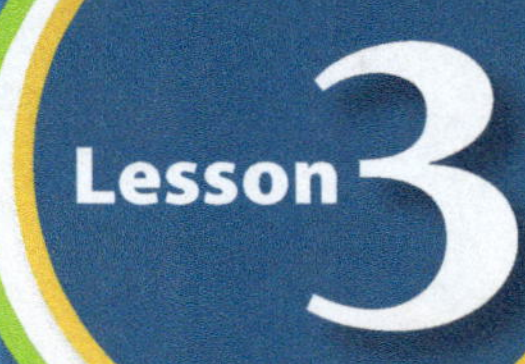

Essential Question
How Are Landforms Created?

The processes we've been studying can happen slowly and imperceptibly or quickly and dramatically during a catastrophe. Name some landforms you see in your region. Think about how the landforms you learn about in this lesson might be related to the worldwide Flood described in the Bible.

Landforms Produced by Water `Explain`

What adjectives would you use to describe water? Is one of them *powerful*? Water is responsible for most of the landforms found on the Earth's crust. Some of these landforms are the result of erosion. Other landforms form as water deposits the sediment it is eroding. Some of the landforms created by water are the results of rivers, streams, and the water that flows under the ground. Others are the work of ocean waves and currents. What landforms where you live or that you have seen were created by water?

Rivers

Streams and rivers form when water flows along the same path time and time again. As the water moves, it erodes the underlying land to form a channel, or **valley**. The valley of a young river is very steep and is shaped like a capital *V* when viewed from the side. Because water flows very quickly through young river valleys, it causes much erosion.

As a river matures, its valley widens. At this stage, the flow of the water is more from side to side than it was when the river was younger. This sideways movement causes a large, flat area called a **floodplain** to form on either side of the river's channel.

The Mississippi River
Where is the floodplain located in this picture?

Objectives

- Describe landforms created by water.
- Identify landforms created by ocean waves.
- Describe landforms created by glaciers.
- Explain how sand dunes form.
- Compare and contrast creep, slump, and talus.

Vocabulary

valley
floodplain
oxbow lake
delta
alluvial fans
caves
cirque
arête
horn
drumlins
eskers
dunes
mass wasting
slump
creep
talus

Floodplains are often marked with curves or bends in the river that are called *meanders*. When a meander becomes totally cut off from the rest of the river, a landform called an **oxbow lake** forms.

Often, long mound-like deposits of sand or gravel called bars form parallel to the flow of the water in a river channel. Crescent-shaped bars often form on the inside of a mature river bend. Why don't these bars form on the outsides of the bends?

As rivers or streams empty into larger bodies of water, a fan-shaped deposit called a **delta** often forms. The coarse sediment grains drop close to the river's mouth. The finer sediments drop farther from the mouth. Rivers that flow down mountain sides deposit sediment when the water reaches a broad, flat area. Like deltas, these deposits are shaped like fans, but are called **alluvial fans**.

Any water that is beneath Earth's surface is called groundwater. Some groundwater is trapped in the pore spaces within rocks. Some of this water even flows like underground rivers. At times groundwater dissolves rocks to form huge holes called caverns, or caves. Some caves are only a meter or so long and high. Others can be very large and form massive underground cave systems.

Have you ever visited a **cave** and learned how it came to be? Scientists often say it took thousands of years for the cave to form. These time estimates are based on uniformitarian assumptions. If the groundwater always dissolved the rocks at the same rate, would these calculations be correct? What if the rate at which the groundwater dissolves the rocks changed? Can you think of an event that might have caused more groundwater to be present at one time?

Rivers deposit sediment when they enter the ocean.

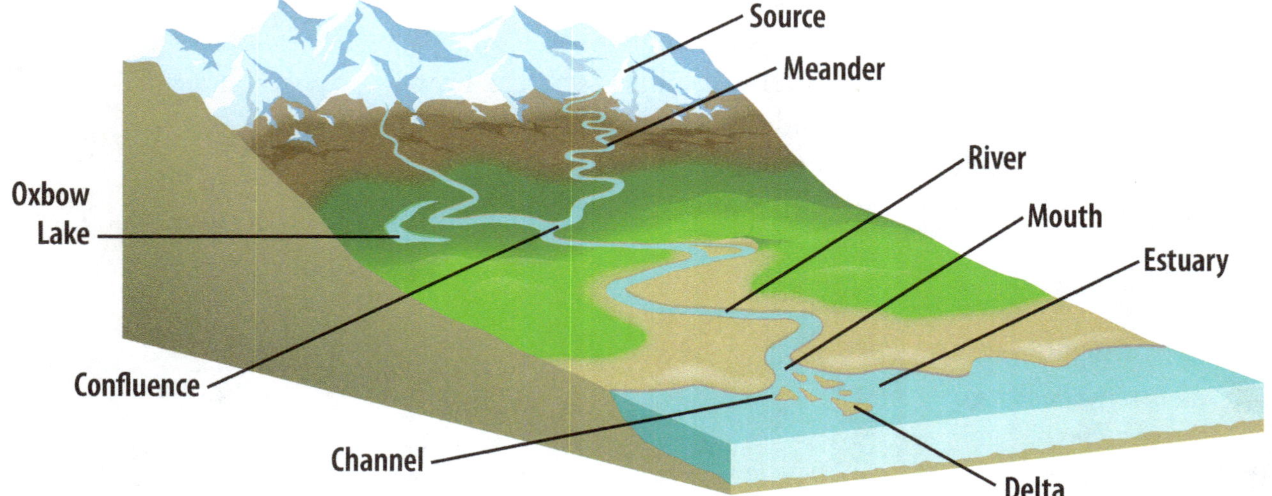

Modeling Sand Dunes

How are sand dunes created?

Procedure

1. Put on your safety goggles and leave them on until everyone is done with this activity.

2. Work with others in your group to set up the box so that it is level and the open end faces you.

3. Put four or five full scoops of sand along the open end of the box. Use your hands to level the sand to form a rectangular deposit.

4. Hold the hair dryer or place the fan so that it is about 20 cm (8 in.) from the open end of the box. Turn the hair dryer or fan on low speed and **observe** what happens to the sand for three minutes. Draw your observations.

5. Collect all of the sand in the box and level it again as you did in Step 3. Repeat Step 4 with the hair dryer or fan on high speed.

6. Collect the sand once again and re-level it. Place some pebbles and a piece of sod in the box, about 10 cm (4 in.) behind the leveled sand.

7. Repeat Step 4 with the model wind on low, then on high. **Observe** what happens. Draw your observations.

Materials
- safety goggles
- large rectangular box
- scoop
- fine- or medium-grained *dry* sand
- metric ruler
- variable-speed hair dryer or fan
- stopwatch
- large pebbles
- piece of sod

Analyze Results

Compare what happened in Steps 4 and 5 with what happened in Step 7. **Record** your results in your *Science Journal*.

Create Explanations

1. How are sand dunes created?

2. How does wind speed affect the formation of a sand dune?

3. How did the pebbles and sod affect dune formation?

Scripture Spotlight

What does **Isaiah 48:17–19** say about waves?

The sea stacks seen here are part of Australian coast have been named the Twelve Apostles.

Are these landforms the result of erosion or deposition? How can you tell?

Ocean Waves and Currents

Rivers and streams are not the only moving water that produces landforms. Oceans waves and currents constantly shape and reshape a shoreline. Waves along a rocky coast, for example, erode rocks to form nearly vertical landforms called *sea cliffs*. Sea cliffs are often eroded to form flat, horizontal surfaces called *platforms*.

When waves erode some rocks along a coast but not others, *sea caves* form. If sea caves are eroded from both sides, *sea arches* are created. Sea arches eventually collapse as waves pound against them. The rock towers left behind are known as *sea stacks*.

Waves also create and destroy beaches. Waves that crash on shore constantly move sand along the beach. During winter, large waves take sand away from the beach, causing the beach to shrink. The beach grows larger during the summer months when smaller waves add sand back to the beach. How do you think large storms like hurricanes might affect beaches?

Waves and currents flow onshore and erode or deposit sediment to change a coastline, producing various types of landforms. *Spits* are long ridges of sand that extend outward from a shore into the ocean. *Barrier islands,* such as those along much of the eastern shoreline of the United States, are deposits of sand that form parallel to the coast. Do you think it is a good idea for people to build homes on barrier islands? Why or why not?

 **How do waves move beach sand?**

Select a sample of fine-grain sand from your teacher. Also, take one sample of glitter (about $\frac{1}{4}$ teaspoon). Gently pour the sample of sand along one edge of a shallow, clear baking pan to produce a beach. Use a measuring cup to gently pour water into the pan to produce a body of water. Gently sprinkle the glitter along the edge of the beach. Use a hair dryer in one location to produce waves in the body of water. Watch where the glitter moves along the beach. What direction does the glitter move?

Landforms Produced by Glaciers `Explain`

You recall that alpine glaciers form high in the mountains and continental glaciers cover large areas of land, usually near the poles. Both types of glaciers change the landscape as they move over the surface.

Alpine glaciers, like rivers and streams, carve enormous valleys as they move downslope. Unlike a river valley, however, a glacial valley is shaped like capital *U* when viewed from the side. In addition to these rounded troughs, alpine glaciers erode the rocks beneath them to form cirques, arêtes, and horns.

A **cirque** is a bowl-shaped landform that forms as an alpine glacier erodes the rocks at the top of its valley. This erosion takes place as parts of the glacier melt, sending water into the cracks in the rocks. When the water refreezes and becomes part of the glacier again, the rocks are pulled, or plucked, from the valley wall and carried away by the ice. The scoop-shaped land left behind is the cirque. Why do you think glacial valleys are shaped differently than river valleys?

Arêtes and horns form when cirques undergo more erosion. An **arête** is a sharp, winding ridge that forms as alpine glaciers erode a cirque. A **horn** forms when three or more adjacent valley glaciers erode adjacent cirques, producing a pyramid-shaped peak.

Scripture Spotlight

Compare **Job 38:29–30** and **37:10**.

Hanging valleys are another landform produced by moving ice. Hanging valleys form when smaller glaciers join the main glacier in a mountainous area. The valley floors of the smaller glaciers are much higher than the valley floor of the main glacier. After the glaciers retreat, the hanging valley lies far above the main valley floor. As a result, water flowing along the hanging valley creates beautiful waterfalls above the main valley floor.

Deposition by glaciers also results in various types of landforms. *Moraines* are irregular hills or mounds of sediment that are deposited directly from the ice. Some moraines form along the bottom of a glacier. Others form along the sides of a glacier. A *terminal moraine* forms at the leading edge of a glacier. Because the sediment in moraines is deposited directly from the ice, it is poorly sorted, so coarse- and fine-grained sediment are mixed together.

Like moraines, drumlins also form from sediment deposited directly from glacial ice. **Drumlins** are long, cigar-shaped hills that are usually tens of meters high and no more than 1 kilometer (.6 mile) in length. Drumlins occur as parallel landforms in areas once covered by continental glaciers. The smaller ends of these landforms point in the direction from the glacier's origin. Some well-known drumlins are Beacon Hill and Bunker Hill in Massachusetts.

Some glacial landforms that are created when sediment is deposited by meltwater. Meltwater, as its name suggests, is simply glacial ice that has melted. *Kames* are small glacial hills with steep sides. Like drumlins, kames often occur in groups. Unlike drumlins, though, the sediment in kames is sorted. **Eskers** are ridges that form within channels of moving water in a glacier. Like the sediment in kames, the sediment in eskers is sorted. Why do you think this sediment is sorted?

Erosion features associated with glaciers.

Which features represent the most active erosion?

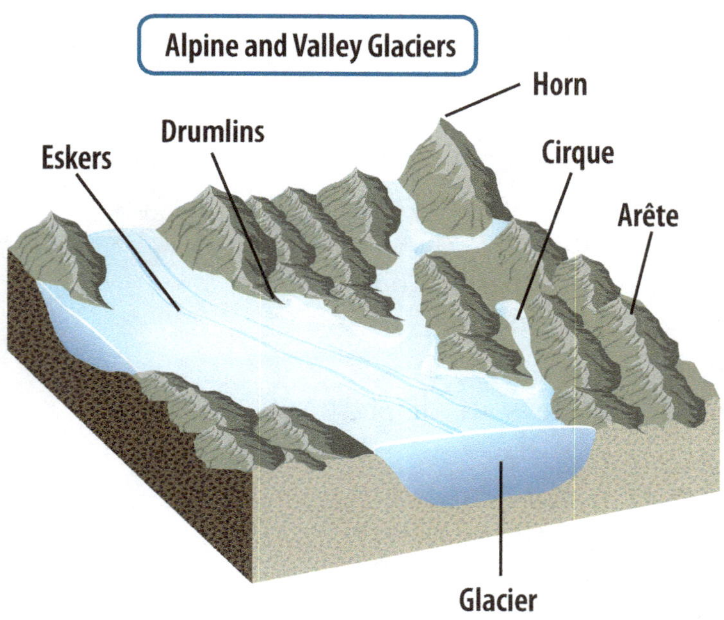

Landforms Produced by Wind Explain

What landforms do you think wind could form? Many of these landforms are found in desert areas. It is important to note, however, that even though wind is responsible for many arid landforms, water is still the most effective agent of change in dry areas. Can you explain why?

Wind typically only erodes sediment no larger than sand. Thus, when wind blows over a desert landscape, the sand is picked up and carried along the ground or just above it. Soils do not accumulate or are very thin in the desert. The angular, interlocking fragments of pebbles, gravel, or boulders on the desert floor forms *desert pavement*.

When sand and silt are dropped by the wind, they often build up to form mounds of sediment called **dunes**. Dunes, as you probably have guessed, change with the wind. Dunes are classified by shape. Does the constant movement of dunes have any impact on people? If so, how?

Another type of wind deposit is called *loess*. Loess is silt-sized material that is often deposited by glaciers. When the material dries out, it is carried by the wind and deposited over broad areas. These deposits can range from a few centimeters thick to over 90 meters (300 ft) thick. In some areas this material can form cliff-like features.

Scripture Spotlight

Which landform on this page sounds like what is mentioned in **Isaiah 40:3**?

Check for Understanding

Think back to the *Structured Inquiry*. How does wind affect the formation and structure of a dune?

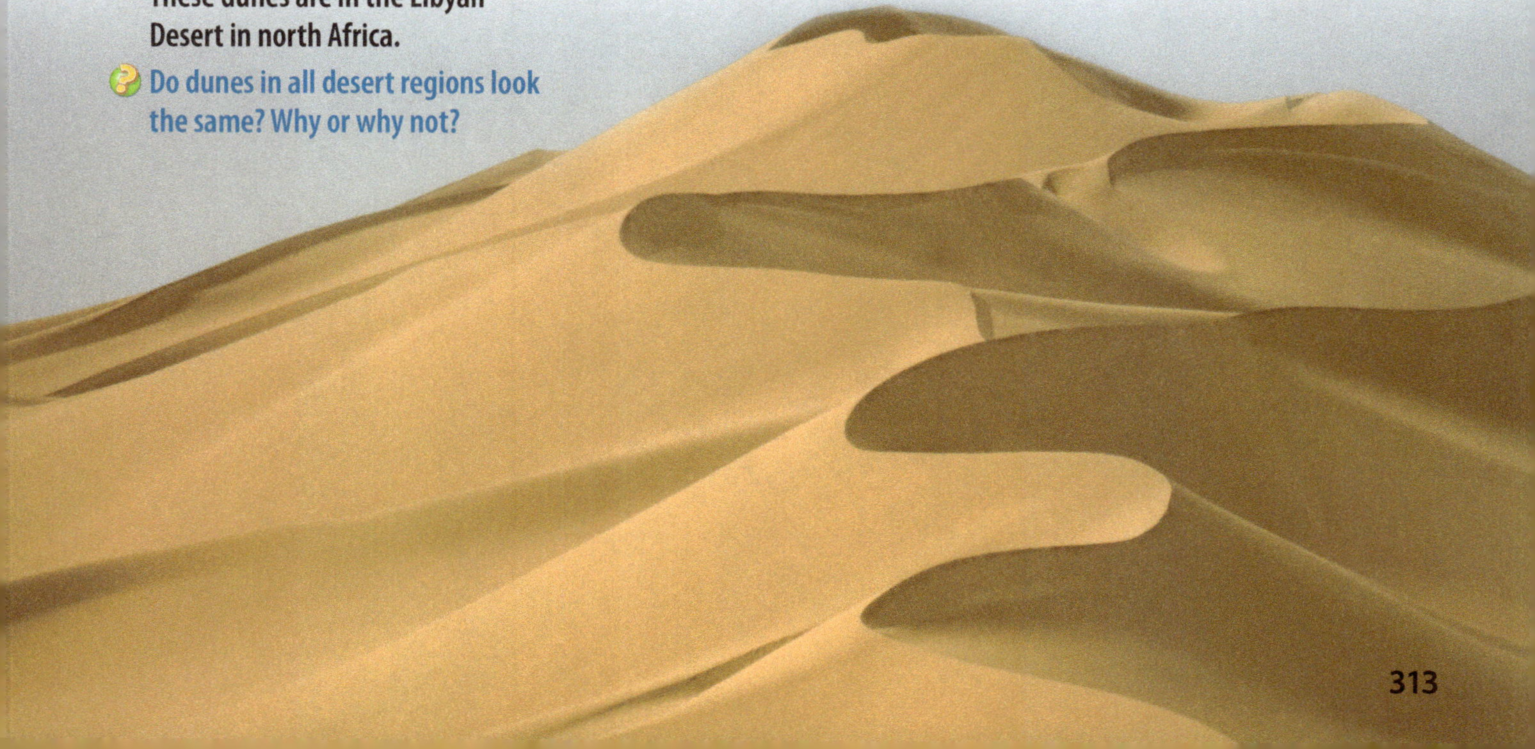

These dunes are in the Libyan Desert in north Africa.

Do dunes in all desert regions look the same? Why or why not?

Landforms Produced by Gravity Explain

Recall from Lesson 1 that some erosion by gravity happens relatively slowly, while other movements of materials happen quickly. Recall, too, that the amount of water mixed in with the rocks and soils on a slope affects their movement. Why do you think gravity takes longer to make changes than other geologic forces we have discussed?

Mass wasting in geology refers to the downhill movement of rocks and loose material under the influence of gravity. One type of mass wasting, **slump**, describes materials that move down a slope as a unit and along a curved surface. The materials in a slump usually do not move very fast or far from the slope. Slump is common along slopes where clay is abundant. Why?

Mass wasting is the movement of loose materials down a slope.

What type of mass wasting is shown here, and how did it happen?

Creep is a type of mass wasting in which materials move down a slope over a long time. Creep is commonly caused when water in the soil on a hillside freezes and thaws many times. This freezing and thawing cause the particles that make up the soil to slowly move, or creep, down the slope.

A **talus** is mass wasting that takes place when rocks fall from a higher elevation to a lower elevation. Most talus is created when rock beds are cracked or fractured. Over time, the beds simply break into large pieces that tumble or drop down a slope. Water, in the form of rain or melting snow, can also cause talus. What kind of human activity encourages these types of erosion?

Concept Check   Assess/Reflect

Summary: How are landforms created? Landforms on Earth are created and shaped by many different forces. Water, ocean waves, glaciers, wind, and gravity are examples of these forces. Landforms such as flood plains, oxbow lakes, deltas, and alluvial fans are created by flowing water. Groundwater creates caves. Ocean waves erode and deposit sand to form sea cliffs, platforms, sea arches, sea stacks, spits, and barrier islands. Other landforms, including eskers, kames, moraines, drumlins, cirques, arêtes, and horns, are created by glaciers. Wind and gravity create dunes, slumps, creeps, and talus.

1. How are arêtes, horns, and cirques the same?

2. Identify the major process—erosion or deposition—that produces each of the following landforms: an oxbow lake, an underground cave, a sea stack, a barrier island, an esker, a sand dune, and a slump.

3. Suppose you are in Wisconsin on a class field trip. You are observing a mound of material that was deposited by a glacier. You think that the mound is a kame. Your friend says it's a moraine. What could you do to easily settle this argument?

4. Describe how dune grasses aid in the formation of sand dunes and why they might be important to maintaining a healthy beach.

5. You are helping your father build a stone wall to hold back a hill on your property. He tells you that the wall will be here for your great-grandchildren. Explain why you think he might be wrong.

Get to Know
Dr. Hartmut Heinrich

What do you know about global warming? Have you heard the theories that Earth's climate has slowly been getting warmer? Scientists have been closely watching the things that may cause global warming. These include air pollution and an increase in greenhouse gases such as carbon dioxide. But what do you know about the evidence that supports global warming?

Dr. Hartmut Heinrich is a marine geologist and climatologist. He is a professor at the Federal Maritime and Hydrographic Agency in Hamburg, Germany. He recognized a connection between melting glaciers, ocean water circulation, and climate change. He thinks that this connection provides evidence that global warming is occurring. Read on and see if you agree with his research. Many scientists already do.

Dr. Heinrich studies what happens when the edges of large continental ice sheets break off and become icebergs. He is interested in the effect that these large pieces of ice have on Earth's climate. Studies show that the edges of glaciers have experienced large breakups several times in history. These breakups are called Heinrich events.

Glaciers are made of frozen freshwater. Many are near the ocean. When they melt, freshwater flows into the ocean. Freshwater is less dense than saltwater, so it floats on top of the saltwater. When large chunks of glaciers break off and become icebergs, the edges of the glaciers move landward. This reduces the amount of fresh meltwater that flows into the ocean. This results in less freshwater floating on top of the saltwater. Dr. Heinrich thinks that warm water from farther south is then able to flow more easily toward the North Pole. Because the surface of the water is in direct contact with the air above it, the warm water causes the air to become warmer as well. This in turn causes the climate of the area to become warmer over time. Evidence in the form of oxygen isotopes from the Greenland Ice Sheet indicates that changes in temperature have occurred at regular intervals throughout Earth's history. Global warming is a real concern. The more we are aware of its causes and effects, the better prepared we can be to help stop it.

Dr. Heinrich's work shows us why we need to be better stewards of God's Creation.

✔ Concept Check

1. **What possible problems can occur if glaciers continue to melt?**
2. **How would a warmer climate near the North Pole affect you and the area where you live?**

Erosion Control Specialist

Erosion control specialists are trained professionals whose work helps to prevent serious erosion from happening. Why do we need them? Have you ever built a sand castle at the beach? You probably worked hard getting it just right, and then the waves washed it away. Imagine that happening on a much larger scale. Homes built on hillsides, roads that run alongside rivers, and farm fields filled with crops are a few examples of locations that are at risk for erosion. Erosion can be a serious problem in many places.

Any time the ground is dug up, the materials that hold the soil together are disturbed or removed. After that, nothing is holding the soil together in that field or on the hillside or along the riverbank. If the water gets too high, or the wind blows too strong, the ground can become unstable. Without protection, the material can begin to move. Erosion control works to prevent or control erosion by wind, water, and even gravity.

The job of erosion control specialists is varied. They visit areas to analyze and identify the erosion risks and hazards. They help plan soil and water conservation practices such as terracing and contour plowing. They set up soil and water management programs. They also may advise farmers and other land users on the best way to protect their land from erosion. They might even be involved in landscape design so that a row of trees or decorative rocks can block strong wind or moving water.

To be an erosion control specialist, individuals must complete at least a bachelor of science degree from an accredited institution in civil, environmental or agricultural engineering, landscape architecture, soil science, geology, or a related field. They must also pass a written exam to become a Certified Professional in Erosion and Sediment Control (CPESC).

Concept Check

1. What types of erosion control can you identify around your home or school?
2. How might the work of an erosion control specialist help prevent pollution in your community?

Study Guide

Lesson 1

1. Mechanical weathering is caused by water, changes in temperature, and living things.

2. Chemical weathering is caused by acidic water, oxygen in the air, and living things.

3. Erosion is the movement of weathered materials, whereas deposition is the dropping of weathered materials.

4. Erosion and deposition are caused by water, glaciers, wind, and gravity.

Lesson 2

1. Soil is a mixture of weathered rocks, air, water, and humus.

2. The different layers of a soil are called horizons.

3. Major soil types are based on soil properties that include color, composition, texture, and permeability, and can determine plant productivity.

Lesson 3

1. Valleys, floodplains, oxbow lakes, deltas, and alluvial fans are landforms produced by rivers.

2. Ocean waves produce spits, barrier islands, sea arches, sea stacks, sea cliffs, and platforms.

3. Moraines, drumlins, cirques, horns, eskers, and kames are a few landforms created by glaciers.

4. Dunes are landforms produced and changed by wind.

5. Gravity causes creep, slump, and talus, which are all types of mass wasting.

Show What You Know

Visualize It Complete the following concept map in your *Science Journal*.

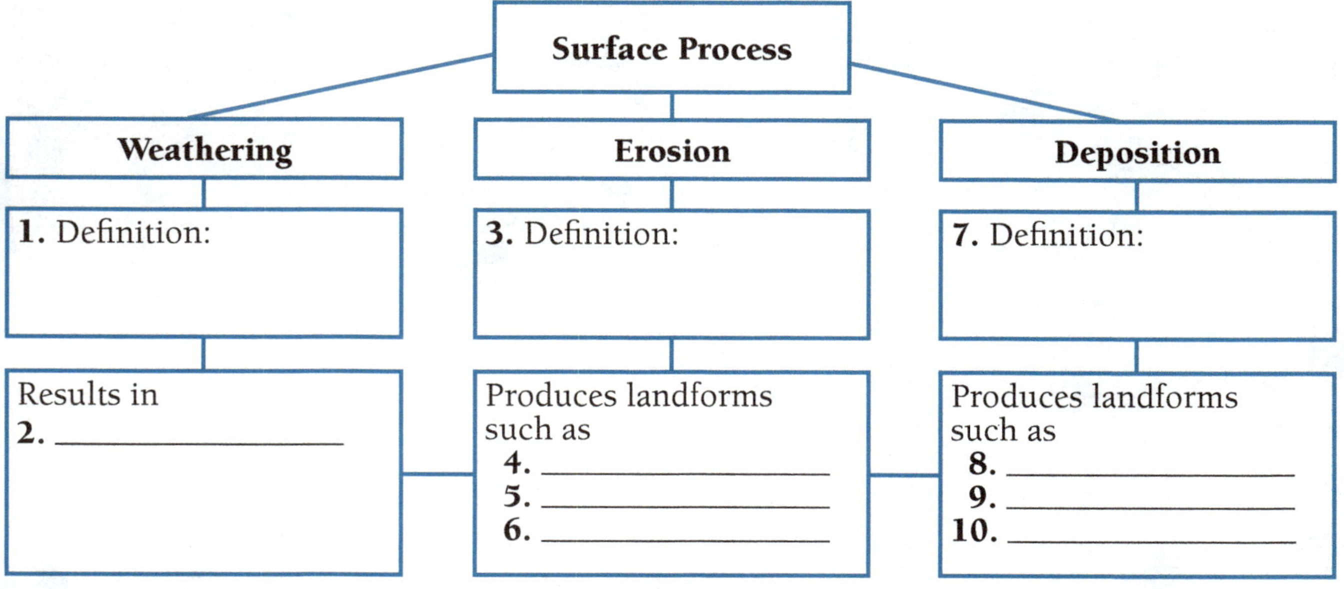

Vocabulary Check

Provide the correct word or phrase to complete each sentence.

11. When a river empties into another body of water, a(n) __________ forms.

12. An arête forms when a(n) __________ erodes a cirque from behind.

13. Sorted sediment dropped by a glacier creates __________ and __________.

14. The horizon of a soil profile that contains the *least* amount of organic matter is the __________.

15. A soil that is rich in clay has __________ permeability.

Multiple Choice

Choose the best answer.

16. Which process causes caves to form?
A. erosion by gravity
B. deposition by glaciers
C. chemical weathering by water
D. mechanical weathering by wind

17. Which process causes dunes to form?
A. mechanical weathering
B. chemical erosion
C. deposition by water
D. erosion by wind

18. Where do rivers cause the most erosion?
A. where they empty into a lake
B. on the outside of curves
C. where they stop or slow down
D. where they run straight

19. Where is the finest sediment in a delta found?
A. in the middle of the deposit
B. on the sides of the deposit
C. near the river's mouth
D. farthest from the river's mouth

20. You are in a large valley with steep sides that is high in the mountains of Alaska. How did this valley likely form?
A. erosion by gusty winds
B. erosion by an alpine glacier
C. deposition by a river
D. deposition by a continental glacier

Check Point

Answer the following questions.

21. **Explain** why rocks weather more quickly in warm, wet climates than they do in cold, dry climates.

22. What kinds of weathering forces might be involved in the wearing down of a large boulder?

23. **Identify** this landform and explain how it formed.

24. What force discussed in this chapter erodes the largest amount of sediments? Why?

25. Describe the permeability of a soil made of equal parts of clay and sand compared to a soil that contains only sand or only clay.

26. **Compare** two of the major soil types.

How Life on Earth Has Changed

Scripture Spotlight

God created Earth and everything on it. Today, Earth is different from how it was in the beginning. You can strengthen your faith by what you are learning about fossils and Earth's past.

You will read the following passages in this chapter.

Genesis 7:21–23 (p. 324)
1 Samuel 23:19–28 (p. 327)
Genesis 6–8 (p. 333)
Genesis 7:17–20 (p. 337)
Genesis 7:4 (p. 344)
Genesis 7:21–23 (p. 344)

Fossilized skeletons of the meat-eating dinosaur Utahraptor were unearthed in Utah's Cedar Mountain formation. Many scientists believe they lived about 125 million years ago. However, scientists who believe in Creation believe that dinosaurs probably disappeared at the time of the Genesis Flood. This species may have hunted and lived in groups. This idea is different from past hypotheses about large dinosaurs traveling and living alone.

The Big Idea

Earth and its life forms have changed over time. Fossils provide clues about many of these changes.

What do these fossilized bones tell you about this organism?

Inquiry Kick-Off Engage

Fossil imprints can provide valuable insights into how an organism lived or about its ecosystem. In your *Science Journal*, you will have an opportunity to draw inferences in an activity about what created a mock imprint fossil.

Science Journal

Objectives

- Describe how rock layers are deposited and change over time.
- Describe the geologic column.
- Distinguish between the major types of fossils.
- Explain the relationship between the geologic column and time.

Vocabulary

strata

lithification

geologic column

fossils

index fossil

Precambrian

Paleozoic

Mesozoic

Cenozoic

eons

eras

periods

relative age

relative dating

radiometric dating

Essential Question

What Is the Geologic Column?

Rock layers found all around the world are filled with clues about creatures that lived in the past. What ideas have you heard to explain why some of these creatures are not here anymore? How does a scientist's worldview influence his or her interpretation of the data in the rock layers?

Rock Layers Explain

Rock layers, or **strata**, usually form when sediment is carried along by moving water. When the water slows down, the largest pieces are deposited first followed by increasingly smaller sediment.

Sediments become cemented together and harden into rock layers during **lithification**. The chart below lists some rocks formed by lithification.

Kind of Rock	Interpretation of Formation
Limestone	Mostly formed from the remains of living organisms (like broken seashells)
Sandstone	Often formed in flowing water
Mudstone and Shale	Formed from fine clay particles settling out in quiet water
Conglomerates	Deposited by high-energy flows

Strata are usually laid down horizontally, but they do not always remain in horizontal layers. As you know, the movement of tectonic plates can cause earthquakes, volcanoes, and mountain building. As these processes happen, along with weathering, erosion, and deposition, rock formations change. The evidence left behind provides clues that scientists can study to solve the mystery of the rock's history. What evidence do you think scientists study, and what do they learn?

The rock layers and the fossils found in them make up what we call the **geologic column**. Most of the geologic column is buried out of sight, but where sections of it are visible, people have discovered millions of fascinating fossils.

Putting It Together

How do scientists use fossil bones to reconstruct animals?

Procedure

1. For this inquiry, imagine you are an archeologist working at a dig site. Get an envelope from your teacher. Remove three fossil bones from the envelope without looking at the remaining bones in the envelope. These are the first three bones you discover at the dig site! Spend five minutes trying various combinations to fit them together. In your **Science Journal**, draw your arrangement of the three bones. **Infer** what the animal might be.

2. Your dig site yields three more bones. Remove three more fossil bones from the envelope. Spend five minutes examining your new finds. Try to incorporate them into your **model** fossil reconstruction of the mystery animal. Draw your arrangement of the six bones.

3. It's the last day of the digging season. A teammate finds three more bones. Take three more bones from the envelope. Incorporate your team's latest finds into the **model**. You may wish to glue or tape the bones of the skeleton in place in your **Science Journal**. Draw your arrangement of the nine bones. **Record** what you think the animal is now.

4. Use colored pencils to make a drawing of your fossil **model** as it might have appeared when the animal was alive.

Materials
- envelope with fossil bones
- glue or tape
- colored pencils

Analyze Results

Compare your **model** with those of two other teams. **Communicate** any differences among the reconstructions. What might account for the differences?

Create Explanations

1. How do scientists use fossil bones to reconstruct animals?

2. What might account for differences in scientists' interpretations of the fossil record?

3. Looking at the assembled fossil, what can you infer about this animal?

Scripture Spotlight

What does **Genesis 7:21–23** describe, and how is it related to what you've been reading?

Fossils (Explain)

Usually when a living organism dies, it decomposes. However, if an organism is buried rapidly before it can decompose, there is a chance it may be preserved. These preserved remains are called **fossils**. Scientists understand that most types of animals are fossilized fairly quickly or not at all. This is especially true if the fossil has preserved soft tissues—like muscle or skin—implying that the animal was buried very rapidly and its remains became mineralized rather than decomposing. In experiments, some types of fossils have been made in the lab in a year or less. Why do you think preservation of a fossil would have to happen quickly?

The following chart details some of the types of fossils scientists find.

amber

Types of Fossils	
Body Fossils	**Trace Fossils**
Preserved remains of animals or plants	Evidence (or traces) of an organism's behavior or activities
Actual Remains	**Footprints and Trails**
Bones and shells are the most common types of fossils.	Footprints give clues about: • The size of the organism that made them. • Whether it walked on two or four feet. • How quickly or slowly it moved.
Woolly mammoth fur, skin, and other body parts are sometimes found preserved in ice.	
Insects can be found preserved in fossilized tree resin called *amber*.	Trails provide similar clues about snakes, worms, and other footless creatures.
A plesiosaur, a marine reptile with a long neck and flippers, was found in tar sands in Alberta, Canada.	
In North Dakota, scientists found mummified hadrosaur bones that they named Dakota.	

Plesiosaur bones

Plesiosaurs were marine reptiles.

What evidence suggests that this reptile may have lived in water?

Types of Fossils

Petrified Fossils	Burrows and Borings
When an organism's soft parts are mineralized, they are said to be petrified. Entire petrified trees were found in Petrified Forest National Park in Arizona. Tiny arthropods are beautifully preserved in 3-D in the Orsten fauna in Sweden.	Burrows are holes or small tunnels made by worms and other organisms that lived in mud or soft sand. Borings are holes or tunnels made by organisms that lived in wood or other hard materials. Burrows and borings provide clues about the size of the organisms that made them and sometimes about their feeding habits.

Fossilized Eggs and Fossilized Nests

These provide evidence about:
- The number of offspring an adult may have produced.
- The size of the young.
- How their parents may have taken care of them.

Fossil Molds and Casts	Fossilized Wastes
A mold is the impression left in sediment by a body part, such as a shell, that is preserved (the body part disappears). A cast forms when a fossil mold becomes filled with minerals or mud.	Coprolites are fossilized dung that often contains bits of other organisms' bones, teeth, or shells. Gastroliths are fossilized gizzard stones that are sometimes found among reptile bones. Coprolites and gastroliths provide clues about an organism's diet.

Carbon Films

Carbon forms very thin sheets that were once a part of a plant or animal.

petrified ammonite fossil

fossilized egg

petrified wood

petrified
fish fossil

Studying the Layers **Explain**

While William Smith (1769–1839) was supervising the digging of the Somerset Canal in England, he observed that the same kinds of fossils always occurred in a specific order in the rock layers. He used his observations to support the *law of superposition,* which states that newer layers form on top of existing layers. Using this approach, he could successfully predict the type of fossils that would be found above or below any layer he was studying. What scientific skills helped Smith in his work?

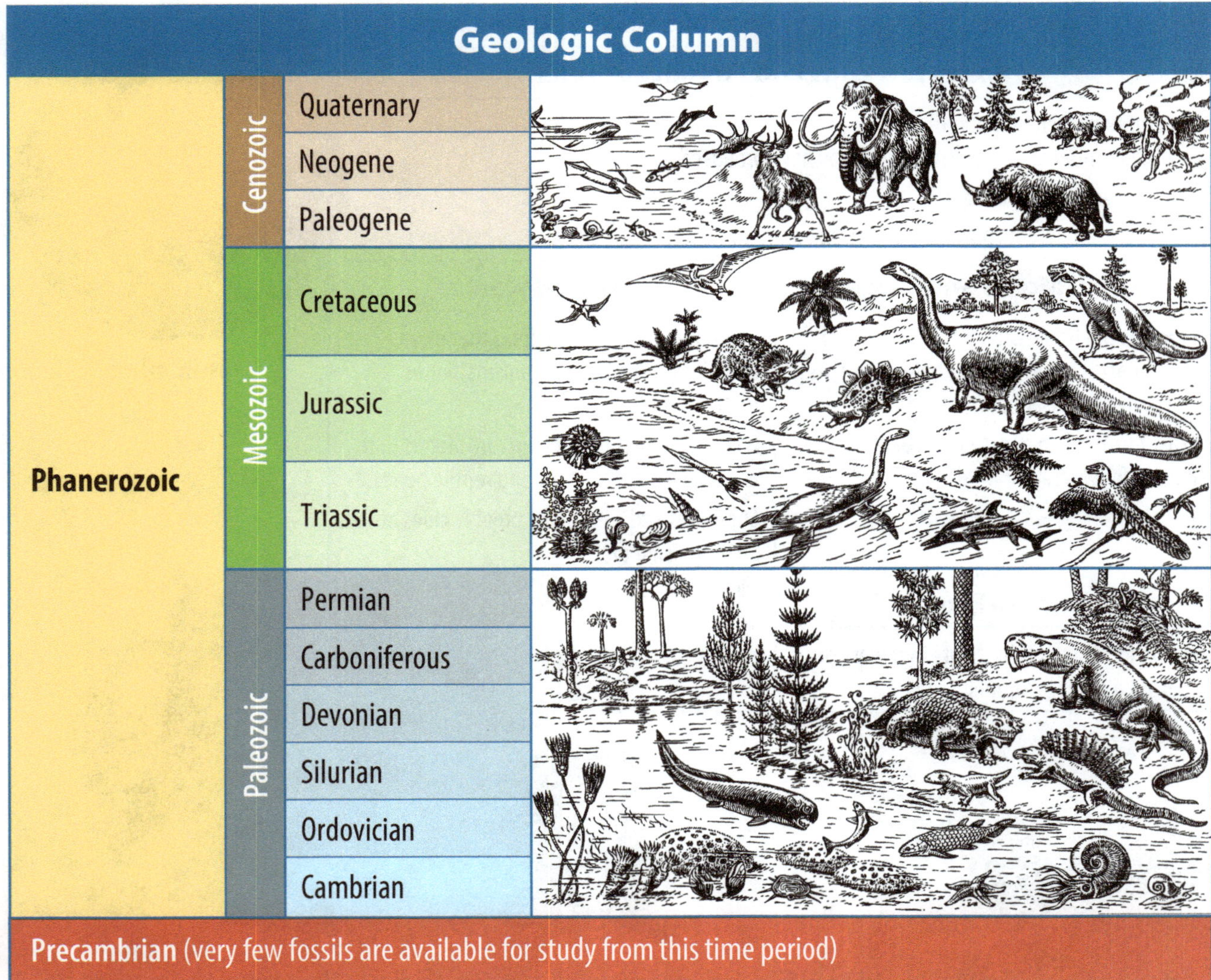

Scientists separate geologic column into intervals based on the appearance and disappearance of fossils from the rock record.

What factors do you think scientists used to separate the Paleozoic, Mesozoic, and Cenozoic eras?

Using Smith's principles, Sir Roderick Murchison (1792–1871) carefully collected fossils from different layers. He found that the fossil communities changed as he worked his way down through the layers of rock in Wales. Later, he was invited by the czar to study the fossil communities in Russia. Murchison began to believe that the same sequence of fossil communities was preserved throughout the world. This specific order of fossils is observable data, and other geologists have confirmed Murchison's idea that the same pattern is consistent in the rock layers throughout the world. Fossils that occur consistently in a limited portion of the geologic column, but are abundant and widespread in that limited portion are called **index fossils**.

The lowest section of the geologic column, the **Precambrian**, contains microfossils, but relatively few macrofossils. *Microfossils* are the remains of small organisms—such as bacteria and protists—that can only be studied using a microscope. *Macrofossils* are larger and visible without a microscope.

Above the Precambrian layers, in a subdivision called the **Paleozoic**, various marine fossils are found. These animals include trilobites and fish. In addition, scientists have found amphibians and huge dragonflies with a wingspan of nearly a meter (3 ft). Near the top of the Paleozoic layers, some fossils of land animals appear. A high percentage of fossils found in these lower layers are creatures that are now extinct.

Above the Paleozoic are the **Mesozoic** layers, which contain a mixture of fossils from marine animals, land animals, and some birds. The most well-known animals in these layers are the dinosaurs.

Above the Mesozoic are the **Cenozoic** layers, which contain fossils from many land animals and birds. Among the interesting animals found in these layers are saber-toothed cats, tiny horses, beavers 2–3 m (6–8 ft) tall, woolly mammoths, and sloths the size of elephants. Why do you think the percentage of fossils of creatures that are still alive today increases as you go up the geologic column?

When scientists speak of the *fossil record,* they refer to the fossils, their placement within the strata, and the information that can be derived from them. Rock layers and the fossils found in them are observable data. Some of the data show very clearly that major changes have occurred during Earth's history. Fossil whalebones have been found in deserts. Fossil seashells have been found on mountaintops all over the world. Fossils of broad-leafed evergreen plants that grow in tropical forests have been found in Alaska.

Time Inferences Explain

The geologic column has come to be closely associated with millions of years of evolutionary history, but it was not that way at first. Most of the scientists who first described and published the relationships between the layers and the fossils believed in the biblical account of Earth's history.

Sometimes the names they assigned to the parts of the geologic column reflected characteristics of the rock layers themselves, like the *Cretaceous* (which means "chalky") or the *Carboniferous* (because of the carbon in the coal found there). Often layers were named after the places where the fossils were first described. The *Jurassic* was named for the Jura Mountains of Switzerland, and the *Permian* was named for the town of Perm in Russia where scientists first described the fossils found in these layers. It was only later that the long ages suggested by some scientists influenced the interpretation of the geologic column. The reason for the fossil sequence and the time span associated with it came to be associated with the newest scientific theory—evolution. As individual layers were grouped into larger categories, they were given names with time connotations.

Now the geologic column is referred to as the geological time scale and uses divisions called eons, eras, and periods.

An **eon** is the longest division of geologic time. There are four eons—Phanerozoic, Proterozoic, Archean, and Hadean. Eons are divided into **eras**, which are still long periods of time, but shorter than eons. For example, the Phanerozoic eon is divided into three eras—Cenozoic, Mesozoic, and Paleozoic. Each of these three eras contain at least three periods. A **period** is the basic unit on the geologic time scale.

Geologic Time Scale		
Eon	**Era**	**Period**
Phanerozoic	Cenozoic 0–65 Ma	Quaternary
		Neogene
		Paleogene
	Mesozoic 65–250 Ma	Cretaceous
		Jurassic
		Triassic
	Paleozoic 250–540 Ma	Permian
		Carboniferous
		Devonian
		Silurian
		Ordovician
		Cambrian
Proterozoic eon	540–2500 Ma	
Archean eon	2500–4000 Ma	
Hadean eon	4000–4600 Ma	

"Ma" is an abbreviation for *mega-annum,* or one million years.

A breakdown of the geologic column with the estimated time periods used by conventional geologists.

How do scientists divide time in the geologic time scale?

Choose two eras and compare them. You may wish to use a graphic organizer, such as a Venn diagram, for this activity. Remember to include details on the types of plants and animals that lived in that time.

How do the fossils of two different eras differ?

shell fossil

ammonite fossils

Study the chart below. Notice how the meanings of the words Cenozoic, Mesozoic, and Paleozoic include the idea of time. Lower layers are considered *older* than higher layers because they were laid down earlier. How does this help in classifying fossils?

	Prefix	**Suffix**	**Meaning**
Cenozoic	Ceno = recent	Zoic = animal life	Recent animal life
Mesozoic	Meso = middle	Zoic = animal life	Middle animal life
Paleozoic	Paleo = old	Zoic = animal life	Old animal life

Science Journal

Check out your *Science Journal* for a Structured Inquiry exploring how fossils can be classified.

Extend

Scientists often refer to the age of one layer in relation to another. **Relative age** is the age of a rock or formation in relation relative to other rocks or formations, usually defined as a zone fossil name. **Relative dating** is the science of determining the relative order of past events, without necessarily determining their absolute age. But just by looking at the layers, we cannot tell how much older one layer is than another. Were the layers deposited millions of years apart or only years, months, or even days apart?

Scientists have attempted to assign actual dates to the rock layers using a process called radiometric dating, also called absolute dating, is a method of dating that compares the relative proportions of particular radioactive isotopes present in a sample. Certain elements that occur in nature decay predictably over time, changing from what we call a parent isotope to what we call a daughter isotope. The more time that passes, the less parent isotope is left and the more daughter isotope there is. Scientists know the half-lives of various elements. They can compare the ratio of parent isotopes to daughter isotopes in an attempt to figure out the age of the rock layers. How does this help in dating rocks and fossils?

While the ratios of parent isotopes to daughter isotopes are actual data, the interpretation of those ratios as millions of years conflicts with both the biblical history of Earth and scientific evidence that is difficult to explain if the layers were really laid down over millions of years. Short-age geology predicts that there are more discoveries to be made about radiometric dating and that these discoveries will shed light on why these ratios indicate time spans that conflict with the chronological information found in the Bible.

The geologic column, which includes both the rock strata and the fossil record, is observable data. The time inferences associated with the geologic column are interpretations of that data, which are influenced by the worldview of the scientists who make them. What can be done to test these interpretations?

Explore-a-Lab

Structured Inquiry

How can you use a sandwich to apply the principles of relative dating?

Work with a partner to make a sandwich from two slices of bread, sliced cheese, tomatoes, mayo or mustard, and some pickles, or any other ingredients you have. Place the first slice of bread on a plate and assemble the complete sandwich. Which layer of the sandwich was laid down first? Which one was laid down third? Record your observations by making a labeled sketch of the sandwich (model rock) layers. Flip the sandwich over. Could you tell which way was up in the sandwich layers? What clues did you use? Return the sandwich to its original position and cut it in half. Discuss with your partner: How does the timing of the cut (a disturbance in the rock layers) relate to the making of the sandwich? How would a scientist who didn't see how you created or cut the sandwich interpret what layer was laid down first? Switch sandwiches with another group. Record the order that you believe the layers in that sandwich were laid down. Why did you decide on that order?

How can you use pennies to model the principles of relative dating used by geologists?

Work with a partner. Cover a cup of 25 pennies and shake it. After shaking, pour out all of the pennies on the table. Heads-up pennies represent "daughter product" and tails-up pennies represent "parent product" that has not changed. Remove all "daughter product" pennies. Put the unchanged "parent product" pennies back into the cup. Use a chart to keep track of the number of pennies put back into the cup. Shake, pour onto the table, and again, remove "daughter product" pennies. How long can you continue this process until there are no more pennies left? Switch places and keep track of the half-life intervals for your partner.

Concept Check Assess/Reflect

Summary: What is the geologic column? The geologic column is made up of many different kinds of rock layers as well as the fossils contained in them. Fossils appear in a predictable order around the world and provide evidence that major changes have occurred during Earth's history. Because time inferences have become associated with the geologic column, it is sometimes referred to as the geologic time scale.

1. Explain how rock layers are formed.

2. Define body fossils and trace fossils and give at least two examples of each.

3. Create a chart that includes the four major sections of the geologic column described in this lesson. In each section, describe the kinds of fossils found there. Be sure your chart shows the correct order of the sections.

4. What is the difference between relative dating and absolute dating? Which one is attempted using radiometric dating?

Essential Question

How Was the Geologic Column Created?

Suppose you are walking in a field and find a fossilized bone lying partially buried in the ground. What can you determine about the fossil? How long ago did the animal live? In the previous lesson you learned about the geologic column, which includes the rock layers around the world, and the fossils found in them. The layers and fossils are observable data that offer clues about the history of our world.

Interpreting the Data **Explain**

By now it should not surprise you that one of the main differences between interpretations of geological data involves time. For example, one interpretation suggests that the geologic column was laid down over hundreds of millions of years. Another approach, however, suggests that most of it was laid down in several thousand years, with a significant amount being deposited in just one year during the biblical Flood.

The origin of the geologic column is a matter of **historical science**. Unlike science experiments that are observable and repeatable, historical science seeks to answer questions about events that no living person witnessed and that cannot be repeated. Historical science has limitations.

Natural processes, like erosion, and human actions, like road building, uncover buried fossils.

What can you learn about the events that preceded its burial by simply observing this fossil?

Biblical Flood Account

As geologists seek to determine which view of Earth's history describes the conditions best suited for creating the geologic column, those who believe in the worldwide Flood of Genesis suggest this **catastrophe**, or sudden disaster, as a reasonable scientific explanation for the data.

The biblical account does not give many geological details about what happened during the Flood, but we do know:

- It lasted a little over a year.
- All Earth was covered with water, which later receded.
- As the Flood began, "the fountains of the great deep were broken up and the windows of heaven were opened," and both were closed after 150 days.

Because of our biblical worldview, we accept the Bible as a source of accurate, reliable information about Earth's history. While it is not a history book or a science book, we believe it to be a revelation of God's dealings with our world. When it does speak of historical or scientific topics, we can trust its accuracy.

The biblical account of the Flood is found in **Genesis 6–8**. As you read it, remember what you have learned in this unit about the structure of Earth and the changes that happen due to plate movement, earthquakes, volcanoes, weathering, erosion, and deposition.

Scripture Spotlight

Read the story of the Flood found in **Genesis 6–8**.

Lesson Activity

Work with a partner. Read the story of the Genesis Flood in the Bible together. List all the details about the events that occurred before, during, and immediately after the Flood described in **Genesis 6–8**. Make a chart. Then add up the days/weeks/months mentioned in this account of these disastrous events in Earth's history.

How long did events surrounding the Genesis Flood last?

Who Goes There?

What can you discover about animals and their environment by studying their tracks?

Procedure

1. Observe the picture of the preserved dinosaur tracks in your *Science Journal*.

2. In the data table, **record** general information about the tracks. Use the hand lens to **observe** the tracks more closely.

3. Use colored pencils to make a drawing of the tracks. Use a different color for each different set of tracks. Label the different sets of tracks with different letters (A, B, C, etc.).

4. Use the dinosaur track key to identify the kind of dinosaur that made each set of tracks you labeled in Step 3. **Record** the information in the data table.

5. For each set of tracks, **infer** the direction the organism was traveling. If possible, tell whether the animal was moving fast or slow.

Materials
- colored pencils
- dinosaur trackway photo
- dinosaur track key
- hand lens

Analyze Results

Compare your data with the data of two other classmates. What similarities and differences are there between your data and your classmates' data?

Create Explanations

1. What can you discover about animals and their environment by studying their tracks?

2. What evidence did you use to infer the direction the dinosaurs were moving?

3. Why is it that even though scientists are looking at the same evidence, they may come up with different explanations of that evidence?

Evidence Explain

A catastrophe as monumental as the Flood described in Genesis would be expected to leave behind significant evidence. What kinds of evidence might it leave? Many features in the geologic record can be explained by a complex event like the biblical Flood.

Burial and Preservation of Remains

As you know, millions of fossils have been found all around the world. The preservation of some fossils provides evidence of sudden death and burial. Sometimes organisms even appear to have been preserved in the act of dying, like an Ichthyosaur that was giving birth at the time of its burial. In other cases, evidence indicates that dead creatures lay exposed for some time or were even washed to other locations before they were buried. Preserved tracks and evidence of burrowing indicate the passage of time while the geologic column was being formed. In many cases, individual fossils are preserved, but sometimes we see evidence of mass extinction. Whether suddenly or slowly, affecting individuals or larger populations, destruction on a worldwide scale is evident in the fossil record.

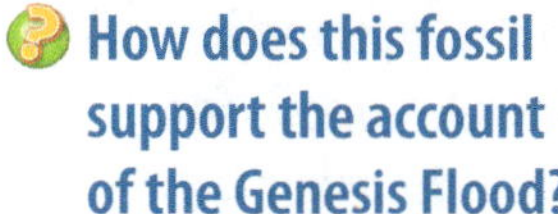

This fossil displays a large number of fish that died at one time.

How does this fossil support the account of the Genesis Flood?

Mesozoic Formations in Northern Utah	
Frontier	300,000 sq km (115,830 sq mi)
Mowry	250,000 sq km (96,525 sq mi)
Dakota	815,000 sq km (314,673 sq mi)
Cedar Mountain	130,000 sq km (50,193 sq mi)
Morrison	1,000,000 sq km (386,102 sq mi)

Widespread Sedimentary Deposits

You may recall that sedimentary rock, which is usually deposited by water, covers 75% of Earth's surface. There are many examples of sedimentary deposits that are relatively thin and so widespread that they cover hundreds of thousands of square kilometers.

One such deposit is the Morrison Formation, which is well known for its abundant dinosaur fossils. It stretches from Texas to Canada, covering one million square kilometers (621,371 square miles). Imagine the amount of water and sediment necessary to create these huge layers. Such large-scale processes could be explained by the Genesis Flood.

Flat Gaps

You learned that the fossils in the geologic column appear in a consistent, predictable order around the world, but that not all layers are found in all locations. Recall also that conventional geologists interpret the layers as representing millions of years. According to that interpretation, the absence of certain layers in a particular location would represent a gap of millions of years of time. If you were a conventional geologist, how would you explain those gaps?

You also learned that weathering and erosion cause layers of rock to change when exposed to the elements over time. Even though rock layers are usually laid down flat, we would not expect them to remain smooth if they were uncovered for a long time. Yet the geologic column contains examples of places where, in spite of missing layers, the rocks above and below the absent layers are in flat contact with each other, showing little or no evidence of erosion. This can be interpreted as evidence that the layers were laid down quickly instead of forming over millions of years. There are other examples of more recent natural events that formed layers quickly.

Turbidites

When mud and sand begin to slide down an underwater slope, they flow downhill like an avalanche in a turbidity current. As the sediment reaches a flat area, it spreads out and stops. The sedimentary layers it forms are called **turbidites**. In 1929, the Grand Banks Earthquake caused this to happen along the east coast of Canada. A large mass of sediment on the edge of the continental shelf traveled toward the abyssal plain at about 80 km/hr (50 mi/hr) and spread out across thousands of square kilometers.

Thousands of these turbidites, which form only underwater, are found on the continents. This could mean that our continents were once under water, which would be consistent with the Flood. Another possibility is that these turbidites were formed on oceanic crust that was somehow moved to its current location on the continent. What type of events might have moved oceanic crust?

Ocean Sediments on the Continents

There is much more sediment on the continents than there is on the ocean floor, and about half of it contains fossils of marine organisms. The presence of marine fossils on land raises some interesting questions, especially since some are located on the top of Mount Everest. How do you think marine animals got to the top of a mountain?

Some marine fossils appear to have been deposited as water flowed across the continents. Others that are found on mountaintops would have been deposited on the ocean floor and then thrust up later as plates collided to form mountains. Both scenarios— flood waters and oceanic crust being thrust up—would be consistent with a worldwide Flood.

Turbidites are massive sea-bottom deposits that often form in response to earthquakes.

How do formations like this on continents support the account of the biblical Flood?

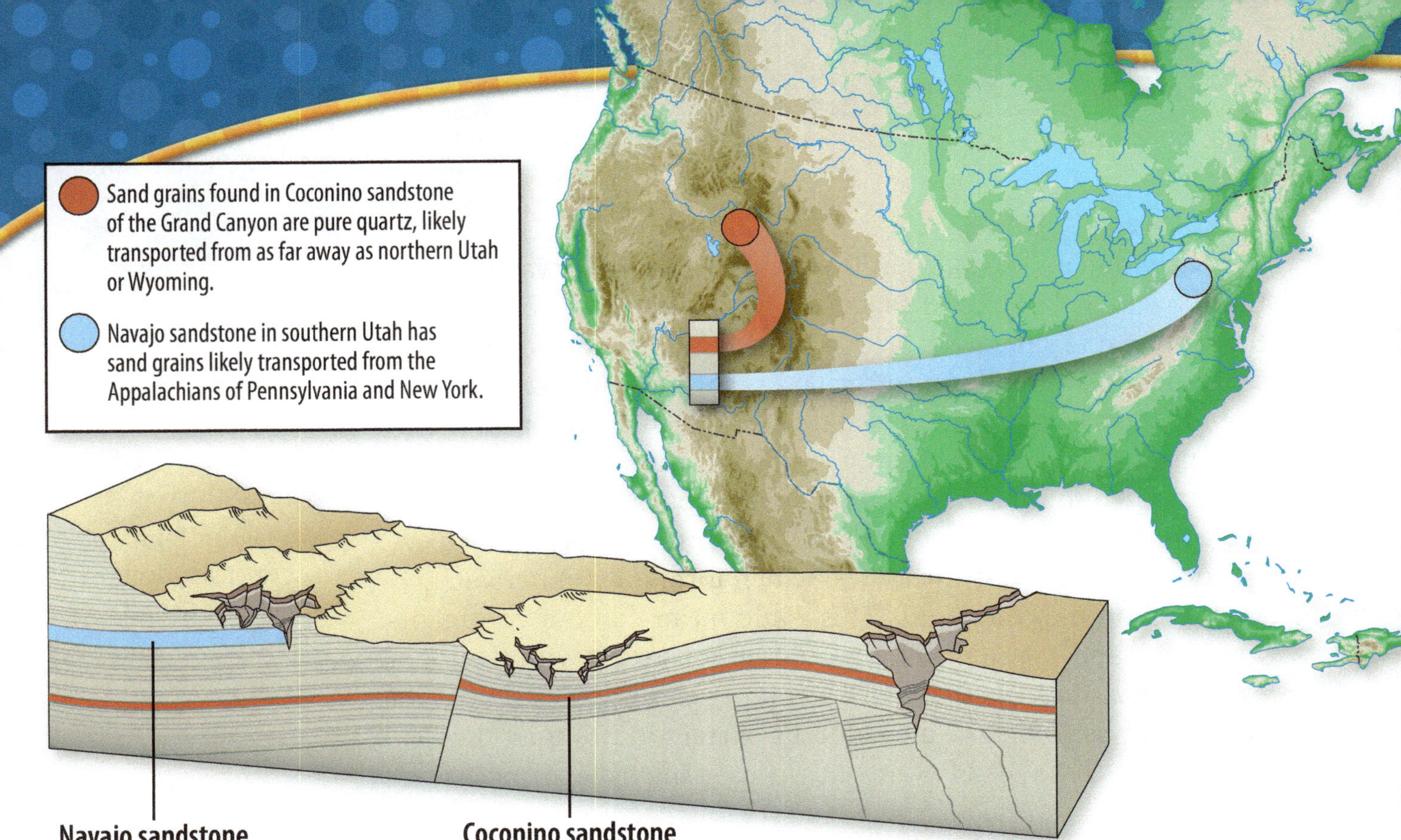

Evidence of massive water flow in North America.

How might this geologic evidence be interpreted?

Continental-scale Currents

Geologists studying sedimentary layers can often tell which way the water was flowing when the sediments were deposited. Today water flows in all directions over the continents as rivers flow to the oceans. However, many sedimentary layers indicate that they were laid down by water that was flowing in the same direction across the continent. A review of 15,000 locations across North America shows a pattern of water flowing toward the southwest in some of the lower layers of the geologic column. How do you think a conventional geologist would explain the water flow patterns? How would a scientist with a biblical worldview explain them?

Incomplete Ecological Systems

In the geologic column scientists sometimes find animal fossils without any fossilized plants nearby. For example, in the Morrison Formation, which is one of the world's richest sources of dinosaur fossils, plant remains are rare. However, in a catastrophic flood, animals could have been washed away from their habitats, explaining why we sometimes do not find them preserved near their food sources.

Ice Ages Explain

In addition to the direct impact, it is possible that the Flood created certain conditions that indirectly changed the Earth in another significant way. You will recall that an ice age is defined as a period when glaciers spread beyond the polar regions and covered more of our planet than they do now. What would cause such glaciers to form?

Recall several facts that you learned about glaciers in Chapter 9:

- Alpine glaciers are found in very high mountains.
- Continental glaciers are found in cold regions near the North and South Poles (like Greenland and Antarctica).
- Because of gravity, both alpine and continental glaciers move, changing the ground underneath.
- Moraines, eskers, erratics, and U-shaped valleys are evidence that glaciers existed in the past.

Although glaciers are only found in very high mountains or in cold regions near the poles today, many geologists believe that was not always the case. Recall that fossils of tropical plants and animals have been found in the Antarctic. Apparently at some point in the past, Earth was warmer and the poles were free of ice. The fact that the poles are now covered with huge ice sheets means that at some time, temperatures dropped and sufficient snow fell to create these glaciers. With evidence of glacial processes in places where no glaciers currently exist, scientists know that glaciers were there sometime in the past. Throughout much of Canada and the northern United States, we see clues left by glaciers, but no glaciers.

Many scientists who believe Earth has been around for millions of years believe that glaciers have advanced and retreated many times, creating multiple ice ages.

The retreat of the glaciers after the end of an ice age left behind debris such as these huge erratics carried to this location by the ice sheets.

Why do many scientists believe glacial activity is responsible for depositing this debris?

Two conditions are necessary to initiate an ice age. First, there must be enough snow to build up an ice sheet. Second, summers must be cool enough to keep the ice sheet from melting. A variety of theories have been suggested by scientists to explain how these conditions could occur. Most of them in some way involve a reduction in the amount of sunlight that reaches Earth's surface. One such theory is that volcanic eruptions, which spew millions of tons of ash into the air, blocked out the Sun's warmth and caused a drop in temperature. Another theory cites the cause of colder weather as a change in the tilt or orbit of Earth, causing the Sun's rays to reach the planet at an oblique angle. What other events could have caused Earth to become so cold?

Creationists have suggested that volcanic eruptions connected with the Flood could have caused the cold weather. Some scientists have suggested that the "fountains of the deep," which the Bible says erupted during the Flood, could have contributed to warming the oceans. Warm water would evaporate rapidly and create many clouds carrying large amounts of moisture. Once these moisture-laden clouds moved to an area of colder temperature, heavy snowfall would result. These conditions, which could reasonably result from the Genesis Flood, are the conditions that could be expected to cause an ice age.

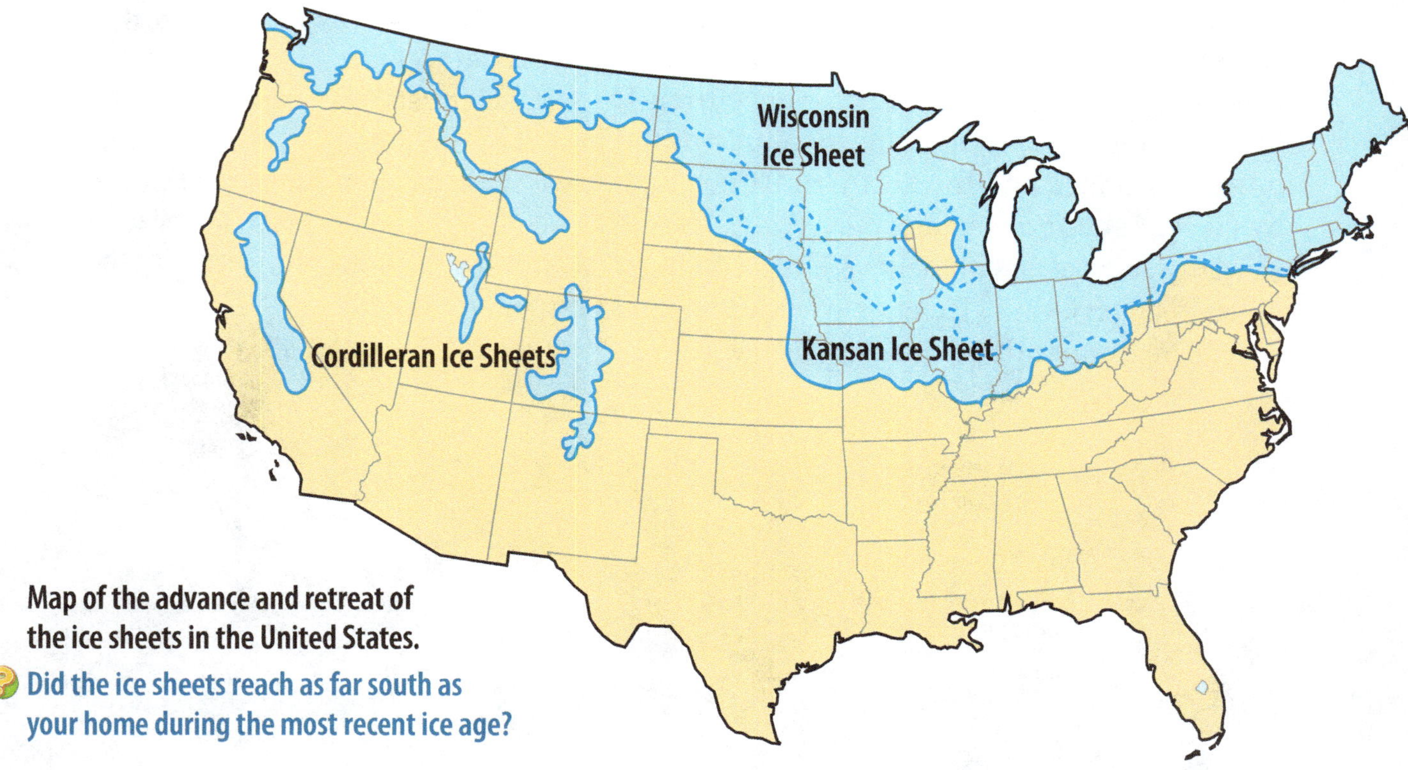

Map of the advance and retreat of the ice sheets in the United States.

Did the ice sheets reach as far south as your home during the most recent ice age?

Drawing Conclusions Explain

Evidence presented in this lesson can be interpreted in ways that are consistent with the Genesis Flood. Our interpretations are influenced by a biblical worldview. Conventional geologists also offer explanations of these data that are influenced by their worldviews. And while geologists may disagree about the interpretation of the evidence, they agree that different conditions existed in the past than what we see now.

Although our understanding is still incomplete, scientists with a biblical worldview have made progress toward understanding how the Genesis Flood impacted our Earth and how it may have contributed to an ice age. By following the scientific method, these scientists who believe in the biblical Flood have made valuable contributions to science concerning the Genesis Flood. From these observations and questions, they develop testable hypotheses, gather data, and analyze that data carefully. These scientists predict that many significant discoveries yet await us and that diligent scientific research will bring us to a fuller understanding of the events of Earth's history.

Concept Check Assess/Reflect

Summary: How was the geologic column created? Historical science is the study of events in Earth's past that cannot be repeated. Because experiments cannot be conducted on the data collected, scientists have developed different theories based on their interpretations about the evidence they have found in the geologic column. Geological evidence that is consistent with the Flood includes fossils, widespread sedimentary deposits, flat gaps, turbidites, ocean sediments on the continents, continental-scale currents, and incomplete ecological systems in the fossil record. The Flood may have created the conditions necessary to produce an ice age.

1. Why is historical science more challenging to study than experimental science?

2. From the evidence presented in this lesson, choose an example that points to destruction on a much larger scale than what we see happening today. Describe what can be observed.

3. From the evidence presented in this lesson, choose an example that points to how rapidly deposition happened. Describe what can be observed.

Objectives

- Distinguish between gradual and mass extinctions.
- Compare natural causes of extinctions with human-caused extinctions.
- Identify species of animals that have become extinct.

Vocabulary

gradual extinction
mass extinction
bird-hipped
lizard-hipped
megafauna

This fish, the Tecopa pupfish, was native to the hot springs of the Mojave Desert in California. The fish's habitat declined because of development of the Tecopa hot springs and surrounding land.

The Tecopa pupfish was the first species removed from the endangered list because it became extinct. How did its limited habitat lead to its extinction?

Essential Question

What Do Fossils Tell about Extinction?

In 1700, there may have been 6 billion passenger pigeons living in North America. By 1900, the passenger pigeon was extinct in the wild. In 1914, the last passenger pigeon on Earth died at the Cincinnati Zoo. The passenger pigeon is one of many species that have become extinct. Some became extinct after populations declined over hundreds of years. Others became extinct more quickly. Still others appear to have become extinct as the result of a catastrophe. What are some catastrophic events that could cause such extinctions? What human actions contribute to the extinction of species?

Types and Causes of Extinction (Explain)

In Lesson 1, you learned about various types of fossilized creatures, many of which no longer live on Earth—they are extinct. Extinction occurs when the last member of a species dies. What do you think can cause a species to completely disappear from the planet?

Extinctions occur at different rates. Extinctions in which only one or a few species disappear are called **gradual extinctions**. Extinctions in which large numbers of species suddenly die out are called **mass extinctions**. Scientists estimate that more than 95% of all the species that have ever lived on Earth have become extinct. Some of these extinctions happened long ago. Others have occurred in the past 500 years. In fact, the extinction of the black rhinoceros occurred in 2011. What do you think leads to a gradual extinction? What causes mass extinctions?

342

ID the Trilobites

What characteristics do all trilobites have in common?

Procedure

1. Study the labeled diagram of the trilobite in your *Science Journal*. Note the names of its body parts.

2. Carefully examine the five trilobite specimen pictures in your *Science Journal*. Use a hand lens to **observe** more detail.

3. **Record the data** you **observe** for each trilobite in the table in your *Science Journal*.

4. Use the information provided in the trilobite key and the characteristics you identified in Step 3 to identify the scientific name of each trilobite.

Materials
- trilobite key
- trilobite picture set
- hand lens

Analyze Results

Compare your results with another classmate. Are there any instances where your classifications are different? What might account for any differences in your interpretations of the key?

Create Explanations

1. What characteristics do all trilobites have in common?

2. Which characteristics were most helpful in identifying the trilobites?

3. Where are trilobites found in the geologic column? How is this information useful?

4. What living group appears to be most similar to trilobites?

Scripture Spotlight

Compare **Genesis 7:4** with **Genesis 7:21–23**. How might this be related to mass extinctions?

Gradual Extinctions

Most gradual extinctions occur when a species' environment changes. Suppose, for example, that a species lives in a very small area that receives much rain. If the amount of rainfall steadily decreases over time, the species will try to adapt by changing its behavior or perhaps migrating to another location. If it cannot adapt to the change in water, it will become extinct.

Although humans are expected to care for God's Creation, some human actions can affect the survival of wild species of plants and animals. Threats of pollution, reducing their habitat or food supply, cutting off wildlife passageways, illegal poaching, or hunting can push species past the recovery point.

Mass Extinctions

Unlike gradual extinctions, mass extinctions are the result of major changes on Earth. Scientists have found evidence of several mass extinctions in the fossil record. For example, in the Paleozoic section of the geologic column, the Ordovician layers contain fossils of many marine creatures. But in the Silurian layers just above, 60% of those creatures no longer exist in the fossil record. Scientists interpret this data to mean that those species became extinct. Evidence of another mass extinction is found in the Cretaceous layers near the boundary between the Mesozoic and Cenozoic. This is probably the most well known extinction event because it involved the dinosaurs.

Recall what you have learned about the difference between experimental and historical science. Into which category do you think the challenge of figuring out the cause of these mass extinctions falls? Because of clues in the fossil record, scientists have suggested that climate change, dropping sea levels, volcanic eruptions, or various combinations of these may have been responsible for the extinctions. How do you think some species would have been able to survive these major events? It is impossible to know for sure.

Geologic Column

Eon	Era	Period
Phanerozoic	Cenozoic	Quaternary
		Neogene
		Paleogene
	Mesozoic	Cretaceous
		Jurassic
		Triassic
	Paleozoic	Permian
		Carboniferous
		Devonian
		Silurian
		Ordovician
		Cambrian
Proterozoic		
Archean		
Hadean		

The geologic column provides clues about changes that have happened in the past.

What do the red arrows marking sections of the geologic column indicate?

Many scientists believe that the dinosaur extinction event involved a meteorite impact that sent massive amounts of dust into the air. The dust would have prevented sunlight from reaching Earth. As a result, plants and other producers could not make food, and the animals that eat plants could not have survived. This hypothesis is based on the presence of unusually high levels of iridium in these rock layers. Iridium is an element that is rare in Earth's crust but is more abundant in meteorites. Some interesting questions exist, though. Why would dinosaurs become extinct but not alligators, crocodiles, and fish? Why would ammonites become extinct but not the chambered nautilus? Why would one kind of clam become extinct but not another? How could the effects of a meteorite impact cause such selective extinctions? Can you think of any other events that could have caused mass extinctions? Scientists with a biblical worldview believe many mass extinction events are linked with the Genesis Flood and the events immediately afterward.

ammonite

chambered nautilus

Explore-a-Lab

Structured Inquiry

How can you model and analyze extinction rates?

Tracking extinction rates for animals that are already long extinct can only be done with modeling. Play this Extinction Game to help you model extinction rates for a herd of mammoth. Gather 30 number cubes. Distribute 20 number cubes among the members of your group. Each cube represents one mammoth. The extra 10 cubes will be used for new births. Each roll of the cubes equals one year. Roll all the cubes at the same time. Adjust your population size by adding or subtracting a number cube for every new birth or death in the herd. Create a table to track your number cube toss results. Play the game for 20 rounds (20 years). Does your herd become extinct? If so, how many years does it take?

1 = **Dies of starvation**	**4** = **Lives another year**
2 = **Killed by a predator**	**5** = **Lives another year**
3 = **Gives birth to a calf**	**6** = **Lives another year**

Look back to the
Structured Inquiry. Now
that you have learned
more about trilobites,
would you change
your interpretation of
the data?

A fossilized trilobite.

 **How do you think
the trilobites used
their eyes and
numerous legs?**

Extinct Animals (Explain)

Although thousands of species have become extinct throughout
Earth's history, you will read about only a few groups of them
in this lesson. The first group, the trilobites, consisted of small
organisms that lived in shallow seas long ago. The other two
groups included some of the largest organisms that ever lived on
land, such as the dinosaurs and woolly mammoths.

Trilobites

Trilobites were marine arthropods that we find preserved
throughout the Paleozoic layers of the geologic column. Although
they are sometimes compared to the horseshoe crab, trilobites
are unlike any creature alive on Earth today. The word *trilobite*
means three lobes, which accurately describes the bodies of these
arthropods. All trilobites had a central lobe, or section, and a side
lobe on either side of the main lobe. Numerous legs were attached
to the lobes. How do you think scientists deciphered all these
details about an extinct organism? Is there any way for scientists to
know for certain the function of these structures? Why or why not?

The existence of trilobites in the fossil record
provides evidence that is consistent with the
biblical view of Earth's history. Their complexity,
their appearance early in the fossil record without
evidence of evolutionary ancestors, and their
exquisite preservation are compatible with the
biblical accounts of Creation and the Flood.

Explore-a-Lab

Guided Inquiry

How can you make a model of a trilobite?

Use clay or some other modeling
material to make a three-dimensional
model of one species of trilobite. Base
your model on information provided in
the text and build on what you learned
in the Structured Inquiry about these
animals' complex structures. Be sure to
add labels to your model identifying
specific structures unique to the species
you decided to model. Compare your
model with a classmate's. Use the
Internet to discover the diversity of
different species. What features did you
use? How could you use fossils like this
trilobite to identify a specific layer of the
geologic column?

Dinosaurs

Dinosaurs have fascinated people for many years. In the nineteenth century, scientific attention turned to new discoveries made in strata along the coast of England. A 10-year-old named Mary Anne Mantell found large reptile teeth in 1822. Her find was named *Iguanodon* because the teeth resemble the teeth of an iguana lizard. Sir Richard Owen studied the fossils found by Mantell and by others. He made models for the Crystal Palace exhibition and first coined the name "dinosaur."

While scientists were studying these new finds, North America had its own dinosaur rush! Two important early dinosaur hunters were O. C. Marsh and Edward Drinker Cope. They began competing to find the biggest and best skeletons to exhibit. Often in their zeal, the same dinosaur was named twice. For example, Apatosaurus and Brontosaurus are two names for the same animal. Since these early discoveries, many dinosaur fossils have been found in every region and continent, even on Antarctica. Many more remain buried, waiting to be found.

These reptiles probably varied greatly in size and in appearance. The fossil bones can provide solid data about the size and body plan of these animals. Scientists generally classify dinosaurs into two main groups based on the shape of their hip bones: the *Ornithschian* (bird-hipped) and the *Saurischian* (lizard-hipped) dinosaurs. Both groups stood erect with their legs held under their bodies.

However, as far as their appearance and behavior are concerned, there is a lot of interpretation involved in what you see in museums or in reconstructions. For example, skin prints can provide clues about skin texture, but scientists must use their imaginations and knowledge about living organisms to fill in missing details like coloring, striping, or camouflage patterns. Footprints can help us figure out how they might have moved, and coprolites and fossilized stomach contents can suggest what foods they ate, but the conclusions remain tentative and a matter of interpretation. Scientists often disagree with the conclusions and interpretations of other scientists.

Scientists continue to study dinosaur evidence.

Why do scientists continue to study dinosaurs?

Ornithischians (Bird-hipped Dinosaurs)

Recall that scientists generally classify dinosaurs into two groups based on skeletal features. The Ornithischians or **bird-hipped** dinosaurs, as their name suggests, had a hip structure similar to birds. Most of these dinosaurs ate plants and many had bony plates on their skin. Stegosaurs were one group of bird-hipped dinosaurs. Stegosaurs walked on four legs, had small heads, and had short necks. Bony plates on their backs or heads are thought to have been used for defense or to perhaps regulate body temperature. Is there any way we can know for sure? *Hadrosaurs* were another group of bird-hipped dinosaurs common in western North America. These animals had duck-like beaks and hundreds of teeth in their mouths. Fossil evidence indicates that hadrosaurs called Maiasaura took care of their young. What type of evidence do you think scientists may have found to indicate this?

Saurischian (Lizard-hipped Dinosaurs)

The Saurischian or **lizard-hipped** dinosaurs were much larger than bird-hipped dinosaurs. Sauropods were among the largest lizard-hipped dinosaurs. One sauropod, Seismosaur, achieved a length of 42 m (140 ft), making it the largest land vertebrate that walked on Earth. About 150 species have been identified. *Theropods* were another group of lizard-hipped dinosaurs. Most of these animals ate meat. All had long tails that were thought to have been used for balance. Theropods are believed to have walked on their strong hind legs and used their front limbs for grasping. Why do you suppose that scientists reached this conclusion? Theropods included the enormous tyrannosaurs as well as the relatively small but ferocious raptors. They were not all giants, and included the turkey-sized Compsognathus, which was one of the smallest dinosaur known, and a diverse group of medium-sized cousins.

Pterosaurs

What advantage did these reptiles have over most dinosaurs?

Marine Reptiles and Pterosaurs

Long-necked marine reptiles called plesiosaurs also appear in the Mesozoic layers. At 15 m (50 ft) long, they were as long as a 5-story building is tall! Ichthyosaurs were even longer, up to 21 m (70 ft) long. Flying reptiles called pterosaurs flew in the skies above them. Some were no larger than a sparrow, while others were the size of an airplane with a wingspan of up to 12 m (40 ft).

Marine reptiles have frequently, and erroneously, been called dinosaurs. Though they have been found in the same part of the geologic column, they differed from dinosaurs in several important anatomical details. No marine reptile had a skeleton with a hip bone shaped like either of the main groups of dinosaurs, and their limbs were modified flippers or paddles, somewhat similar to the fins and flippers of marine mammals such as seals and whales.

Pterosaurs are not considered dinosaurs either. They had hollow lightweight bones, long necks, short bodies, and forelimbs that took the form of wings. Their wings were featherless and made of a skin membrane similar to the wings of a bat. Whether they flew by flapping their wings or gliding through the air is uncertain. How could scientists test their theories about pterosaur flight? On the ground, pterosaurs may have walked or crawled, or they may have been bipedal, walking on only their hind limbs. What do you think caused these marine reptiles and pterosaurs to become extinct?

Called to Serve

The "DinoDig," sponsored by Southwestern Adventist University, set a new standard of excellence in fossil fieldwork. The data gathered there since 1997 is consistent with a catastrophe like the biblical Flood. This project demonstrates how a biblical worldview can be a catalyst for rigorous, productive science done with excellence. You will read more about the director of the DinoDig at the end of this chapter.

Megafauna

Another group of extinct animals, called the **megafauna**, includes large animals that lived during the Cenozoic Era. This group of animals included mammals, birds, and reptiles.

Birds called moa were nearly twice as tall as an average human. Although they could not fly, fossilized footprints suggest that moa could walk up to 5 km/h (3 mi/h). Another flightless giant bird belonged to the *Aepyornis* species. These birds were more than 3 m (10 ft) tall and could have weighed nearly 500 kg (1100 lbs). The eggs laid by these birds measure at least 1 meter (3 ft) around and were more than a third of a meter (3 ft) long. How could scientists tell that these birds could not fly?

The giant ground sloth was about the size of a black bear, but had claws that could be 50 cm (20 in.) long. These sloths did not have front teeth, but were still able to eat various types of plants, including yuccas, willows, and cacti. Australia's giant marsupials included a giant kangaroo and marsupial bear, 2 m (6.5 ft) tall at the shoulder.

Mastodons and woolly mammoths are two other megafauna. Mastodons had large tusks and teeth. They stood about 3 m (10 ft) tall and weighed about 550 kg (1200 lb). Woolly mammoths were much larger and probably weighed around 8 000 kg (17,600 lb). At the shoulder, they measured 5 m (16 ft) tall and had long, curved tusks, which were often 4–5 m (13–16 ft) long. What do you think may have caused these animals to become extinct?

Math in Science

Use a metric tape measure to find the circumference and length of an extra-large chicken egg. Compare this to the size of the eggs of the extinct bird *Aepyornis*. How much larger, in terms of percent, were the eggs of *Aepyornis*?

Several extinct rodents belonged to the megafauna. Two species of giant beavers that lived in North America measured nearly 2.5 m (8 ft) long and weighed close to 100 kg (220 lb). The teeth on these rodents were up to 15 cm (6 in.) in length. Another mega-rodent was the North American giant capybara that lived in Florida, South Carolina, Texas, and Arizona. Capybaras were the world's largest rodents. This extinct animal weighed about 90 kg (200 lb), which is nearly twice that of capybaras living on Earth today.

The largest animal in the megafauna was a hornless rhinoceros called *Paraceratherium.* Commonly known as *Indricotherium,* it was probably the largest mammal to have ever lived on land. Based on fossil finds, an adult could have been 5.5 meters (18 ft) tall at the shoulder and 12 m (39 ft) long! Scientists believe it had a neck like a giraffe that would have allowed it to eat the leaves from treetops. Why do you think these large animals could not survive?

The fossil record is filled with a fascinating variety of animals that are now extinct. It is likely that many of those became extinct as a result of the Flood. Many other animals have become extinct since then as a result of human activities. Although we cannot control what has happened in the past, we can work to prevent the loss of more species in our time.

Faith Connection

On page 112 of *Patriarchs and Prophets,* Ellen White indicates that animals many times larger than those we see now used to exist on Earth.

Concept Check   Assess/Reflect

Summary: What do fossils tell about extinction? Gradual extinctions occur when a species' environment changes. Extinction events that involve large numbers of species dying out at the same time are mass extinctions. Some scientists have concluded that these events are the result of catastrophes such as meteor impacts or volcanic eruptions. But other scientists believe most of the mass extinction events found in the geologic column could have happened during and immediately after the Genesis Flood. Subsequent extinctions involving plants and animals have occurred since that time.

1. What is meant by the word *extinction*?
2. Compare gradual extinctions and mass extinction events.
3. Some scientists believe dinosaurs became extinct as the result of a meteorite impact. What data is this interpretation based on?
4. According to the interpretation described in question 3, how might a meteorite impact have affected the plant-eating dinosaurs?
5. Compare an animal from the megafauna with an animal alive today.

Get to Know
Dr. Arthur Chadwick

Dr. Arthur Chadwick is a biologist and geologist. He is currently studying many dinosaur fossils that are preserved in a quarry in Wyoming. Dr. Chadwick uses the latest technology to understand how these dinosaur fossils ended up where they did.

After fossils are discovered, scientists follow steps to understand the relationship between each fossil and the surrounding rock. They describe the fossil and its condition. This might involve drawing pictures, making a map of the location, and taking photos. Finally, the scientists begin the task of carefully removing the fossil from the surrounding rock. This can take a very long time. The scientists have to be careful not to damage the fossil. Once it is removed, the fossil is carefully wrapped for transport to the laboratory where it will be studied.

After a fossil is collected, the relationship between the fossil and the rock is lost. Enter Dr. Chadwick and his high-precision GPS and GIS equipment.

You might be familiar with the global positioning system, or GPS. Most people use this technology every day while driving or walking around town. Dr. Chadwick uses GPS to mark the exact location of each dinosaur bone before it is removed from the rock. He enters this information into another computer system called GIS (the geographic information system). This powerful program combines all of the data to produce a three-dimensional map of the quarry. This information has given Dr. Chadwick and his research team a more complete picture of what the area looked like before the dinosaurs became fossilized.

Called to Serve

Dr. Chadwick believes that making a detailed map of a quarry where numerous fossils are found will best preserve the data for current and future scientific study. It makes good sense and it is good science.

Concept Check

1. Why do scientists record so much information about the fossils before they are removed from the rock?
2. How can studying an area in three dimensions help scientists understand the area's history?

Virtual Paleontology

When paleontologists find the remains of a dinosaur or other fossilized organism, they usually have to dig it up to study it. This means chipping carefully at the area around the organism and freeing the fossil from the rock that surrounds it. If the fossilized remains are in several pieces, the pieces must be carefully put back together exactly as they were found so they can be studied. Sometimes, this means that fossils are damaged or destroyed. In most cases, the process is painstaking and long.

Thanks to new technology, this process is no longer the only option. "Virtual paleontology" is a way that paleontologists can examine the fossils of long-dead organisms without removing them from the rock. The idea of virtual paleontology occurred to researchers who were studying reptile fossils in stone. The fossils and the rock around it were the same color, so it was hard to tell the rock sample from the fossils it contained. In addition, the fossils were very thin and fragile.

The researchers took the rock samples to a medical research lab and used the computed tomography (CT) scanner to look at the fossils. A CT scanner uses X-rays to produce images of the insides of objects or people, based on the different densities of the materials inside. The researchers were able to see the fossils using the scan. Later, they used a more powerful CT scanner to examine the fossils. Researchers believe that this technique will become a common way that paleontologists can learn about fossils.

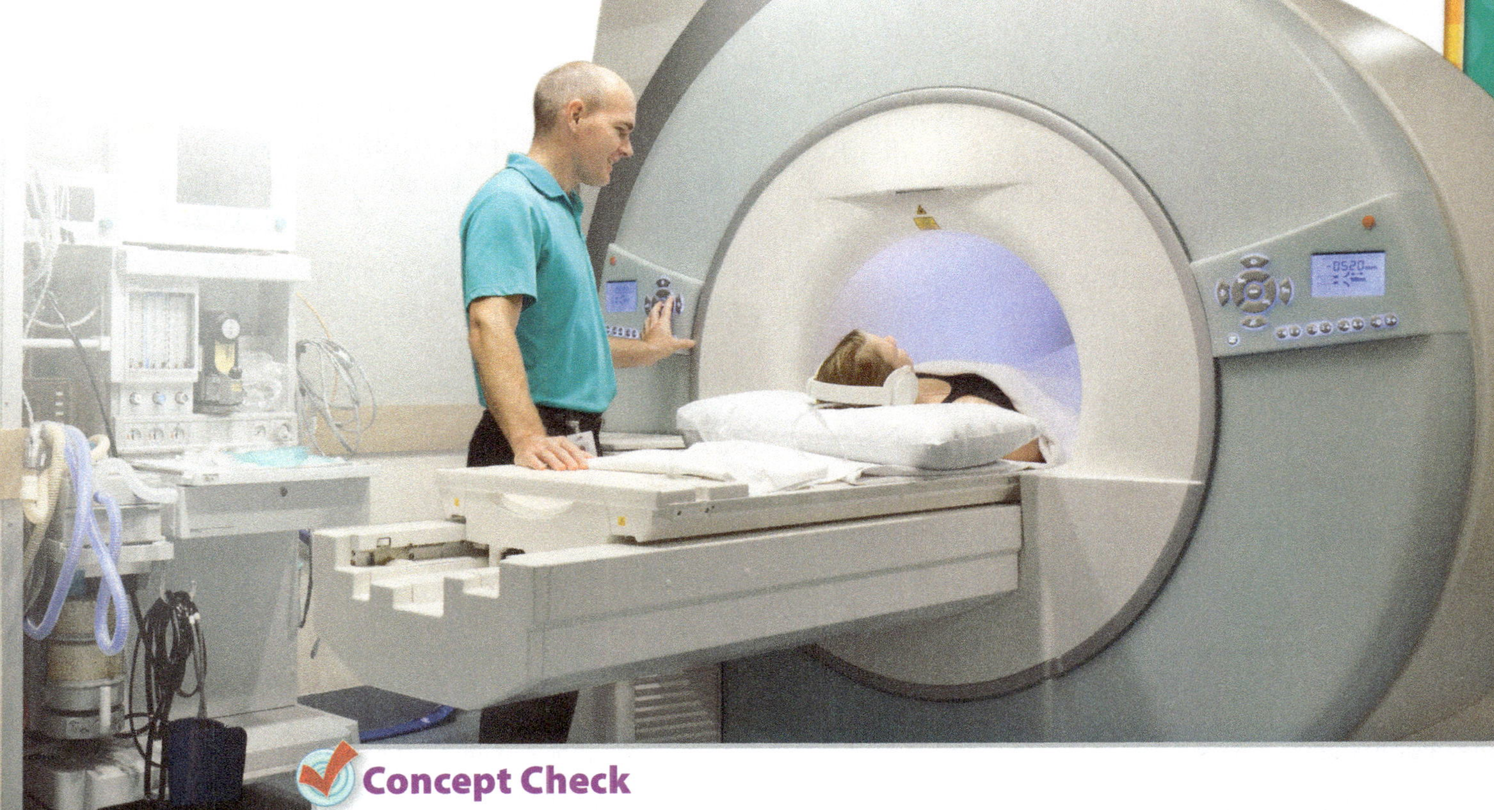

Concept Check

1. How did advances in medical technology help paleontologists?
2. What are the advantages of virtual paleontology?

Study Guide

Lesson 1

1. Rock layers and fossils found within the strata make up the geologic column.

2. Strata are laid down horizontally as sediments are deposited by moving water, but they do not always remain horizontal.

3. The rock strata and fossils in the geologic column are data; the times associated with the layers are interpretations of the data.

Lesson 2

1. Historical science seeks answers to questions about events that are not observable or repeatable.

2. Interpretations of the geological column suggest that it was either laid down over a long period or as a result of the death and burial of organisms during and after the Genesis Flood.

3. Evidence consistent with the Flood includes flat gaps, turbidites, ocean sediments on continents, and incomplete ecological systems.

4. Some scientists believe that there were several ice ages; others think conditions necessary for an ice age could have resulted from the Flood.

Lesson 3

1. Many extinction events have been identified in Earth's history. Some of these events were gradual, occurring over hundreds of years. Others were mass extinctions where a large number of species died at the same time.

2. Some extinction events occur because of human actions. Others result from organisms' inability to adapt.

3. The Paleozoic and Mesozoic layers of the geologic column contain evidence of mass extinction events. Many kinds of animals, such as trilobites, dinosaurs, and megafauna, are examples of extinct species.

4. Some scientists have concluded that these events resulted from catastrophes, such as meteor strikes or volcanism. Others believe these events can be explained by the Genesis Flood.

Show What You Know

Visualize It Complete the following concept map.

Earth's Past

Geologic Column

Contains:

1. ______________________
2. ______________________
3. ______________________

Genesis Flood

Evidenced by:

4. ______________________
5. ______________________
6. ______________________

Mass Extinctions

Evidenced by:

7. ______________________
8. ______________________

Contrast each pair of terms.

9. experimental science—historical science
10. fossil mold—fossil cast
11. gradual extinction—mass extinction
12. relative dating—radiometric dating
13. geologic column—geologic time

Multiple Choice

Choose the best answer.

14. Which of these is the actual remains of an organism?
    A. trails or burrows
    B. a fossil mold
    C. insect in amber
    D. a coprolite

15. What kind of information do fossilized footprints or trails provide about the organisms that made them?
    A. body plan
    B. age
    C. presence of exoskelton
    D. size and movements

16. What is one way in which the Genesis Flood may have provided the conditions needed for an ice age?
    A. It caused ocean water to cool rapidly.
    B. It caused volcanic eruptions that could have cooled Earth.
    C. It caused continents to move.
    D. It caused sea levels to rise.

17. What is the standard explanation of why fossils of marine life are found on mountains?
    A. The sea level once reached that high.
    B. They were deposited on the ocean floor, and later thrust up.
    C. Faults caused the rock layers to shift.
    D. Glaciers moved them to the mountaintops.

18. What evidence leads some scientists to think the dinosaur extinction is related to a meteor impact?
    A. dropping sea levels around the globe
    B. signs of climate change
    C. high levels of iridium in rock layers
    D. incomplete ecosystems in the fossil record

Check Point

Answer the following questions.

19. **Infer** why jellyfish are rarely fossilized.

20. **Sequence** the events that suggest how the Genesis Flood and the ice age could be related.

21. Describe at least three lines of geological evidence that support the account of the biblical Flood and explain their significance.

22. **Compare** the fossil organisms found in the Paleozoic layer with those of the Cenozoic layer of the geologic column.

23. Identify the type of fossil shown here and explain how it formed.

Physical Science

Computers and other modern electronic devices provide immediate access to information. Electronic devices are complex systems with many parts that work together.

Unit Overview

In this unit, you will explore the characteristics of matter, which makes up all of the objects in the world around you. In addition, you will explore the kinds of changes that take place in matter and how to tell one kind of change from the other. You will also learn about the subatomic particles that make up all matter in the physical world. These tiny parts combine to form materials you see and use every day.

From there, you will focus on electricity. You will compare different kinds of electricity, explore sources of electricity, and learn how electricity is used in electronic devices. Because electricity and magnetism are related, you will learn to identify the characteristics of magnets and explore how electricity and magnetism affect each other.

Properties and Changes of Matter

Scripture Spotlight

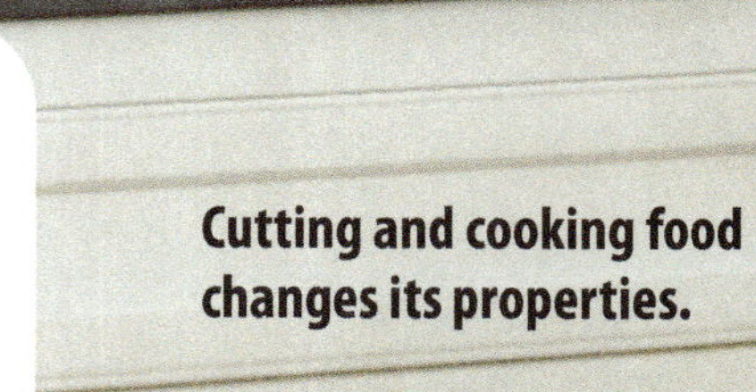

You will study the properties of matter and the changes matter can experience. You can strengthen your faith by what you are studying about matter.

You will read the following passages in this chapter.

Numbers 11:7 (p. 364)

Ecclesiastes 10:1 (p. 364)

Matthew 11:8 (p. 364)

Revelation 4:6 (p. 364)

Exodus 30:23 (p. 365)

Daniel 5:27 (p. 367)

Daniel 3:22 (p. 369)

Revelation 3:15–16 (p. 369)

Psalm 97:5 (p. 373)

Exodus 16:23 (p. 378)

1 Samuel 28:24 (p. 383)

Cutting and cooking food changes its properties.

The Big Idea

Objects are made of matter and have physical properties that can be observed and measured. Matter can be changed physically or chemically according to laws created by God.

How do the properties of foods change during preparation?

Inquiry Kick-Off Engage

Different types of matter have different properties. How would you describe custard? Could you walk on it? What makes quicksand hazardous? What is the best way to get out of it? In this inquiry, you will investigate the properties of an unusual form of matter.

Science Journal

Essential Question

What Are Physical Properties?

What do all the objects in your classroom have in common? How are they different? God created matter with many different characteristics to allow humans to make almost any object they could think of. Think about the tools that were available in biblical times and the tools available now. With God's thoughtfulness to give us so many different kinds of materials—hard and soft, liquid and solid, rigid and flexible, and many others—humans have been able to invent many tools that make life easier.

Choose one object from your desk to examine. Is it made of a single substance or several materials mixed together? How would you test it to prove your answer? Why do you think God gave matter different characteristics that allow mixing of materials?

Classifying Properties Explain

How would you classify properties of matter? Refer to the flow chart on the next page. The chart will help you determine if a property is a physical property or a chemical property. You will learn much more about chemical properties later in this chapter. How can you use this flow chart to classify the properties of an object in your classroom?

These pieces of exercise equipment are made of different kinds of matter. They differ in physical properties, such as size, shape, elasticity, and mass.

What are some properties of this equipment that help you when you exercise?

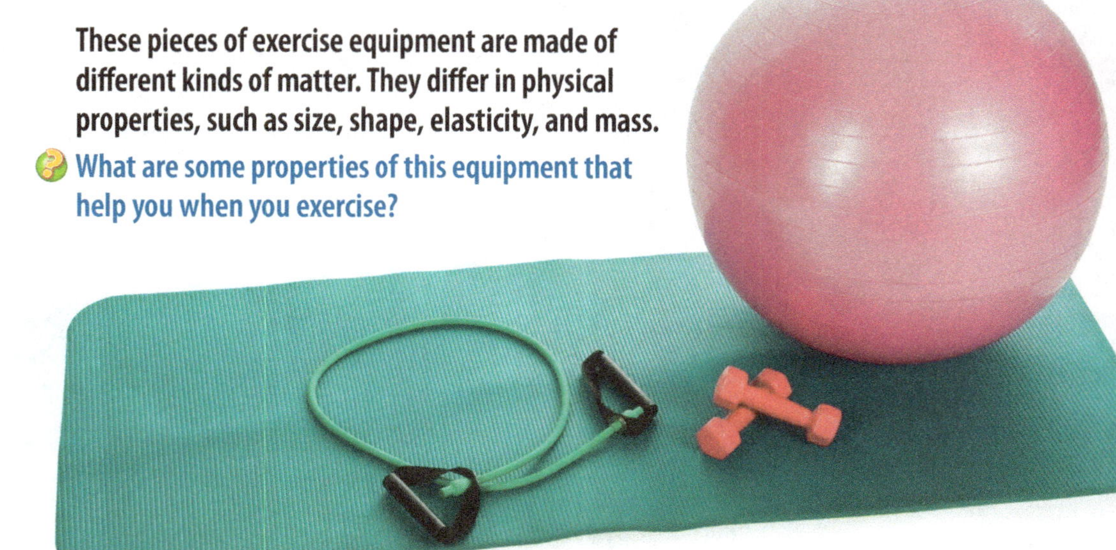

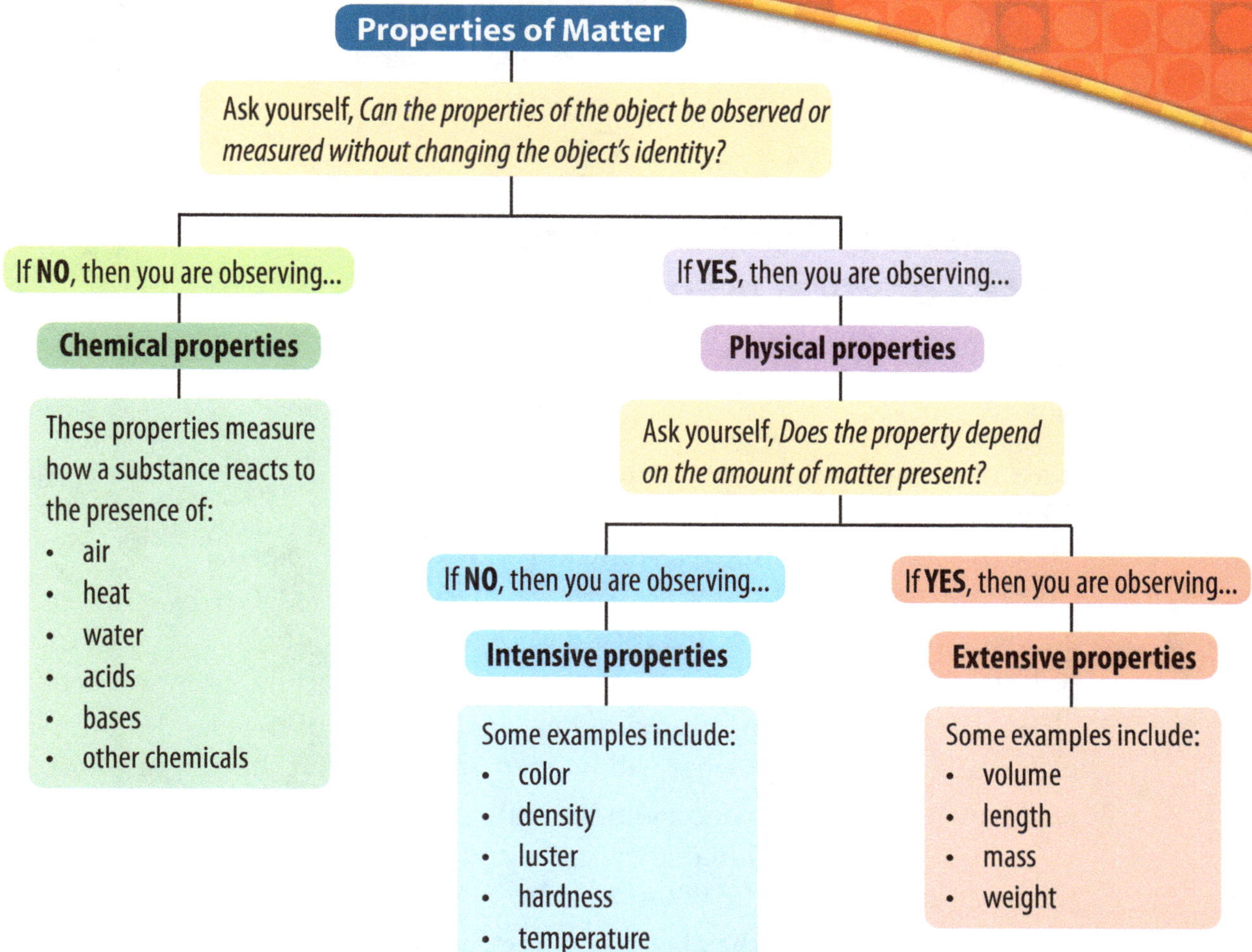

Physical Properties

A **physical property** is a characteristic of matter that can be observed and measured without changing the matter. You can use your senses to observe physical properties, such as color, shape, odor, taste, and texture. You can also learn about matter by measuring physical properties, such as size, mass, weight, length, and temperature. Other physical properties can be tested. Physical properties also determine the uses of different types of matter.

Consider the properties of a bike helmet. The strength of a bike helmet is a property of the lightweight plastics. The comfort of the helmet is a property of the cloth in the band; its tiny holes allow air to flow and sweat to evaporate.

Other properties of the matter in the helmet are color, odor, hardness, state (solid, liquid, gas, or plasma), temperature, and luster (how shiny or dull). These properties are characteristics of the kind of matter. They are the same whether the matter the helmet is made from is large or small.

Float That Boat!

How does physically changing a clay ball affect its physical properties?

Procedure

Materials
- triple beam balance
- 100-mL (3-oz) graduated cylinder
- 50-mL (1.7-oz) beaker
- 20-cm (8-in.) square plastic pan, 5 cm (2 in.) high
- plastic bowl, about 15 cm (6 in.) diameter
- $\frac{1}{2}$ stick of clay
- water

1. Use the balance to **measure** and **record** the mass of an empty graduated cylinder. Pour 50 mL (1.7 oz) of water into the cylinder. **Measure** its mass again. Determine the mass of the 50 mL (1.7 oz) of water. **Record** the mass and volume of water. Use the formula: D = m/v to find and **record** the density of the water.

2. Next, use a balance to **measure** and **record** the mass of the clay. Roll the clay into a ball. **Predict** what will happen when you put the ball in water.

3. Add water to the graduated cylinder and **record** the volume of the water used. Tilt the graduated cylinder and gently slide the clay ball to the bottom. **Measure** the volume again.

4. To find the volume of the clay ball alone: subtract the volume of the water from the volume of the clay and water. Then, use the density formula to calculate and **record** the density of the clay ball.

5. Remove the clay ball. Shape it into a clay boat. **Experiment** with the boat until you find a shape that floats the best.

6. Place the bowl into the pan. Fill the bowl completely to the top with water. Place the boat into the bowl of water. Let the water from the bowl spill out into the pan.

7. Use the beaker to **measure** the amount of water that spilled into the pan. **Record** this volume. Calculate and **record** the density of the clay boat.

Analyze Results

Examine your data table. **Compare** the density of the water with the density of the clay ball and the density of the clay boat.

Create Explanations

1. How does physically changing a clay ball affect its physical properties?
2. When you made the clay ball into a clay boat, what properties stayed the same and what properties changed?

States of Matter Explain

There are four main states of matter—solid, liquid, gas, and plasma. Can you name one example of each? The atoms that make up matter are constantly in motion, which means the atoms have kinetic energy. The states of matter are defined by their kinetic energy. Solids have the least amount of kinetic energy. Plasma has the greatest amount of kinetic energy. How could you reduce the kinetic energy of milk molecules to make ice cream?

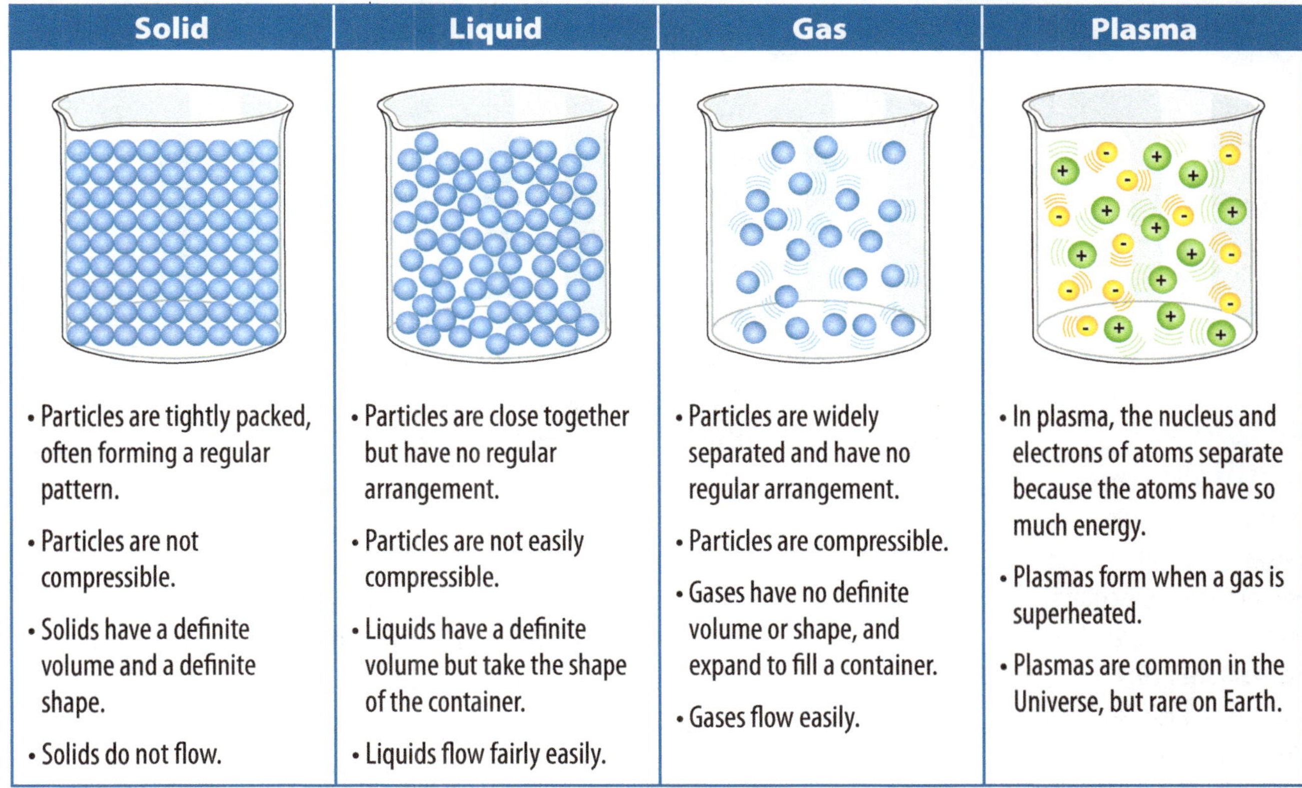

Each state of matter has its own distinctive properties.

What is one type of matter that can exist as a solid, liquid, and gas at temperatures found naturally on Earth?

Solids

The atoms of solids move back and forth and up and down. Scientists describe this movement as vibration. Why is this vibration not felt or seen? Wood, sugar, rocks, and most metals are examples of solids. Many solids form crystals arranged in regular repeating patterns. Some of these are quartz, table salt, and diamond. Solids that do not form regular patterns are called amorphous solids. Glass, wax, and plastics are amorphous solids.

Solids can be amorphous, without crystal structure, like obsidian on the right. Solids can also have a definite crystal structure, like fluorite on the left.

Draw what you think diamond molecules might look like. Now, draw what wax particles might look like. What is alike or different about the drawings? Why did you draw them this way? Do you think either of these substances could be compressed into a smaller space? Explain.

Liquids

Like the atoms of solids, the atoms of liquids also vibrate. But the atoms of liquids vibrate more than atoms of solids. As a result, the atoms of liquids spread apart more, and this allows them to slide past each other and flow. When poured, liquids spread out, so they lack a definite shape. Common liquids include milk, water, juice, and oil. What do you think are some other factors that might affect how a liquid spreads out?

Draw what you think molecules of water would look like when they are flowing in a stream. What would those molecules look like if the stream started to freeze?

Gases

Gas particles are in constant, rapid motion. Many gases are invisible, colorless, and odorless. They are easily compressed or squeezed into smaller spaces. Common gases include oxygen, helium, and carbon dioxide. What properties do you think gases might have that make them easier to compress than a solid or liquid?

Scripture Spotlight

Choose two of these verses and identify what properties are described there. **Numbers 11:7**, **Ecclesiastes 10:1**, **Matthew 11:8**, and **Revelation 4:6**.

Plasmas

Plasma is made up of free electrons and ions of different elements. It forms when a gas is heated enough to separate the electrons from the atoms, producing the electrons and positively charged ions. Plasmas are very common in the Universe, but rarely found on Earth. Why do you think naturally occurring plasmas are rare on Earth?

The Northern Lights are a natural type of plasma. But most plasmas at Earth's surface are created by people. What might be some uses for plasma? Look around your home. What electronics might contain plasma?

Scripture Spotlight

In which state does myrrh exist in **Exodus 30:23**?

Lesson Activity

Obtain a group of coins from your teacher. Identify different properties that you use to describe the coins. Group the coins based on some of the properties. Select one coin and describe it to a friend.

 **What is the coin and what properties helped you to identify it?**

Measuring Matter (Explain)

Regardless of the state of matter, it can always be measured. Using various methods, you can determine the volume, mass, weight, density, and temperature. What tools would you use to measure these properties? Can you use the same tools for solids, liquids, gases, and plasmas?

Common properties of matter include mass, weight, volume, shape, and size. These properties depend on the amount of matter in an object. **Mass** refers to the amount of matter in an object. We often use the terms *weight* and *mass* interchangeably, but they are different properties. **Weight** is a measure of the effect of Earth's gravity on an object. If you went to a planet with less gravity than Earth, your weight would be less, but your mass would be the same. **Volume** is the amount of space that matter occupies.

Volume

Recall that volume is the amount of space that matter occupies. Besides science class, when might you need to know the volume of some type of matter? Ask your family members if they ever need to know the volume of something for work or school.

Measurement	States of Matter	Equipment	Units
Volume	• solid • liquid • gas	• metric ruler • graduated cylinder • beaker	• cubic centimeters (cm^3) • milliliters (mL)

==Displacement== is the volume of fluid that is pushed aside by the object. Recall that you measured displacement to find the volume of the clay ball in the Structured Inquiry. Displacement provides a way to measure irregular solids. Conduct independent research to determine how Archimedes used displacement to understand buoyancy.

The water in this measuring cup curves up at the edges.

 How much liquid is in this measuring cup? How accurate is your measurement? Explain.

Math in Science

Determine the volume of your science book. What measurements do you need to take to determine the volume?

Mass

A balance measures the mass of objects. Balances come in several forms. A simple pan balance consists of two pans set at equal distances along a lever arm. An unknown mass is placed in one pan. Standard masses are placed in the other pan until the two pans balance. How was this type of balance useful in biblical times? Is it still used today?

A triple beam balance uses three beams with known masses, which slide along the beam. The middle beam reads in 100-gram units. The rear beam reads only in 10-gram units. The front beam reads from 0 to 10 grams.

To find an unknown mass using this kind of balance, follow these steps:

1. Zero out the balance by moving all the masses to the left, and adjust the balance screw if the pan is not balanced.

2. Set the unknown mass on the pan.

3. Slowly slide each mass to the right until the pointer points directly at the center mark.

4. Add up the values of all three sliders to find the mass of the object.

Do you think a triple beam balance is more accurate than a pan balance? Explain. Which is easier to use?

An electronic balance provides a digital display of mass. It contains a single pan. The reading on the balance is set at zero. The object is placed on the pan and load cells within the device sense the mass of the unknown sample. A digital readout is displayed on the scale. Electronic balances are used to measure chemicals and biological samples. How is an electronic balance used in the home. Why is it important for us to know how to use all three types of balances?

Scripture Spotlight

Daniel 5:27 is a message to someone important about a balance. Read the verses around it to learn who the person was and what the message was.

The photo shows a triple beam balance.

What is the mass of the unknown object on this balance?

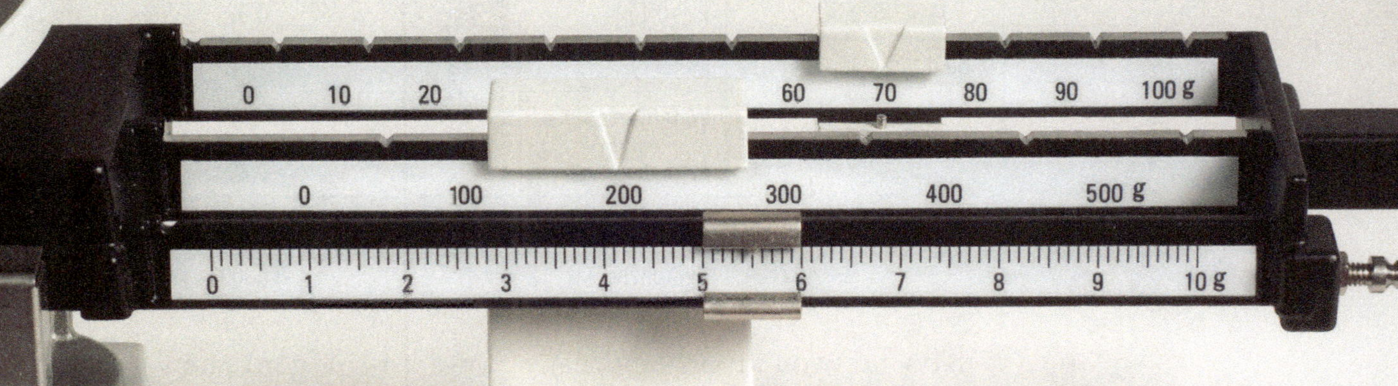

Measurement	States of Matter	Equipment	Units
Mass	• solid • liquid • gas	• simple pan balance • triple beam balance • electronic balance	• grams (g)

Weight

Recall that weight is a measure of the pulling force of gravity on objects. A spring scale is a useful tool for measuring weight. The object to be weighed is attached to a hook. The movement of the spring is read on a scale or dial that shows how much the object weighs. Spring scales measure objects in newtons, units of force. The amount of force acting on an object will vary depending on the object's location and its mass. How would your weight on the Moon compare to your weight on Earth? Why?

Density

Density is a measure of mass per unit of volume, or how closely the particles in an object are packed together. For example, a rock and a foam block of the same volume have different masses. The rock has a greater mass, so its density is greater. Density is found by dividing mass by volume ($D = m/v$). Common units for density are g/cm^3, kg/m^3, and g/mL. Density allows us to compare the same volume of different kinds of matter. Why is it important for an engineer to know the density of substances? The table gives the densities of some common substances. How could a chemist use this table?

Density of Common Types of Matter		
State	**Kind of Matter**	**Density (g/cm^3 or g/mL)**
Solid	Gold	19.3
	Carbon steel	7.85
	Ice (0°C)	0.92
Liquid	Mercury	13.6
	Water (4°C)	1.0
	Alcohol	0.81
Gas	Air	0.0013
	Hydrogen	0.00009

Now that you know the density of water is 1.0 g/mL, how can you test the density of objects? Think back to the clay ball you used in the Structured Inquiry. Based on your observations, do you think its density is greater or less than 1 g/mL? An object floats in a liquid if the object is less dense than the fluid. For example, ice floats in water because it is less dense than water. Objects that are denser than water sink. Why does a hot air balloon rise in the atmosphere?

Temperature

The temperature of matter—how hot or cold something is—can also be measured. Thermometers are tools used to measure this.

There are several types of thermometers available that use the fact that matter expands when heated in order to determine temperature. One type of thermometer has a long glass tube with a bulb at the base. The marks on the outside of the tube indicate the degrees of temperature. The liquid inside the glass tube expands when heated and contracts when cooled. Bimetallic strip thermometers use two different metals that expand at different rates as they heat. Liquid-crystal thermometers have heat-sensitive crystals that change color to show different temperatures. A digital thermometer uses sensors to detect changes in heat. What careers might require the use of a thermometer?

The Celsius and Fahrenheit scales are the two most common scales for measuring temperature. Both systems use the freezing and boiling points of water as reference points. The Fahrenheit temperature scale is divided into 180 degrees. Anders Celsius divided his scale into 100 degrees. Do you think one temperature scale might be better to use than another in certain situations? Which scale are you most familiar with?

Scripture Spotlight

Daniel 3:22 and **Revelation 3:15–16** involve temperature. Which one is about literal heat and which is about symbolic heat?

Concept Check Assess/Reflect

Summary: What are physical properties? Physical properties of matter can be observed using the senses. Properties of matter include, color, mass, hardness, and state (solid, liquid, gas, or plasma). Tools, such as graduated cylinders, balances, and thermometers, are used to measure physical properties. Mass is the amount of matter in an object, whereas weight is a measure of the force of gravity pulling it. Density measures how close together particles are in an object. Volume is the amount of space that matter occupies. Temperature, a measure of how cold or hot something is, uses either the Celsius or Fahrenheit scale.

1. Compare the physical properties of a block of wood with the physical properties of milk.

2. How are mass and weight related?

3. A metal cube measures 2 cm on each side. The cube has a mass of 40 grams. What is the density of the cube? Is the cube made of gold or another mineral? Explain.

4. Explain how a liquid thermometer measures temperature.

Objectives

- Explain what a physical change is.
- Identify examples of physical changes in matter.
- Explain how thermal energy affects the states of matter.
- Identify factors that affect the freezing point and boiling point of matter.
- Describe why dissolving substances affect the properties of mixtures.

Vocabulary

physical change
dissolve
melting point
sublimation
boiling point

Essential Question

What Are Physical Changes?

Think for a moment of a hot sunny day in the summer. Now think of getting an ice cream cone to enjoy. As you walk outside enjoying your cone, what is likely to happen? Why does this happen? What might be different if you were enjoying the ice cream cone on a cold day? Now think of sharpening your pencil, or watching frost form on a window pane, or seeing powdered drink mix dissolving in water. What do all of these events have in common? These are all physical changes. What physical changes can you see happening around you right now?

Physical Changes `Explain`

A **physical change** is a change in the physical properties of a substance. Freezing a liquid is a physical change. Physical changes only change the form of matter. They change properties of the size, shape, or state of matter. The matter itself remains the same. How can you physically change wheat or an aluminum can? A physical change is usually reversible and does not create a new substance. For example, if you melt an ice cube, you can change it back into ice again. Other physical changes are not so easy to reverse. If you crack a raw egg, can you reverse this physical change? Why or why not?

Cars are designed to buckle and absorb energy during a collision.

How could the physical changes to this crushed car be reversed?

Physical changes can be classified into three different
kinds of changes based on the condition or appearance of
matter. These include: (1) changes in the shape or size of
objects, (2) changes that result from mixing matter together,
and (3) changes in the form or state of matter.

Changes in Shape or Size

All changes in shape or size are physical changes. Take three
sheets of paper. Change each sheet of paper without changing
the nature of the paper fibers. What did you do to each sheet of
paper? How do you know these are physical changes? What could
you do to reverse the changes?

Look around your classroom. Identify some objects that are
made by making a change in the shape or size of a substance.
What metal objects did you identify? Heating metals can change
them into thousands of useful objects. Melting a substance is one
way to change its shape.

Changes in Mixtures

Using a clean spoon and paper cup, mix a spoonful of salt in some
water. The salt seems to disappear. It spreads and mixes evenly in the
water, or **dissolves**. You have made a mixture—a substance made by
mixing two or more substances together. How do you know the salt
is still there? How could you reverse this physical change?

Solids, liquids, and gases can all form mixtures. The air we
breathe is a mixture of gases, such as oxygen, nitrogen, and water
vapor. A tuba is made of brass, which is a mixture of copper and
zinc. Some of the most enjoyable mixtures involve food. What
are some ways you can combine foods to produce a mixture and a
tasty treat? You will study mixtures in more detail in Chapter 12.

How can you change a piece of paper so you can step through the middle of it?

Your group will be given several sheets of construction paper and pairs of scissors. Work together as a team to develop a way to cut one sheet of construction paper so that it forms a complete circle large enough for you to step through. You are not allowed to use tape or glue. If you need more construction paper, ask your teacher. What type of change have you made to the paper? Is it still paper? Can you reverse the change?

Dissolving Sugar

How does temperature affect the amount of sugar that can be dissolved in water?

Materials
- graduated cylinder
- two glass beakers
- thermometer
- sugar
- plastic spoons
- balance
- weighing paper
- cold water
- warm water
- hot water
- ice water

Procedure

1. **Measure** 100 mL (3 oz) of cold water and pour it into a beaker. **Measure** and **record** the temperature of the water.

2. Place the weighing paper on the balance. The mass of the paper is not needed in this activity, so re-zero the balance before moving on to Step 3.

3. **Measure** 250 g (8 oz) of sugar using the balance. Use weighing paper so the sugar can be easily removed from the balance.

4. Using a plastic spoon, add the sugar to the beaker, a little at a time. Stir until the sugar is dissolved. Stop adding sugar when it will no longer dissolve.

5. **Measure** the amount of sugar on the weighing paper that you did not use. **Use numbers** to find how much sugar was added to the water. **Record** the result.

6. Repeat Steps 1–5, using the warm water, the hot water, and the ice water.

Analyze Results

Compare how much sugar dissolved in cold water, warm water, hot water, and ice water. Graph your results.

Create Explanations

1. How does temperature affect the amount of sugar that can be dissolved in water?

2. How did you know when each mixture was saturated with sugar?

3. What do you think will happen as the mixture of sugar and hot water cools?

4. How did the amount of sugar dissolved in ice water compare to the amount of sugar dissolved in hot water? Explain.

Changes of State

When a substance changes from one state of matter to another, it has undergone a change of state. Recall that there are four common states of matter: solid, liquid, gas, and plasma. After a rainstorm, you might see a puddle. After a day, you notice the puddle gets smaller. Where did the water go? Did a physical change occur? How do you know? Will the water return again? If so, how? All matter changes from one form or state into another under the right conditions. For example, at room temperature, hydrogen is a gas, iron is a solid, and alcohol is a liquid. However, if the right conditions are created, each of these samples will change. Hydrogen gas will change to a liquid under pressure at low temperatures. Iron will melt and become a liquid if it is heated to a high enough temperature. And at room temperature, given enough time, the alcohol will evaporate away as a gas.

Solid-Liquid Changes

Thermal energy causes the physical changes you just learned about. For example, iron will begin to melt when it is heated to 1535°C (2795°F). Recall that adding energy increases the kinetic energy of the atoms. The atoms begin to bump into one another faster. Eventually the atoms move fast enough to break free of the rigid structure of the solid, melting into a liquid. The temperature at which a solid melts is called its **melting point**. For most substances, the melting point and freezing point are the same temperature. Melting and freezing points are physical properties of matter. Why is knowing the melting point of a substance important? How do engineers use this information?

Because melting is only a physical change, the iron can be changed back to a solid state. The freezing point of iron is the same as its melting point and is also a physical property of matter. By removing the thermal energy and letting the liquid iron cool, the atoms begin to move slower. This allows the atoms to once again form a rigid pattern and become a solid. What materials can you change from a liquid to a solid and then back to a liquid?

Why is this change occurring?

The melting point and freezing point of distilled water is 0°C (32°F). When you freeze liquids, they lose energy. The particles begin to slow down because they contain less thermal energy. The particles begin to take on the regular crystal structure of a solid. They vibrate in place instead of sliding over each other. How can a substance freeze and melt at the same temperature? How can a meteorologist use this information to predict frost or snow?

In which setting will the change from solid to liquid occur the fastest?

Place an equal number of ice cubes into each of three plastic cups. Place a thermometer into each cup and record the starting temperatures. Then, place one cup in the Sun, one in the shade, and one indoors. Observe the cups and record the temperatures every three minutes for at least 30 minutes.

Solid-Gas Changes

Sometimes, a solid changes directly into a gas. This change of state is called **sublimation**. Frozen carbon dioxide is commonly known as dry ice. Carbon dioxide gas changes to a solid under low temperatures and pressure. Dry ice *sublimes,* or changes directly to a gas, at −78.5°C. Therefore, when it is exposed to room temperature it bypasses the liquid state entirely. As the particles of the solid gain thermal energy, they move faster and spread apart.

The formation of frost on a windowpane in winter is the reverse of sublimation. It is called *deposition.* Water contains enough kinetic energy to change directly from a gas, water vapor, to the solid state, frost. When frost forms on the window, water vapor has lost thermal energy. As energy is released by the gas particles in this process, they slow down. They form crystals of ice. What conditions are necessary for frost to form on a window?

You can see dry ice "fog" when it sublimes. Because of this property, dry ice is often used in theater productions to produce fog-like effects.

Why might dry ice be more useful than water ice for transporting refrigerated goods?

Liquid-Gas Changes

When water is heated on the stove, it transitions to a gas during *evaporation*. The heat imparts energy to the individual water particles. These higher-energy particles can escape from the surface of the liquid and evaporate in the gaseous form, water vapor.

If the temperature increases enough, boiling occurs. During boiling, tiny bubbles of water vapor form in the liquid. These tiny bubbles of gas combine to make the larger bubbles you see rise to the surface. The temperature at which boiling occurs in a liquid is called its **boiling point**. Boiling point is a physical property of matter. The normal boiling point of distilled water is 100°C (212°F) at sea level.

Condensation is the reverse of evaporation. As water vapor cools, the molecules slow down. The molecules meet and collide with other water particles. They slow down even more and change back to the liquid state, or condense. How does this explain the fog that sometimes appears over water? Is it always possible to get water vapor to condense on the outside of a glass? Explain your reasoning.

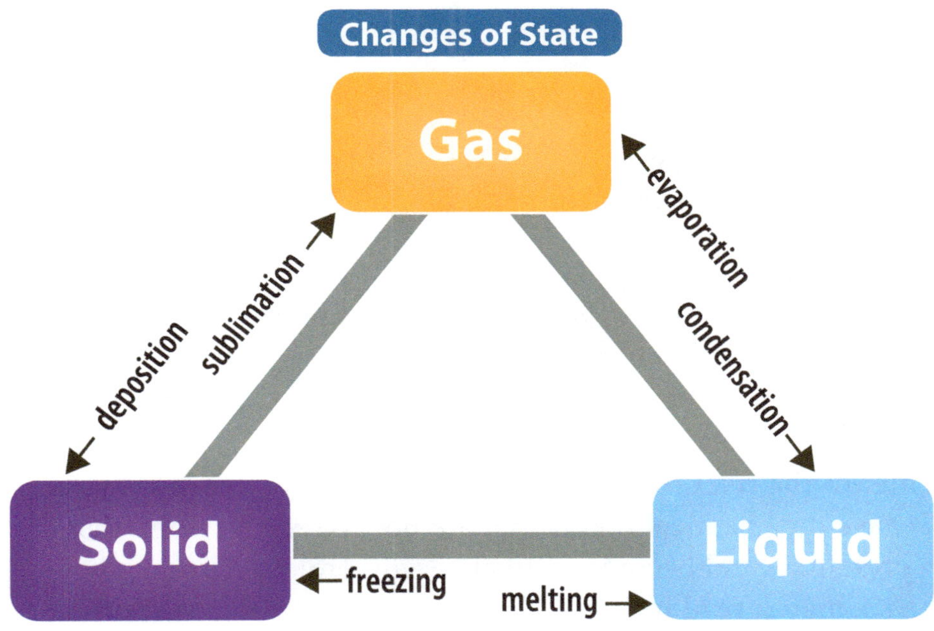

Matter moves between different states as energy is added or removed from a substance.

What are some examples of things that change state for each part of the diagram?

Visitors to the Holy Land today often swim in the Dead Sea. Do some research to see what is unique about the Dead Sea, and then explain what would happen to it in cold weather.

Factors That Affect Freezing and Boiling Points (Explain)

While the freezing points and boiling points of matter are generally specific for the particular type of matter, it is possible to affect these two properties of matter.

Factors That Affect Freezing Point

Solutes

When a solute is added to a liquid, the freezing point of the liquid becomes lower. If a solute, such as salt or sugar, is mixed into water and allowed to completely dissolve, the temperature at which the water mixture freezes will be lower than the usual 0°C (32°F). This happens because the foreign particles interfere with the attractive force between the water particles. The particles have to slow down more so that the regular crystal pattern of solid ice can form. In cold climates, salt is used on sidewalks and roads to lower the freezing point of water and prevent ice formation. Do you think this works at all temperatures? Go online and learn about the limitations of the effect of salt in lowering the melting point of water.

Pressure

While changing pressure does not have a great effect on the freezing point of matter, it does have some effect. Increasing pressure tends to decrease the freezing point.

Movement

If you live where you get freezing temperatures in the winter, you have probably noticed that rivers can freeze. The edges of the rivers where the water is slow moving and shallow will cool and freeze faster than the deeper, quickly moving water in the center of the river. The fast moving water won't freeze as easily because the chemical bonds that are necessary for the particles to crystallize and become solid are easily broken. Eventually, the ice may spread from the edge across the width of the river. The thickness of the ice is usually not consistent. Because the river water under the ice is still moving and often turbulent, warmer water from the river bottom flows upward and melts the underside of the surface ice.

How can you use the physical change in a mixture to make a frozen treat?

Make ice cream in a bag. Fill a zip-top gallon bag with crushed ice. Measure and record the temperature of the ice. Add 120 g (4 oz) of rock salt. Put 120 mL (0.5 cup) of whole milk, 1 mL (0.03 oz) of vanilla, and 20 g (0.7 oz) of sugar into the zip-top quart bag. Seal this bag securely. Place the smaller bag inside the larger one, and seal it. Wear gloves or use a towel to protect your fingers. Shake or knead the mixture for about 10 minutes or until the mixture thickens. Measure and record the temperature of the ice again. How did the temperature of the ice change? Explain. Wipe off the top of the small bag. Open the bag carefully. Enjoy the ice cream with a cup and spoon.

SAFETY: Students who have allergies to milk or vanilla should not eat the product of this experiment.

This family is making homemade ice cream.

How does adding salt to the ice help freeze the milk mixture inside the ice cream maker? Why?

At higher elevation, water reaches a boiling point at lower temperature.

How would the cooking time of the potatoes be different at sea level compared to a very high elevation?

Factors That Affect Boiling Point

Solutes

As with the freezing point, solutes also affect the boiling point. When a solute such as salt is added to water, the water's boiling point is increased. Generally, the more solute that is added to a liquid, the greater is the increase in its boiling point.

Pressure

Pressure is the factor that has the greatest effect on the boiling point of a liquid. This pressure is usually the outside air pressure created by Earth's atmosphere. As pressure increases, the boiling point increases, and as pressure decreases, the boiling point decreases. For example, at sea level, where the full weight of the atmosphere presses down on the water, water boils at 100°C (212°F). As elevation increases, water boils at a lower temperature. At the top of Mount Everest, water boils at around 72°C (162°F). Why might it take longer to boil an egg or cook pasta when camping in the mountains than it does when cooking these things at the seashore?

The table lists some physical changes. Work with a partner to complete the second column of the table. Identify a way that each change described could be reversed. Share your ideas with the class. Which solutions were the most creative?

Why do you think some physical changes are harder to reverse than others?

Physical Change	Reversal Method
Crushing a can	
Cutting up fruit to make fruit salad	
Boiling water	
Molding clay into a different shape	
Tearing paper into small pieces	

Concept Check   Assess/Reflect

Summary: What are physical changes? A physical change is a change of matter from one form to another without producing any new substances. Tearing paper, chopping wood, and crushing an aluminum can are physical changes. Changes of state are common physical changes. In a state change, particles gain energy or lose energy, causing the particles of the substance to speed up or slow down. As temperature changes, matter changes from one state to another.

1. Name three kinds of physical changes and give an example of each.

2. Describe the relationship between atoms of water in a solid, liquid, and gas.

3. Explain what happens to the particles of a liquid when it freezes.

4. Describe the energy changes that accompany evaporation and condensation.

What Are Chemical Properties and Changes?

Objectives

- Explain what a chemical property is.
- Identify the chemical properties of matter.
- Explain the law of conservation of mass.
- Identify common examples of chemical changes in matter.

Vocabulary

chemical property

flammability

corrosion

chemical change

law of conservation of mass

chemical reaction

reactivity

reactant

product

chemical equation

Some properties of matter cannot be observed and described until a change takes place. To observe chemical properties, the matter must actually be changed. For example, what types of changes occur when a cake is baking?

Chemical Properties Explain

A **chemical property** of matter describes the ability of a substance to change into new forms of matter. Listed below are a few examples of chemical properties. When have you observed some of these properties?

Some Chemical Properties	Examples
reaction with oxygen	rusting iron
flammability; able to burn or ignite	flammable substances include oil, gasoline, dry wood, and matches
reaction with acids or bases	structures damaged by acid rain

Chemical properties help us understand how substances will behave under different conditions. For example, Frederick Bertholdi, the sculptor who designed the Statue of Liberty, gave it a copper skin over an iron support structure. The copper skin helped reduce the effects of **corrosion**, the weakening of metals caused by chemical action, on the statue. Copper combines with water and oxygen to form a green coating, or patina, that prevents further damage to the metal. For many years, this blue-green patina protected the statue.

Why does the Statue of Liberty have a blue-green color instead of the reddish-brown color of copper metal?

Reaction in a Bag

What happens to mass during chemical changes?

Procedure

1. Wear goggles. Use the graduated cylinder to **measure** 20 mL (0.7 oz) of water. Pour the water into the vial.

2. **Measure** 2.5 g (0.08 oz) of baking soda, and pour it into the bag. Add 2.5 g (0.08 oz) of citric acid to the bag, and mix the two powders together. **Observe** and **record** what happens.

3. Put the bag inside the bowl. Place the bowl and its contents on the balance. Put the container of water inside the bag. Be careful not to spill the water. **Record** the total mass.

4. Seal the bag, and carefully remove it from the bowl. Tip the vial to mix the water with the powders in the bag.

5. Hold the bag gently in your hands, but do not squeeze the bag or open it. **Observe** and **record** the chemical reaction.

6. Once the reaction has stopped, place the sealed bag in the bowl. Place the bowl and its contents back on the balance. **Measure** and **record** the mass again.

7. Often, scientists will calculate the percent error in their measurements. **Use numbers** to determine any difference between the mass of the reactants and the mass of the products. **Record** the percent error. To find percent error, find the difference between the mass of the reactants and the mass of the products. Divide that difference by the starting mass of the reactants and multiply by 100.

Materials
- 50-mL graduated cylinder
- 20 mL (0.7 oz) water
- 50-mL vial
- safety goggles
- zip-top quart bag
- 2.5 g (0.08 oz) baking soda
- 2.5 g (0.08 oz) citric acid
- bowl
- balance

Analyze Results

Compare the mass at the beginning of the chemical change with the mass at the end.

Create Explanations

1. What happens to mass during chemical changes?

2. What evidence shows that a chemical change took place?

3. How would you explain any differences in mass in your results?

Conservation of Mass Explain

In 1789, a French scientist named Antoine Lavoisier figured out that when wood burns, it combines with oxygen from the air. What type of experiment do you think Lavoisier devised to prove his discovery? What Lavoisier observed was a **chemical change**. This type of change in matter forms new substances with different physical and chemical properties from the original matter. When the wood burns, it changes to carbon (ash). It may seem that the resulting ash has less mass than the original log, but burning also produce gases that have mass. During the chemical change, these gases are released into the air.

When Lavoisier conducted his experiments, he made sure none of the products of the reactions was lost. He carried out his experiments in sealed containers. Lavoisier measured the mass of matter present at the beginning of each experiment and compared that with the total mass of products at the end of the experiment and found that the masses remained the same.

The results of his experiments have become known as the **law of conservation of mass**. This law states that whenever a chemical change occurs, the total mass of substances in a chemical reaction remains the same throughout the change. It also tells us that matter cannot be created or destroyed in an ordinary chemical change.

Explore-a-Lab

Structured Inquiry

 **Why did the mass of the steel wool change?**

Wear goggles and gloves during this experiment. Tear off a small piece (the size of an egg) of steel wool. Place the steel wool in a 250-mL beaker and add white vinegar until the entire piece is covered. Soak for about 5–7 minutes. Remove the steel wool from the vinegar and wring dry. Place the steel wool in a clean, dry 250-mL flask and cover the opening with a balloon. Measure the mass of the flask, steel wool, and balloon on a balance and record the mass. Allow the flask to sit for about 45 minutes until changes can be observed in the steel wool. Measure the mass of the flask, steel wool, and balloon again. What happened to the steel wool? How did the steel wool and the vinegar react?

Chemical Changes

What happens to metals left outside? Most metals are susceptible to oxidation, like your bike or your parent's car. What is rust? When oxygen interacts with iron, a chemical change occurs and a new material called iron oxide, or rust, forms. Iron oxide is brittle and orange in color. Rust can be very costly because it destroys the metal layer by layer. Acid in rainwater increases the effect. What other factors can increase the rate at which rust forms?

Matter combines, separates, and changes in countless ways. You have learned about physical changes in matter, but you also encounter chemical changes every day. Many chemical changes occur during cooking. Chemical changes are usually permanent. Baking cookies produces new matter with very different properties from the ingredients that you started with. Would it be possible to change baked cookies back into their original ingredients? Explain.

Scripture Spotlight

What examples of chemical change are described in **1 Samuel 28:24**?

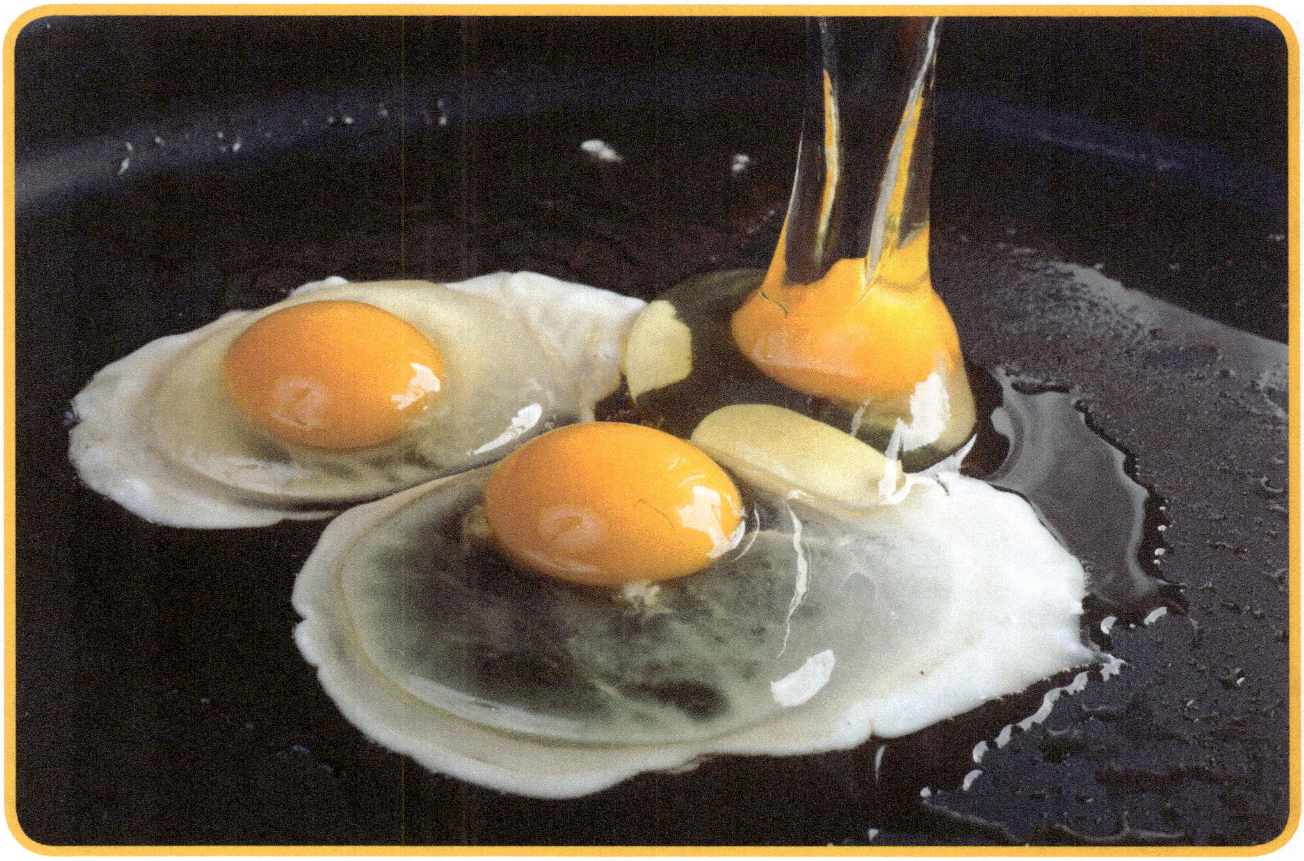

Cooking food involves many chemical changes.

How do the properties of the cooked eggs differ from those of the eggs before cooking?

Other examples of chemical changes include tarnishing metal, burning fuel in a furnace, setting off fireworks, and forming yogurt from milk. Chemical changes even occur inside you. When your body digests food, the food undergoes a chemical change. What are some other examples of chemical changes you can think of?

The list below describes some indications that a chemical change has taken place. Which of these happen when you light charcoal in a grill and cook vegetable kabobs? Is there a time when a change in color might be an example of a physical change instead of a chemical change? You will learn more about the signs of a chemical change in the next lesson.

- formation of a *precipitate* (a solid)
- formation of a gas
- change in color
- change in smell
- change in energy

Another name for a chemical change is a *chemical reaction.* A **chemical reaction** occurs when two or more kinds of matter interact. You are already familiar with some chemical reactions. For example, *photosynthesis* is the reaction plants use to make food. *Cellular respiration* is the reaction in all cells that changes sugar into energy. Some chemical reactions break matter apart. An electric current breaks water into hydrogen and oxygen. Others combine two or more kinds of matter, such as a rusting nail. What other chemical reactions have you seen today?

Some substances react more easily than others. **Reactivity** refers to the ability of matter to undergo chemical reactions with other kinds of matter. You have already learned that the ability to burn easily, or flammability, is a common chemical property. How would a chemist use knowledge of reactivity to make a fire extinguisher? Why do you think it is necessary to have different kinds of fire extinguishers?

Some substances do not react easily with other materials because they are more stable. For example, gold does not combine readily with oxygen or acids. It does not tarnish easily. Such materials are said to have a low reactivity. They are less likely to produce chemical changes. What would life be like if God designed all substances to be stable and not react, or what if all substances were reactive? What would change in your life?

Writing Chemical Equations

The substances used in chemical reactions are called the **reactants**. The new substances produced during the chemical change are called the **products**. Chemical reactions can be described with simple **chemical equations**.

The reactants, or starting materials, are written on the left side of the equation. The products are written on the right side of the equation. They are separated by an arrow that shows the direction of the reaction. Scientists sometimes denote the state of each reactant and product in a reaction, such as solid (s), liquid (l), or gas (g). We use (aq) to describe an aqueous solution that is formed when a substance is dissolved in water. For example, the equation for the rusting of iron is written this way:

$$\text{iron (s)} + \text{oxygen (g)} \rightarrow \text{iron oxide (s)}$$

When solids are dissolved in liquids, it is easy to believe they have disappeared completely. The reaction in the Explore-a-Lab on the next page shows the combination of two liquids containing dissolved matter. A *precipitate* is the solid that sometimes forms when two solutions are combined and react chemically. This reaction produces a white solid precipitate: calcium chloride (aq) + sodium bicarbonate (aq) → sodium chloride (aq) + calcium bicarbonate (s).

Rust has damaged this metal gate.

How does the reaction between iron and oxygen change the metal?

Math in Science

Write equations to describe each of the following chemical changes:

(1) Magnesium metal burns in oxygen gas to produce a white powder called magnesium oxide; (2) baking soda (sodium bicarbonate) and vinegar (acetic acid) form carbon dioxide gas and water; (3) hydrogen and oxygen gases unite to form water. The equations should follow this pattern:

reactant 1 ( ) + reactant 2 ( ) → product 1 ( ) + product 2 ( )

Explore-a-Lab

Structured Inquiry

What changes take place when sodium bicarbonate and calcium chloride solutions combine?

Wear goggles and gloves during this experiment. Label two plastic cups: "baking soda solution" and "calcium chloride solution" with masking tape. Use a graduated cylinder to add 20 mL (0.7 oz) of water to each cup. Add 2 g (0.07 oz) of calcium chloride to the water in its labeled cup. Stir until the solid dissolves. Add 2 g (0.07 oz) of baking soda (or sodium bicarbonate) to the water in its labeled cup. Stir until the solid dissolves. Carefully pour the baking soda solution into the calcium chloride solution. Do not pour in any undissolved baking soda. Observe what happens. Do you think a chemical reaction took place? How could you tell for sure? How could you have used this experiment to demonstrate the law of conservation of mass?

Concept Check   Assess/Reflect

Summary: What are chemical properties and changes? Chemical changes are changes in matter in which new substances form with different physical and chemical properties from the original matter. The changes are usually permanent. The total mass remains the same.

1. Identify two examples of chemical changes. Tell how the matter changes.

2. What chemical property do substances like gold display?

3. You fry pancakes for breakfast and cut up an apple for a snack. Compare and contrast the changes to matter in these two actions.

4. What does the law of conservation of mass tell us about what happens to matter during physical changes?

What Are Chemical Reactions?

When you cut up veggies and other toppings while making a pizza, are you making physical or chemical changes? When the pizza bakes, is it a physical change or chemical change that is happening? How do you know? How do you know when the pizza is done? You have just learned that during a chemical change, matter is rearranged and new substances form. But how do you know if or when a chemical reaction happens?

Recognizing the Signs Explain

The burning of a candle on a birthday cake and the ripening of fruit involve a chemical reaction. What changes do bananas undergo as chemical reactions ripen and rot them? Several signs indicate that chemical reactions are happening. These include a change in color, a change in odor, the production of a gas, the production of a solid, and a change in energy.

Color Changes

Color change in the leaves of deciduous trees is a sign of a chemical reaction taking place in the leaves. The change in color of a roasted marshmallow indicates that a chemical reaction has taken place on the surface of the marshmallow. What other color changes have you observed that would indicate that a chemical reaction has occurred? Do all color changes mean that a reaction has occurred?

Objectives

- Identify the signs that indicate a chemical change has occurred.
- Identify the main types of chemical reactions.
- Explain the difference between endothermic and exothermic reactions.
- Identify the factors that affect the rate of chemical reactions.

Vocabulary

endothermic reaction

exothermic reaction

catalyst

Autumn leaves are a beautiful sign of a chemical change.

What chemical reactions are happening in these leaves?

Record your work for this inquiry. Your teacher may also assign the related Guided Inquiry.

Pondering Plaster

How can you tell that a chemical reaction has occurred?

SAFETY: Wear goggles and gloves at all times. Do not stir the plaster with your bare hands. Never pour plaster of Paris down the drain.

Procedure

1. Use masking tape and a pen to label three sandwich bags and three cups "Vinegar," "Saltwater," and "Water."

2. Put on goggles and gloves. In one cup, mix 20 mL (0.7 oz) of warm water with 5 g (0.2 oz) of salt. **Measure** and **record** the temperature of the saltwater.

3. In a second cup, pour 20 mL (0.7 oz) of water. In a third cup, pour 20 mL (0.7 oz) of vinegar. **Measure** and **record** the temperature of each.

4. Use the teaspoon to slowly sprinkle three scoops of plaster of Paris into the saltwater. Slowly stir the mixture. Pour it into the "Saltwater" bag. Seal the bag, pressing out as much air as you can. Attach a thermometer with tape to the outside of the bag. Dispose of the spoon and cup in the garbage.

5. Repeat Step 4 with the "Water" and the "Vinegar" cups and zip-top bags.

6. **Record** the temperature of each plaster of Paris mixture every five minutes for 35 minutes. At the same time, **observe** the hardness of the mixtures by gently pressing on the bag. **Record** your observations.

7. After 35 minutes, dispose of the plaster of Paris mixtures in the garbage.

Materials

- water
- salt
- vinegar
- plaster of Paris
- 3 plastic disposable teaspoons
- 3 zip-top sandwich bags
- marking pen
- 3 small plastic disposable cups
- 3 thermometers
- masking tape
- balance
- 25-mL graduated cylinder
- goggles
- disposable plastic gloves

Analyze Results

Compare the hardness and temperature changes that occurred in each bag. Graph the temperature data. Place time on the *x*-axis and temperature on the *y*-axis.

Create Explanations

1. How can you tell that a chemical reaction has occurred?

2. Which mixture hardened more slowly? How would this knowledge be beneficial to an artist making a plaster of Paris sculpture?

Changes in Odor

Have you ever forgotten about a glass of milk on the counter for a few days? When milk goes through a chemical change, it produces a sour odor. You know that this smell is a sign that the milk is spoiling. Why do you think the odor changes over time?

Production of Gases

Fizzy antacid tablets contain dry chemicals. When the tablets are dropped in water, the bubbles of gas produced are signs of a chemical reaction. Does the gas produced also give off an odor? Chemical smells are often related to the production of gases. Bacteria, decaying plants, and chemical changes give off gases that produce a rotten-egg smell.

Faith Connection

You are learning to recognize the signs of chemical changes. What outward signs indicate that God is making spiritual changes in people's hearts?

Explore-a-Lab

Structured Inquiry

What is being produced during the reaction? How does this reaction make yeast beneficial to baking?

Add 5 g (0.2 oz) of yeast to one cup of warm water. Add 5 g (0.2 oz) of sugar. Gently stir the mixture. Observe and record what happens. Record observations every 5 minutes for 15 minutes.

Production of Solids

The appearance of a solid that will not dissolve when mixing two liquids is another sign of chemical change. The liquid might turn cloudy because the solid stays dissolved. Or, the solid may settle to the bottom of the container. How does this information relate to the last Explore-a-Lab you performed in Lesson 2?

Changes in Energy

A change in temperature is also a sign that a chemical reaction has taken place. Some reactions take in energy. For example, instant cold packs contain chemicals that absorb heat when mixed with water. Other chemical changes give off energy. A burning magnesium ribbon releases heat and light.

What signs tell you fireworks involve chemical changes?

Types of Chemical Reactions Explain

Chemical reactions are constantly happening. What chemical reactions can you identify happening around you right now? Chemical reactions are organized into main types of reactions. Study the chart below to learn the types of chemical reactions.

Some Common Types of Chemical Reactions		
Type of Reaction	**Description**	**Example**
Synthesis	two substances combine to make a new substance	hydrogen and oxygen forming a water molecule
Decomposition	a complex substance breaks down to form two separate substances	release of carbon dioxide from calcium carbonate treated with an acid
Combustion	occurs when oxygen combines with another compound to form water and carbon dioxide	burning of fossil fuels
Single displacement	one compound takes a substance from another compound	recovering silver from unwanted X-ray film
Double displacement	occurs when two compounds trade substances	an antacid neutralizing stomach acid

Endothermic Processes

Chemical reactions that absorb energy are called **endothermic reactions**. Photosynthesis is a familiar endothermic reaction. The light energy absorbed by the plant's leaves is needed to change carbon dioxide and water into sugar and oxygen.

Instant cold packs used to treat injuries also make use of endothermic chemical reactions. When you strike the cold pack, water in the pouch mixes with a dry chemical. This reaction absorbs energy and the temperature of the pack drops. Do you think endothermic reactions take place in your body? Why or why not?

Exothermic Processes

Chemical reactions that release energy are called **exothermic reactions**. The energy given off in these reactions may be in the form of heat, light, or sound. Many reactions between acids and bases are exothermic. Combustion reactions, or burning reactions, are familiar exothermic reactions.

Cellular respiration is an exothermic reaction that takes place in your body. Your body uses the energy produced by cell respiration to power your muscles and other life processes. What happens to your body that shows that cellular respiration is an exothermic reaction?

Hand warmers are another example of exothermic reactions. These packets contain several different chemicals. When the plastic covering is removed and these chemicals are exposed to air, the resulting reaction releases heat energy.

Check out your *Science Journal* to further explore what happens during exothermic and endothermic processes.

Extend

Explore-a-Lab

Structured Inquiry

How quickly will temperatures change during endothermic and exothermic reactions?

SAFETY: Wear goggles.

Work in small groups to study temperature changes in two chemical reactions.

Part I Put on goggles and gather materials. You will need a 250-mL beaker, 100 mL (3 oz) of white vinegar, 30 g (1 oz) of baking soda, a plastic spoon, a thermometer, and a stopwatch. Pour 100 mL (3 oz) of vinegar into the beaker. Measure and record the temperature. Do not remove the thermometer from the liquid. Slowly add 30 g (1 oz) of baking soda, stir briefly with a spoon, and measure the temperature change every five seconds until the temperature stops changing. Graph the results. Clean your work station.

Part II Put on goggles and gather materials. You will need a 250-mL beaker, 100 mL (3 oz) of 3% hydrogen peroxide, 30 g (1 oz) of yeast, a plastic spoon, a thermometer, and a stopwatch. Pour 100 mL (3 oz) of hydrogen peroxide into the beaker. Measure and record the temperature. Do not remove the thermometer from the liquid. Add 30 g (1 oz) of yeast, stir briefly with a spoon, and measure the temperature change every five seconds until the temperature stops changing. Graph the results. Clean your work station.

How do the temperature changes compare? Which reaction is an example of an endothermic reaction and which is an example of an exothermic reaction? When did the temperature change most drastically in each example?

Rate of Chemical Reactions Explain

Chemicals have the ability to react when exposed to other chemicals or to certain physical conditions. Exothermic reactions release heat and can result in fires or even explosions. Such out-of-control chemical reactions can cause serious injuries, health hazards, and damage to property and the environment. Sometimes a catalyst is added to a reaction. A **catalyst** does not participate in the reaction, but increases the rate of the reaction. You can take actions at home or at school to prevent, slow down, or reduce the risks of these hazards.

Reactions occur at different rates or speeds. Several factors can increase or decrease the rate of a chemical reaction.

Ways to Control a Chemical Reaction	
Factors	**Rate of Chemical Reaction**
Increase concentration of reactant	Increases rate
Increase temperature	Increases rate
Add catalyst	Increases rate
Decrease temperature	Slows rate

Explore-a-Lab

Structured Inquiry

 **How does adding a catalyst affect a chemical reaction?**

SAFETY: Wear goggles. Work outdoors.

Work in small groups. Wear goggles throughout this investigation. You will need 118 mL (4 oz) of 3% hydrogen peroxide, dishwashing detergent, 5 g (0.2 oz) of yeast (a catalyst) dissolved in 30 g (1 oz) of very warm water, a funnel, a 0.5-L (16-oz) empty water bottle, an aluminum cake pan, and a thermometer. Using the funnel, pour the hydrogen peroxide into the plastic bottle. Add 5 mL (0.2 oz) of dishwashing soap to the bottle. If you prefer, add a few drops of food coloring as well. Swirl the bottle to mix the liquids together. Place the bottle in the pan. Measure and record the temperature of the mixture. Then, pour the yeast and warm water mixture into the bottle. Step back quickly. Record your observations of the reaction. When the reaction slows or appears to be over, measure and record the temperature once more. Was there a difference in temperature before and after the reaction?

What factor increases the rate of rust formation the most?

Compare the effects of air, water, acid, and salt on the rate of the rusting reaction. Put equal amounts of fine iron wool into four test tubes. Label the test tubes: *dry, water, acid rain,* and *saltwater*. Add 10 mL (less than 1 oz) of water to the second tube, 10 mL (less than 1 oz) of white vinegar to the third, and 10 mL (less than 1 oz) of saltwater to the fourth tube. Observe and record what you see in each test tube for three days.

After having studied this chapter, think about different physical and chemical changes that take place every day. How many changes do you think you observed already today? In Chapter 12, you will look further into the study of chemicals. You will study atoms, elements, compounds, and mixtures. While you study the next chapter, think about how the information relates to physical and chemical properties and changes.

Concept Check Assess/Reflect

Summary: What are chemical reactions? Changes in color, odor, and energy and the production of gases and solids are signs of chemical change. A variety of chemical reactions can occur, including synthesis reactions, decomposition reactions, combustion reactions, single displacement reactions, and double displacement reactions. Physical and chemical processes can give off energy (exothermic) or absorb energy (endothermic). Catalysts can increase the rate of a reaction without participating in the reaction.

1. Name three signs of chemical change. Give an example of each.
2. During a holiday celebration, a friend hands you a glow stick. You crack the stick and it glows. Provide a general explanation for the chemical reaction occurring.
3. Describe the role of energy in chemical reactions.
4. What factors affect the rate of chemical reactions?

Get to Know
Charles Stine

Charles Stine (1882–1954) was a chemist who worked for many years as the Director of Research for the E.I. DuPont Company. Stine had several degrees and honors. From his research, he developed many new products and patents for the company. He was a well-respected scientist in his field, and he was a Christian. He spoke to fellow scientists about his faith and wrote *A Chemist and His Bible*.

As a chemist, Stine's job was to find ways to make chemicals useful to people. Stine hired a young chemistry professor from Harvard University named Wallace Carothers. Before he left Harvard, Carothers had been experimenting with the chemistry of polymers. A polymer is a large molecule that is made of repeating units or chains connected to each other by chemical bonds. Plastics, silicon putty, and rubber are some examples of polymers.

One day, as Carothers was experimenting with a molten polymer in the lab, he accidentally touched a glass rod to the polymer. He was able to pull "fibers" out from the molten polymer. A man-made fiber had been discovered. However, testing showed that the fiber melted at high temperatures and softened in water and cleansers. This fiber product could not be used to manufacture fabric or clothing.

Called to Serve

Charles Stine said, "The world about us, far more intricate than any watch, . . . marvelous beyond even the imagination of the most skilled scientific investigator, this beautiful and intricate creation, bears the signature of its Creator, graven in its works."

Under Stine's leadership, the DuPont laboratory continued to mix polymers, looking for a better result. Finally, a fiber was made that was not affected by water and most cleaners. Several names were suggested for this new product. DuPont settled on "nylon."

During World War II, the production of nylon provided the military with parachutes, airplane tire cord, ropes, and tents. After the war, nylon was used to make other products, such as clothing, ropes, and carpets. What is a nylon product you use?

Concept Check

1. In addition to working as a research chemist, what else did Charles Stine do?
2. How was nylon invented?

Chemical Engineer

Chemical engineers use knowledge about the properties and reactions of substances (chemical knowledge) to help fix problems, such as world hunger or the pollution of our environment. Chemical engineers help to create new materials and find ways to supply our energy needs. They use their chemistry knowledge to improve processes, machines, buildings, and products.

The term "chemical engineer" does not fully describe the work chemical engineers do. This term separates them from other kinds of engineers. Chemical engineers use their chemical know-how in a variety of fields. Chemical engineers work in so many different jobs that they are sometimes called "universal engineers."

Many chemical engineers work as leaders of private companies, for government agencies, and for colleges and universities. They have important influence in the development of many products, such as microwavable foods, vehicle tires, biofuel, bullet resistant vests, and items made of plastics like carpeting, toys, clothing, and bottles. Can you think of other products chemical engineers created?

Chemical engineers work toward a brighter future by solving problems with pollution, finding new energy sources, creating new medicines, and creating new buildings. They are responsible for creating life-saving devices, such as the artificial kidney. Chemical engineers are working hard to find new ways to recycle plastics. Engineers try to find the best answer to a question or problem. They help people by fixing things that are broken, improving things that work, and working to discover new inventions. If you were a chemical engineer, what would you study?

The difference between chemical engineers and other types of engineers is that they use their knowledge of chemistry as they work on solving problems, creating new materials, and finding ways to supply our energy needs. Chemical engineers typically obtain a bachelor's degree in engineering with classes focused on chemistry and math. Many chemical engineers obtain higher levels of education, such as a master's degree or doctoral degree, to advance in the particular field or teach at a university.

Concept Check

1. Why are chemical engineers sometimes called universal engineers?
2. What are some things chemical engineers have developed to improve your life?

Study Guide

Lesson 1

1. Physical properties can be observed and measured without changing matter.

2. There are four states of matter: solid, liquid, gas, and plasma.

3. Common physical properties include size, shape, color, texture, mass, volume, and density.

Lesson 2

1. Physical changes change the form, shape, size, or appearance of matter.

2. Thermal energy causes changes of state.

3. For most substances, the melting point and freezing point are the same temperature.

Lesson 3

1. Chemical properties describe the ability of matter to change into new substances.

2. Chemical changes produce new substances with different properties.

3. When a chemical change takes place, the mass of the reactants equals the mass of the products.

4. Physical changes are often reversible. Most chemical changes are permanent.

Lesson 4

1. Signs of a chemical change include changes in color, odor, and energy.

2. Types of chemical reactions include synthesis, decomposition, combustion, and displacement.

3. Changes in matter generally involve an exchange of energy with the surroundings.

4. Exothermic processes give off energy and endothermic processes take in energy.

5. A catalyst can speed up a reaction.

Show What You Know

Visualize It Complete the following concept map.

Changes in Matter

1.	Chemical Changes
↓	↓
The physical form of the matter may be changed but its essential 2. _________ remains the same.	The new matter has different 3. _________ than the original matter.
↓	↓
The form or 4. _________ of matter is changed by adding or removing 5. _________.	Energy is used to change the 6. _________ of the matter.
↓	↓
Examples: ice 7. _________ and sugar 8. _________.	Examples: iron 9. _________ and the 10. _________ of food.

Explain how each pair of terms is related.

11. mass—density
12. volume—displacement
13. dissolve—physical property
14. single displacement—combustion
15. endothermic reaction—exothermic reaction

Multiple Choice
Choose the best answer.

16. Which tool could you use to measure the mass of objects?
    A. digital thermometer
    B. graduated cylinder
    C. spring scale
    D. triple beam balance

17. How could you test the magnetic properties of a piece of metal?
    A. Put it in an electric circuit.
    B. Scratch it with your fingernail.
    C. Bring it near a magnet.
    D. Stretch it into thin wires.

18. What is the change of state called when a solid changes directly into a gas?
    A. condensation
    B. evaporation
    C. melting
    D. sublimation

19. What indicates that the Statue of Liberty went through a chemical change?
    A. The original matter is still present.
    B. The color of the statue changed.
    C. The total mass stayed the same.
    D. The changes are reversible.

20. What is a clear sign of an endothermic reaction?
    A. Heat and light are given off.
    B. There is a combustion reaction.
    C. There is a drop in temperature.
    D. There is a change of state.

Check Point
Answer the following questions.

21. Explain the signs showing that cooking broccoli produces a chemical change.

22. A graduated cylinder was filled to the 25-mL (1-oz) mark. Placing a solid object with a mass of 75 g (2.6 oz) raised the water level to 46 mL (1.6 oz). **Use numbers** to calculate the density of the object.

23. **Compare** the properties of mass and weight. Describe how each is measured.

24. Explain what the law of conservation of mass says about chemical changes.

25. Describe why adding salt affects the boiling point of water.

For items 26–28, refer to the photo.

26. Does the photo show a physical change or a chemical change? How can you tell?

27. Describe what happens to the particles of matter during this change.

28. Is this change an endothermic process or an exothermic process? Explain your reasoning.

The Atomic Theory

Scripture Spotlight

By studying atoms, elements, compounds, and mixtures, you can become closer to God by understanding His Creation. You will read the following passages in this chapter.

Daniel 2:31–35 (p. 401, 430)   Ephesians 4:3 (p. 421)
Colossians 1:16 (p. 406)   Colossians 3:14 (p. 421)
Proverbs 11:21 (p. 419)   Ephesians 2:13–15 (p. 425)
Isaiah 56:6 (p. 419)   Matthew 25:31–46 (p. 427)

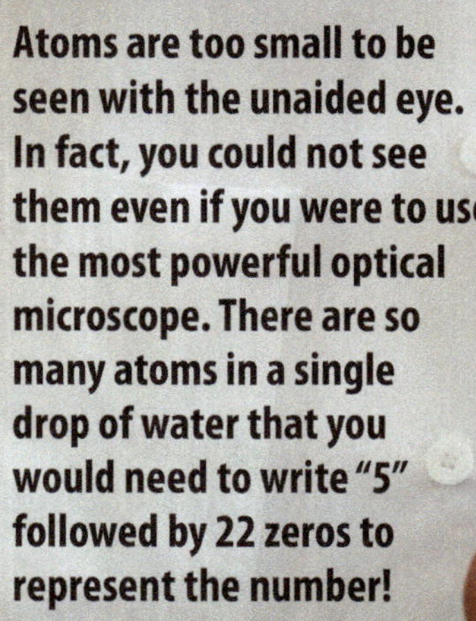

Atoms are too small to be seen with the unaided eye. In fact, you could not see them even if you were to use the most powerful optical microscope. There are so many atoms in a single drop of water that you would need to write "5" followed by 22 zeros to represent the number!

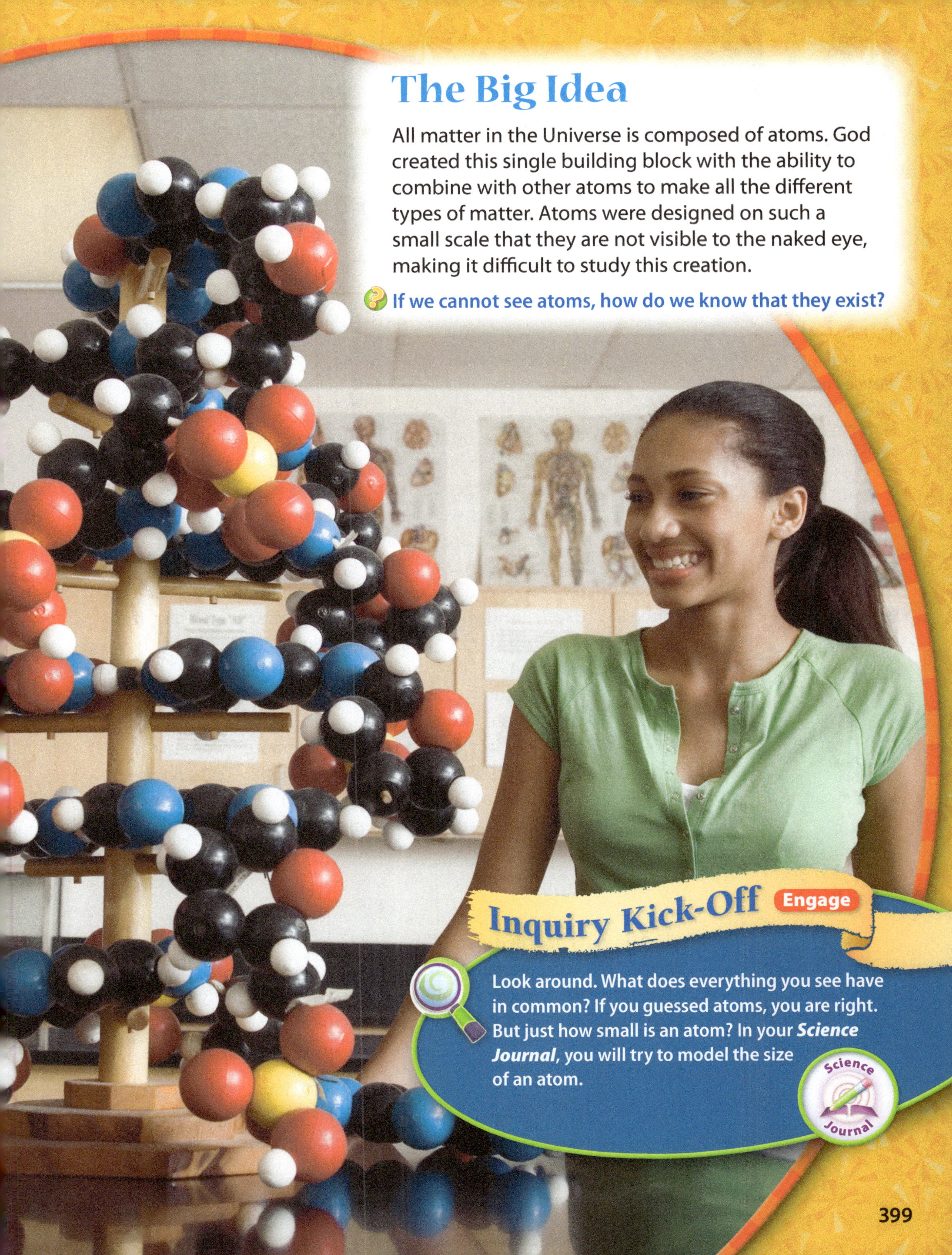

The Big Idea

All matter in the Universe is composed of atoms. God created this single building block with the ability to combine with other atoms to make all the different types of matter. Atoms were designed on such a small scale that they are not visible to the naked eye, making it difficult to study this creation.

If we cannot see atoms, how do we know that they exist?

399

? Essential Question

What Are the Building Blocks of Matter?

Have you ever wondered how an electronic toy, engine, or other device is made? How do all of the small pieces come together to make the device work? If you took apart your favorite device and separated it into pieces, and then broke each piece into the smallest pieces, you would still have pieces many thousands of times larger than the smallest particles of matter.

The ancient Greeks were among the first people in history to think about the nature of matter. They wondered what made up matter. The Ancient Greek philosophers Aristotle and Empedocles (em•ˈpe•də•ˌklēz) used reason alone to decide that there were four kinds of matter in the Universe: earth, air, fire, and water. They thought all matter could be made out of these four "elements." Why do you think they thought this?

In the fifth century B.C., Greek philosophers Democritus (di•ˈmä•krə•təs) and Leucippus (luːˈsɪpəs) developed the idea of elements. They determined that elements were made of indivisible particles, which they called *atoms*. Years later, English scientist John Dalton used the term "atom" to explain how related elements may have some similarities, but their atoms are not identical.

Atoms `Explain`

You have learned that all matter is made up of these tiny particles called **atoms**. These particles are too small to be seen with an unaided eye. Atoms, grouped in particular ways, make up all the matter that exists in the Universe.

Elements

The simplest form of matter is called an **element**. The modern science concept of an element is very different from the four elements of the ancient Greeks. Why do you think that concept changed?

You are already familiar with many of the common elements—metals, such as iron, gold, silver, and copper, and gases, such as oxygen and helium. These substances are found on Earth and throughout the Universe.

Elements cannot be broken down into anything simpler without losing their chemical properties. Atoms are the smallest component of an element. Are there things smaller than atoms? Even if you were able to isolate a single atom of iron, it would still be iron. However, if you changed the basic components of the atom, it would become a different element. The atoms of one element are different from the atoms of every other element. For example, the atoms in the element gold are not the same as the atoms that make up oxygen. What do you think is different in atoms from different elements?

Elements in Ancient Cultures

Ancient cultures were familiar with about nine of the elements we now know today. Some of the earliest known studies of the elements were done by the Persian scientist Jabir ibn Hayyan (ˈja•ˌbir•ˌi•bən•hī•ˈyan). While the names he used to identify these elements were different from the names we use today, the elements are the same. These nine elements—gold, silver, copper, iron, lead, tin, mercury, sulfur, and carbon—were the first elements to be discovered because they exist naturally in their pure form or can be easily purified.

Scripture Spotlight

King Nebuchadnezzar of Babylon dreamed about a statue. What elements are named in the description of this statue in **Daniel 2:31–33**?

copper

iron

Nine elements, such as these samples of native copper and iron ore, were known to many ancient cultures.

How do you think your ancestors separated an element from the other rock it was imbedded in?

Atomic Models

How can you build a three-dimensional model of an atom?

Materials
- modeling clay (three colors)
- plastic knives
- large bag of candy or jelly beans (two colors)
- large bag of smaller candy (one color)
- metric rulers
- large round paper plates
- atomic structure display chart or projection
- periodic table of the elements

Procedure

1. Create a size scale for the particles you will make to model atoms. Then, using colored clay, make small balls of two different colors to represent the protons and neutrons in the nucleus of three different kinds of particles. Gently push the protons and neutrons together to form a sphere to represent the nucleus.

2. Cover the "nucleus" with a layer of the third color of clay to represent the electron cloud region where the electrons are found. Use the plastic knife to cut the "atom" in half to **observe** its cross-section.

3. In your **Science Journal**, draw your atom's "cross-section" and label the parts where you expect to find the nucleus, a proton, a neutron, and an electron.

4. Next, separate at least eight large candies into two color groups—one color to represent protons and another color to represent neutrons. Arrange the number of large candies in the center of the paper plate to represent the nucleus of a carbon atom. Decide how you want to show the electrons in the cloud region.

5. **Record** your carbon atom by drawing a diagram of it in your **Science Journal** and use the periodic table of the elements to find the chemical symbol for carbon. Repeat Steps 4 and 5, drawing and labeling models of hydrogen, nitrogen, and oxygen atoms.

Analyze Results

What kinds of shapes do you see in your atomic models?

Create Explanations

1. How can you build a three-dimensional model of an atom?

2. How does an atom of one element differ from that of another element?

3. What is the hardest thing to show accurately in any model of an atom? Explain.

Parts of Atoms Explain

Scientists continue to explore elements and their atoms. They discovered that while atoms are the smallest part of an element, atoms are, in fact, made of smaller parts. All atoms are arranged in the same basic way. Most of the mass of any atom is at its center, or **nucleus**. In Chapter 2, we discussed the structure of a cell. The cell nucleus is the control center. Similarly, in atoms the nucleus is at the center of the atom. What are some other things that atoms and cells have in common?

Protons

The nucleus of an atom contains extremely tiny particles called **protons**. A proton has a mass of 1.7×10^{-24} grams. It also has a positive charge. The number of protons in the nucleus identifies the element. For example, the element copper contains 29 protons in its nucleus. We say that copper has an atomic number of 29. What is the atomic number of an element?

Neutrons

In addition to protons, an atom's nucleus also contains particles called **neutrons**. Neutrons have a mass nearly the same as protons. However, neutrons have no electrical charge. They are neutral. What kind of effect do neutrons have on protons?

Electrons

The nucleus of every atom is surrounded by a cloud of negatively charged particles called **electrons**. These particles are much, much smaller than protons or neutrons. The mass of an electron is only 9.1×10^{-28} grams, which is 1/2000 of the mass of a proton. The number of electrons present in an atom depends on the element of the atom. Different elements will have different numbers of electrons in their atoms. An electronically neutral atom contains the same number of electrons as protons. So, one copper atom has 29 electrons surrounding the nucleus. If an atom has less protons than electrons, what is its charge?

In the image, the blue and yellow balls represent neutrons and protons, while the green balls are electrons that travel around the nucleus of the atom.

Do you think the term *nucleus* is appropriate to describe the center of an atom? Explain.

Atomic Models Explain

As their understanding of the structure of atoms grew, scientists used models of atoms to illustrate what they learned. As new information was discovered, the existing models were refined or abandoned entirely and new models were developed to reflect the new information.

Dalton Model

John Dalton proposed that all matter was composed of atoms. Atoms of each element were identical in atomic mass and their properties. Dalton considered the atom to be a solid unbreakable sphere. Because of this description, Dalton's atomic model is often referred to as the "billiard ball" model. What did Dalton not consider in his model of the atom?

Thomson Model (Blueberry Muffin Model)

In the late nineteenth century, most scientists agreed on the existence of atoms. In experiments with hydrogen atoms, J. J. Thomson knew that atoms were electrically neutral. However, he also knew that atoms contained particles of positive charge—protons. In order to account for the mass of a hydrogen atom and to maintain its neutral charge, Thomson realized that atoms must contain smaller, negatively charged particles. At this time, the structure of atoms was still very much unknown. Thomson's model of the atom can be compared to a blueberry muffin. According to his explanation, the negatively charged electrons were the "blueberries" suspended in the "muffin" of positive charges. His model was discredited upon the discovery of the nucleus of atoms. How do you think other scientists proved Thomson wrong?

In Thomson's model the negatively charged electrons were scattered through the positively charged atom.

How does this differ from the model of an atom on the previous page?

How can you tell if a material has an attractive charge?

Inflate a balloon and press it against the classroom wall. The balloon should fall to the floor. This means the balloon is neutral. Rub the balloon against your hair. Your action has created positive and negative charges. Press the balloon against the wall again. What happened? Recall that when it comes to electron particles interacting, "opposites attract." This means that positive and negative charges are attracted to each other. Develop a hypothesis that will explain the effects on the charged objects.

Rutherford Model

In 1911, Ernest Rutherford, a student of Thomson's, published a model of the atom with a central, positively charged nucleus orbited by negatively charged electrons. How do you think Rutherford determined there was a nucleus in the atom? Why didn't other scientists discover this sooner?

Rutherford proposed that most of an atom is made up of empty space. The space between the nucleus and the electron cloud is quite large compared to the size of the nucleus and the atomic particles. In fact, over 99% of an atom's total volume is empty space. This is equivalent to a marble, representing the nucleus, in the center of a large football stadium with the rim of the stadium representing the path of the orbiting electrons. What was still wrong with Rutherford's model?

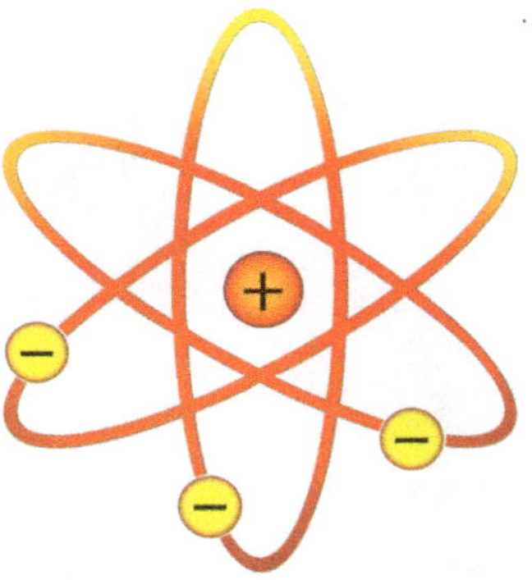

Rutherford atom

Solar System Model

Niels Bohr's experiments using Rutherford's atomic model led him to realize that electrons orbit the nucleus at specific distances. These distances depend on how much energy the electrons have. This model was compared to the model of our Solar System and is often referred to as the solar system model because electrons orbit the atom's central nucleus.

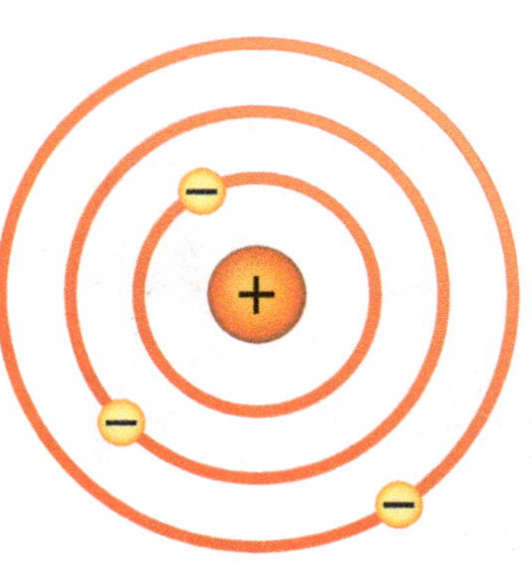

Solar system model

Scripture Spotlight

Do you think the Bible talks about atoms or electrons? Read **Colossians 1:16**.

Electron Cloud Model

Erwin Schrodinger refined the atomic model to take into account that we cannot exactly know where an electron will be at any given moment. He was able to determine at what level of energy an electron might be located. The likely locations of the electrons were represented by a collection of dots resembling a cloud. The amount of energy an electron has determines where in the electron cloud it is most likely to be found. As an electron gains energy, the distance it orbits from the nucleus increases. As it loses energy, the distance between it and the nucleus decreases. How does an electron get more energy? Can we ever draw or create a perfectly correct model? Why or why not? Do you think the model of the atom will continue to change? Why or why not?

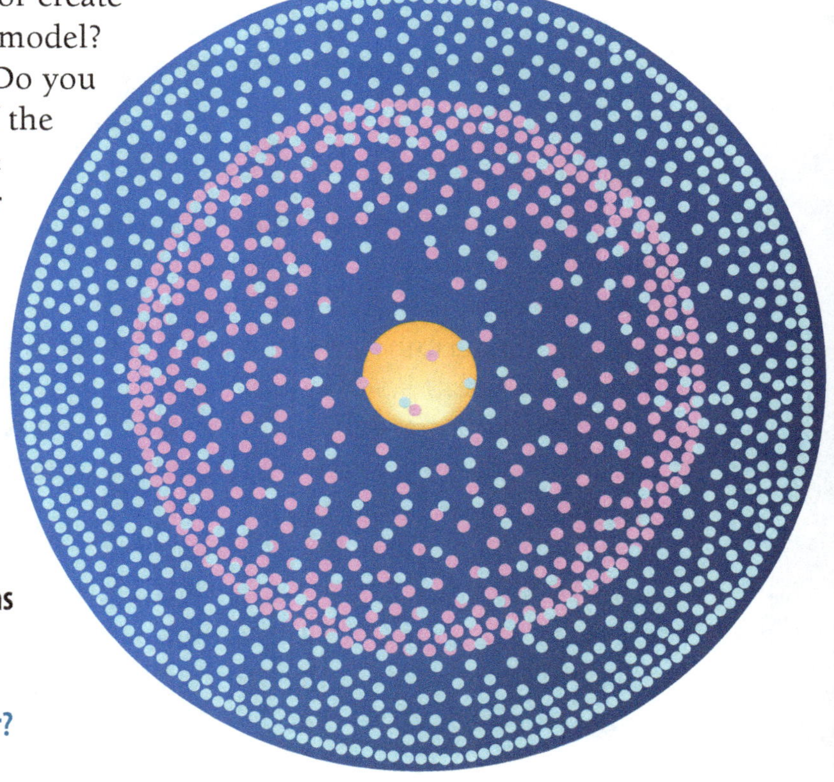

The cloud model is used to represent current ideas on atomic structure.

 How are the five models alike? How do they differ?

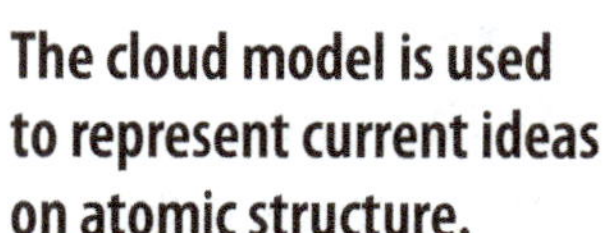

Math in Science

Atoms are so small that scientists use a special unit of length to measure them called the *angstrom* (Å). This unit is equal to one ten-billionth of a meter (1×10^{-10} meter). Atoms range in size from less than 1Å to more than 4Å in diameter. How many meters across is the widest atom?

The Electron Clouds

Unlike protons and neutrons that stay inside the atom's nucleus, electrons move very quickly in specific electron clouds or levels around the nucleus. The electron clouds are similar to planetary orbits around the Sun in our Solar System. The amount of energy an electron has determines the path it takes in its orbit. Electrons with the same levels of energy are most likely to be found in the same electron level around a nucleus. As an electron gains energy, the distance it orbits from the nucleus also increases. How does an electron get more energy? The electron cloud provides a visual of where the electrons orbit, but they do not have fixed movements like the planets.

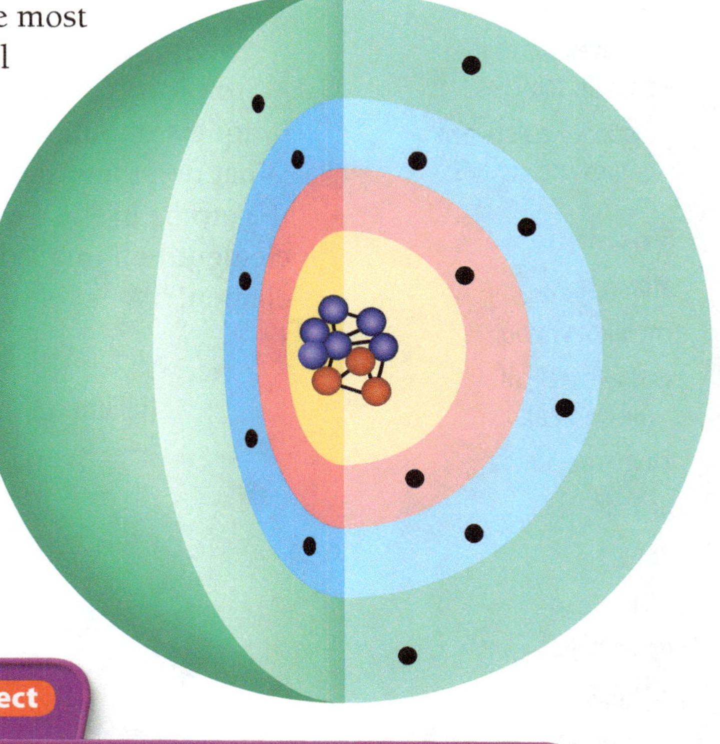

The model illustrates the multiple levels that electrons follow in an atom of mercury.

Scientists have compared atoms to our Solar System. What parts of the Solar System represent the nucleus, protons, and electrons?

Concept Check

Summary: What are the building blocks of matter? All matter is made of atoms. A substance composed of one type of atom is called an element. It cannot be broken into anything simpler or it would lose its chemical properties. Atoms are composed of positively charged protons and neutral neutrons, which are in the nucleus. Negatively charged electrons travel in a cloud-like orbit of the atom's nucleus. Scientists, such as Dalton, Thomson, Rutherford, Bohr, and Schrodinger, have all contributed to our understanding of the structure of the atom by creating models.

1. What is the simplest form of matter called?
2. What are the parts of an atom?
3. Compare and contrast electrons and protons.
4. Does any atomic model accurately show what an atom really looks like? Explain.
5. How are atoms and elements different?

Essential Question
What Is the Periodic Table of the Elements?

You learned that elements are composed of atoms with unique combinations of protons, neutrons, and electrons. What do atoms and elements have in common? How are the atoms in the element carbon different than those in the element helium? What are the physical properties of each element? What is it that makes these elements so different?

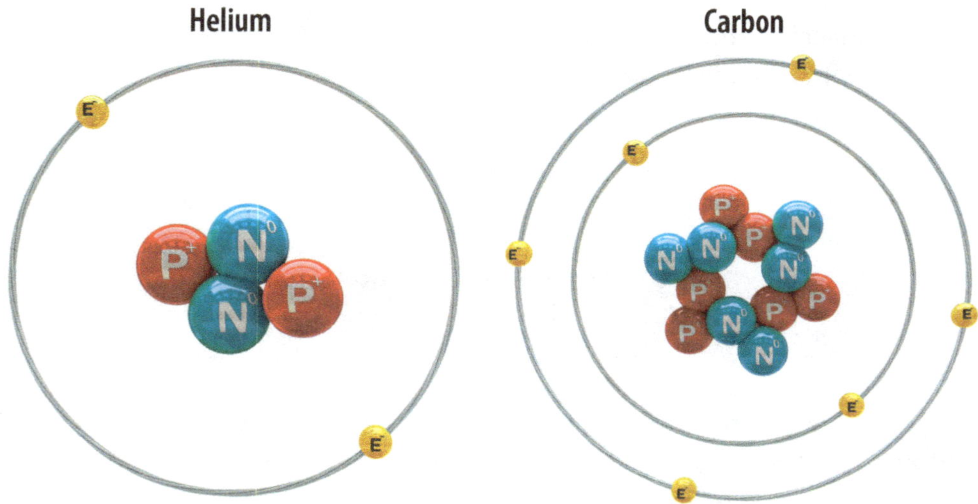

These images show the atomic structure of helium and carbon.

What would happen to these elements if you were able to remove some of their protons, neutrons, or electrons? Would the elements remain the same?

Scientists have now discovered 118 distinct elements in the Universe. Each of these elements is composed of atoms with their own unique combination of protons, neutrons, and electrons. Knowing that there is only a few elements, what does that tell you about the matter around you? Do you think there could be more elements that haven't been discovered? Why or why not?

Discovering and Classifying Elements (Explain)

Long ago, craftspeople made tools and jewelry with gold, silver, copper, iron, and other materials. They came to know the properties of each material very well. They understood how each material reacted with other materials. They knew the temperatures at which materials melted and evaporated and whether or not they could be bent or stretched.

These ancient craftspeople were not chemists as we think of them today. They did not use the scientific process to study the materials they were using. Many of them did not share their discoveries. Why do you think many craftspeople kept new discoveries of reactions as trade secrets? Could some of these discoveries have been dangerous? If so, can you think of any examples?

Before the advent of modern science, *alchemists* were people interested in the properties of certain materials. They studied how materials changed in certain conditions. Their ultimate goal was to change one material, such as the inexpensive element lead, into another element, like the precious metal gold. We now know that this would actually require changing the atomic structure of these elements, a feat far beyond the technology of that time or even of our own time. Do you think scientists will ever be able to change one element to another element? Explain your answer.

Lesson Activity

Alchemists wanted to understand how materials changed under certain conditions. Think about substances you use regularly, like water, salt, and soap. Make a list of five or more of these substances. Think about how each substance can change depending on conditions. Record the change in condition next to the substance on your list. Predict what will happen to each substance. Try some tests to see if your predictions were correct.

Why is it important to understand what causes a substance to change?

Copper Coat a Nail

How can you tell that a chemical reaction has occurred?

Materials
- safety goggles
- 100 mL ($\frac{3}{4}$ cup) white vinegar
- clear plastic cup
- 5 mL (1 teaspoon) salt
- 20 copper pennies
- paper towels
- 2 steel nails
- masking tape
- plastic spoon

Procedure

1. Wearing goggles, **measure** 100 mL ($\frac{3}{4}$ cup) of white vinegar and put it into the clear plastic cup. Add 5 mL (1 tsp) of salt. Stir the salt until it dissolves in the vinegar.

2. Dip a penny into the vinegar solution about halfway for approximately 10 seconds.

3. Remove the penny. **Record** what you **observed**.

4. Record how the pennies look before putting them in the solution. Then, place all the pennies into the plastic cup. Let them sit in the vinegar solution for 10 minutes. Use the plastic spoon to occasionally stir the pennies. While the pennies are soaking in the vinegar solution, look for any changes that you can see happening to the pennies. **Record** what you observe.

5. Use the plastic spoon to remove the pennies and put them on a paper towel. **Record** how the pennies look now.

6. Place the two nails into the plastic cup of vinegar. Let the nails sit for at least 30–40 minutes.

7. Remove and rinse the nails. **Observe** and **record** the condition of the nails.

Analyze Results

What did you **observe** happening to the nails that were submerged in the vinegar? Why?

Create Explanations

1. How can you tell that a chemical reaction has occurred?

2. What happened when you placed the pennies into the vinegar solution?

3. Why would you want to coat a steel object with another metal like copper or zinc?

4. What signs of a chemical reaction did you observe in this investigation?

Seeing Patterns and Making Groups Explain

Although alchemists were not scientists, they did observe that certain materials had similar properties and could be grouped together.

Scientific methods of chemistry introduced by Jabir ibn Hayyan in the seventh century helped bring about the discovery of the element phosphorus in 1649 by Henning Brand. He kept it secret because he was looking for a way to change iron and lead into gold. Can you think of any other scientific experiments that had unexpected, yet important, results? In 1680, Robert Boyle discovered phosphorus and made the news public. In 1680, what types of scientific methods do you think scientists used to prove their theories? It was Robert Boyle who first defined an element as a substance that cannot be broken down into a simpler substance by a chemical reaction. Robert Boyle is often considered the father of modern chemistry.

By 1789, when Antoine-Laurent de Lavoisier published the first chemistry textbook, he was able to list 33 elements. He grouped the elements as gases, metals, nonmetals, and earths.

As more and more elements were discovered and their chemical properties were compared, chemists continued to see patterns in their properties. They noticed a relationship between the atomic structure of elements and their chemical properties. It was observed that certain elements could be grouped into triads, or groups of three, based on their chemical properties. For example, iron, cobalt, and nickle make up the iron triad. Copper, silver, and gold are metals. They are shiny, malleable, and good conductors of heat and electricity. These shared properties are one reason the elements are ordered by **atomic mass**, which is the total mass of all of the protons and neutrons in the nucleus.

Faith Connection

Besides his great contributions to both physics and chemistry, Robert Boyle was a humble Christian and diligent Bible student. He was interested in missions and gave much of his own money to translate the Bible into other languages. In his will he founded the "Boyle Lectures" for proving the Christian religion.

Robert Boyle (1627–1691) followed the scientific method and published his results, even if they were negative.

Why might scientists be interested in publishing results that are contrary to their predictions?

Math in Science

Lithium, beryllium, and potassium are one of the triads. Lithium has an atomic mass of about 7.0. Potassium has an atomic mass of about 11.0, and beryllium has an atomic mass of about 9.0. What is the average mass of the triad?

A Periodic Table

Russian professor Dmitri Ivanovich Mendeleev created a table of elements organized by atomic mass. Mendeleev's table showed the elements and the relationships between them. It even allowed him to predict the existence of elements that had not yet been discovered.

Mendeleev's table and today's use **chemical symbols**. A chemical symbol may be either one capital letter or a combination of one capital and one lowercase letter for the name of an element.

Some chemical symbols, like hydrogen (H), are easy to figure out. Other symbols are harder to understand because they refer to older names or a different language, like Greek or Latin, for the element. For example, Fe, the symbol for iron, is derived from the old Latin name for the element iron, *ferrum*. Look at the box on the left. What information does the box provide about the element hydrogen?

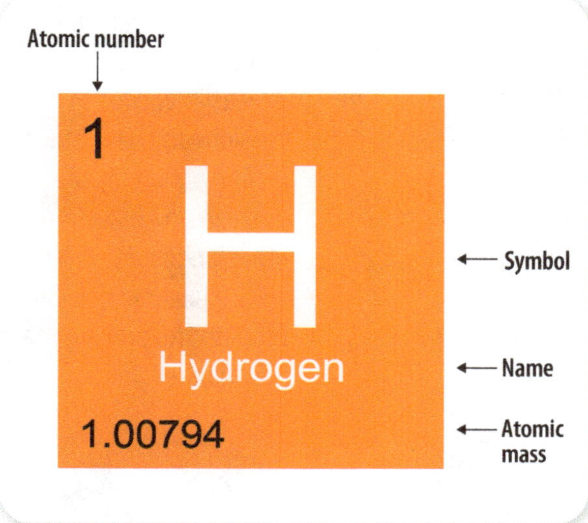

Atomic number

1

H

Hydrogen

1.00794

← Symbol

← Name

← Atomic mass

The periodic table provides information about the element.

What information can the table tell you about an element?

Lesson Activity

Learn more about chemical symbols by copying and completing the following chart. Do research and use the periodic table of the elements on pages 414 and 415 to help you find the names and chemical symbols.

Atomic Number	Current Name	Original Name	Symbol
13		Natrium	
26		Ferrum	
29		Cuprum	
47		Argent	
79		Aurum	
80		Hydrargyrum	
82		Plumbum	

How do the chemical symbols relate to the names of the elements you identified?

The Periodic Table of the Elements Explain

The **periodic table of the elements** that we use today organizes the elements by repeating patterns in their properties. Mendeleev and others ordered the elements by their atomic mass. However, in 1913, Henry Moseley ordered the elements by **atomic number**. The atomic number of an element is the number of protons in its nucleus. Each element has a unique atomic number that never changes. The atomic number is different from the atom's atomic mass. Atomic mass is an average of the masses of both the protons and neutrons in the nucleus. An average number is used because an element's number of neutrons can change. For example, carbon is normally referred to as carbon-12 (C-12). Carbon-12 has 6 protons and 6 neutrons in its nucleus, equaling an atomic mass of 12. Two extra neutrons result in an isotope of carbon, carbon-14.

As you read further, refer to the periodic table of the elements on pages 414 and 415. The atomic numbers and masses are used with other characteristics to organize the elements into *periods* and *groups*. Rows of elements are called **periods**. These run from left to right, with some skips in between. The period number for a row indicates the highest stable energy level of electrons. The period number increases as you move down the periodic table because there are more sublevels where the energy of the atoms increases. The **groups** make up the columns of the table, and these elements have the same number of electrons in their outer levels. Beryllium, magnesium, and calcium all have two electrons in their outer levels. The elements in a group have similar chemical properties. How many electrons do you think barium has? How many electrons to you suppose aluminum and boron have?

Today's periodic table is an updated version of that created by Mendeleev.

How has the periodic table changed since Mendeleev's time?

Lesson Activity

Use the periodic table of the elements on pages 414 and 415 to write the chemical symbols for potassium, cobalt, boron, and helium. Include the atomic number, atomic mass, name, and symbol.

What two things does the atomic number of an element tell you about the atomic structure of that element?

Check out your *Science Journal* for a Structured Inquiry that explores organizing elements into a periodic table.

Extend

Atomic Number

An element's atomic number indicates the number of protons in the atomic nucleus and usually tells the number of electrons in the atom. An element's position on the periodic table will also reveal some of its chemical properties.

For example, you are likely familiar with iron. Iron (Fe) on the table has an atomic number of 26. However, you may not be familiar with the element cobalt (Co). Its atomic number is 27. What might you infer about cobalt based on its atomic number and what you know about iron? If you study the elements and their properties you will realize why some are grouped as they are. What elements on the table do you think were easiest to discover? Why did it take so long to discover the relatively few elements that are on the chart?

Periodic Table of the Elements

The elements in the periodic table are arranged by their atomic masses.

What sort of patterns can you spot in the table?

Color Key

Alkali metals	Alkaline earth metals	Transition metals	Post-transition metals	Metaloids
Other nonmetals	Halogens	Noble gases	Lanthanide	Actinide
Radioactive element	Synthetic element	H Gas	Hg Liquid	Li Solid

The various elements on the periodic table of elements are further identified using the color code key provided.

Reference complete periodic table on pages 514–515 of your student edition.

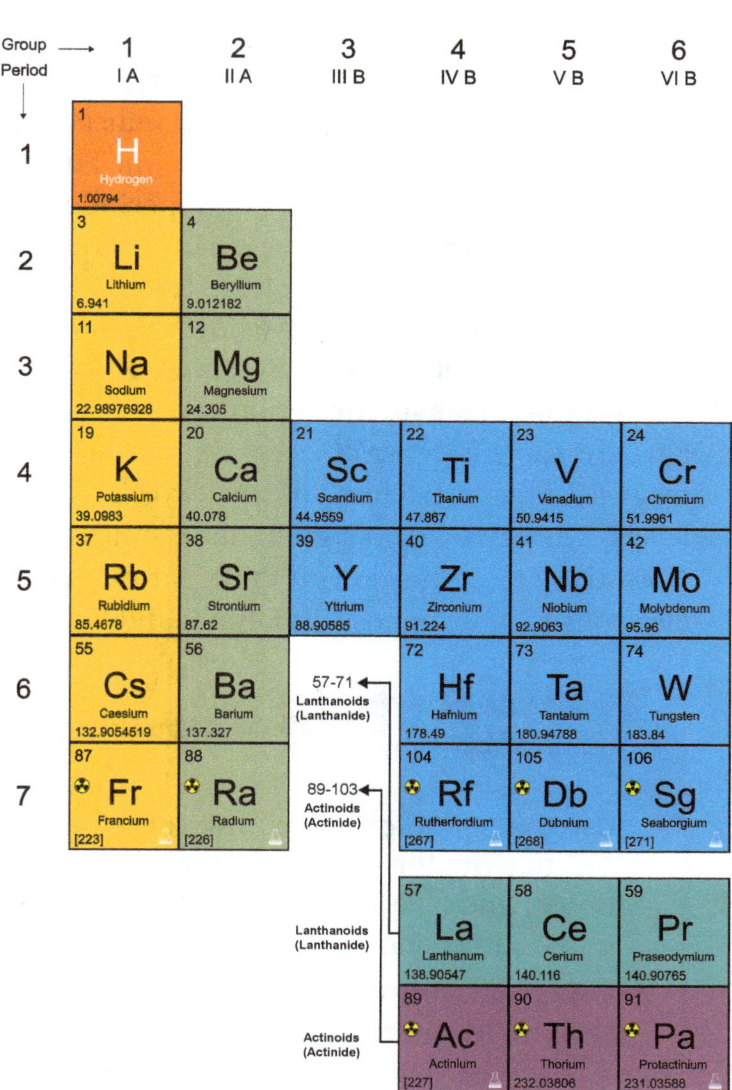

414

Table of Many Colors Explain

The colors in the periodic table below show how elements can be organized into groups. Study the characteristics of three groups of elements. Use the table to name some metals, nonmetals, and metalloids. Which ones are you familiar with?

Metals	Nonmetals	Metalloids
• Lustrous (shiny appearance) • Good conductors • Ductile (can be made into a wire) • Malleable (can be bent and shaped) • Usually solids at room temperature (mercury is the exception) • Very reactive	• Not lustrous (dull appearance) • Poor conductors • Brittle • May be solid, liquid or gas at room temperature • Less reactive	• Have properties of both metals and nonmetals • Lustrous (shiny appearance) • Semi-conductors • Form alloys with metals

7 VII B	8 VIII B	9 VIII B	10 VIII B	11 I B	12 II B	13 III A	14 IV A	15 V A	16 VI A	17 VII A	18 VIII A
											2 **He** Helium 4.002602
						5 **B** Boron 10.811	6 **C** Carbon 12.0107	7 **N** Nitrogen 14.0067	8 **O** Oxygen 15.9994	9 **F** Fluorine 18.9984032	10 **Ne** Neon 20.1797
						13 **Al** Aluminium 26.9815386	14 **Si** Silicon 28.0855	15 **P** Phosphorus 30.973762	16 **S** Sulfur 32.065	17 **Cl** Chlorine 35.453	18 **Ar** Argon 39.948
25 **Mn** Manganese 54.938045	26 **Fe** Iron 55.845	27 **Co** Cobalt 58.933195	28 **Ni** Nickel 58.6934	29 **Cu** Copper 63.546	30 **Zn** Zinc 65.38	31 **Ga** Gallium 69.729	32 **Ge** Germanium 72.64	33 **As** Arsenic 74.9216	34 **Se** Selenium 78.96	35 **Br** Bromine 79.904	36 **Kr** Krypton 83.798
43 **Tc** Technetium [98]	44 **Ru** Ruthenium 101.07	45 **Rh** Rhodium 102.9055	46 **Pd** Palladium 106.42	47 **Ag** Silver 107.8682	48 **Cd** Cadmium 112.411	49 **In** Indium 114.818	50 **Sn** Tin 118.71	51 **Sb** Antimony 121.76	52 **Te** Tellurium 127.6	53 **I** Iodine 126.90447	54 **Xe** Xenon 131.293
75 **Re** Rhenium 186.207	76 **Os** Osmium 190.23	77 **Ir** Iridium 192.217	78 **Pt** Platinum 195.084	79 **Au** Gold 196.966569	80 **Hg** Mercury 200.59	81 **Tl** Thallium 204.3833	82 **Pb** Lead 207.2	83 **Bi** Bismuth 208.9804	84 **Po** Polonium [209]	85 **At** Astatine [210]	86 **Rn** Radon [222]
107 **Bh** Bohrium [272]	108 **Hs** Hassium [270]	109 **Mt** Meitnerium [276]	110 **Ds** Darmstadtium [281]	111 **Rg** Roentgenium [280]	112 **Cn** Copernicium [285]	113 **Uut** Ununtrium [286]	114 **Fl** Flerovium [289]	115 **Uup** Ununpentium [288]	116 **Lv** Livermorium [293]	117 **Uus** Ununseptium [294]	118 **Uuo** Ununoctium [294]

60 **Nd** Deodymium 144.242	61 **Pm** Promethium [145]	62 **Sm** Samarium 150.36	63 **Eu** Europium 151.964	64 **Gd** Gadolinium 157.25	65 **Tb** Terbium 158.9253	66 **Dy** Dysprosium 162.5	67 **Ho** Holmium 164.93032	68 **Er** Erbium 167.259	69 **Tm** Thulium 168.93421	70 **Yb** Ytterbium 173.054	71 **Lu** Lutetium 174.9668
92 **U** Uranium 238.02891	93 **Np** Neptunium [237]	94 **Pu** Plutonium [244]	95 **Am** Americium [243]	96 **Cm** Curium [247]	97 **Bk** Berkelium [247]	98 **Cf** Californium [251]	99 **Es** Einsteinium [252]	100 **Fm** Fermium [257]	101 **Md** Mendelevium [258]	102 **No** Nobelium [262]	103 **Lr** Lawrencium [262]

Locate the grey colored elements on the periodic table once more. These are the noble gases. **Noble gases** are colorless and odorless gases that do not react easily with other elements. They are located on the far right of the periodic table. These elements have all of the electrons they need, so they rarely combine with other elements. They are used in light-up signs, light bulbs, headlights, balloons, and they are also used to cool things.

How Has the Table Changed? (Explain)

As scientists learned more about the composition of different elements, they expanded on the concept of the periodic table. In 1939, francium was discovered. It is the last naturally occurring element on the periodic table. Since then, new elements have been synthesized in the laboratory using nuclear reactors and particle accelerators. All of the elements with atomic numbers greater than 92 (uranium) have been discovered this way. Do you think more elements will be discovered in the future? It has been some time since scientists discovered a naturally occurring element, so why do you think it is important for them to continue their research?

Scientists use giant particle accelerators, such as the Tevatron device at Fermilabs in Batavia, Illinois, with a 6 km (4 mi) circumference to discover new elements and learn more about the structure of atoms.

Why do we need such large devices to study things as small as atoms?

These super-heavy elements are not stable and decay almost instantly into other elements. Examples of these new elements are flerovium (Fl), which has an atomic mass of 114, and livermorium (Lv), with an atomic mass of 116. Four other super-heavy elements—113, 115, 117, and 118—are waiting for their permanent names and addition to the periodic table.

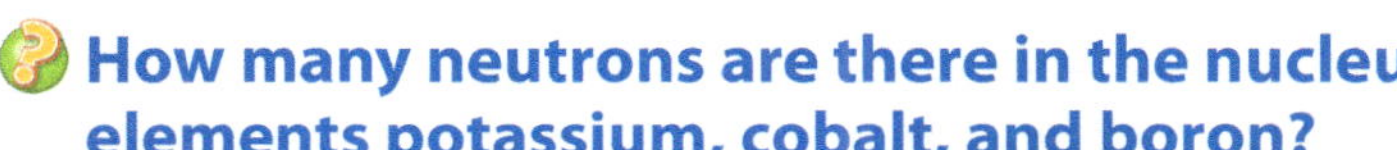

Lesson Activity

Determine the number of neutrons in the elements potassium, cobalt, and boron. Find the number of neutrons by rounding the atomic mass of each element into a whole number. Subtract from that whole number the number of protons (the same as the element's atomic number) in each of the elements.

How many neutrons are there in the nucleus of the elements potassium, cobalt, and boron?

Concept Check Assess/Reflect

Summary: What is the periodic table of the elements? The atomic mass of an element is the total of all of the protons and neutrons found in the nucleus of that element. Elements are identified by their symbols and arranged by their properties in the periodic table of the elements. The atomic number of an element is the same as the number of protons in the atom's nucleus and is usually the same as the number of electrons in the atom. Elements are identified and grouped on the periodic table as metals and nonmetals. The noble gases are elements that have all the electrons they need.

1. Which large group of elements has the properties of conducting heat and electricity and the ability to be bent and stretched?

2. What makes the noble gases different from most of the other elements?

3. How are elements in the same group on the periodic table alike?

4. How were Mendeleev and other chemists able to use the periodic table to organize elements?

How Are Compounds and Molecules Related to Elements?

Objectives

- Define molecules and compounds.
- Describe the relationship between atoms, elements, molecules, and compounds.
- Explain what a chemical formula is.
- Distinguish between ionic and covalent bonds.

Vocabulary

chemical bond

molecule

compound

valence electron

ion

ionic bond

covalent bond

Most matter is only made of a few of the elements you learned about in the previous lesson. What elements do you think are most common? What elements combine to form your body? If you found just one of these elements on its own, would you be able to identify it as being from a human? Why or why not?

Atoms and Molecules Explain

Due to their chemical nature, most elements do not exist in pure form. They form bonds with one or more atoms of different elements. A **chemical bond** forms when two atoms either share or exchange electrons in a way that links them together.

A **molecule** is formed when two or more atoms join together chemically. A **compound** is a molecule that contains at least two different elements. All compounds are molecules, but not all molecules are compounds.

Molecular hydrogen (H_2), molecular oxygen (O_2), and molecular nitrogen (N_2) are not compounds because each is composed of a single element. Water (H_2O), carbon dioxide (CO_2), and methane (CH_4) are compounds because each is made from more than one element. The smallest bit of each of these substances would be referred to as a molecule. For example, a single molecule of molecular hydrogen is made from two atoms of hydrogen, while a single molecule of water is made from two atoms of hydrogen and one atom of oxygen.

These are atomic diagrams of hydrogen (H) and oxygen (O).

Compare each of these to their display on the periodic table. What does an atomic diagram tell you?

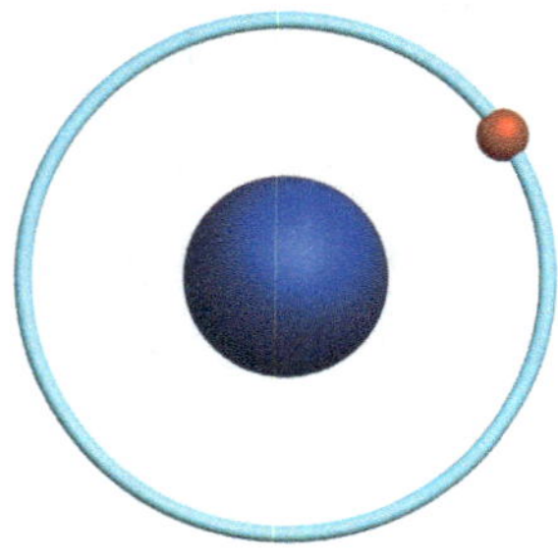

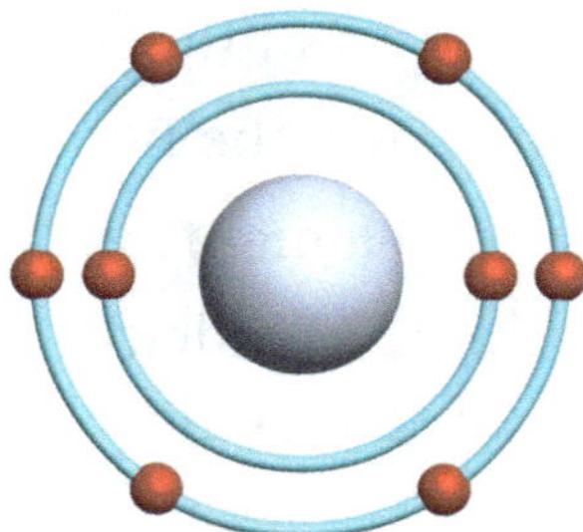

Compounds and Properties

When elements combine to form a compound, they form a new substance. The properties of this new substance are usually very different from the properties of either of the original elements. For example, hydrogen (H) is a gas that burns very easily. Oxygen (O) is a gas that helps other things to burn. However, when you combine these two elements, the hydrogen and oxygen atoms form a water molecule (H_2O). Water is non-flammable. If you then pass electrical energy through water, it will split into hydrogen and oxygen again. Can you think of another element that has different properties when it combines with a second or third element?

Chemical Formulas

What is a chemical formula? A chemical formula is used to express the way that atoms of various elements form bonds to become molecules and compounds. The chemical formula identifies each element by its chemical symbol and by the number of atoms found in a molecule of that compound. If there is more than one atom, a subscript is used to indicate that number. For example, the chemical formula of a water molecule is H_2O. Another simple molecule, methane (natural gas), is made of one carbon atom bonded to four hydrogen atoms. The chemical formula for methane is CH_4. Glucose (sugar) has 6 carbon atoms, 12 hydrogen atoms, and 6 oxygen atoms. What is its chemical formula?

What happens if you change the number of atoms in a compound's chemical formula? For example, if you add one more atom of oxygen to H_2O, do you still have a water molecule? If the number of atoms of either element in a molecule changes, then the compound they form is different.

Water and hydrogen peroxide are made of the same elements.

How are molecules of water and hydrogen peroxide different?

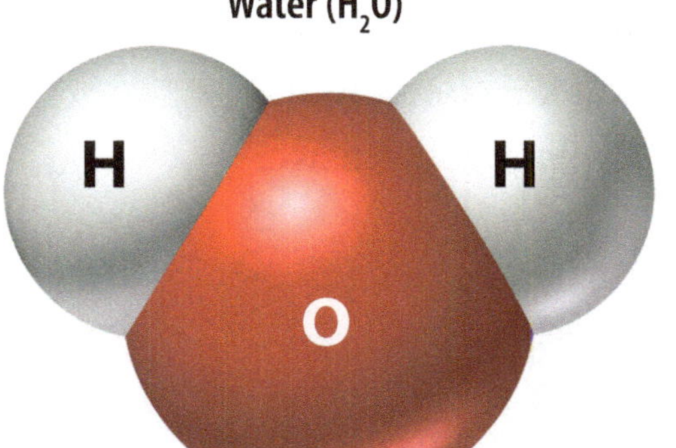

Water (H_2O)

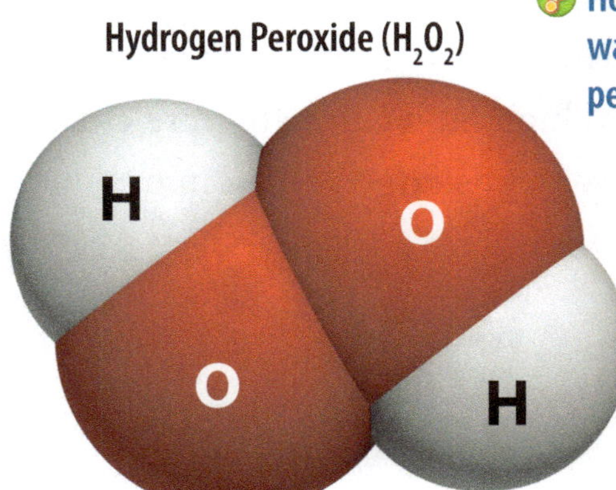

Hydrogen Peroxide (H_2O_2)

Molecular Models

How can you build a three-dimensional model of a molecule?

Procedure

1. Fill in the table in your **Science Journal** using your periodic table.

2. Pick a colored gumdrop to represent each of the elements listed in your **Science Journal**. Color the circle with the color you chose to represent the element.

3. Using the gumdrops and toothpicks, build three molecular models using the following configurations and the chemical formulas provided in your **Science Journal**:

   a. Molecules with two atoms are linear:

   b. Molecules with two atoms on either side of a center atom are bent:

   c. Molecules with three atoms around a center atom are shaped like pyramids:

4. Fill in the chart in your **Science Journal** with the correct number of atoms for each element in the molecule.

5. Make a diagram of each molecule listed in your **Science Journal**.

Materials
- colored pencils
- small gumdrops
- toothpicks

Analyze Results

How do your models show the structure of the molecules?

Create Explanations

1. How can you build a three-dimensional model of a molecule?

2. Which molecules are pure elements? Which are compounds?

3. Which is larger, an atom or a molecule? Explain your answer.

Find the formulas for copper sulfate, calcium chloride, household bleach, and toilet bowl cleaner. Then design a chart that identifies the common name (if there is one), the elements that are included in the formula, and the number of atoms of each element in the compound.

How are you able to determine the chemical formula for each?

Chemical Bonds **Explain**

A chemical bond is the binding that occurs when electrons from different atoms combine. The electrons in chemical bonds are **valence electrons**, electrons in an atom's outermost shell. These electrons repel each other and are also attracted to the protons within the atoms. This interaction forms bonds between atoms, sticking them together.

Ionic Bonds

When an atom gains or loses an electron, the balance between the positive and negative charges is lost and the atom obtains a charge. An **ion** is an atom that gains or loses an electron. An **ionic bond** forms from the attraction of the oppositely charged ions.

When the metal sodium (Na) loses an electron, it forms a sodium ion. When the non-metal chlorine (Cl) gains an electron, it forms a chlorine ion. When the two ions combine, they form a compound called sodium chloride. This compound is more commonly called table salt.

Scripture Spotlight

What kinds of bonds are mentioned in **Ephesians 4:3** and **Colossians 3:14**?

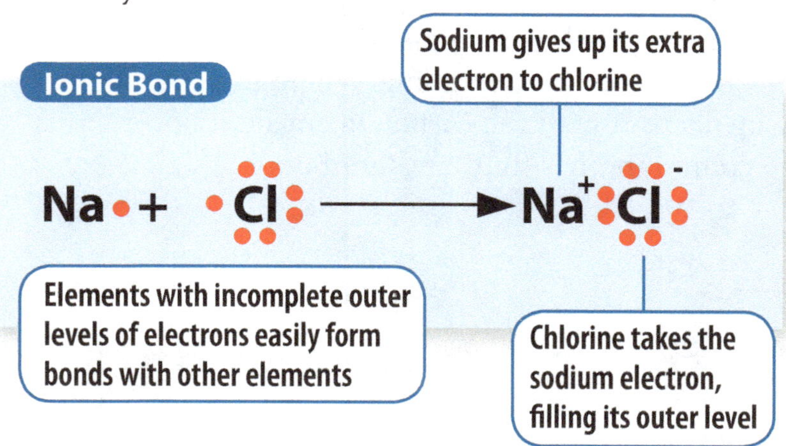

The chemical bonding that forms the compound sodium chloride.

How does the bond between sodium and chlorine in salt form?

Depending upon the ions involved, compounds may possess different properties and uses. Besides cooking, seasoning, and preserving foods, we use salts in many different ways. Can you think of other ways we use salt?

What determines how atoms combine to form molecules?

Gather four gumdrops of each color red, green, and white, toothpicks, paper, and colored pencils. Each gumdrop is an atom and each color represents a different element. Draw and color your elements. Name your elements and create a chemical symbol for each. Use the toothpicks to connect the gumdrops to make three different molecules. The red gumdrops only have one place where another atom can attach; the green gumdrops can attach to two atoms; and the white gumdrops can attach four atoms. All combinations of atoms have to be balanced and geometric in nature. Draw and color your creations and write the chemical formula for each molecule. Remember to account for multiple atoms of the same element in your molecule. How many different molecules can you create with your elements? How does the number of atoms an element can bond with limit the molecules that can be formed?

Covalent Bonds

When two atoms share electrons to form a chemical bond, that bond is known as a **covalent bond**. Water is an example of a compound formed by covalent bonds. Each hydrogen atom shares one electron with the oxygen atom to form a pair of covalent bonds. Another example is, methane (CH_4), composed of one carbon atom and four hydrogen atoms. Each hydrogen atom shares one electron with a carbon atom. Where the hydrogen atom shares an electron with the carbon, a covalent bond forms. The methane compound has four covalent bonds.

The chemical bonding that forms the molecule hydrogen.

How does this bonding differ from that for sodium chloride?

When an atom shares two electrons, it forms a double bond. If an atom shares three electrons, it forms a triple bond. For example, the compound carbon dioxide (CO_2) contains one carbon atom and two oxygen atoms. Each oxygen atom shares two electrons with the carbon atom. Carbon dioxide has two double bonds.

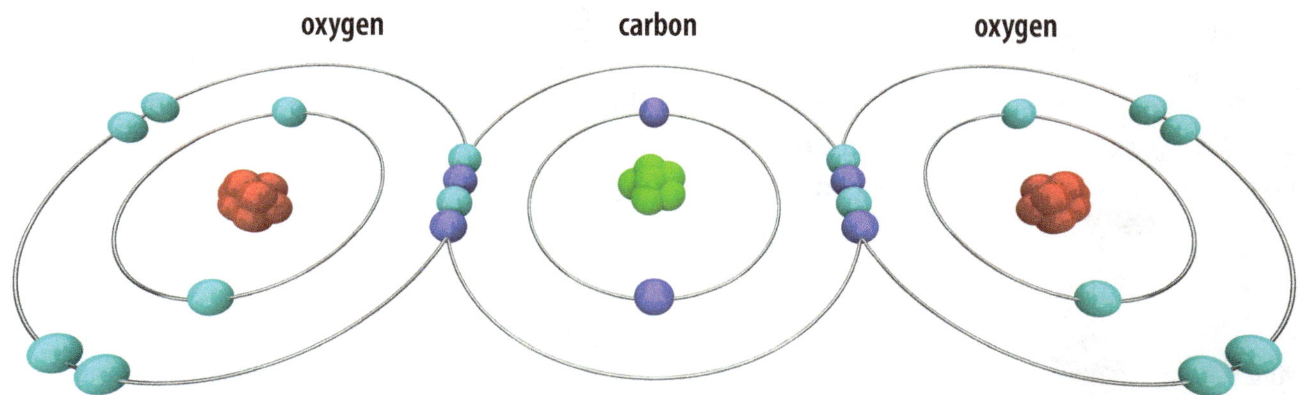

Two atoms of oxygen share electrons with an atom of carbon to form carbon dioxide.

 **Once the bonds are formed, how many electrons does carbon have in its outermost orbit?**

Concept Check Assess/Reflect

Summary: How are compounds and molecules related to elements? A chemical bond forms when two atoms either share or exchange electrons in a way that links them together. A molecule is formed when two or more atoms join together chemically. A compound is a molecule that contains at least two different elements. Chemical symbols and formulas are used to show the bonding of atoms to form molecules. An ionic bond forms when an atom gives up an electron that another atom needs to fill its outer level. Atoms that gain or lose electrons are called ions. Covalent bonds form when atoms share outer-level electrons.

1. What is the smallest unit of a compound called?

2. How does a covalent bond form?

3. How does an atom become an ion?

4. Why are elements in column 1 of the periodic table able to react chemically with many other elements, while those in column 18 do not?

5. Does a compound share the same chemical properties as its elements? Explain.

Essential Question

How Do Compounds and Mixtures Differ?

Think about vegetable soup. What ingredients might be in your vegetable soup? Could you separate the ingredients? What is in tomato soup? How is it different than vegetable soup? How are they the same? Both vegetable soup and tomato soup are mixtures. In what ways are these mixtures different from compounds?

Compounds or Mixtures **Explain**

In a previous lesson, you learned about elements, molecules, and compounds. All matter can be classified as either a pure substance or a mixture. Elements and compounds are pure substances; they have a constant composition and consistent properties throughout. A **mixture** forms when two or more substances combine without changing the nature of the component substances. Unlike compounds, which change chemically and result in a new substance, mixtures result from *physical* changes and retain their original properties. The compounds in a mixture do not bond to form a new substance.

In some mixtures, it is easy to see the individual substances. A collection of coins from around the world is this type of mixture. You can pick out types of coins by size, color, or unique designs that have been stamped on them. Can you think of other examples of mixtures made of substances that can easily be identified?

Both types of soup represent examples of a mixture.

How can you tell what substances are in the soup?

Some mixtures do not have a uniform composition, and contain tiny pieces of one substance mixed within the other substance. Some examples of these mixtures are as follows:

- **emulsions**, which have two or more liquids, where one liquid contains the other, such as mayonnaise and many salad dressings
- **colloids**, which are liquid mixtures containing small particles that do not separate out
- aerosols, such as room fresheners that come out of spray cans
- foams, such as whipped cream and shaving cream
- gels, such as toothpaste

In many mixtures, the substances are blended so evenly that the mixture appears to be the same throughout. Examples include the air we breathe and the sodas and juices we drink.

Separating Mixtures

Mixtures do not have chemical formulas like compounds. They do not have fixed compositions. You can mix different amounts of the substances and still have the same type of mixture. For example, air is a mixture of mostly nitrogen and oxygen with a blend of carbon dioxide and other gases. Smoke is a mixture of particles of carbon, ash, and air.

Because no chemical reaction occurs when you blend substances in a mixture, each compound keeps its original physical properties. You can use those physical properties to separate the individual substances from the mixtures. Think about the soup discussed earlier. How would you separate the different ingredients? Think of another mixture and explain how you would separate it. Do you think it is possible to separate all mixtures? Why or why not?

Explore-a-Lab

Structured Inquiry

How can you separate a mixture of sand and water?

Pour some sand into a glass of water and stir it to make a mixture. Once you have a mixture, predict how you can separate the sand from the water. Test your prediction and revise your plan as needed.

Dissolving

What factors affect the rate at which solids dissolve?

SAFETY: Handle hot water with care to avoid scalds. Wipe up spills immediately to avoid slipping.

Materials
- hot water
- room-temperature water
- cold water
- clear plastic cups
- thermometer
- stirring rods
- sugar cubes
- metal spoon
- masking tape (for labeling)
- stopwatch

Procedure

1. **Control the variables** by observing one variable at a time. Label cup 1 "solid" and cup 2 "crushed." Fill each about halfway with room-temperature water and **record** the temperature. Place one sugar cube in the cup labeled "solid." Crush the other sugar cube with the spoon and put it into the cup labeled "crushed." **Observe** the time it takes for the sugar to dissolve in each cup separately. **Record** your observations in a table.

2. Label cup 3 "stirred" and cup 4 "unstirred." Fill each cup about half-full with room temperature water and record the temperature. Crush two sugar cubes. Place one crushed cube in each cup. Use the rod to stir one of the solutions for one minute. Do not stir the other solution. **Observe** how long it takes the sugar to dissolve in each cup.

3. Label cup 5 "hot." Fill it halfway with hot water and record the temperature. Label cup 6 "room temperature." Fill it halfway with room-temperature water and record the temperature. Label cup 7 "cold." Fill it halfway with cold water and record the temperature. Place a sugar cube into each cup. **Observe** how long it takes for the sugar to dissolve in each cup.

Analyze Results

Examine the data you collected. Summarize how each variable may have affected each solution.

Create Explanations

1. What factors affect the rate at which solids dissolve?

2. How does crushing the solute affect the rate of solution? Why?

3. How does stirring and water temperature affect the rate of solution?

4. Which variable resulted in the fastest time for the sugar to dissolve? Why?

Ways to Separate Mixtures

Process	Examples
Magnetism	Separating iron from sand
Filtration	Separating sand from water
Evaporation	Separating salt from seawater
Distillation	Separating pure water from seawater
Paper Chromatography	Separating out the colors in black ink

Density

You can also use density to separate materials in a mixture. Homemade salad dressings are made with oil, vinegar, water, and spices. Each of these substances has a different density. If you let the homemade salad dressing sit on a tabletop, the different substances will separate into layers based on density. The less dense substances will float on top of the denser substances.

Combinations

Sometimes you need to use a combination of properties to separate the substances in a mixture. Soil is a mixture of many different substances. Most of them you can separate mechanically by using different sizes of sieves or screens to separate the larger particles from the smaller ones. Another way to separate some of the substances is to pour the soil into water. The least dense substances will float while the substances with greater density will sink. Think about a mixture of iron filings with salt and sand. How would you go about separating the different substances?

Scripture Spotlight

You are learning how different mixtures can be separated. Read about a great separation in **Matthew 25:31–46**. Who is doing the separating, and how is it being accomplished?

Explore-a-Lab

Structured Inquiry

 **Which property are you using to separate the salt and pepper in each test?**

You can use the different properties of salt and pepper to separate them from a mixture. Mix a spoonful of salt and pepper into a pile on a small plate. Try each method below and observe the results.

- Blow up a balloon. Rub it on your hair and then bring it close to the mixture.

- Place some of the mixture into a cup of water.

- Place some of the mixture into a cup of vegetable oil.

- Repeatedly tap on the side of the plate hard enough to shake the particles.

Which properties allow you to separate the mixture?

Use what you have learned about separating mixtures. Devise a way to separate out the iron filings, salt, sand, and water from the mixture given you. How many processes did it take to separate everything?

Practical Uses

Being able to separate materials from mixtures using their different physical properties has many practical uses in the real world. Recycling plants are designed to separate all the materials mixed together in our recycling bins so they can be recycled into new materials. Here are some of the techniques a typical plant uses:

- Sieves and screens sort items mechanically by size.
- Magnets separate metal objects, such as steel cans.
- Rapidly spinning drums separate lighter paper and plastic from denser metal and glass.
- Optical scanners trigger blasts of air to separate materials by color.
- Shredded plastic can be separated from paper using static electric energy.

A recycling plant is one place where mixtures are separated.

What are some ways materials are separated out of this mixture?

Solutions `Explain`

A **solution** is a mixture in which the particles of each substance are mixed evenly. When you place salt into water and stir, what happens? A substance that dissolves in another substance is called a **solute**. The substance that a solute is dissolved into is called a **solvent**. When you make a saltwater solution, which substance is the solute and which is the solvent? When you make hot chocolate, which substance is the solute and which is the solvent?

Salt dissolves in water, forming a salt solution that freezes at a lower temperature than pure water. The salt here has caused the snow closest to it to melt because it lowers the freezing point of water.

What does this tell you about icebergs in the ocean?

Usually when you make hot chocolate you stir in the cocoa mixture. What would happen if you didn't stir the mixture? The stirring helps change your mixture to a solution faster by evenly spreading out the solute. How quickly one substance dissolves in another is known as the rate of solution. Can you think of other ways to increase the rate of solution?

To be called a solution, substances must mix evenly but not form new substances. There is no chemical reaction. This means each substance keeps its properties. This does not mean that the properties of the solution are exactly the same as those of the substances that were mixed. For example, the freezing point of a salt solution is lower than the freezing point of pure water. What properties change when you add cocoa to your warm milk to make hot chocolate?

Making a solution by dissolving a solid, like cocoa, into a liquid, like milk, is common. However, solutions can be made by mixing gases, solids, or liquids. Dissolving gases in liquids is another way to form solutions. Carbonated drinks are a solution of gaseous carbon dioxide dissolved in liquid water. Other ingredients are dissolved to add flavors. Oxygen, carbon dioxide, and other gases dissolve in nitrogen gas to form the gaseous solution you are breathing right now—air. Solids can also form solutions when they are in a liquid state.

Check out your *Science Journal* for a Structured Inquiry that explores chemical separation.

Extend

Check for Understanding

Think back to your observations in the *Structured Inquiry*. What factors increased the rate of solution in your experiment?

Math in Science

The concentration of a solution is a measure of the amount of solute dissolved in a solvent. The more solute that is dissolved, the higher is the concentration. If no more solute can be dissolved, the solution is saturated. One way to express concentration is as a percentage of the mass of the solute over the volume of the solution (g/mL $\times$ 100). If there are 35 grams of salt dissolved in 1000 mL of seawater, what is the concentration of the solution?

An alloy is made by melting metals and mixing them to cool again into a solid.

Alloys

Alloys are solid solutions made of two or more metals. Early metalworkers discovered that they could mix a metal with other metals or nonmetals when they were melted to form alloys. The different combinations produced desirable properties, like strength. What do you think these metalworkers made with the alloys they created?

One of the alloys these metalworkers discovered early on was bronze. Bronze is a mixture of copper and tin. It is harder than either copper or tin alone and was used for weapons and other implements. Other common alloys are steel and brass.

Steel is made when carbon is mixed with iron, forming a very strong alloy. Brass is an alloy of copper and zinc. It is much harder than either copper or zinc alone. Brass was made as a substitute for gold. Gold and silver items are rarely made with pure gold or silver, which are too soft to hold a shape. To lower costs and strengthen the metal, gold and silver are mixed with less expensive metals, such as copper or nickel. What other alloys can you think of? Why are these alloys useful?

Solubility

The ability of one substance to dissolve in a solvent is the **solubility** of the substance. Recall the Structured Inquiry in this lesson. Which of the factors investigated had the greatest effect on how fast the sugar cube dissolved?

Factors That Affect Rate of Dissolving

- **Temperature**—As particles in a solvent and solute speed up, they mix together faster.
- **Agitation**—Stirring or shaking a solution to break up, dissolve, and mix the solvent and solute together.
- **Surface area**—Dissolving occurs on the surface of a solute. Smaller pieces dissolve faster than large pieces because there is more surface area exposed to the solvent.

Separating Solutions

Even though a solute seems to disappear when it dissolves into a solvent, it is still there. Remember that solutions are mixtures. No chemical reaction has taken place to change either the solute or the solvent into new substances. You can use the properties of the solute and solvent to separate them.

You can use the differences in the boiling points of salt and water to separate a saltwater solution. Salt has a much higher boiling point than water. If you heat a saltwater solution enough, the water will evaporate and leave the salt behind. Most of us get our salt from salt deposits that formed when ancient oceans evaporated. Some countries get their freshwater by using desalination plants to remove salt from seawater.

Explore-a-Lab

Structured Inquiry

How does the rate at which carbon dioxide solute escapes the solvent differ?

Work with a partner. Compare the rate at which gas escapes from four bottles of seltzer water. Open the cap of one bottle at room temperature. Test one bottle directly from the refrigerator, leave one bottle at room temperature after a gentle shake, and give one bottle (directly from the refrigerator) a gentle shake. Using a stopwatch, record how much time it takes from when the lid is removed to when the bubbles no longer appear in each bottle. Record the data in a chart.

Concept Check   Assess/Reflect

Summary: How do compounds and mixtures differ? A mixture is a combination of two or more substances that retain the properties of the original substances. Mixtures are formed by physical changes, not chemical changes. Mixtures can be separated into their original substances by magnetism, density, or other means. Solutions are created when a solute dissolves in a solvent. Alloys are solutions of molten metals, such as steel, bronze, and brass. Solubility is the ability of one substance to dissolve in a solvent and is affected by the size of the solute, temperature, and pressure.

1. Two or more substances blended without changing the substances makes what?

2. Which are the solute and solvent when carbon dioxide and water mix?

3. How could you separate a mixture of iron filings, salt, sand, and water?

4. If a solute does not dissolve in a solvent, what can you try to help make it dissolve?

5. What is the benefit of an alloy?

Extend

Get to Know
John Dalton

John Dalton (1766–1844) was born into a modest Quaker family in Cumberland, England. Dalton grew up farming and working in the family cloth shop. The family was poor, and John did not get much formal education. However, he learned reading, writing, and arithmetic at the nearest Quaker school. When Dalton was 12 years old, he started teaching at the school, and by age 16, he had become principal.

He remained there until he became a math and philosophy tutor at the New College in Manchester. For one of his first research projects, Dalton followed his interest in meteorology. He kept daily logs of the weather, paying special attention to details such as wind speed and barometric pressure—a study that Dalton would carry on for the rest of his life. He began to keep a meteorological diary. Over the next 57 years, he entered more than 200,000 weather observations.

Dalton's interest in weather and air pressure led him to study gases. Dalton's experiments with gases led to the scientific principle that we know as Dalton's Law of Partial Pressures. Dalton's fascination with gases led him to state that all forms of matter (solid, liquid, or gas) were made up of small individual particles.

Called to Serve

The English Royal Society asked Dalton to become a member in honor of his distinguished work. He turned them down because Quaker beliefs did not allow him to seek glory. But glory came to him regardless of his faith. He was quietly elected to the Society without his knowledge.

Reading ancient textbooks, he found that the philosopher Democritus had described water as mostly empty space with smooth balls gliding over each other. Dalton called the "balls" that make up matter *atomos,* or *atoms*. Each type of atom was different from the next by its atomic mass. In an article he wrote in 1803, Dalton created the first chart of atomic masses. This was the beginning of the development of the atomic theory.

Concept Check

1. John Dalton was very interested in the study of air pressure. How did his interest lead to the atomic theory?

2. How is each atom different from another?

Food Chemist

A food chemist researches the chemical properties of the foods and beverages we consume. This work helps improve the quality, safety, and reliability of our food. Food chemists study proteins, fats, starches, and carbohydrates, as well as additives and flavorings, to figure out how well they work in a food system.

One role of a food chemist is to work in quality control to make sure all foods from a food production facility meet certain standards. This is critical for public safety, as contamination of foods with chemicals, bacteria, and other unwanted additives can be dangerous. People who consume contaminated foods could become sick. A food chemist works to prevent contamination and identify its sources when it does occur.

If a food chemist is working on the delivery of food, he or she may study the changes food may go through as it is exposed to different environmental conditions. The chemist's research helps determine the best way to prepare, package, and keep foods fresh for transportation. Many foods in their natural states do not hold up well in storage. For example, when tomatoes are harvested, the food chemist must decide the best way to keep them fresh so that they will continue to ripen and arrive at the supermarket at their peak of ripeness.

A food chemist may work on ways to develop foods to meet nutritional needs. Nutrition bars, for example, may involve the research of a food chemist to figure out how to add the right nutrients and keep the bars fresh until the consumer eats them. The chemist may work with nutritionists, health-care providers, and other businesspeople interested in making this product.

Food chemists work for businesses, government agencies, and research facilities as they focus on food quality and safety. Some chemists may choose to specialize in certain areas, like dairy products or starches, and become experts in these areas.

Many food chemists begin their educational training with an undergraduate degree in chemistry or biology and enter a food sciences program at a master's level degree. A Ph.D. may be required for food chemists wishing to pursue work in teaching or conducting fundamental food chemistry research.

 **Concept Check**

1. How does the work of a food chemist help to make our lives better?
2. What kind of education does a food chemist need to have?

Study Guide

Lesson 1

1. All matter is made of atoms. All atoms have the same basic structure. They have a nucleus made of neutrons and protons that is orbited by a cloud of electrons.

2. The simplest form of matter is called an element. Elements are pure substances; they have a constant composition and consistent properties throughout. Elements cannot be broken down into anything simpler without losing their chemical properties.

3. The atoms of one element are different from the atoms of every other element. Atomic models include the Thomson model, Rutherford model, Solar System model, (Bohr model), and electron cloud model.

Lesson 2

1. The periodic table of the elements shows all of the elements arranged by atomic number.

2. Metals are the largest group of elements and the most reactive. They are on the left side of the periodic table and share common characteristics.

3. Nonmetals have properties that are generally opposite those of metals.

4. Nonmetals are found on the right side of the periodic table and are brittle and transparent; they may be solids, liquids, or gases at room temperature.

Lesson 3

1. A molecule forms when two or more atoms join chemically. A compound is made of two or more elements chemically combined into a single substance.

2. The number and kind of atoms in a molecule determine what compound forms and is expressed in a chemical formula.

3. Covalent bonds form when atoms share electrons. Ionic bonds form when atoms give up an electron. Ions form with the gain or loss of electrons.

Lesson 4

1. A mixture can be a combination of molecules, compounds, or other mixtures.

2. The substances in a mixture can be separated using their physical properties. A solution forms when one or more substances known as solutes are dissolved in another substance called a solvent.

Show What You Know

Visualize It Complete the following concept map in your *Science Journal*.

Element ⟶ Compound ⟶ Mixture

is   is   is

1. ______________________

2. ______________________

3. ______________________

Explain how each pair of terms is related.

4. atom—element
5. molecule—compound
6. nucleus—neutron
7. group—period
8. proton—atomic number
9. electron—chemical bond
10. solute—solvent
11. alloy—solution

Multiple Choice
Choose the best answer.

12. In an atom, this particle has a positive charge.
    A. electron
    B. ion
    C. neutron
    D. proton

13. What is the ability of a substance to dissolve in another substance?
    A. density
    B. atomic number
    C. solubility
    D. boiling point

14. What has to happen for an atom to become an ion?
    A. It loses or gains an electron.
    B. It forms a chemical bond
    C. It dissolves.
    D. It shares an electron with another atom.

15. Which is not an example of something made from an alloy?
    A. a bronze medal
    B. a gold ring
    C. a steel nail
    D. a copper wire

16. Which decreases the rate at which a solid dissolves in a solvent?
    A. higher temperatures
    B. stirring
    C. larger particle size
    D. higher pressure

Check Point
Answer the following questions.

17. What makes the atom of one element different from those of other elements?

18. How does the periodic table of the elements help us **classify** elements?

19. What **sequence** do you go through to separate a mixture of water, oil, salt, sand, and steel filings?

20. Why do people often prefer to use alloys rather than pure metals?

21. How does a solution compare to a compound?

22. Identify the process shown in the diagram below and explain what has changed because of the process.

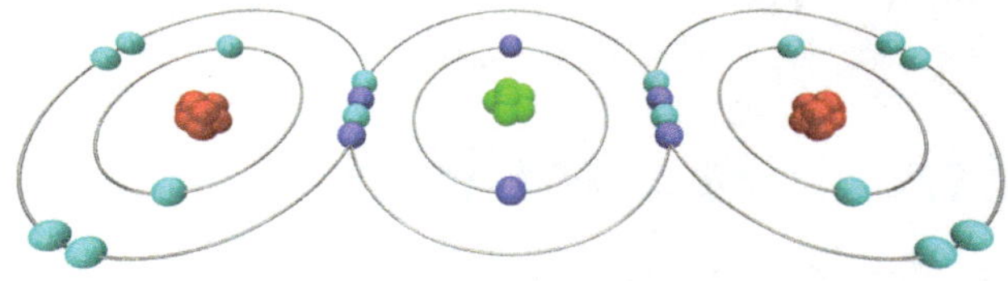

23. Explain how the concept of the structure of the atom changed through new scientific discoveries.

24. What is the difference between a covalent and ionic bond? How is each significant?

Chapter 13

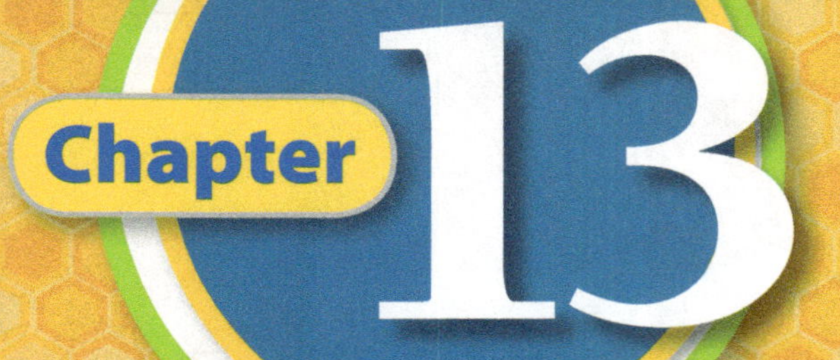

Electricity

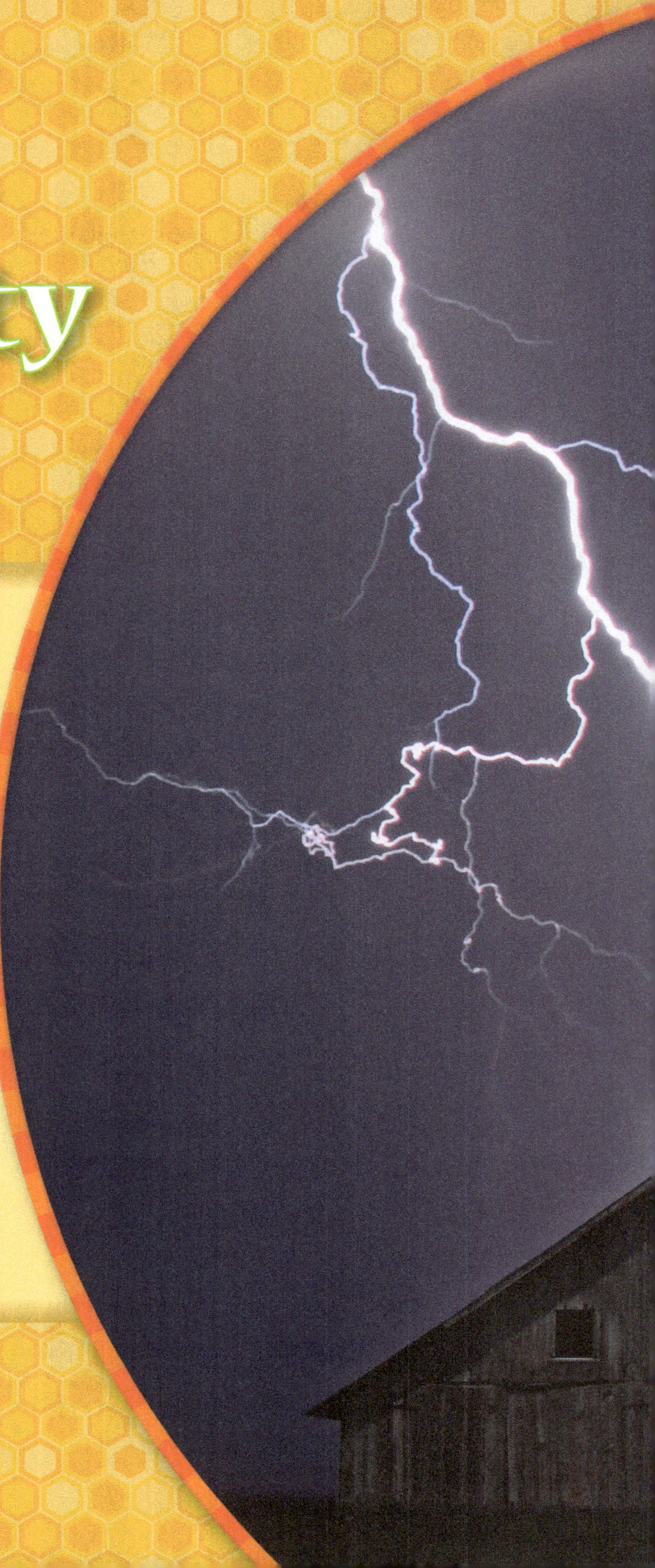

Scripture Spotlight

As you study about electricity, you will read the following passages from scripture:

Exodus 19:16 (p. 442)
Matthew 24:27 (p. 442)
Psalm 119:11 (p. 455)
Romans 10:17 (p. 456)
2 Peter 1:3 (p. 459)
Luke 11:13 (p. 464)
John 3:34 (p. 464)

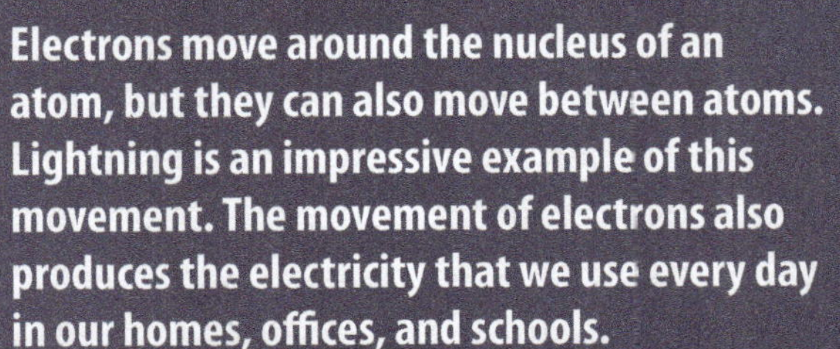

Electrons move around the nucleus of an atom, but they can also move between atoms. Lightning is an impressive example of this movement. The movement of electrons also produces the electricity that we use every day in our homes, offices, and schools.

The Big Idea

Electricity is a form of energy produced by charged particles. In a continuous flow, or stored in a battery, electricity provides power for our technology.

How are lightning and the electric current that you use at home related?

Objectives

- Explain the causes and effects of electric charge.
- Explain what causes an electrostatic discharge.
- Explain how lightning forms.
- Describe the practical uses of static electricity.

Vocabulary

electrostatic discharge

lightning

? Essential Question

What Is Static Electricity?

Have you ever walked across a carpet and reached for the doorknob, only to be "zapped" by an unexpected electric shock? What causes the spark? Why does hair mysteriously stand on end after pulling off a wool cap on a winter day? Have your clothes ever made strange crackling noises and clung together when you pulled them out of the dryer? Have you ever wondered why lightning leaps out of a thundercloud? Is it possible for all these events to be related? If so, how?

Electric Charges and Forces Explain

These seemingly unrelated events are each caused by *static electricity*, the buildup of electric charges on the surface of objects. Recall that protons and neutrons are located at the center of an atom and electrons orbit the nucleus. Neutrons do not carry a charge. Protons have a positive charge and electrons carry a negative charge.

Normally, atoms have a neutral charge because they have equal numbers of protons and electrons. The outer electrons in some atoms are held together very loosely. These electrons have the ability to move from one atom to another. Do you think there are any benefits to losing electrons? Does losing an electron change the type of substance the atom is? Explain.

Lesson Activity

Sprinkle a little salt and some finely ground pepper onto a paper plate. Mix them together. Rub a plastic pen back and forth very quickly with a piece of wool cloth. Hold the pen very close to the plate and move it slowly over the salt and pepper mixture.

? **How is this similar to and different from separating things with a magnet?**

When electrons jump from one material to another, the atoms gain or lose electrons. The electric charges are no longer neutral. When an object loses electrons, it gains a positive charge. When an object gains electrons, it gains a negative charge. Rubbing is one way to transfer electrons. How can you tell if an object has gained or lost electrons? When can it be dangerous for an object to become charged?

Electric Forces

What happens when charged objects are brought near each other? The charged objects exert forces on each other. This force can be attractive (pulling objects together) or repulsive (pushing them apart). The effects of these forces can be summarized in this way:

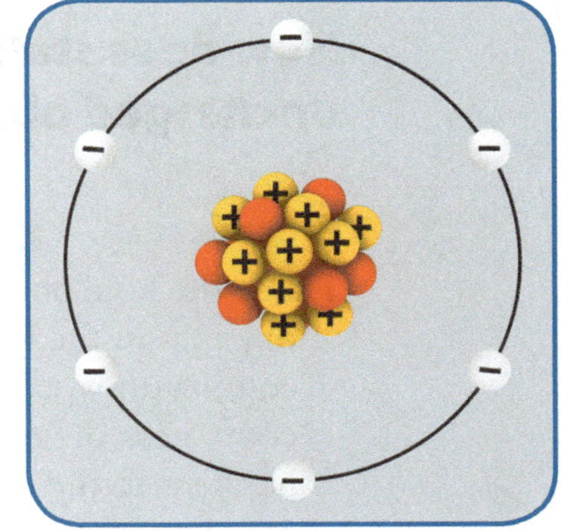

Gaining electrons gives an object a negative charge.

Which kind of particle do you think would be easiest to rub off an atom? Why?

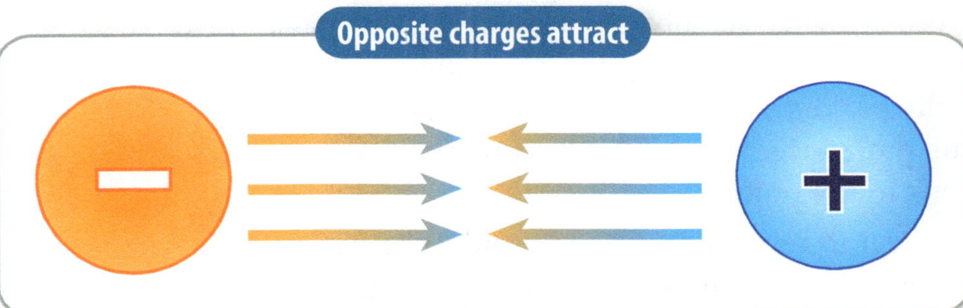

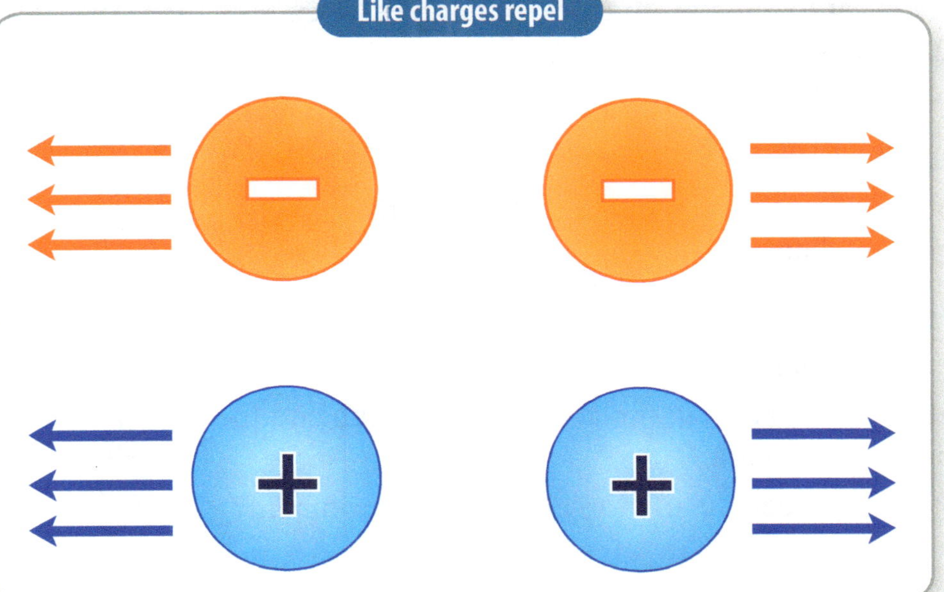

Opposite charges attract each other, while like charges repel each other.

Taking Charge!

How does static charge affect different uncharged objects?

Procedure

1. Use the hole punch to create about 30 circles of paper in the Petri dish. Rub the plastic comb with wool cloth. Bring the charged comb near the paper circles. **Observe** what happens to the paper. **Use numbers** to count and **record** how many pieces of paper the comb picks up.

2. **Predict** what will happen if you bring the comb near other uncharged objects. How will the numbers compare? Bring the charged comb near the broken pieces of packing pellets and then the puffed rice. **Observe** and **record** what happens to the objects.

3. Use a dryer sheet to remove any static buildup between tests. Repeat Steps 1–2 using nylon cloth instead of wool. **Observe** and **record** your findings.

4. Use a dryer sheet to remove any static buildup between tests. Repeat Steps 1–2 using the glass stirring rod and then the rubber rod instead of the comb.

Materials
- tracing paper
- hole punch
- Petri dish
- wool cloth
- plastic comb
- dryer sheets
- nylon cloth
- glass stirring rod
- rubber rod
- foam packing pellets
- puffed rice cereal

Analyze Results

Examine your data table. **Compare** the effects of the electric force produced by each charged object.

Create Explanations

1. How does static charge affect different uncharged objects?

2. Why do the various particles first stick to the comb and then "jump" off the comb after a while?

3. How could this process be used to clean the environment?

Electrostatic Discharge Explain

Recall what happened when you brought your finger near the charged aluminum pie plate during the Chapter Inquiry Kick-Off. How does that relate to what happens when you rub your feet across the carpet on a dry winter day, and then touch a metal doorknob? What causes the shock?

An **electrostatic discharge** is the sudden release of static electricity between two differently charged objects. The electric force is so great that electrons jump the gap between the objects. Often this discharge results in a spark and a popping or crackling sound. When you touch the doorknob, an electrostatic discharge produces the spark that gives you a small electric shock. You can sometimes see, feel, and hear the sound of the spark when it jumps. These discharges can damage or even melt small parts inside fragile electronic equipment, such as computers. Can you think of other times when these small sparks might cause damage or be dangerous? What can you do to prevent them?

A Van de Graaff generator is a device that produces a static charge. Touching the aluminum globe causes all the strands of a person's hair to have the same electric charge. This causes the hairs to repel one another. Based on information that you learned in the chapter, how might the device produce a static charge? In fact, the generator uses a moving rubber belt driven by a felt-covered pulley. This interaction results in the belt carrying away electrons from the felt. The electrons are drawn from the rubber belt to the globe at the top.

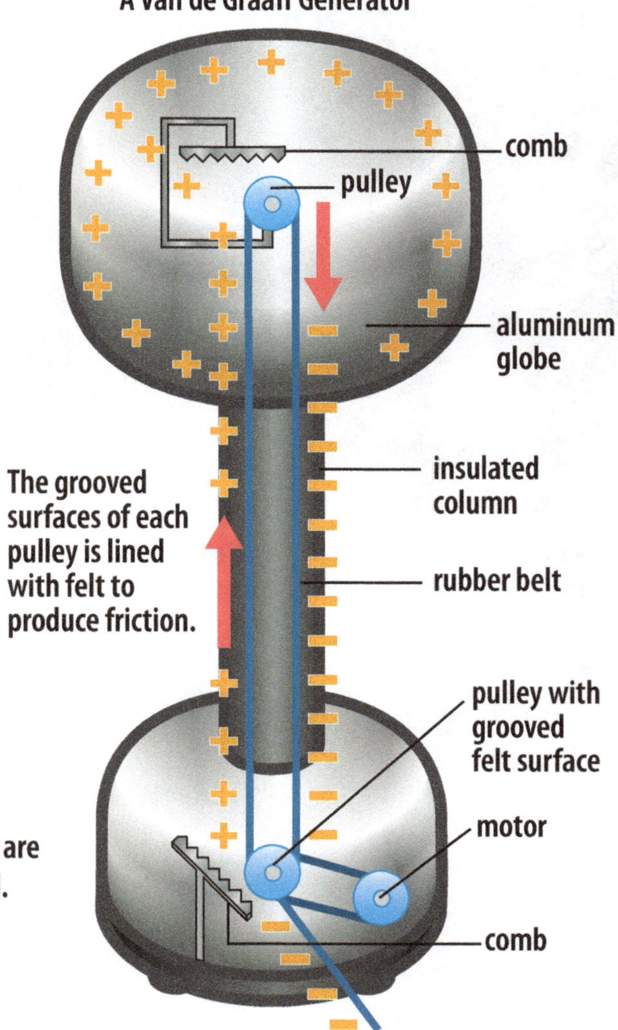

The combs in the Van de Graaff generator are made of a conductive material, like metal.

How does the charge travel to the globe?

Electricity in the Air (Explain)

The most spectacular example of a natural electrostatic discharge occurs in a thundercloud. **Lightning** bolts occur because of a difference in charge between a cloud and the ground, between two clouds, or between parts of the same cloud. Why do you think some lightning bolts are brighter than others or last longer?

What do you think causes areas of negative and positive charge to develop and separate within the cloud? Positive charges are more common at the top of the cloud where ice and snow form. The heavier, negatively charged water molecules drop lower in the cloud. The negative charges at the base of the cloud are attracted toward the ground below.

When the difference in the charge becomes great enough, a giant spark is released through the air. Once a complete path forms, a surge of electricity moves back up the cloud. The flash we call lightning is actually this return stroke. This results in an explosive release of energy. The surrounding air is heated and expands rapidly, producing thunder. Lightning can reach temperatures of more than 28,000°C (50,400°F) and can release 200 million volts of electricity. Now that you understand lightning, how would you describe what you observed in the Kick-Off Inquiry?

The chart below outlines some ways you can stay safe during a thunderstorm.

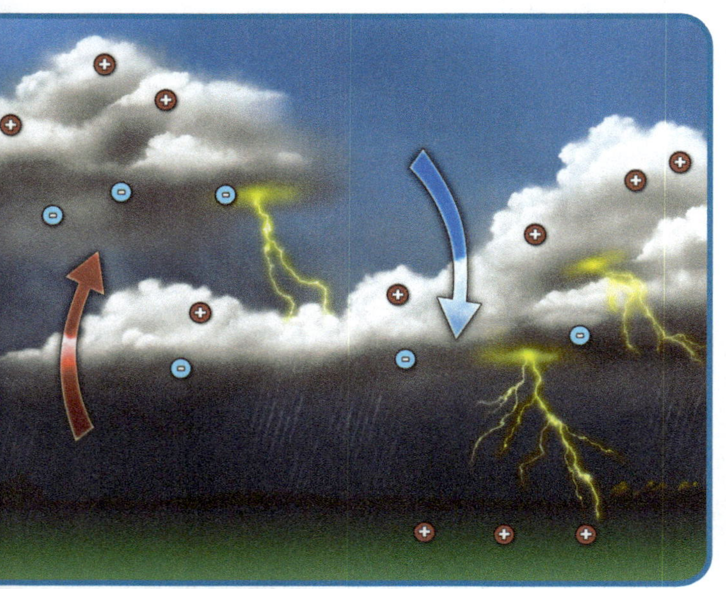

Charge separation within storm clouds

Why does cloud-to-cloud lightning travel from the top of one cloud toward the bottom of another cloud?

Lightning Safety	
Indoors	**Outdoors**
• Get indoors during thunderstorms and stay away from windows and doors. • Avoid plumbing, electrical appliances, and corded phones.	• If outdoors, reduce your risk of injury by avoiding tall objects, isolated trees, metal structures, open areas, and water. • Crouch down to reduce your height.

Math in Science

Light travels at 300,000 km/sec (186,000 mi/sec) and sound travels at about 340 m/sec (or 1 km per 3 sec) (1115 ft/sec). To calculate how far away a thunderstorm is, count the number of seconds between the lightning flash and the related thunderclap. Then divide by 3. If you count 18 sec between the lightning and the related thunder, how far away is the storm?

Using Static Electricity

You know static electricity can be a nuisance when it makes your clothes stick together or when you get a shock from touching a doorknob. But the attraction between charged objects can be used for a variety of practical modern devices. How do you think people use static electricity to make the world better?

Pollution Control

Many industries use static electricity to help control pollution. Study the diagram of how pollution can be controlled in a smoke stack.

1. Dirt and smoke particles move through a smoke stack.

2. The particles pass through a negatively charged metal plate and receive a negative charge.

3. The charged particles are attracted to positively charged collecting plates.

4. Collecting plates are knocked to remove the dirt and smoke particles.

5. Clean waste gas continues through the smoke stack.

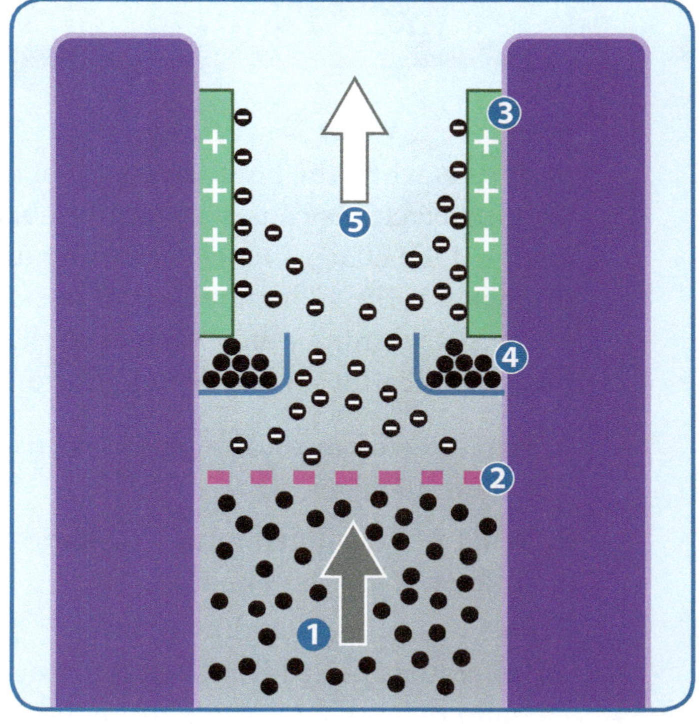

Diagram of pollution control equipment in a smoke stack.

Printing and Advertising

Photocopiers and printers use electric charges, dry ink (toner), and light to copy print onto paper. The drum is given an electric charge. An image of the original is projected onto the drum. The charge disappears where the light strikes the metal surface. The toner ink is attracted to the dark image on the drum. As paper with the opposite charge moves past the drum, the ink transfers to the paper. Heat fixes the image onto paper.

Advertisers use static-cling vinyl for brightly colored, removable signs, decals, and stickers. The plastic is positively charged and requires no glue to stick to negatively charged glass windows and metal surfaces. If a static-cling decoration no longer sticks, can you correct the problem? If so, how?

Spray Painting

A metal car body is treated so that it has a negative charge. As paint is sprayed from a specially designed sprayer, it becomes positively charged. The positively charged paint is attracted to the negatively charged metal. This ensures an even coating of paint on the car, which will protect the car from weather damage.

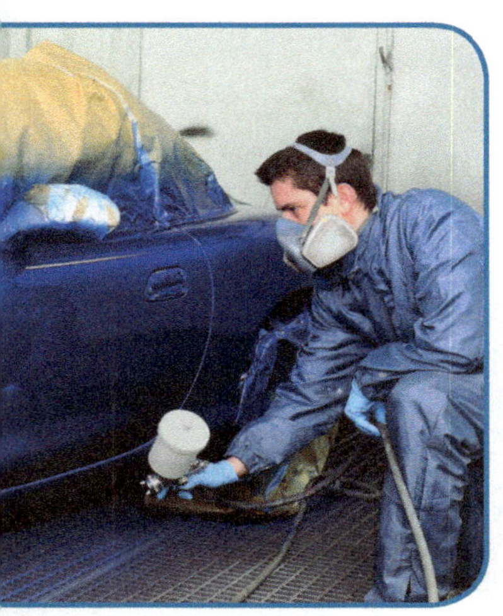

How is the auto painter protecting himself from the paint and paint fumes?

Concept Check Assess/Reflect

Summary: What is static electricity? Static electricity builds up on the surface of objects when the outer electrons of atoms jump from one material to another. Gaining electrons causes a negative charge. Losing electrons causes a positive charge. Like charges repel each other. Unlike charges attract. An electrostatic discharge is the sudden release of static electricity between objects with unlike charges. Lightning is an example of electrostatic discharge. Static electricity has many applications, such as in pollution control and photocopying.

1. Explain why most objects are electrically neutral. Describe one way to charge a neutral object.

2. What is the sudden release of static electricity between two objects with different charges called?

3. Use your understanding of electric charges and electric force to explain why: (a) clothes pulled from a dryer cling together, (b) a wool cap causes hair to stand on end, and (c) lightning flashes in a thunderstorm.

4. Name three devices that make use of static electricity. Explain how static electricity is involved in each process.

What Is Electric Current?

Think for a moment what it would have been like this morning when you were getting ready for school if there was no electricity. How would this have affected your morning routine? Now think of what it was like in your town 150 years ago, before electricity. Back then, people relied on daylight and fire for light. All the work that is done today by electrical tools and machines was done by hand or with the aid of animals. Can you imagine life without electricity? We depend on electricity to power our modern world. Do you know how the electricity gets to your home? How is it generated? If your home lost power for a week, what would you miss most?

Current Electricity **Explain**

A flow of electrons produces electricity. **Current electricity** refers to the continuous flow of electrons. Electric current powers all electrical appliances. It can be used to heat and cool your home, cook your food, and provide you with information. What else can current electricity do for you?

Electric current travels to your home through a system known as a power grid. Have you seen power cables and poles in your neighborhood? These are visible parts of power grids in most cities, towns, and rural areas. Much of the electricity that is transported through the power grid comes from power plants. The power plant's generators are likely powered by a steam turbine, running water, or a diesel engine. From the

New York City was one of many cities along the East Coast of the United States affected by a major electric power blackout.

What vital services would be affected by a major power failure?

Objectives

- Describe how the electric power grid distributes electricity.
- Explain the difference between alternating and direct current.
- Compare voltage, current, and resistance.
- Distinguish between a conductor and an insulator.
- Identify materials that are conductors and materials that are insulators of electricity.

Vocabulary

current electricity

direct current

alternating current

voltage

amperage

ampere

conductivity

resistance

insulator

conductor

superconductor

power station, this electricity is fed into a network of wires. Its power is boosted to allow it to be transmitted over long distances. Then, it is reduced to safer levels for use in homes, schools, and businesses.

Flowing Electric Charges Explain

When an electric current flows, the electrons all move in a continuous, unbroken path. There are two main types of electric current. **Direct current** (DC) flows in one direction. This current is used in batteries. The electric current that we obtain from outlets is called **alternating current** (AC). This current changes direction because it is produced at a power station using a rotating current generator. You will learn more about this later in the chapter.

The flow of electricity through wire can be thought of as water flowing through a pipe. The electrical generator that produces the electricity is like the water pump that brings water into the pipe. Just like water in a pipe is under pressure, electric current also has pressure; this pressure is known as **voltage** and it is what drives the electrons through the wire. The rate at which the electrons flow through the wire in a given period is the **amperage**. Continuing with the water pipe analogy, the amperage is similar to the rate (or amount) at which water flows through the pipe. The unit for measuring the flow of electric current is the **ampere**, or amp for short. Remember, the voltage of the current is a measure of the force that pushes the electricity through the wire. The amperage is a measure of the rate at which electricity flows through the wire.

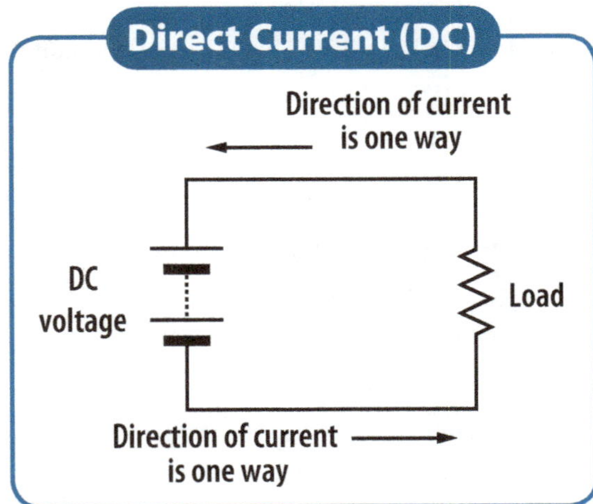

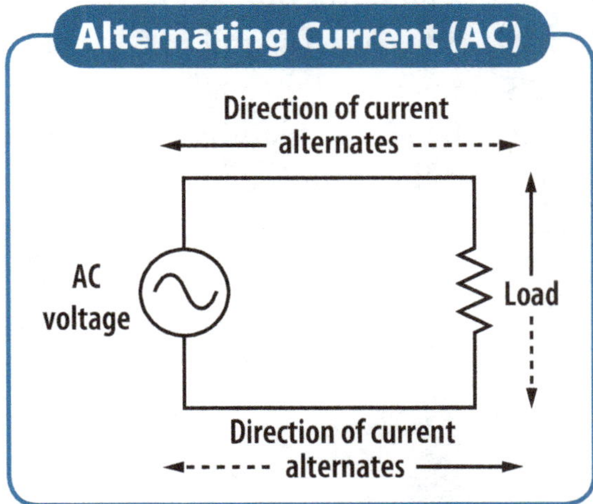

The diagram compares direct and alternating current.

How do the diagrams explain the difference between these two currents?

Does It Resist?

What is the difference between conductors and insulators?

Procedure

1. Attach the first wire to the negative (−) end of the battery holder with the alligator clip. Attach the other end of the wire to one end of the bulb holder.

2. Attach a second wire lead to the positive (+) end of the battery.

3. Insert the bulb into the bulb holder.

4. Attach the third wire lead to the other end of the light-bulb holder.

5. Your conductivity tester should look like the drawing in your **Student Journal**.

6. Briefly touch the two alligator clips at the gap. If the bulb lights, you have a complete circuit and you may continue. If the bulb does not light, check all connections and test again.

7. Place the first object between the two free alligator clips. Make sure there is good contact between the clips and the object. **Observe** the bulb. Does it light? Is it dim or very bright? **Classify** the material as a conductor or an insulator. **Record** your observations in the data table.

8. Repeat Steps 6 and 7 for each of the objects you are testing.

Analyze Results

Examine your data table. **Compare** how well each object resisted or conducted an electric current.

Create Explanations

1. What is the difference between conductors and insulators?

2. What do the materials that are the conductors have in common? What about the materials that are insulators?

3. What is the handle of a screwdriver made from? Why?

Materials
- 1.5-volt battery in a battery holder
- 3 wire leads with alligator clips at each end (or insulated wires with stripped ends)
- 1.5-volt bulb in a bulb holder
- variety of test materials such as aluminum foil, craft stick, shoelace, penny, paperclip, rubber band, nail, plastic game piece

Conductivity and Resistance Explain

You have learned that electric current moves in direct or alternating paths. But, how does that electricity move through different material to power your computer? What substances allow electricity to pass through? What materials stop electricity?

Matter reacts in various ways to move electric charges. **Conductivity** refers to a material's ability to allow electrons to flow. When a material prevents electrons from flowing, it is called **resistance**. A scientist named George Ohm discovered that shorter or thicker wire provides less resistance to charge flow. Ohm also found that resistance is related to the kind of material the charge flows through. The unit for electrical resistance was named the ohm in his honor. Why do you think the length and thickness of a material affect its conductivity?

Insulators and Conductors

An **insulator** is a material that resists or blocks the flow of electrons. A **conductor** is a material that allows electrons to move freely from atom to atom in a continuous path. What kind of interruptions might cause electrons to stop flowing as expected, or flow too fast?

The conductivity of materials ranges from those that are strongly conductive to those that are the least conductive and make the best insulators. Where are some places you would want to use material that is a good insulator?

This diagram allows you to compare the conductivity of different materials.

How could you use this information to select a conductor for wiring a house?

Math in Science

George Ohm discovered the relationship between voltage, current, and resistance. He expressed it in an equation that is known as Ohm's Law: $V = I \times R$, where V = voltage, I = current, and R = resistance.

To solve for current, rewrite the equation as $I = V/R$.

To solve for resistance, rewrite the equation as $R = V/I$.

To help you remember the relationships, write them in a triangle:

Then, cover the quantity you want to solve for with your finger.

Given a 12-volt battery and a bulb with a resistance of 3 ohms, what would be the amount of the current flowing through the conductor?

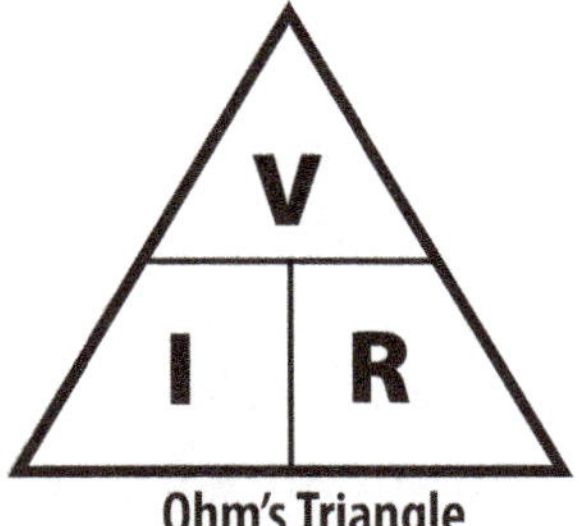
Ohm's Triangle

Faith Connection

Imagine that the Holy Spirit wants to flow through you like an electric current to bless the people around you. Are you more like an insulator or a conductor?

Explore-a-Lab

Structured Inquiry

 **How does water conduct electricity?**

You will need distilled water, tap water, salt, a 9-volt battery, three wires with alligator clips on each end, and a light bulb. Build a conductivity tester by connecting one alligator clip to the negative battery terminal; leave the other end loose. Connect a second clip to the positive battery terminal and clip the opposite end onto the light bulb. Attach the third wire to the light bulb and leave the opposite end loose. Place the two loose alligator clips into a beaker of distilled water. Record your observations. Now make a saltwater solution and repeat the test. Add more salt to see how it affects your results.

The Shanghai maglev train system uses superconducting materials to travel at speeds of nearly 500 km (310 mi) per hour.

What are the advantages of using superconducting materials?

Liquid solutions, especially those that contain salts and acids, also make good conductors. Even objects that are normally good insulators, such as a dry tree branch, will conduct electricity when wet. You have probably seen or read the warning labels on electric appliances warning you not to use them near water. Pure water is generally considered to be a nonconductive material. Tap water, and most freshwater, on the other hand, is a good conductor. Why is pure water a nonconductor and tap water a conductor?

The impurities, dissolved minerals and salts, that are present in tap water, but absent in pure water affect conductivity. The dissolved minerals make tap water and most freshwater a good conductor. Thus, touching an appliance, such as an electric hair dryer, near a bath or shower could put you in real danger. The water could conduct an electric charge to your skin and cause serious electric shock.

Superconductors are materials that conduct electricity without any loss of energy due to resistance. The low-resistance quality of superconductors is because their electrons are bound together in pairs and flow with extremely low energy loss. This creates an electric current that moves with little resistance. Japan, Germany, and China have already built maglev (magnetic levitation) train systems that use these superconductors and the technology associated with them. Another use of superconducting materials is in cross-sectional magnetic resonance imaging (MRI) of the inside of the human body.

Insulators are materials that have tightly bound electrons. These materials strongly resist the flow of charges. Common insulators include ceramics, glass, plastics, rubber, and wood. Insulators are often used to protect us from the dangerous effects of electricity flowing through conductors, including electric transmission lines. For the same reason, a plastic coating is often placed around electric wires on appliances used to conduct electricity.

You will wrap up your study of electric current in this lesson by creating a list of common conductors and insulators. As you work on the lesson activity, think about the following questions: Why is it dangerous to use an appliance with a damaged cord? How can birds sit on electrical lines and not get electrocuted?

Lesson Activity

Use the Internet or other available resources to research conductors and insulators. Make a data table that lists 10 conductors and 10 insulators. List them in order of effectiveness. Then, include everyday uses for each of the conductors and insulators that you listed in the data table.

 **What conductors and insulators do you use on a daily basis?**

Concept Check   Assess/Reflect

Summary: What is electric current? Current electricity is the flow of electrons. Electric current runs all the electrical devices you use. There are two kinds of electric current—alternating current and direct current. Alternating current is supplied to homes and businesses. Batteries supply direct current. Voltage is the electric force that pushes electrons through a wire. The rate at which electrons flow through a wire is the amperage. Some materials prevent the flow of electrons through them. These materials are called insulators. Other materials allow electrons to move freely through them and are called conductors.

1. Explain how electric current reaches the outlets in our homes.

2. What is the electric force called that drives electric current through wires? What unit is used to measure it?

3. Explain how the width of a wire affects conductivity.

4. Silver is a better conductor than aluminum. If a silver wire and an aluminum wire have the same length and the same thickness, which will have a higher resistance? Explain.

Essential Question

What Are Types of Batteries?

What items in your home run on batteries? Batteries are not connected to wall outlets, so where do they get their energy? Why is the power supply of a battery limited?

Electricity from Chemicals Explain

Batteries are a portable source of electrical energy stored in chemical form containing two or more cells. How is the energy transferred to power electronic devises?

An **electrolytic cell**, also called an electric cell, is a device that converts chemical energy to electrical energy. It is the working part of a battery. An electrolytic cell consists of three parts: two different metals or a metal and some carbon called **electrodes**, or terminals, and an electrolyte. If you look at an ordinary electric cell, you will notice that one terminal is marked positive ($+$) and the other is marked negative ($-$). These two electrodes are separated by the **electrolyte**, which is either a solid or a liquid solution that contains *ions* from chemical reactions occurring at the electrodes. It conducts the current inside the electrolytic cell.

How Does an Electrolytic Cell Work?

When you connect the electric cell, a series of chemical reactions occurs inside. The chemical reactions cause electrons to collect on the negative electrode and flow through an outside conductor back to the positive terminal. The positive electrode gains electrons due to the reactions. The electrolyte and the outside conductor form a complete path that allows the current to flow. The chemicals inside the electric cell gradually change into different chemicals. When the original chemicals in a single-use electric cell are used up, the electric cell stops working. How do rechargeable electrolytic cells keep working?

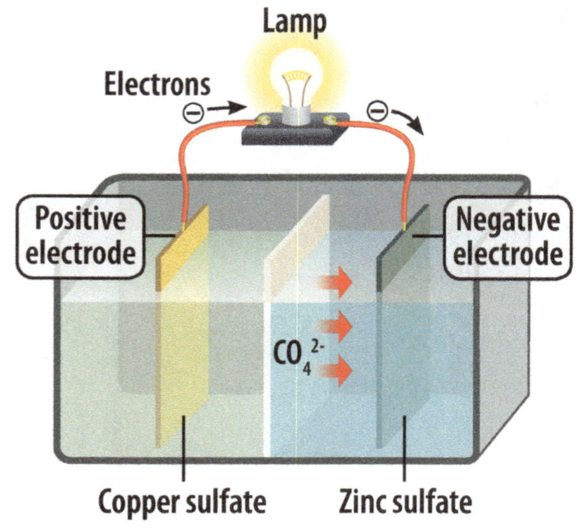

Electrolytic cell

What is the purpose of an electrolyte?

Different Kinds of Electrical Cells

What is the function of an electrode? What might be some good qualities of an electrode to fulfill this function? What materials do you think might make a good electrode?

All electrolytic cells use chemical changes to produce electrical energy. Electrical cells are commonly classified as wet cells or dry cells. Certain metals can use electrons better than other metals. For example, aluminum, zinc, and iron all give up electrons more easily than copper. Two metals that have different abilities to give up electrons are used in wet cells.

Wet Cells

A **wet cell** contains a liquid electrolyte, which is free to flow and move. The wet cell in the diagram shows the metal strips, which are the electrodes, and the acid solution, which is the electrolyte. That is why the wet cell in the diagram uses copper and zinc as electrodes. An automobile uses a storage battery, which is made up several individual wet cells connected together. Storage batteries contain lead/calcium plate electrodes suspended in a solution of sulfuric acid as the electrolyte. Advancements in battery technology have produced batteries that can power electric vehicles. These batteries are also rechargeable.

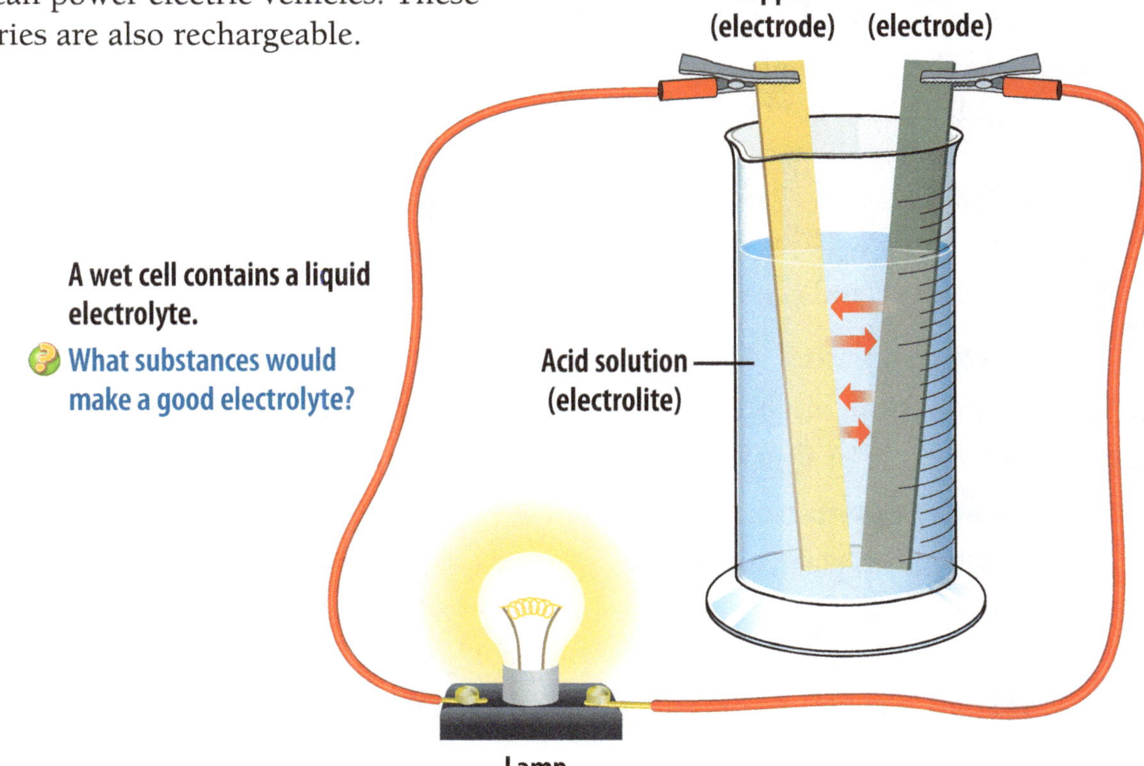

A wet cell contains a liquid electrolyte.

What substances would make a good electrolyte?

Saltwater Cell

Why is an electrolyte necessary in an electrolytic cell?

Procedure

1. Pour 400 ml (14 oz) of water into the beaker.

2. Use a balance to **measure** 200 g (7 oz) of salt. Add the salt to the water and stir until most of the salt is dissolved.

2. Screw the light bulb into the base and connect one wire to each of the contacts on the base.

3. Test the light bulb by touching the alligator clips to opposite ends of the AA cell. **Observe** whether the bulb lights or not. If it does not light, replace it with another bulb.

4. After testing the bulb, use the alligator clips to connect one wire to the stainless-steel nail and the other wire to the magnesium strip.

5. At the same time, insert both electrodes into the saltwater at opposite sides of the container. Be sure the electrodes are mostly in the saltwater.

6. **Observe** and **record** what happens to the light bulb.

7. Have a partner pour some hydrogen peroxide into the saltwater. **Observe** and **record** what happened to the light bulb.

Materials
- salt
- balance
- water
- 600-ml (20-oz) beaker
- long-handled plastic spoon
- stainless-steel nail
- magnesium strip
- 2 wire leads with alligator clips
- low-voltage light bulb (1.2 V) with socket base
- AA cell
- 3% hydrogen peroxide

Analyze Results

Compare the brightness of the bulb before and after adding the hydrogen peroxide. Why do you think this happened?

Create Explanations

1. Why is an electrolyte necessary in an electrolytic cell?

2. If electrons collect on the stainless-steel nail terminal, how does electricity flow in your saltwater cell?

3. Draw a diagram that shows the parts of your electrolytic cell and use arrows to show the flow of electrons.

Dry Cells

Wet cells tend to be large, so it is impossible to slip a wet cell into a flashlight. If flashlights don't use wet cells, how are their electric cells different? How can the electric cell produce a current without a liquid electrolyte solution? Flashlights and other small electrical devices use dry cells. **Dry cells** work like wet cells, but use a combination of thick, paste-like chemicals for the electrolyte, including carbon/zinc, alkaline, and lithium. Why do you think some moisture is needed inside a dry cell battery? In a dry cell, the negative electrode contains the electrolyte paste. Dry cells are available in a wide range of sizes. Why do you think so many sizes are necessary?

There is a difference between a dry cell and a dry cell battery. Dry cells are single units such as the C, D, AA, and AAA dry cells commonly purchased to run small electrical devices. A battery is a series of dry or wet cells connected together, such as a 9v battery (made up of 6, 1.5 volt cells) and 12v car battery (made up of 6 2 volt cells).

One kind of dry cell contains a carbon rod set in the middle of a zinc can. A moist paste fills the can. Chemicals in the paste react with the zinc and release electrons. These electrons flow through a conductor, such as a flashlight bulb, connected to the dry cell. The current flows from the zinc can, through the bulb, and back to the carbon rod. What would happen if the moisture in the paste dried up?

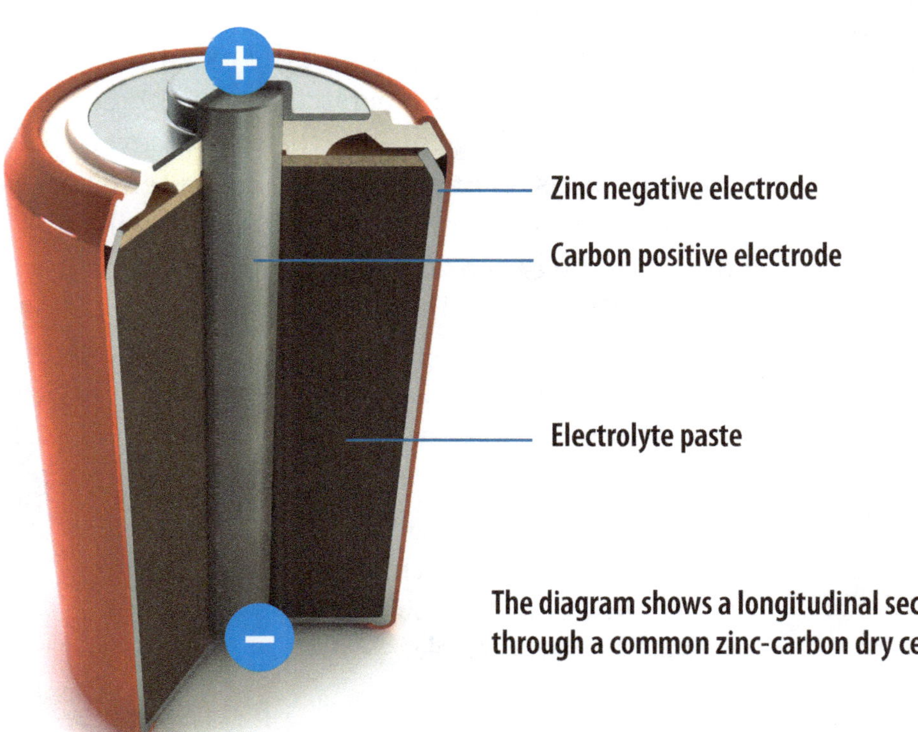

The diagram shows a longitudinal section through a common zinc-carbon dry cell.

Science Journal

Check out your *Science Journal* for a Guided Inquiry re-creating one of the first batteries.

Extend

Scripture Spotlight

Which kind of battery best describes your faith—rechargeable or non-rechargeable? What does **Romans 10:17** say about where faith comes from?

Recycling Batteries

Keep in mind that when most people talk about batteries, they are actually talking about single electric cells. What do you do with an electric cell or battery when it has stopped working? Is it safe to place a battery in the garbage? What happens to the chemicals in the battery when the battery structure breaks down? Is that safe for the environment?

Batteries contain hazardous heavy metals such as lead, cadmium, zinc, mercury, and lithium. It can be very damaging to the environment to bury them in landfills. The harmful metals can leak out and pollute land and water resources. Metallic lithium will react with moisture to cause landfill fires. Burning batteries causes air pollution.

To avoid the landfill, the first choice is to use **rechargeable** batteries, which can be reused repeatedly. The first rechargeable battery system was invented in 1960. Introducing an external electric current to rechargeable batteries causes the chemical changes that occur during normal battery use to be reversed. Examples of rechargeable batteries include nickel cadmium, nickel-metal hydride, and lithium ion batteries. They are more environmentally friendly, but eventually even rechargeable batteries do wear out.

Many recycling centers will accept used batteries. Collection centers start by separating the batteries by the chemicals in the electrolyte. After combustible material is removed, the cells are chopped into smaller pieces and the metal is heated until it melts. The metals can then be separated by density. Some recyclers ship the solid metal in large masses to metal recovery plants for reuse in new products. While it is expensive in terms of energy use to reclaim metals, so is mining new raw materials. In addition to the expense, what are other reasons to recycle this material? After some of these materials are recycled several times, do they still contain the same original properties? Explain.

Lithium ion batteries are often found in mobile devices, laptops, tablets, and hand-held game consoles.

Single-Use Batteries		Rechargeable Batteries	
Advantages	Disadvantages	Advantages	Disadvantages
• Less expensive first cost than rechargeable batteries • Ready to use; do not need to wait for charging • Dependable amount of charge • Less likely to explode or start a fire	• Must continuously repurchase • Produce more waste; the batteries themselves and the packaging • Contain toxic chemicals	• Long-term cost savings • Produce less waste • Less toxic than single-use batteries	• Must wait for charging; must plan in advance for charged batteries • Lose ability to hold a charge over time • More likely to explode or start a fire • Contain toxic chemicals

Explore-a-Lab

Structured Inquiry

 **What supplies the electrolyte in a potato battery?**

Use a galvanized (zinc-coated) nail and a 5-cm (2-in.) piece of 14-gauge copper wire as electrodes and a fresh potato to make a potato battery. Sandpaper the electrodes until they are shiny. Insert them into the flatter side of the potato about 2.5 cm (1 in.) apart. Make sure they do not touch. Attach wire leads with alligator clips to each electrode. Use a voltmeter. Attach the clip of the wire lead from the nail to the negative (−) slot on the voltmeter. Attach the clip of the wire lead from the copper electrode to the positive (+) slot on the voltmeter. Observe what happens.

Concept Check Assess/Reflect

Summary: What are types of batteries? Electrolytic cells are portable sources of electrical energy that is stored in chemical form. An electrolytic cell is made from an electrolyte and two different metals, called electrodes. Chemical reactions in the cells cause electrons to flow from the negative electrode, through an outside conductor, and back to the positive electrode. Wet cells have a liquid electrolyte, while dry cells have thick, paste-like electrolytes. Batteries are two or more electrolytic cells used together. Batteries contain hazardous chemicals and are harmful to the environment when disposed of improperly. Many people now use rechargeable batteries, which are friendlier to the environment. All batteries should be taken to recycling centers.

1. Explain how an electrolytic cell produces electric current.

2. Explain the function of an electrolyte.

3. Name two kinds of electrolytic cells and describe how they differ.

4. In what ways are rechargeable batteries better than single-use batteries?

Essential Question

What Are Electric Circuits?

What happens when you turn on the light switch in your room? Where does the power come from? How does an electric current flow? An electric current flows only when it can follow a complete path back to its starting point. Such a path is called a *circuit*. You depend on electric circuits all the time in your everyday life. How does a circuit turn the light on or off?

Electric Pathway Explain

An electric circuit consists of a complete pathway connecting a conductor to the power source and passing through an object that needs the current to operate. There are three main parts to a circuit:

1. The power source, such as a battery, solar cell, or a generator
2. A **load** or electric device that uses current, such as a bulb or motor
3. A conductor or wire to connect these parts together

Many electric circuits also have a fourth part: a switch, which opens or closes a circuit. In a **closed circuit**, electrons can flow from one pole of the power source to the other in a continuous path. An **open circuit** happens when the switch is open and there is a break in the circuit. Because the path is incomplete, and electrons cannot flow. Why do some circuits have switches while others do not?

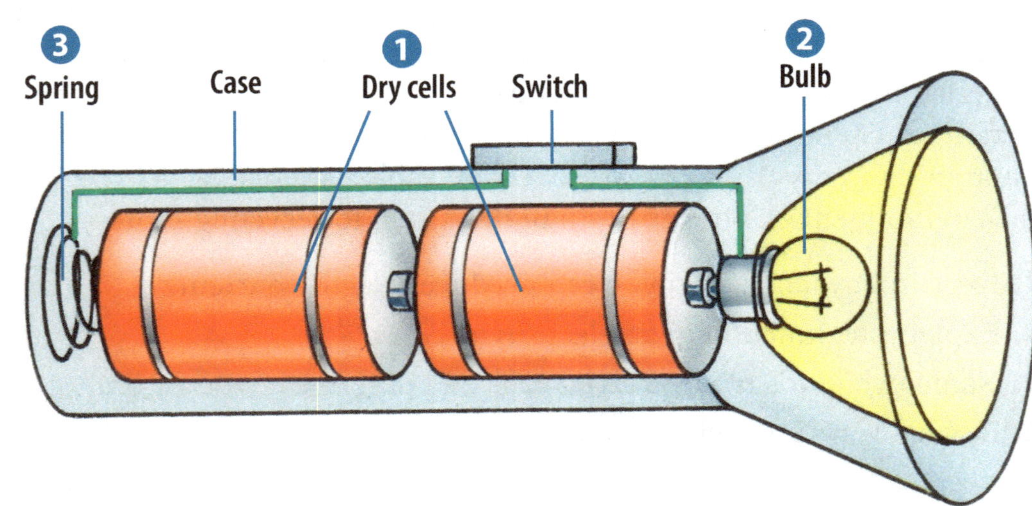

The circuit inside a flashlight

? What are three ways you could open the circuit in the flashlight?

A flashlight is a good example of a basic electric circuit. It has batteries as a power source, a switch to open and close the circuit, a bulb as the load, and metal springs and strips that act as conductors between the parts. Can you take away any of these parts and still have a working flashlight? If so, what parts?

Series Circuits

There are two different ways of designing circuits. Both require a complete path. In the first design, all the parts (the power source, the loads, and any switches) are connected in a line. When the electric current flows in a single path, the circuit is a **series circuit**. The picture shows a series circuit.

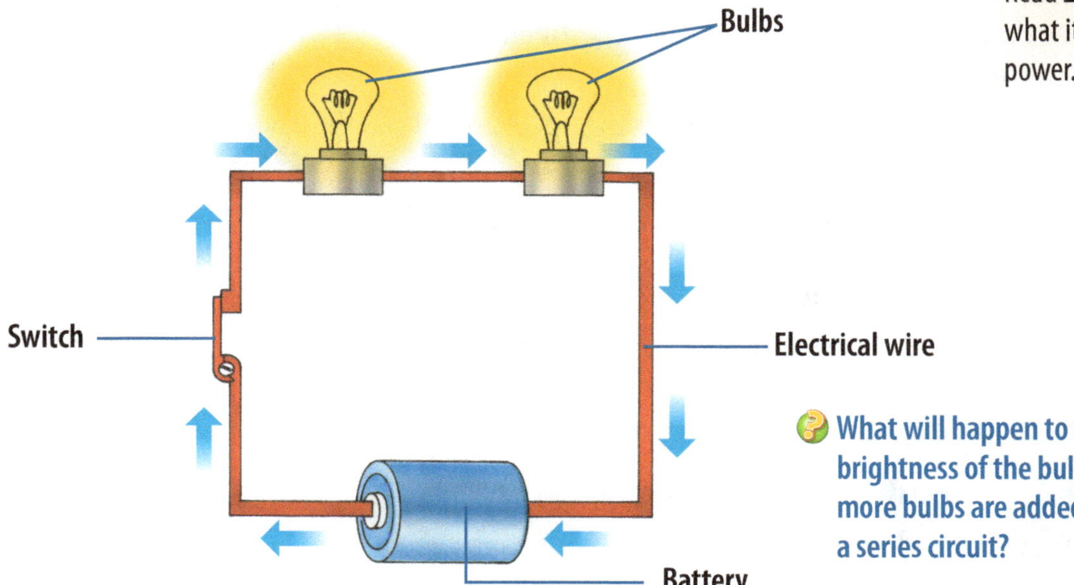

What will happen to the brightness of the bulbs if more bulbs are added to a series circuit?

With two bulbs connected in series, the charge passes through each bulb in turn. If you unscrew one of the bulbs or disconnect one end of the battery, the circuit will be broken. The entire circuit goes out because the circuit is open. Where have you seen a single burned-out light bulb make other lights not work?

When more light bulbs are added to a series circuit, the resistance goes up. For this reason, the term *resistance* is often used as a synonym for *load*. Each bulb in the circuit shares the available energy. This means if you add more bulbs to the circuit, the overall current is reduced. The bulbs will decrease in brightness. You must add more voltage to the circuit for the bulbs to shine equally brightly.

Making Connections

How does the brightness of bulbs compare in the series circuits?

Procedure

Materials
- 4 D cells
- 2 battery holders
- 15 wire leads with alligator clips
- 6 bulbs
- 6 bulb holders
- 4 switches

1. Design a simple circuit with two D cells, one switch, one bulb, one holder, and connecting wires. **Use models** by sketching the layout for the circuit you plan to build. Then, build the circuit. If the bulb does not light, check your circuit and its connections.

2. **Observe** what happens when you open and close the switch. **Record** your results.

3. Design a series circuit by adding one or more bulbs and holders to your simple circuit. **Use models** by first sketching the layout of the new circuit. Then, build the circuit.

4. **Compare** the brightness of the bulbs in the new circuit. **Predict** what will happen if you unscrew one of the light bulbs in this circuit. Test your prediction and **record** your results.

Analyze Results

Review your sketches and the results of each test.

Create Explanations

1. How does the brightness of bulbs compare in the series circuits?

2. What problems did you encounter in building the circuits? How did you fix them?

3. How do the circuits compare to the circuits that wire your home? Explain.

Parallel Circuits Explain

The second way to design a circuit is to split the current into separate paths. A **parallel circuit** is formed when the circuit is broken into several branches so the current flows into multiple paths. The picture shows a simple parallel circuit with two branches. Because the electric current travels in several different paths, if you unscrew one bulb in a parallel circuit, the current continues to flow, because it is able to follow an alternate path. It does not affect the other bulb. Do you think a series circuit or parallel circuit would be better to use? Explain your answer.

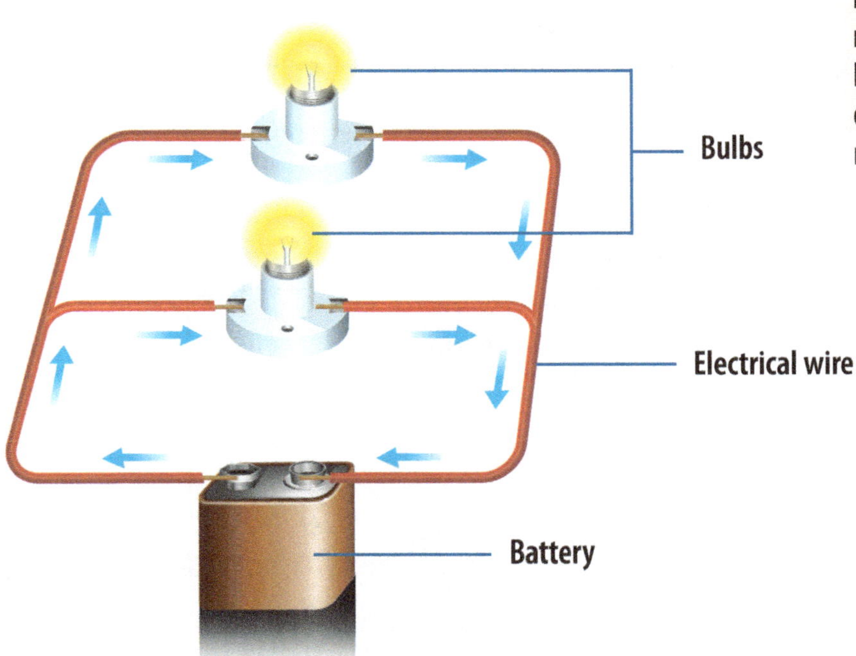

Parallel circuit with two branches

What are the advantages of connecting a circuit in parallel?

What happens if you add an additional load to the parallel circuit? A parallel circuit is like a multilane highway. Because there are multiple paths that the current can flow on, more charges can travel in the same period. The electric charge only flows through one of the three branches, so the overall flow rate through the circuit, or current, increases. Each individual bulb receives the same amount of current, so the bulbs shine just as brightly as in the original circuit. But the total resistance in the circuit actually decreases. How could you keep the same number of light bulbs, but increase resistance?

Circuit Diagrams `Explain`

Engineers draw the path of a current using a circuit diagram. They use a series of special symbols to represent each part of the circuit. The chart provides common symbols that identify different parts in simple circuits. For example, the longer bar on the battery represents the positive terminal, while the shorter bar represents the negative terminal. Why do you think common symbols needed to be created?

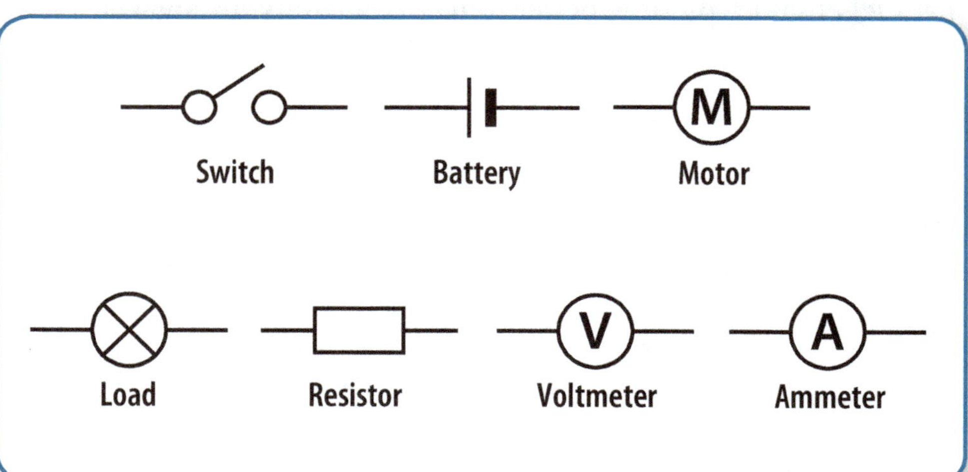

How would you draw a parallel circuit diagram using circuit diagram symbols?

Lesson Activity

On a sheet of paper, draw a circuit diagram for the main floor in your house. Identify locations of all electrical outlets and wall and ceiling fixtures. Include the location of the circuit breaker box. Are there any locations where you feel a new outlet or fixture would enhance the living area? Add them to your diagram using a different color of pencil.

Why are standard symbols important in a circuit diagram?

It is a good idea to draw the circuit you plan to build before you start assembling the parts. When you try to analyze existing circuits, draw a circuit diagram to help you understand how the current flows through the circuit. How might this be helpful to an electrical engineer?

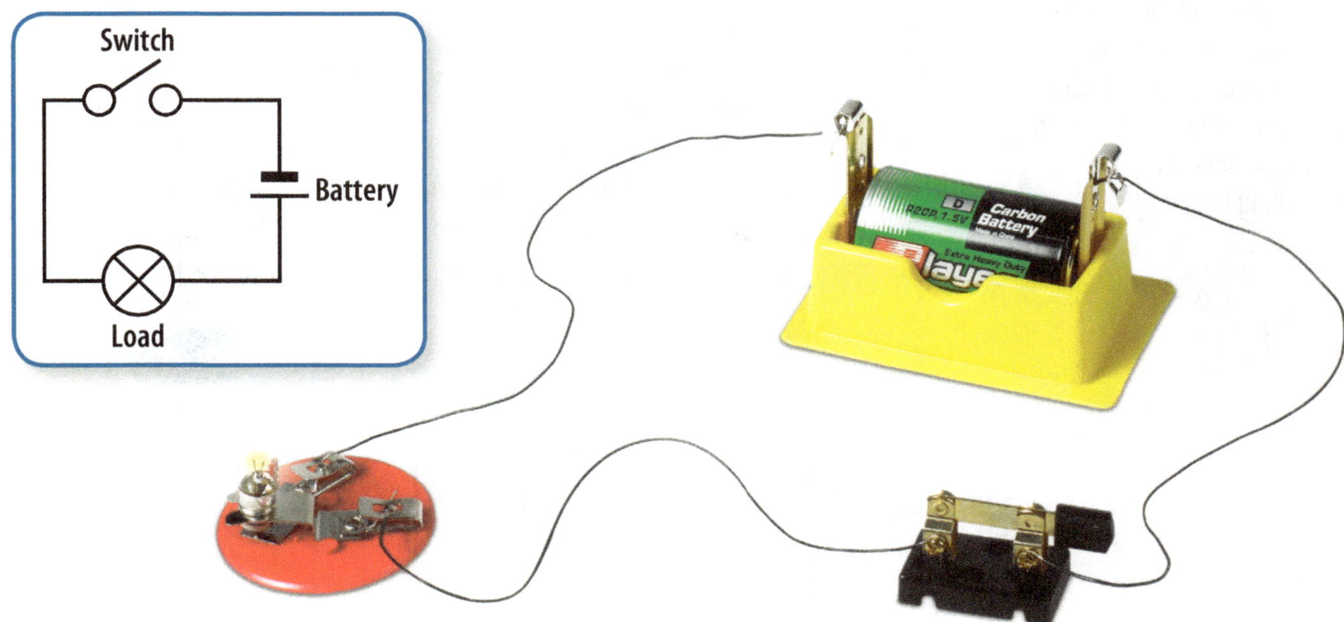

Compare the sample circuit with the circuit diagram.

Do you think the bulb will light? Explain.

Explore-a-Lab

Structured Inquiry

Without breaking the circuit, what do you need to know to figure out where to put the switches in each circuit?

On a separate sheet of paper, draw a circuit diagram for each of the following circuits.

1. A single D cell, a light bulb, and a switch are connected by wires in a circuit so the switch can turn the light bulb on and off.

2. Three D cells are placed in a battery pack to power a circuit containing two light bulbs wired in parallel that are controlled by a switch.

3. Three D cells are placed in a battery pack to power a circuit containing three light bulbs wired in series.

After you draw each diagram, trade diagrams with a classmate and build the other person's circuits.

Measure Current, Voltage, and Resistance Explain

Measuring quantities like current, voltage, and resistance in an electric circuit makes it easier to understand what is happening in that circuit. This is especially helpful when troubleshooting electrical problems, such as why an electric dryer does not produce any heat.

Meters are devices used to measure electrical quantities. An *ammeter* measures current, a *voltmeter* measures the voltage across two points in a circuit, and an *ohmmeter* measures resistance. A *multimeter* combines these three functions in a single meter.

These meters are connected in different ways. The diagrams below show how a voltmeter and ammeter are connected in circuits.

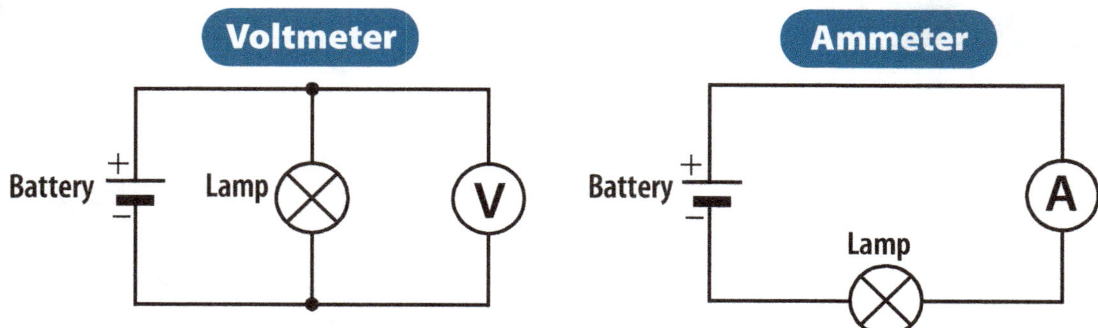

A voltmeter and ammeter are connected in a circuit.

To determine how much current is flowing, an ammeter is connected in series to test the power source and the load. To measure voltage, a voltmeter is connected in parallel to the device being tested. The circuit does not actually need to be broken to measure the voltage. The measurement probes can just be touched to the parts of the circuit that are to be measured. An ohmmeter is not connected directly to the power supply. To test a load, one component in the circuit is removed and tested separately.

Breaking the Circuit

Several types of switches are used to control the flow of energy through a circuit. You know that a switch is used to open and close a circuit or to change the path of electricity in a parallel circuit. Switches are generally used to turn various devices on or off. What might be another reason to have a switch in a circuit?

How much voltage is in a lemon electrical cell?

You will need a lemon, a D-cell, a steel paperclip, uninsulated copper wire, a multimeter with probes, and two leads with alligator clips.

1. Touch the multimeter probes to each end of the battery to determine the voltage. Record the data on a sheet of paper.

2. Soften the lemon by pressing down and rolling the lemon on your desk.

3. Insert the paperclip into the lemon, leaving a portion exposed.

4. Coil the copper wire, leaving about 3 cm (1.2 in.) uncoiled. Insert the coiled end of the wire into the lemon, about 3 cm (1.2 in.) from the paperclip.

5. Attach one lead with alligator clips to the paperclip and one to the copper wire. Attach the other ends of the leads to the probes on the multimeter. Read the voltage of the lemon. Record the voltage on your paper.

6. Gather data from the other student groups and calculate the average voltage of the lemon batteries.

How does the average voltage of the cell compare to that of the lemon? What would be some pros and cons of powering a LED bulb with a lemon electric cell?

Electrical Hazards and Safety

It is important to use care when handling electrical circuits and respect the dangers they present. What electrical safety rules do you know? Can you think of any safety precautions you or your parents take when working around electrical equipment?

Overloads

Most of the circuits in homes or businesses are connected in parallel. This means that unplugging or using a switch to turn off one device will not affect the other devices in the circuit. How would you turn off all the items on a circuit without disconnecting each one individually?

Every device we plug into the circuit gets the same voltage no matter how many there are. This allows them to work properly. Recall that each load we add to a parallel circuit increases the current and lowers the resistance. This makes it very easy to overload a circuit. Overloading a circuit can cause it to overheat and cause fires. What devices does your home have that help prevent these problems? What can you do to avoid this situation in your home?

Short Circuits

A short circuit is a problem that occurs when insulation wears away from electric wiring. The unprotected wiring provides a low-resistance path for current flow.

The current may jump to another material and cause sparks. The exposed wires may become hot enough to start a fire or cause a shock. What can you do to prevent this from happening in your home?

Electric Shock

People can be severely injured by touching exposed wires in electric circuits. Our bodies have high resistance. Electric current that passes through our bodies will slow down and release thermal energy. The electrical energy can cause burns or cause muscles to jerk violently, which is harmful for the heart. If someone is receiving an electric shock you should not touch the person until the power source is turned off. Why is this so?

Electrical Safety

The list below describes some ways to prevent electrical shocks and burns. How will you share this information with loved ones?

- Never touch electrical outlets with your fingers or with objects.

- Never play with electrical cords, wires, or switches.

- If you are in the bathtub, shower, or standing on a wet floor, never touch anything electrical such as a light switch or hair dryer.

- When playing outdoors, never play around electrical wires or equipment.

- Stay away from areas marked DANGER: HIGH VOLTAGE.

- Never climb utility poles, transmission towers, or fences around substations.

- Do not climb trees near power lines.

- Never throw objects at wires or utility poles.

- Only fly kites in dry weather and in open spaces, away from power lines.

A **fuse** is a device that limits the amount of electrical current flowing through a circuit. A fuse is connected so that the current runs through it. If too much current moves through a fuse, a piece of metal melts or in a circuit breaker a switch is flipped. This breaks the circuit and shuts down the flow of electrical power. Fuses have to be replaced after they have blown.

Many older buildings continue to use fuses and fuse boxes to limit current flowing through home wiring. But many modern buildings use circuit breakers instead. **Circuit breakers** are special switches that are designed to open when too much current passes through them. Why is it important to check safety equipment regularly?

The advantage of circuit beakers is they can be reset instead of needing to be replaced. It is important to correct the problem or the electrical hazard that set off the circuit breaker first before resetting it. Why do you think this is important?

The electrician is installing a circuit breaker.

Why would it be useful to have each of the breakers labeled?

 ## Concept Check — Assess/Reflect

Summary: What are electric circuits? An electric circuit provides a path through which electric current flows. Three key parts of a circuit are the power source, the load, and the conductor. Switches allow circuits to be opened and closed. In a series circuit, all the parts are connected in a line. In a parallel circuit, the current is split into separate paths to two or more loads. Circuit diagrams use symbols to represent the parts of a circuit. Current is measured with an ammeter, voltage is measured with a voltmeter, and resistance is measured with an ohmmeter. Safety devices help prevent dangerous, overloaded circuits.

1. What is a closed circuit?
2. Describe how to design a series circuit and a parallel circuit.
3. Compare what happens to energy when you add loads to each kind of circuit.
4. Describe how a switch is used in a circuit.

Essential Question

What Are Electronics?

What is your definition of electronics? What kind of devices do you put into that category? You probably use several types of electronic devices every day. Have you ever wondered what they look like on the inside or who came up with the idea to build such devices? During the last century, electronics have been evolving at a breathtaking, exciting pace. What new electronics do you think could be built in the next 100 years?

Electronics `Explain`

Electronics are devices, such as televisions, radios, and computers that operate using small electrical parts to control the flow of electrons. What makes electronics is a device called a transistor. The **transistor** is a small sandwich of two different types of semiconductors. It changes its conductivity as it works. Transistors act mainly as switches or gates and use a binary code. The switches open to let electricity through, which represents the digit 1. The switch closes and stops the flow of electricity, which represents the digit 0.

The Power of Electrons

Many of the electronic devices and machines that we use every day contain tiny electric circuits printed on thin chips of silicon called **microchips**. Metals, such as copper, aluminum, and gold, are added to increase silicon's ability to conduct current. The circuit design is etched onto the chip using ultraviolet light. Smaller than a grain of rice, microchips contain **binary code**, a series of 1s and 0s. These microchips control and direct the flow of electric currents. They are the "brains" behind modern electronic devices such as video games, mobile devices, digital watches, microwave ovens, and tablets. Microchips do the same work as a transistor, only they are smaller.

Semiconductor Devices

Solid elements, such as silicon or germanium, are **semiconductors**. These elements have electrical conductivity between that of a conductor and an insulator.

There are a number of different types of semiconductor devices, including *diodes, solar cells, photocells,* and transistors. A **diode** is a semiconductor device with two electrodes that only lets current flow in one direction. Diodes are used to convert AC power to DC power. They are commonly used in radios, televisions, mobile devices, computers, and many other electronic devices. Solar cells are electronic devices that change solar radiation from the Sun directly into electricity. Photocells are sensitive to varying light levels. They are used to generate or control an electric current in such devices as burglar alarms and smoke detectors.

This assembled circuit board contains a number of semiconductor devices.

What type of device might contain this kind of circuit board?

As a class, develop a code. Use the code to send a message to one of your classmates. Show that data needs to be converted in order to be transmitted from one place to another.

How is this code similar to how a computer functions?

Diodes

How do diodes affect electrical circuits?

Procedure

1. Attach a flat connector to each wire lead of the battery holder. See diagrams in your *Science Journal*.

2. Place the 4 AAA batteries into the battery holder.

3. Refer to the diagram. Use the bicolor LED and touch the longer of the two LED wires to the positive wire (red) coming from the battery holder and the shorter LED wire to the negative wire (black) from the battery holder. Try reversing the positive and negative wires. **Record** your observations in your *Science Journal*.

4. Attach a connector (B) to each of the LED wires. Mark which LED wire is positive (+).

5. Repeat Steps 3–4 with the other LED (one-color).

6. Attach a flat connector (A) to one end of the 1000-ohm resistor and a connector (B) to the other end of the resistor.

7. Refer to the diagram and connect the positive wire of the one-color LED to the positive wire from the battery. Next connect the negative LED wire to the connector on one side of the resistor. Attach the free end of the resistor to the negative wire from the battery. **Observe** what happens and **record** your data in your *Science Journal*.

Materials
- 6v battery holder (4-cell)
- pliers
- 4 AAA batteries
- quick connects—4 flat
- quick connects—6 receptacle
- LED bicolor diode (6v)
- resistor (1000 ohms)
- LED diode (6v)
- wire strippers

Analyze Results

Draw a circuit diagram to **compare** each set-up in this investigation.

Create Explanations

1. How do diodes affect electrical circuits?

2. How does the LED assembly show the direction of the current?

3. How do you think an LED would light up if it was connected to an alternating current? Why?

Energy Transformations

You have learned that there are many different forms of energy. You hear sound energy when you listen to music. Energy can also change or transform into other forms of energy. What other energy transformations do you use every day?

Some forms of energy are more useful than others. For example, a toaster gives off thermal energy and light energy. The wire inside the toaster is hot, and that thermal energy is used to toast bread. Do you make use of the light energy the wires also give off? Some energy is lost during energy transformations. Often this energy loss takes the form of thermal energy. What other examples of lost energy can you think of?

Another example of energy change is solar cells that change solar energy into electrical energy. Batteries store potential energy in the form of chemical energy. This chemical energy changes to electrical energy when it is connected into a circuit. What are signs that a device is giving off energy that is not being used?

Lesson Activity

Find and cut out pictures of modern electronic devices or appliances from magazines or catalogues. Make a list of different forms of energy each device uses, such as thermal, mechanical, chemical, light, sound, and so on. Work with a partner to classify the pictures of the devices by their main or final energy transformation. Next, make a list of the appliances that you use daily either at home or at school. Make a chart listing seven of the most important electronic devices or appliances you use, and identify the final energy transformation in each one.

Which devices play the biggest part in your everyday life?

Computing Devices and the Internet Explain

Transistors, microchips, and processors made the development of personal computers possible. Advances in microchips and liquid crystal displays led to the development of portable laptop computers. These advances transferred the computing power that once fit in the size of a room to the palm of your hand.

Although you might not realize it, you interact with hidden computers in your everyday life. For example, all cars made today have at least one computer in them. Most cars have multiple computers. Computers control a car's air bags, keyless entry system, stereo system, and cruise control system. Automatic parking systems made up of computers and sensors on some newer models can park the car with little input from a driver. What features would you expect to see on cars in the near future?

Smart homes controlled by applications on smartphones are a reality. You can turn on and off lights right from your phone. There are also applications that can unlock doors, adjust the thermostat, control a sprinkler system, and check a web cam through a smartphone.

A smartphone is a computer that fits in the palm of your hand.

What are some of the ways you use computers?

Explore-a-Lab

Structured Inquiry

What similarities and differences are there in the construction of a CD and a DVD?

Compact discs (CDs) and digital video discs (DVDs) are small pieces of plastic that can hold music data or computer software. Observe a CD and a DVD. Record your observations in a data table. Fold each disc in half while wrapped in a thick towel. Use your fingers and forceps to pry apart any noticeable layers of the discs. Record your observations for each type of disc. Then, use packing tape or duct tape to peel off the reflective coating from the discs. You will end up with a clear piece of plastic. Record your observations. How do the CDs and DVDs compare to other forms of electronic storage? Is any one type of storage better than another? Explain your reasoning.

The Internet Network

The Internet is a part of everyday life, but its use was a revolutionary idea—a unique way to link non-connected computers together to share information. It provides a way for transferring data, gathering information, conducting trades, and holding group discussions. E-mail allows messages to move from one computer to another using telephone lines, cable, or a wireless connection. Information in the form of images, sounds, and video are also communicated.

The World Wide Web (www) is a collection of information, business, social networks, and entertainment spots called websites. With just a click, you can link to other sites, participate in discussions with other students, and virtually tour the world without ever leaving the classroom. How do you use the Internet? How much time do you spend on the Internet in an average day?

Called to Serve

For decades, Seventh-day Adventists have been using the latest technology to spread the gospel. In 1929, H.M.S. Richards began a radio ministry called "Voice of Prophecy."

In 1956, George Vandeman began a television ministry called "It Is Written." Visit their websites to see how they are still using the latest technology to share Jesus.

Concept Check Assess/Reflect

Summary: What are electronics? Vacuum tubes and transistors were some of the earliest electronics developed. Both inventions allowed electrical energy to be used in new ways. Many of today's electronic devices, including cell phones, microwave ovens, and computers, depend on microchips which contain millions of transistors on small chips. In electronic devices, electrical energy can be transformed into other forms of energy, such as light, sound, thermal, and mechanical energy. The Internet allows data, images, and sounds to be exchanged worldwide using personal computers.

1. What are the tiny electric circuits printed on thin pieces of silicon that control and direct the flow of currents in many electronic devices?

2. What is the significance of transistors in the development of computers?

3. Name and describe the energy transformations that take place in three electronic devices. Why are these energy changes never 100% efficient?

4. How would your day be different without the Internet?

Lightning

Lightning is a natural electrostatic discharge of energy produced during a thunderstorm. The discharged energy comes from a buildup of positive and negative charges within a cloud. Lightning is spectacular, powerful, and sometimes deadly.

Inside a thundercloud, small bits of ice (frozen raindrops) rub together as they move around in the cloud. Each time they rub together, a small electric charge is created. After a while, the cloud becomes filled with electrical charges. The positive charges gather at the top of the cloud. The heavier, negative charges gather at the bottom of the cloud.

The negative charges at the bottom of the cloud cause a buildup of positive charges on the ground below. The positive charges on the ground will move up taller objects, such as a person, a tree, a large building, or a mountain. The positive and negative charges exert the force of attraction. The negative charges move down a narrow channel in the air, while the positive charges move up in their own narrow channel. The channels will connect. The result is a sudden release of electrical energy that we see as lightning.

Lightning can be 3–90 miles long, or even longer if there are many bends and twists in the channels. The temperature of a lightning bolt is about five times hotter than the surface of the Sun and can contain 200 million volts of electricity! The intense heat of the lightning causes the air around the channel to heat up, too. The air expands very rapidly and, while doing so, compresses the air in front of it. A shock wave forms; this is thunder. We see lightning before we hear thunder because the light from the lightning bolt travels at the speed of light. The sound waves from thunder travel much slower.

Studying natural lightning is difficult; the bolts are unpredictable. Scientists create lightning to study. While scientists stand in an insulated, grounded trailer, they launch small rockets into thunderclouds. A rocket contains copper wire and is grounded with a wire to a strike point on the ground. The wire alters the electrical charges under the cloud, and the hopeful result is a bolt of lightning to hit the strike point. Around the strike point are instruments that collect data about the lightning. Scientists also test electronic devices, such as surge protectors and airplane insulation, to see if they can withstand a lightning strike.

✓ Concept Check

1. Why do we see lightning before we hear thunder?
2. When creating lightning, will the bolt always hit the strike point? Explain.

Electrician

When you walk into your home or workplace, you flip a few switches and the lights come on. Your computer, television, and the appliances you use come to life. Before this can happen, the wiring structure that delivers your electrical power must be correctly installed. Electricians are the skilled technicians who are responsible for installing and connecting the structures of an electrical system.

When you have problems with an electrical system, it can be dangerous to try and fix it without electrical training. Call an electrician who is trained to install, operate, repair, and maintain electrical systems. Many electricians study for a license for as long as five years, though some courses can be finished in as little as three years. Most states require electricians to be licensed or certified. To receive an electrician's license or certification, you usually have to pass a state exam.

Becoming an electrician means you get hands-on training as well as classroom instruction. Even after being certified, electricians must continue to learn about the constantly changing technology that relies on electricity. Some electricians even choose to specialize in a certain type of electrical work.

Some electricians work on interior wiring in homes, offices, and other buildings. There are also electricians who work outside on the electrical lines that bring power into our homes and businesses. They fix any downed lines or any that have been damaged by storms or wind. This is especially dangerous work because the power currents are high and any accidents can be fatal.

Electricians are also part of the crew when a new building is being constructed. They install the wiring as well as fixtures for light bulbs and sockets to plug in appliances. Electricians may also place special circuits for stoves and heaters, or install voltage meters and breaker boxes.

Other electricians work solely on ships, jets, or automobiles, connecting radios, global positioning systems, radar, and other electronics. Some electricians specialize in research to develop the newest technology.

There are many types of electricians and they all play an important role in keeping our electronics running.

✔ Concept Check

1. What kind of training does an electrician need before he or she is licensed to work?
2. Describe some of the different jobs an electrician might have.

Study Guide

Lesson 1

1. Loosely held electrons in atoms jump from one atom to another, resulting in charged objects.

2. Electrostatic discharge is the sudden release of electricity between two differently charged objects.

3. Lightning is a natural electrostatic discharge.

Lesson 2

1. Electric current travels to homes from generating stations over a power grid.

2. Electric current can be direct or alternating.

3. Conductors allow electricity to flow through them. Insulators resist the flow of electricity through them.

Lesson 3

1. Chemical reactions in electrolytic cells produce electric currents.

2. The two kinds of electrolytic cells are wet cells and dry cells.

3. Batteries are recycled to protect the environment and reclaim useful resources.

Lesson 4

1. An electric circuit includes a power source, switches, and a conductor to connect the parts together.

2. Electric current needs a closed circuit to flow. An open circuit is incomplete and prevents electricity from flowing.

3. Circuits can be connected in either series or parallel.

4. Switches, fuses, and circuit breakers are used to control the flow of electricity.

Lesson 5

1. The power behind modern electronic devices comes from microchips that conduct electric current.

2. Microchips are tiny chips of silicon that control electric currents and store information in binary code.

3. Electronic devices transform electric energy into other forms.

Show What You Know

Visualize It Complete the following Venn diagram.

Static Electricity

Like charges **1.** _______ and unlike charges **2.** _______.

3. _______ is a natural example of an electrostatic discharge.

Used in printing, advertising, spray painting, and in **4.** _______ _______.

Forms of energy produced by **5.** _______ _______.

Electric Current

Sources include generators, **6.** _______, and solar cells.

Travels in both series and parallel **7.** _______.

Measurable quantities include current, **8.** _______, and resistance.

Semiconductors that only let current flow in one direction are **9.** _______.

Explain how the paired terms are related.

10. static electricity—lightning
11. voltage—amperage
12. insulator—conductor
13. battery—electrolytic cell
14. load—series circuit
15. switch—transistor
16. electronics—microchip

Multiple Choice
Choose the best answer.

17. Which tool could you use to measure the resistance in a circuit?
    A. ammeter
    B. resistor
    C. ohmmeter
    D. voltmeter

18. What causes advertising decals to stick to windows without glue or tape?
    A. attraction between charged objects
    B. electrostatic discharge
    C. electric current
    D. repulsion of like charges

19. Which of the following materials would make the best conductor of electricity?
    A. distilled water
    B. glass beaker
    C. tap water
    D. wood block

20. What happens to the other bulbs in a parallel circuit if you disconnect a switch in one of the branches?
    A. They get brighter.
    B. They get dimmer
    C. They will go out.
    D. They will be unaffected.

Check Point
Answer the following questions.

21. Explain what causes lightning between a cloud and the ground.

22. **Compare** alternating current with direct current.

23. Describe how electricity gets from the power station to our homes.

24. **Infer** why you should be especially careful of the buildup of static charges around electronic computing devices.

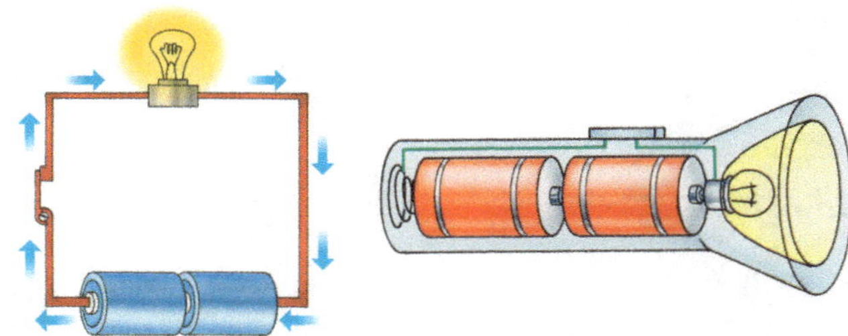

25. Identify the main parts of a circuit and tell how these parts are connected in a flashlight.

26. Choose one electrical device you use every day and **predict** how technological advances in electronics will improve the device.

Electricity and Magnetism Are Related

Scripture Spotlight

As you study these forces that God created, think about what spiritual lessons you can learn from electricity and magnetism. You will study the following passages in this chapter.

John 12:32 (p. 481)   Psalm 104:3 (p. 504)
John 13:35 (p. 484)   Psalm 135:7 (p. 504)
Matthew 5:16 (p. 484)   Psalm 74:16 (p. 505)
James 1:17–18 (p. 499)

The Big Idea

Magnets draw certain materials to themselves. They can be used to generate electricity, and electricity can be used to create magnets. Electricity and magnetism working together have been used to create generators and motors, which power many things we use every day.

What are some devices you use that make use of electricity and/or magnets?

Technicians use a giant magnet to sort and move material at the junkyard.

Inquiry Kick-Off — Engage

How do you sort and move a mountain of metal? Technicians use giant magnets to move large metal pieces to machines that crush and shred the metal into small pieces that can be sorted. In your *Science Journal*, you will use a magnet to move and test objects.

? Essential Question

What Are Magnets?

Lodestone is a natural magnet. Have you ever encountered a lodestone? How might you use a lodestone? Would these stones be worth mining? Why or why not? People have used magnets for more than 2000 years. However, scientists did not discover the links between magnetism and electricity until the 1800s.

Magnetic Materials Explain

Why do you think magnets stick to the door of a refrigerator? What are ways that you use magnets in your home? The metal most people think of when they think about magnets is iron. Other common metals, such as aluminum, copper, silver, and gold, are nonmagnetic.

Iron is not the only magnetic metal. Nickel and cobalt also make good magnets. The best magnets, however, are made of alloys, combinations of two or more metals. What two metals do you think would be the best combination for a magnet? Why? Magnets made from alloys containing iron, boron, and neodymium are the most powerful magnets. These are used in medical electronic devices.

Magnets are used in:

- computer hard drives,
- audio speakers and headphones,
- bicycle generators, for lighting,
- motors in cordless tools, and
- flashlights that are shaken to generate electricity.

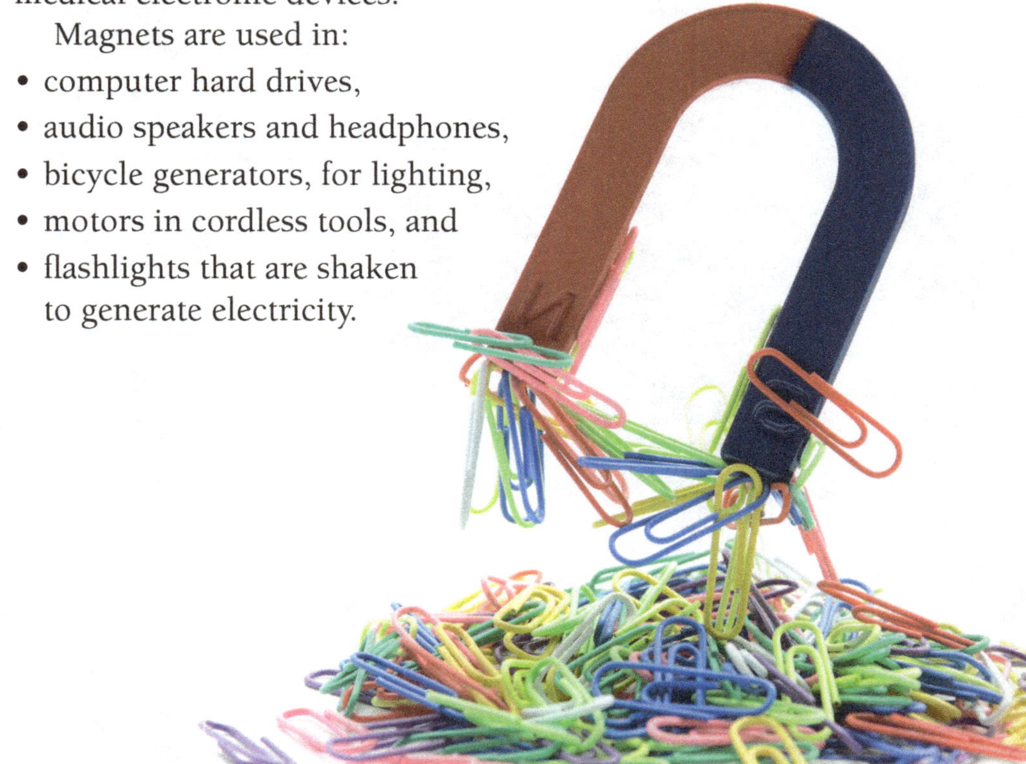

Permanent Magnets

When magnets are put in equipment like computers, we need them to maintain their magnetic properties. A magnet that keeps its magnetic properties for a long time is called a **permanent magnet**. What would happen if the magnets inside your computer lost their magnetic qualities?

The natural mineral magnetite, also called lodestone, is a permanent magnet. All other permanent magnets are made in factories. Most permanent magnets are made as bar magnets or U-shaped horseshoe magnets. These magnets are used in schools. Flexible magnets are made by combining magnetic metals with plastics. Some refrigerator magnets are made this way. What type of magnet do you think is used in your electronic devices? Why?

Temporary Magnets

Metals like iron are made into **temporary magnets** when you put a magnet near them. You can make a temporary magnet by stroking a magnetic material with a permanent magnet. However, temporary magnets will lose all or most of their magnetism when the magnet is removed. Every object that is lifted, moved by, or sticks to a magnet acts as a temporary magnet.

Magnetic Forces and Fields

Magnets can be made from several different types of materials in a variety of shapes. You have just learned that some magnets are permanent while others are temporary. Regardless of their shape or composition, all magnets share two characteristics. They all exert a magnetic force and produce a magnetic field. How have you used this force? What can you see in the classroom that uses a magnetic force? Most often magnetic force is helpful. What are some times that this property might not be helpful?

Magnetic Poles

The places on a magnet where magnetism is the strongest are the **magnetic poles**. Each magnet has a north pole and a south pole. The poles on a bar magnet are at either end. The poles of a donut magnet are around the outside edge of the magnet. Where do you think the magnetic poles are on a horseshoe magnet?

Record your work for this inquiry. Your teacher may also assign the related Guided Inquiry.

Creating Magnets

What controls the strength of a temporary magnet?

Procedure

Materials
- large iron or steel nail
- strong bar magnet
- small paperclips

1. See how long you can make a chain of paperclips using the bar magnet. **Record your data**.

2. Stroke the nail 10 times with the bar magnet. Be sure the nail is stroked in the same direction each time.

3. **Observe** if the nail will pick up a paperclip. If it does, touch a second paperclip to the first one that is hanging from the magnet. The second paperclip will be attracted to the first one because the first clip has become a magnet.

4. Keep adding paperclips in this way to see how long you can make the chain. **Observe** how many paperclips the nail picks up. **Record your data** in a table like this in your *Science Journal*.

Number of times nail is stroked	10	20	30
Number of paperclips in chain			

5. Repeat Steps 2–4, stroking the nail 20 times, and then 30 times. How many paperclips does the nail pick up each time?

Analyze Results

Review the data you collected. Make a graph to compare your results.

Create Explanations

1. What controls the strength of a temporary magnet?

2. How does the strength of the magnetic nail compare to the strength of the bar magnet? Why?

3. What do you think will happen to the nail magnet after a day or two? Explain.

The two poles of a magnet respond differently to the poles of a second magnet. The north pole of one magnet will pull on the south pole of another magnet. However, the north pole of the first magnet will push away the north pole of the second magnet. This simple experiment shows that the two poles of a magnet are different.

The north poles of two magnets push each other away, or repel each other. The same thing happens with the south poles of two magnets. They act like two positive or two negative electric charges. To summarize, like poles repel. Unlike poles attract.

What happens if you cut a bar magnet in half? Each smaller magnet has both a north pole and a south pole. Cutting each small magnet in half produces four smaller magnets. Each magnet still has a north pole and a south pole. You could continue this process until the magnets are very tiny. Is it possible to have a magnet with only a north pole? Why or why not?

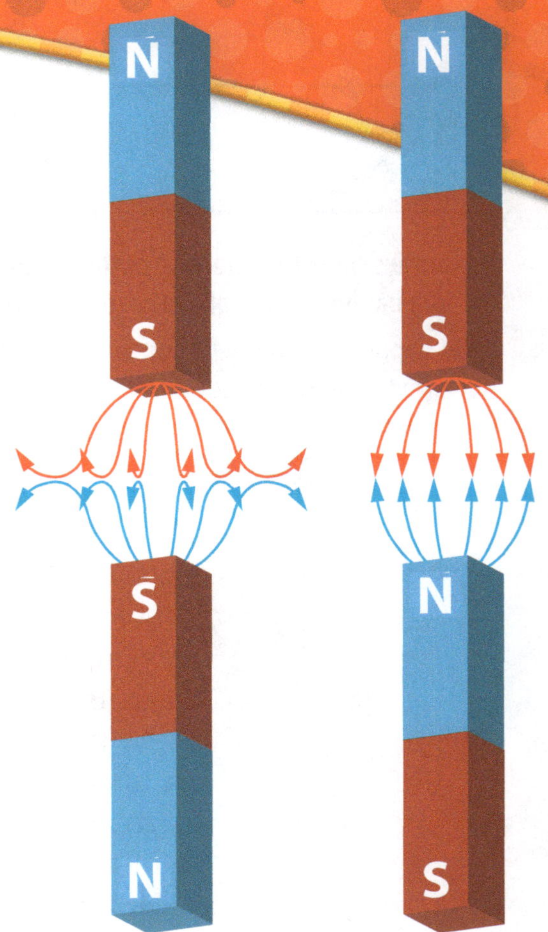

Where are the magnetic poles found on different types of magnets?

Working in pairs, select four different types of magnets from your teacher. Place a sheet of paper or transparent plastic over the first magnet and draw the outline of the magnet on the sheet. Then gently sprinkle iron filings onto the sheet. Observe what happens. Where are the poles of this magnet located? What form do the magnetic field lines take? Draw your observations. When you have completed the experiment, gently brush the iron filings off the sheet back into the storage bottle. Switch roles and repeat with another magnet. Continue until each magnet has been examined.

Magnetic Domains

The domain model explains properties of a magnet. In this model, a magnet is made up of a collection of tiny **magnetic domains**. Each domain acts like a tiny magnet. In a magnet, the domains are all lined up so that their north poles all point in the same direction.

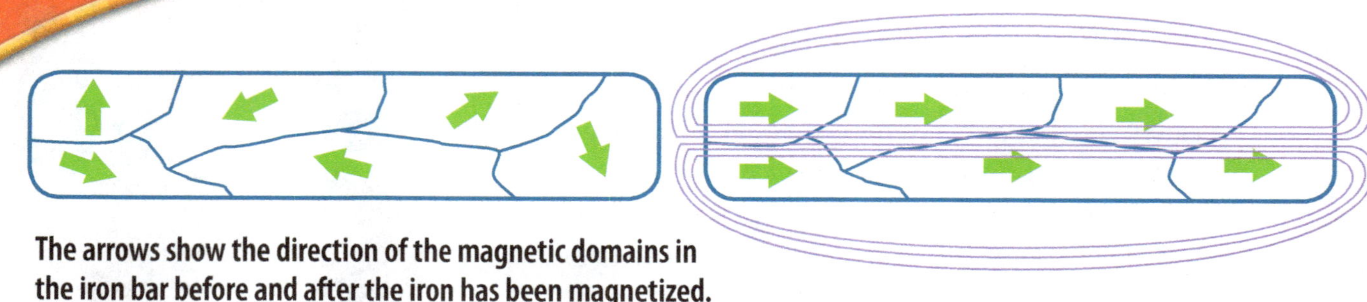

The arrows show the direction of the magnetic domains in the iron bar before and after the iron has been magnetized.

What happens to magnetic domains when a magnetic material is magnetized? How does this relate to the magnetic field?

Nonmagnetic materials are also made of magnetic domains. Their domains are mixed up in all different directions, so these materials do not act as a magnet.

Some materials will have their domains lined up temporarily if they are near a strong magnet. These materials become temporary magnets when stroked many times with a magnet. Each stroke aligns the magnetic domains in the material. Temporary magnets are made just by touching a permanent magnet. Temporary magnets can also be made by running an electrical current through or around some metal objects.

Heating or pounding can destroy the magnetic properties of a magnet by moving the domains out of order. What may happen if a magnet on your refrigerator door gets knocked to the floor a lot? Why?

Magnetic Fields

A magnet creates an invisible area of magnetism all around it called a **magnetic field**. If you slowly push a magnet toward a nail, at some point the nail will jump and stick to the magnet. The magnetic field acted on the nail. How can you use the knowledge that a magnet can make an object jump to help you?

Iron filings can show the direction of the magnetic field of a magnet. The filings are closest together where the magnetic field is strongest at the poles and farther apart where the magnetic field is weaker.

Magnetism is sometimes described as a force that "acts at a distance." This means that a magnetic field can cause a pulling or pushing force on an object without touching it. Push the north pole of a bar magnet toward the south pole of a second magnet. What will happen? Try it and see. What happens if you push the north pole of a magnet toward the north pole of another magnet?

Scripture Spotlight

Think of a Christian as a magnet with an invisible magnetic field around him or her that helps attract others to Jesus. What might that spiritual magnetic field be made of? See **John 13:35** and **Matthew 5:16.**

Earth as a Magnet (Explain)

You have likely heard of the North and South Poles on Earth. We call these two locations poles because Earth acts like a magnet. Earth's liquid core rotates, producing a magnetic field. This magnetic field is almost the same as the one produced by a bar magnet. Just like a bar magnet, Earth has a magnetic North Pole and a magnetic South Pole. A magnetic field surrounds the planet. Do you think Earth's magnetic North Pole is in the same location as its geographic North Pole?

Compasses

For at least 1000 years, people have used Earth's magnetic field to find direction. People in China discovered that lodestone always pointed in the same direction when it could swing freely. Later they discovered that a floating, magnetized needle also always points in the same direction. The invention of the *magnetic compass* was based on this discovery. At first, compasses were used exclusively on land.

The *north-seeking pole* of a compass points toward Earth's north magnetic pole. The *south-seeking pole* of a compass points toward Earth's south magnetic pole. The compass made long-distance ocean travel possible without getting lost.

Zheng He was a Chinese admiral. Zheng He was the first recorded person to use a compass to navigate. He made seven ocean voyages between 1405 and 1433. Later, European explorers used magnetic compasses to travel across the Atlantic Ocean. Early navigational compasses were often suspended in water or hung from a silk cord. On land, the compass also was a valuable tool for early explorers. Why might knowing how to use a compass be a valuable skill?

Check out your *Science Journal* for a Guided Inquiry that explores how Earth's magnetic field works.

(Extend)

The pointing needle on the early Chinese compass was a lodestone spoon-shaped device, with a handle that would always point south.

How is this different from magnetic compasses you have used for finding directions?

How does Earth's magnetosphere protect the planet?

Check for Understanding

Compare and contrast how the magnetic field around a bar magnet and the magnetosphere are alike and different.

The Magnetosphere

The Sun ejects millions of charged particles into space. These particles, which are harmful to us, make up the solar wind. Fortunately, these dangerous particles encounter the magnetosphere. The **magnetosphere** is a region of Earth's magnetic field that extends 58,000 km (about 36,000 mi) into space. The magnetosphere deflects high-speed, charged particles in the solar wind and keeps them from entering Earth's atmosphere and striking Earth's surface. This protective feature is another demonstration of God's design to protect us from harm. If the magnetosphere did not exist, what would happen to our power grid? The electronics we use? How would it affect our health?

Auroras

When charged particles of the solar wind reach the magnetosphere, they travel along magnetic field lines. The particles travel to the strongest part of Earth's magnetic field—the North and South Poles. At the poles, the particles travel down into Earth's upper atmosphere. There they collide with molecules of oxygen and nitrogen in the air. Energy is transferred from the charged particles to the oxygen and nitrogen molecules. As these molecules return to a natural state, they give off their extra energy. Oxygen molecules give off green light or red light at different heights. Nitrogen molecules give off blue or purple/violet light.

Math in Science

Magnetic field strength can be measured in milligauss (mG) units. At 30 cm (1 ft) away, common home appliances had the following measurements of magnetic field strength: refrigerator, 1.5 mG; blender, 11.1 mG; microwave, 60 mG; can opener, 115.5 mG. Which had the weakest magnetic field, and which had the strongest? How much stronger was the strongest than the others?

These different colors of light, which produce glowing "streamers" or "curtains" of light in the night sky, are called **auroras**. In the northern hemisphere, they are called the Aurora Borealis. In the southern hemisphere, they are called the Aurora Australis. Because the light given off is much dimmer than sunlight, the auroras cannot be seen during the daytime.

Auroras occur at both poles at the same time. They most often occur in ring-shaped areas around Earth's magnetic poles. However, the complete rings can only be seen and photographed from space.

The best places to see the Aurora Borealis (Northern Lights) are in Alaska, Canada, and Scandinavia. Auroras are sometimes visible further south during periods of greater sunspot activity.

What kind of molecules produced the green aurora in this picture?

Concept Check Assess/Reflect

Summary: What are magnets? Iron, nickel, cobalt, and a few other metals are magnetic. Most other materials are nonmagnetic. Temporary magnets lose their magnetism easily. Permanent magnets keep their magnetism for a long time. All magnets have a north pole and a south pole. Like magnetic poles repel, while unlike poles attract. A magnet contains magnetic domains that line up in the same direction. The region around a magnet where its magnetism acts is its magnetic field. The magnetic field of a magnet is strongest at its poles. A magnetic field surrounds Earth.

1. How is Earth's magnetic field related to auroras in the night sky?

2. How does a compass work?

3. How can a material contain magnetic domains, yet not be a magnet?

4. What is one thing that would be different on Earth if our planet did not have a magnetic field?

? Essential Question

How Are Electricity and Magnetism Related?

How do like and unlike electrical charges behave? How does this concept relate to magnetic poles? There are many similarities between electricity and magnetism. In this lesson, you will study how the two are related.

Electric Current from Magnets (Explain)

How can a magnet be used to produce an electrical current? The English scientist Michael Faraday conducted experiments to figure this out. He moved a magnet in and out of a coil of wire. When the magnet was pushed into the coil the electrons moved in one direction. The electrons moved in the opposite direction when the magnet was pulled out of the coil. This movement back and forth is important because it produces an electric current that flows through the wire. This process of getting electrons to move through a wire by changing the magnetic field is called **induction**.

Michael Faraday used an apparatus similar to this to induce an electric current by moving a magnet inside a wire coil. The current can be detected with a meter such as a galvanometer or ammeter.

? What would happen to the meter dial if you moved the magnet in and out of the coil at a slower rate? What if you stopped the motion of the magnet entirely?

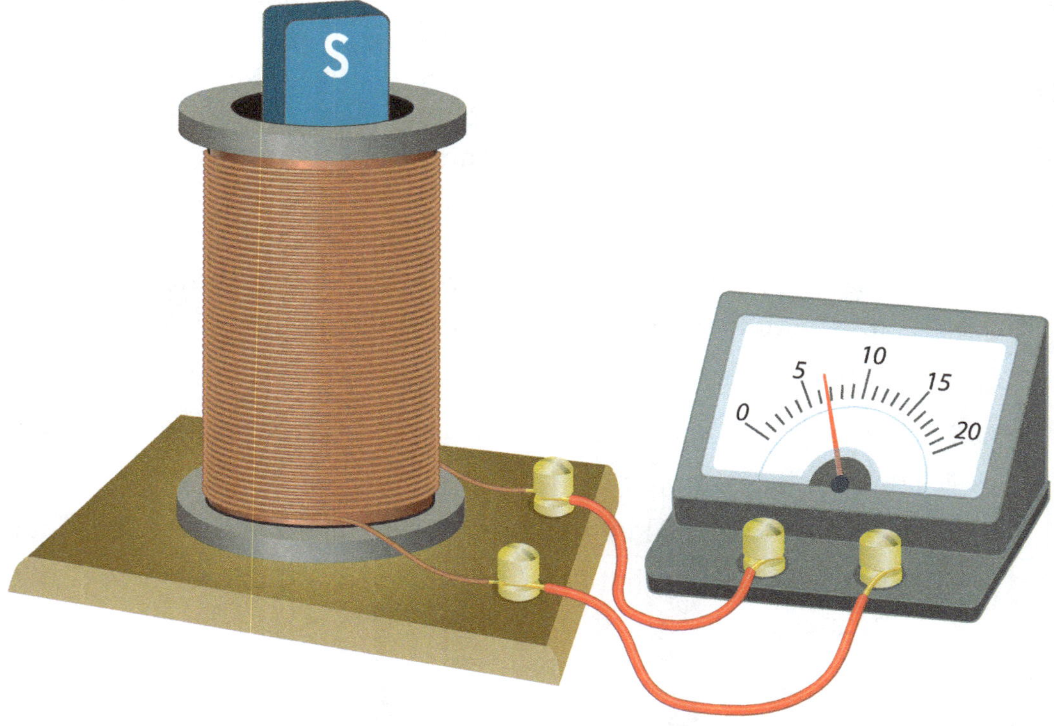

The movement of the current Faraday discovered can be measured with an *ammeter.* An ammeter has two terminals—one for incoming current and one for outgoing current. Inside the ammeter is a conducting coil with a magnet that rotates when current passes through the ammeter. Ammeters can be either mechanical or electronic.

One type of mechanical ammeter is the *galvanometer,* which is named after Italian physicist Luigi Galvani. The galvanometer uses the effect of a magnetic field on a spring coil to indicate the direction and scale of an electric current. In this way, the needle on a galvanometer will also show the direction an electric current is moving. Why would an electrician need to know what direction electric current is moving?

 **How can you make electrons move?**

You will need two empty beverage cans, two foam cups, an inflated balloon, transparent tape, and fake fur. Tape the bottom of one cup to the side of a can. Repeat for the second cup and can. Place the cups upside down on a table with the cans sitting on top of the cup bottom. Arrange the cups so that the ends of the cans touch one another. Rub the balloon with the fake fur to build a negative charge in the balloon. Move the balloon near the end of one can, but do not touch the can. Move the balloon away and separate the cans by moving the cups. Do not touch the cans, which are now charged. Test the cans to see if they are positively or negatively charged by moving the negatively charged balloon closer. What happened when the balloon was near each can? What charge did each can have? How do you know?

Generating Electric Current

How can a magnet generate electric current?

Procedure

1. **Measure** out a length of 50 cm (20 in.) of the wire.
2. Leave several centimeters free at each end of the wire and make a coil by wrapping the wire around the cardboard tube eight times.
3. Remove the cardboard tube from the coil. Attach both ends of the wire to the ammeter.
4. Take the bar magnet and move it close to the coil. **Record** what you **observe** on the ammeter in your *Science Journal*.
5. Push the bar magnet slowly into the coil and **observe** the ammeter as the magnet is at various locations. **Record** what you **observe**.
6. Slowly remove the bar magnet from the coil and record what you **observe** on the ammeter. Repeat Steps 4–6 four more times.

Materials
- tape measure
- 50 cm (20 in.) insulated copper wire
- cardboard toilet paper tube
- ammeter
- bar magnet

Analyze Results

Interpret your observations. Use your observations to **draw a conclusion** about the relationship between the magnet and ammeter.

Create Explanations

1. How can a magnet generate electric current?
2. How did the ammeter help in your observations? How did the ammeter reading compare to your movement of the magnet?
3. What was happening to the electrons in the wire when you moved the magnet into the coil?

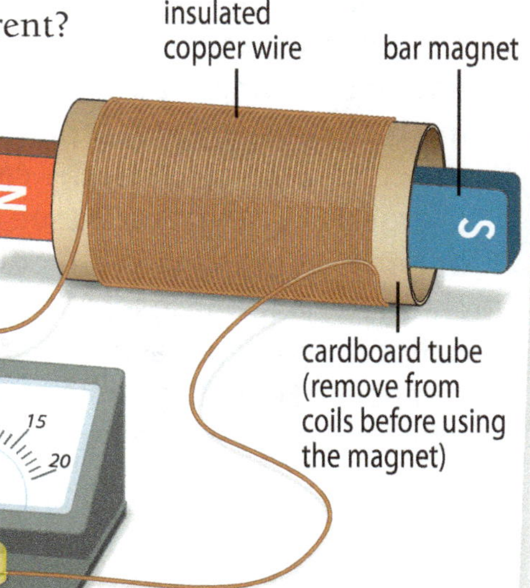

Magnetism from Electric Current Explain

If an electric current can be made using a magnet, can a magnetic field be made using an electric current? Look at the diagram on the right. It shows a simple magnetic field that is produced by an electric current flowing through a straight wire. Notice that the wire carrying an electric current runs straight up and down. Look at the circles going around the wire at right angles. The lines represent the magnetic field produced by the electric current. The direction of the magnetic field created by an electric current is always at right angles to the flow of electricity. To help remember this, remember that the magnetic field is in the direction the fingers of your right hand would curl if you wrapped them around the wire with your thumb pointing in the direction that the current flows. How would the magnetic field change if the direction of the current changed? Why? What if the electric current was stronger—how would this change the magnetic field?

Using this method, your thumb points in the direction of the current.

If a current were flowing toward the tip of your pencil, what direction would the magnetic field be going?

Explore-a-Lab

Structured Inquiry

How can you demonstrate that there is a magnetic field around an electric circuit?

(1) Place a battery in a battery holder and connect two wire leads to the two terminals of the battery holder. (2) Make a hole and place a vertical wire through a box lid or a piece of cardboard. (3) Connect one wire lead to the bottom of the vertical wire. (4) Tape four small compasses on the card around the vertical wire. Observe and record the direction of the compass needles before the electric current is connected.

(5) Connect the other wire lead to the top of the vertical wire. Note the direction in which each compass needle points. (6) What happens to the magnetic field if you reverse the direction of the current in the wires? Try it. Observe and record the direction of the compass needles when the electric current is reversed. Do not leave the wires connected for more than a few minutes.

You have learned that an electric current flowing through a wire causes a magnetic field around the wire. There are several parts to an electromagnet. An electromagnet requires a source of electricity, a conductor, and an iron or nickel core. What happens when the wire is shaped into a coil? An electric current flowing through a coil of wire wrapped around a solid iron core makes an **electromagnet**. The coil of wire has a north pole and a south pole, just like a bar magnet. Increasing the number of turns in the coil makes the magnetic field stronger. The fields of all the turns add together to make up the magnetic field of the whole coil. However, if the electric current is turned off, the electromagnet is also turned off. The coil is no longer magnetic.

The magnetic domains in the iron core line up in the magnetic field and the core becomes a temporary magnet. The magnetic field of the iron core is added to the magnetic field of the wire coil. The electromagnet becomes stronger. You can also make an electromagnet stronger by increasing the current in the coil. Thinking back to the Structured Inquiry, how could you increase the current supplied to the coil?

Explore-a-Lab

Structured Inquiry

How can you make and test the strength of an electromagnet?

Work with a partner. To make a simple electromagnet, you will need one D cell or other dry cell, copper wire, and an iron nail. Take a 1-m (3-ft) length of coated copper wire. Leave about 20 cm (8 in.) of the wire free at one end and wrap it around the iron nail. Cut the wire so that about 20 cm (8 in.) is free at the other end. Remove the coating from about 2.5 cm (1 in.) at both ends of the wire. Place a D-cell battery in a battery holder and attach one wire to one end of the battery holder and the other wire to the other end of the battery holder. To test your electromagnet, bring it near some paperclips or iron filings. How can you increase the strength of the electromagnet? Try increasing the number of loops of wire or using a stronger source of electricity.

When both wires are connected to the dry cell, electric current flows through the wire.

What happens to the nail when current is flowing?

Some large electromagnets on cranes can lift more than 1000 kilograms (2200 pounds) of scrap iron and steel.

What is the benefit of using a magnet rather than other methods to move materials?

Uses of Electromagnets

Both motors and generators operate through *electromagnetic induction*. Electromagnetic induction forms when a magnetic field produces a voltage across the conductor. You observed this when you moved the magnet in and out of the coiled wire.

Electromagnets are used to control switches in relays. They are used in some telephones and computers to perform basic functions. There are also industry uses for electromagnets. Minerals, such as iron, cobalt, or their alloys, respond to a magnetic field. This characteristic allows for sorting and lifting of metal objects, like you saw in the chapter opener. The crane uses a strong electromagnet to move or separate scrap metal when the current is switched on. Once the load is moved and the current is turned off, the scrap is dropped. Some electromagnets can lift more than 1000 kilograms (2200 pounds) of scrap iron and steel.

Engineers use electromagnets in transportation. As you learned earlier, in Europe and Asia, many trains run on electromagnets. Why might these countries use electromagnets to power trains rather than fossil fuels? How would electromagnets help a train run faster or more efficiently? Other uses of electromagnets include security systems, switches, systems for spacecraft propulsion, and solid-state memory computing systems.

Check for Understanding

Think about the *Structured Inquiry* you completed earlier and the Explore-a-Lab you completed on the previous page. How are the induction device and the electromagnet similar? How are they different?

When the doorbell button is pushed, the electromagnet is magnetized. It attracts the iron bar pulling on the contact arm to break the circuit. This shuts off the electromagnet.

What causes the noise you hear?

The doorbell button is a switch that opens and closes an electric circuit connected to an electromagnet. When you push the button of the doorbell, you close a circuit that sends the current through a coil. The coil becomes an electromagnet that repels a permanent magnet inside the coil. How does this make the chime and the doorbell ring?

The next time you ring a doorbell or use a blender, remember that you are not just pushing a button. Instead, you are using electricity and magnetism together to get things done.

Concept Check Assess/Reflect

Summary: How are electricity and magnetism related? Moving a magnet through a wire coil causes an electric current to flow in the coil. A magnetic field exists around every wire carrying an electrical current. An electromagnet is a device made by using electricity to create a magnetic field. It is only a temporary magnet because the magnetic field only exists when the electric current is flowing. Electromagnets are used in doorbells, telephones, and any device that has an electric motor.

1. How does electromagnetic induction work? What can it be used for?
2. What are three devices that use electromagnets?
3. What happens when the direction of a magnet moving in a coil is reversed?
4. How can you increase the strength of an electromagnet?

Essential Question
What Is a Generator and a Motor?

For hundreds of years, people knew that some materials, such as magnetite, had magnetic properties. But it wasn't until people learned about the interrelatedness of magnetism and electricity that important discoveries were made. Why do you think it took so long for people to figure out how electricity and magnetism could work together? The discoveries associated with these concepts led to the development of two of the most widely used applications of magnetism and electricity—the electric generator and electric motor. Do you think there are more discoveries to be made relating to magnetism and electricity? Explain your answer.

Generators Explain

A **generator** is a machine that turns mechanical energy into electrical energy. Generators do this by turning a coil of wire in a magnetic field to produce an electrical current. The two main parts of a generator are a magnet and the wire coil. The generator will not work without a source of mechanical energy. The mechanical energy is needed to move the magnet or the coil. Some sources of mechanical energy are steam produced from coal, gas, nuclear power, or some other fuel. Other sources of mechanical energy include a hand crank, wind, and moving water.

Generators are used for many functions around the house.

❓ **Why might it be good to have a generator in the home?**

Faith Connection

Think of a spiritual analogy for a generator. If the electrical current produced represents spiritual power or growth, what might the coil of wire and magnetic field represent?

Generators supply electric current when there is a power breakdown or where there are no electric lines. Portable generators are available in different sizes for different functions. What businesses might need generators as a backup electricity supply?

In the diagram of the simple generator, the yellow arrow shows the direction that the coil rotates. The red end of the magnet is the north pole. The blue end of the magnet is the south pole. The blue lines represent magnetic field direction from the north pole to south pole. The red arrow shows the direction of the current. The direction of current flow switches back and forth as the coil is turned with the handle.

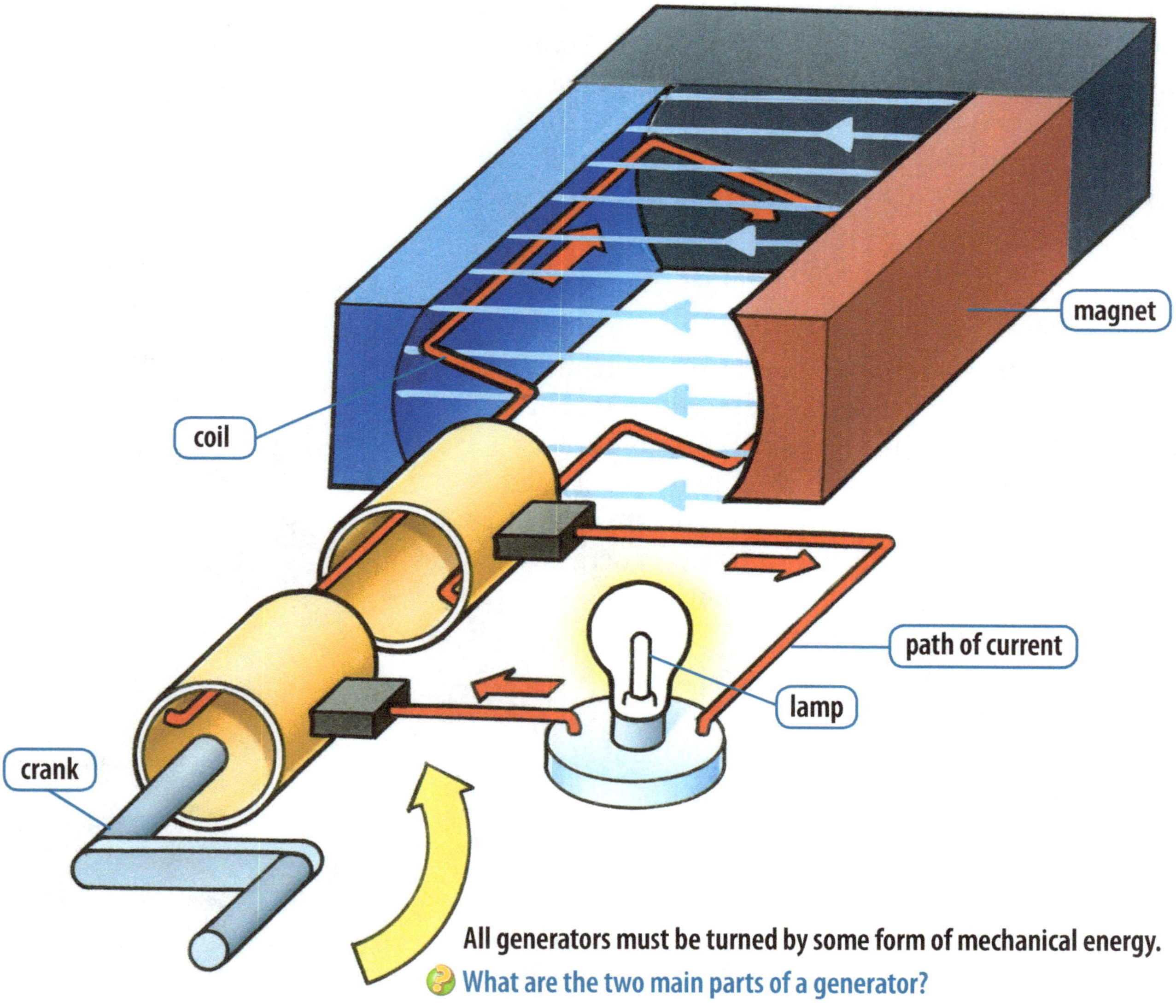

All generators must be turned by some form of mechanical energy.

 What are the two main parts of a generator?

Making a Generator

How does a generator work?

Procedure

1. Use a metric ruler and the drawing compass to **measure** and draw two circles about 5 cm (2 in.) in diameter on the poster board. Cut out the circles.

2. Place the film canister in the center of each circle and trace around the bottom. Your inner circle should be about 3 cm (1.25 in.) in diameter. Cut out the inner circles so you have two donut-shaped pieces of poster board.

3. Take the lid off the film canister. Slide both poster board "donuts" onto the film canister. Position them about 2 cm (0.8 in.) apart. Each should be equal distance on each side of the canister.

4. Use transparent tape to hold each "donut" in place.

5. **Measure** about 7 cm (3 in.) of wire to leave free when you begin winding the coil. Wind the wire onto the film canister between the two "donuts" until about 7 cm (3 in.) of wire is left. Secure the wire with insulation tape.

6. Use sandpaper to scrape off the insulation on the two free wire ends. Fasten each wire end to an LED light. Secure the light to the side of the canister with insulation tape.

7. Put the disk magnet into the canister flat on the bottom. Put on the lid. **Observe** and **record** what happens when you hold the canister between your thumb and finger and shake it up and down.

Materials
- empty plastic 35-mm film canister
- strong neodymium disk magnet
- 3.5 m (11.5 ft) of thin, enamel copper wire (29 AWG or 0.30 mm dia.)
- LED light
- poster board, 12 cm × 12 cm (5 in. × 5 in.)
- scissors
- insulation tape
- fine-grit sandpaper
- drawing compass
- metric ruler
- transparent tape

Analyze Results

Communicate what happened when you shook the film canister.

Create Explanations

1. How does a generator work?

2. How is the generator being powered?

3. What happens when you stop shaking the canister?

4. Explain how a generator that can provide electricity to your home is powered compared to the small generator you made.

The Electric Power Grid Explain

Electricity travels a long distance to get to your house. Electricity is made at a power plant by huge generators, which produce almost all of the electricity that we use. Power plants all across the country are connected in a system called the **electric power grid**. The picture shows the electric power grid and how the electric current generated at these power plants gets to your home.

Many power plants use coal, but some use natural gas, nuclear energy, solar energy, waterpower, wind energy, or even energy from inside Earth to generate electricity. Inside the generators, this energy is converted into another form of energy (usually mechanical or steam). This energy turns fan-like blades inside a **turbine**. The blades are attached to a pole, called a shaft. Wire is coiled around the shaft and the shaft is surrounded by a magnet. When the blades inside the turbine begin to turn, the wire-covered shaft also turns, and an electric current is produced.

This electric current is sent through devices called **transformers** that can alter the voltage to push the electric current over long distances. The current travels over high-voltage transmission lines that stretch across the country.

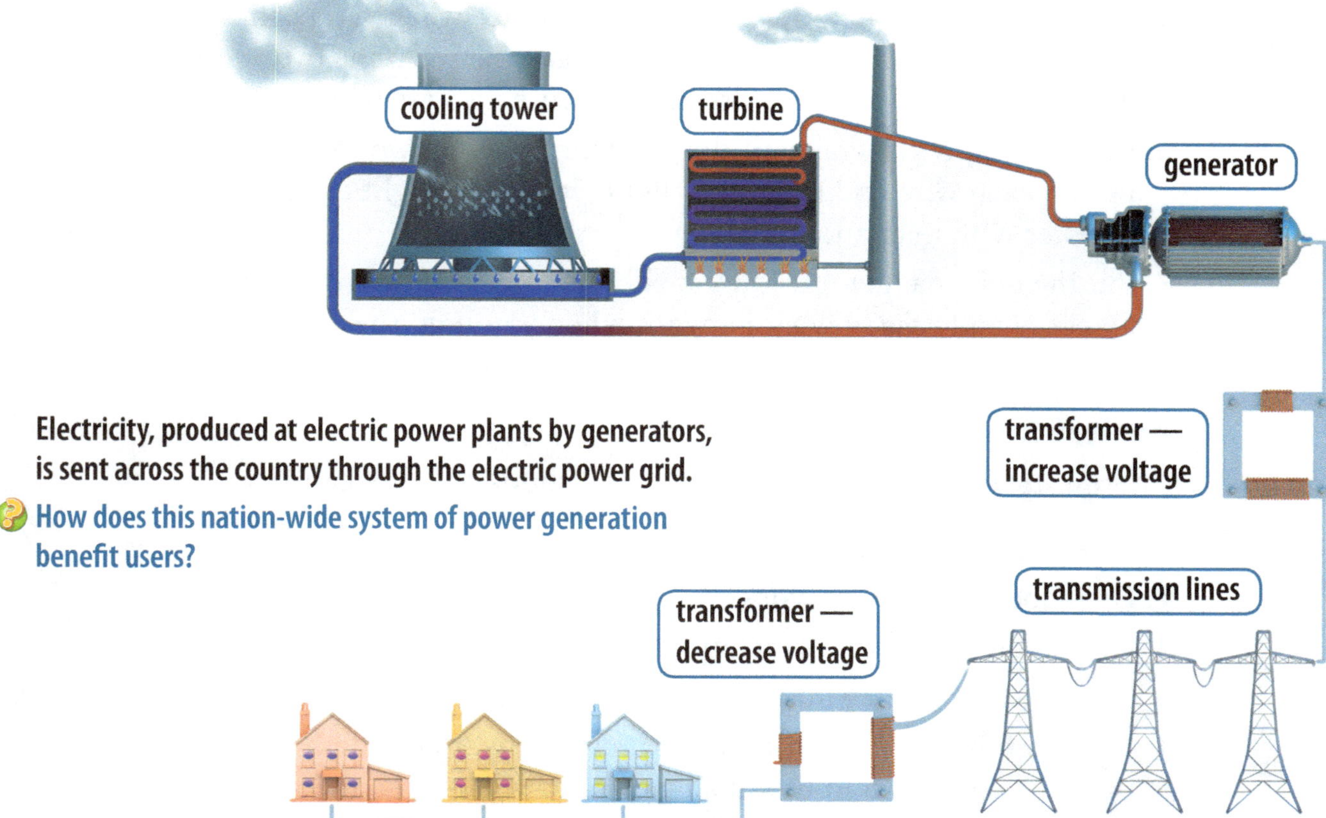

Electricity, produced at electric power plants by generators, is sent across the country through the electric power grid.

How does this nation-wide system of power generation benefit users?

When the current reaches a substation, the voltage is lowered so it can be sent over lower-power lines to your neighborhood. Transformers again lower the voltage so that the power is safe to use in our homes. At your home, the current moves through a meter that measures how much electricity your family uses. This electricity travels through wires in the walls to the outlets and switches in your home. Breakers and fuses protect the wires from being overloaded.

Electric Motors

Generators are used to generate electricity, which can be used to power many devices, including electric motors. An **electric motor** uses an electromagnet to change electrical energy into mechanical energy. The mechanical energy produced can then be used to do work. What devices in your room contain an electric motor? Electric motors can be very large to run machines in factories or commuter trains. They can also be very small to run refrigerators, clocks, and blenders. What appliance has the smallest electric motor in your home?

Electric motors and generators are two different devices that work in opposite ways. Generators change motion into electricity by turning a coil of wire in a magnetic field. Motors use magnets to change electrical energy into mechanical energy. Electric motors are of two basic types: alternating current (AC) and direct current (DC). The names indicate the kind of electricity they consume. AC current changes direction or alternates. DC current flows in one direction through the electric circuit.

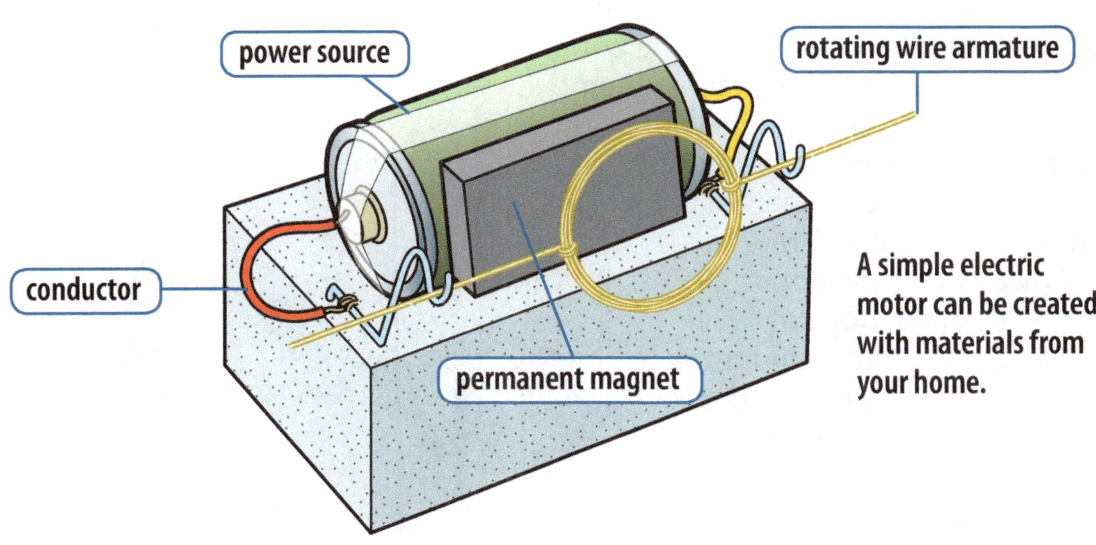

A simple electric motor can be created with materials from your home.

AC Motors

An AC motor consists of four parts:

(1) the rotating wire coil or armature (the rotor),

(2) an alternating electric power source,

(3) brushes that transfer the electric current to the armature, and

(4) permanent magnet(s) that use the electric current to cause the wire armature to turn.

Electric motors function on the two key principles of magnetism—like poles repel, and opposite poles attract. When current passes through the wire armature it turns into an electromagnet with the same polarity to the permanent magnet. Repulsion between "like" magnetic poles causes the armature to turn. After half a turn, the opposite poles attract each other. The current then flips to keep the armature rotating.

Many common appliances, such as fans and hair dryers, use the alternating current from a wall outlet to operate their motors. The current passes through the coil inside the motor, turning the coil into an electromagnet. A permanent magnet inside the device is attracted to the poles of the coil. When the current reverses, it is repelled. The rotating coil turns a shaft that operates the fan.

Explore-a-Lab

Structured Inquiry

How does an electric motor work?

Build a simple electric motor with a C cell, a coil of copper wire, a neodymium magnet, a thick rubber band, and two safety pins. Wrap the wire around the battery to make a coil and remove from the battery. Remove the insulation from both ends of the wire with sandpaper. Attach the heads of two safety pins tightly to the terminals of the battery with the rubber band. Mount the wire coil by pushing the ends through the holes in the bottom of the pins. Place the magnet under the wire coil between the two pins and give the wire coil a spin. Draw a diagram of your motor and label the parts.

DC Electric Motors for Cordless Devices

Cordless devices are designed to run on direct current from batteries. Direct current from a battery passes through a coil, making the coil an electromagnet. A motor made to run on direct current has a reversing switch called a **commutator**, which reverses the current. Reversing the current flips the poles of the coil. The poles of one magnet repel the poles of the other magnet. The coil keeps turning in the same direction until its two poles find the opposite poles of the magnet.

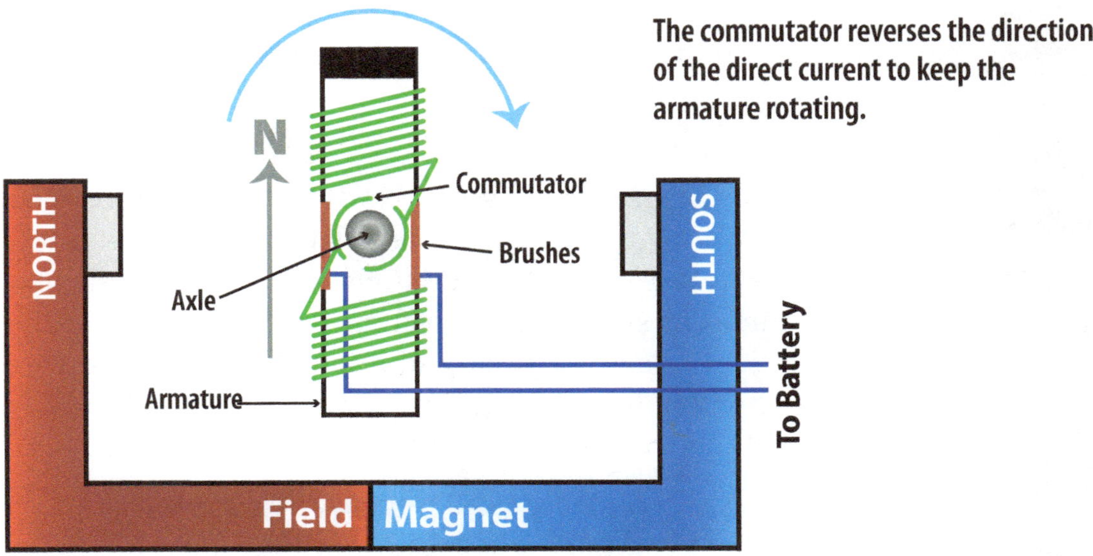

The commutator reverses the direction of the direct current to keep the armature rotating.

Concept Check Assess/Reflect

Summary: What is a generator and a motor? A generator changes mechanical energy into electrical energy. Generators produce almost all the electricity we use. Electric current travels from power plants to homes and businesses through a system called the electric power grid. An electric motor uses an electromagnet to change electrical energy into mechanical energy. Factories use motors to run machines, and people use electric motors of all sizes in their homes and electric devices.

1. Explain how an electric current is produced using a wire and a magnet.

2. Describe how the electricity produced in power plants gets to your home.

3. Explain the role of a commutator in a DC electric motor.

4. Why can a generator be considered the opposite of an electric motor?

Essential Question

How Is Electricity Used and Conserved?

Look around you. Name the first five items you see that use electricity. Identify one task you do every day that requires electricity. How did your ancestors complete this same task without electricity? Have you ever thought about where electricity comes from? What created the current, and how did it get to the light switch in your room?

How Is Electricity Made? Explain

Most of the electricity used in North America is produced in power plants that burn fossil fuels such as coal, gas, or oil. Some use heat from nuclear reactions to boil water and change it to steam. The steam turns a turbine connected to a generator that produces electric current. Additional sources of electricity include hydropower and other *renewable* sources. Why are fossil fuels and nuclear power not renewable?

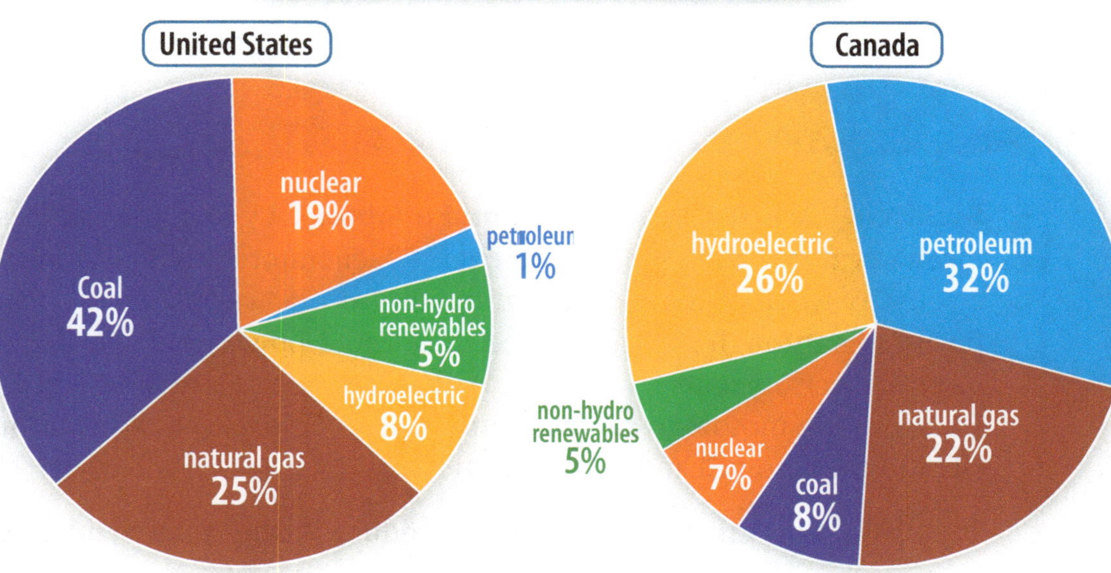

The circle graphs show the sources of electricity used in North America.
How does energy use differ between the two countries?

Hydroelectric Energy

Hydroelectricity is electricity that is produced by moving water produced by dams that are built on or near large rivers. The water stored behind these dams can produce tremendous power that is used to turn the turbines of large generators. Look at the numbers on the diagram as you read the text below to understand how the power plant works. What are the advantages of generating power this way? Can you identify any downsides to the process? Where do you think these dams are used in the U.S. and Canada?

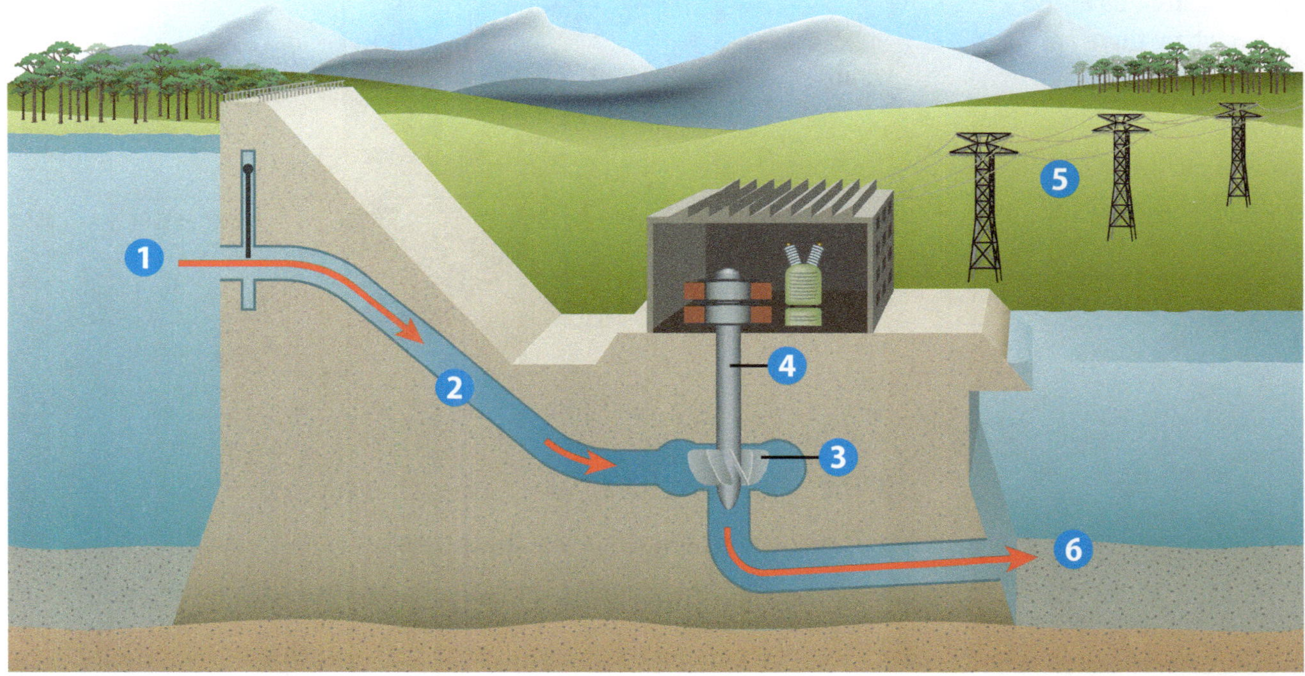

1. Water stored behind the dam pushes forward under great pressure.

2. When the intake gates are opened, water flows at high speed down through the **penstock**, a large tube that carries water to the turbines.

3. Turbines are like big propellers. As the water rushes past, the blades spin. The turbine is attached to a shaft, which also spins.

4. The spinning shaft is connected to magnets in the generator, which produces electricity. The electricity flows to power lines that are connected to the generator.

5. Power lines are connected to the transformer. They carry electricity to homes and businesses.

6. Water empties out into the river below the dam.

Over time, sand and silt can build up in the reservoir behind the dam.

How could the accumulating sediment affect the plant's ability to produce electricity?

Wind Energy

Wind can provide another source of clean and safe electricity. Much like electricity generated at a dam, wind energy is produced using turbines. A wind turbine uses large propeller blades attached to a tower. Look at the diagram as you read about the parts of the wind turbine and how it generates electricity. What locations in the United States and Canada would be best suited to gather wind energy? Why did you pick those locations?

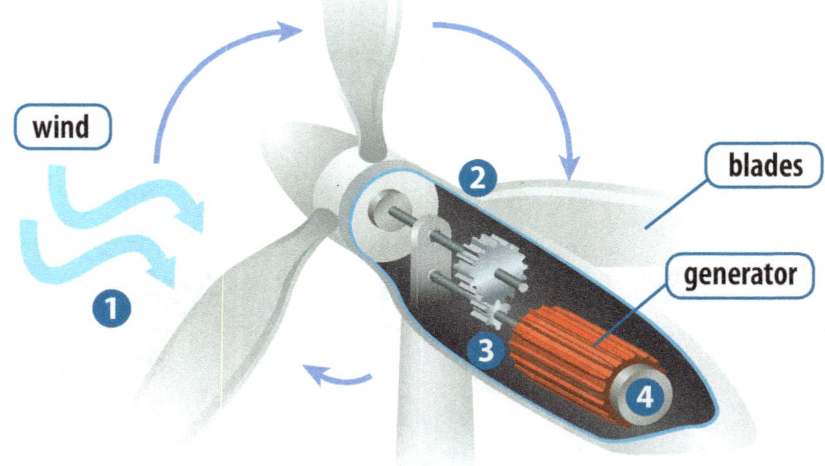

1. The wind turns the blades of the wind turbine.
2. The rotor attached to the shaft turns.
3. As the shaft turns it causes the gears in the gear box to turn.
4. The gears cause the generator to turn, producing electricity.

Wind farms are located on the top of smooth, rounded hills; on flat, open plains; along shorelines; or in mountain gaps, where wind is funneled through the pass. Ideal locations regularly have winds with a velocity of at least 23 km/h (14 m/h). Why would wind speeds that exceed this value be dangerous for the windmills? Why are winds of less velocity than this not desirable?

In 2012, Texas, Iowa, and California produced the most electricity by operating wind farms. In some countries, offshore wind turbines are being used. Offshore winds usually blow at higher speeds than winds on land. These turbines produce more electricity. In some places, people have set up small wind turbines in their own yards or on their farms to generate some of their own electricity. What are the advantages of wind turbines? Can you think of any concerns that wind turbines might raise?

To work efficiently, groups of wind turbines must be situated in areas where there are steady winds.

Do you think the area where you live would be a good place to generate electricity with wind turbines? Why or why not?

Scripture Spotlight

Compare **Psalm 104:3** and **Psalm 135:7**. What do they say about the wind?

Solar Energy

About 30% of the incoming solar radiation is reflected back into space, but the rest of the energy is absorbed by various materials on the planet's surface, including clouds, plants, soil, and water. The absorbed solar energy heats the surface of the planet.

Humans have harnessed sunlight since ancient times for light and warmth and are now using improved technological methods to capture, store, and use solar energy. In Chapter 13, you learned that solar cells produce electricity directly from sunlight. Solar cells are joined together to make **solar panels**. People often place the panels on the roofs of buildings to capture as much sunlight as possible. How might solar-powered traffic signs on roads be useful in an emergency?

Geothermal Energy

Thermal energy from within Earth is called **geothermal energy**. Wells are drilled into geothermally active areas to access very hot water. Geothermal heat is best mined along hot water springs, where water penetrates deep subterranean faults or cracks that reach miles into Earth's interior. Boiling water and steam come to the surface. They are used to turn turbines to generate electricity. Geothermal sources are primarily found in the western and southwestern United States and in British Columbia, Alberta, Northwest Territories, and Yukon in Canada.

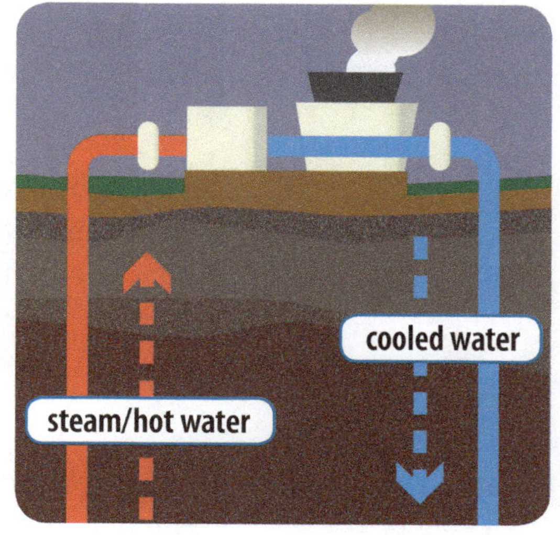

Power plants that use geothermal energy use steam or hot water heated by Earth to turn turbines to generate electricity. Then, the cooled water is pumped back below ground.

Which alternative energy source would be most practical where you live?

Tidal Energy

Studies show that waves and tidal currents have become viable renewable sources of electricity. Tides are more predictable than wind. New technologies are being developed to harness this clean, renewable energy source and efficiently add this electricity to the grid for consumer usage. These systems work on the principle of kinetic energy produced by the flowing water in the tides. Tidal energy has strong potential along coasts in Alaska and Hawaii in the United States. Significant opportunities for wave energy also exist along the East Coast. The most promising locations in Canada include the Bay of Fundy between New Brunswick and Nova Scotia or the Cumberland Basin, a site intersected by the New Brunswick-Nova Scotia border.

Measuring Electricity Explain

How would your life be different if you could not use electricity? Suppose you wanted to do homework after dark. You would have to use a candle or lantern. You could not watch television, listen to a radio, or use a computer. Think about other ways your life would change.

Electric Power

Electric power is the amount of electrical energy used during a certain amount of time. Another definition is the rate at which a device changes electrical energy into another form of energy. Power is measured in units called **watts** (W). What did you learn about watts in previous lessons? How does this relate to the discussion about energy in this chapter? How do you measure energy use in daily appliances? A **kilowatt** is equal to 1000 watts. What devices are plugged in to outlets in your classroom? How much power do they use?

Calculating Electric Energy Use

If you use your computer more during the summer vacation than you use it during the school year, how would this affect the cost to use your computer? You can use the formula below to estimate an appliance's energy use.

$$(\text{Wattage} \times \text{Hours used per day}) \div 1000 = \text{Daily Kilowatt-hours (kWh)}$$

Next, you multiply this by the number of days you use the appliance during the year. Then you can multiply the kWh by your local electric company's rate per kWh. This equation lets you calculate the annual cost to run an appliance.

Math in Science

Follow the steps above to calculate how much it costs to use a particular device for a year.

1. Suppose you have a desktop computer that you use to do homework. It is rated at 60 W. Your computer monitor is rated at 30 W. How many total watts does your computer system use?

2. If you use your computer and monitor for 4 hours each day, what is the computer's daily energy use?

3. If you use your computer every day, and your electric company's rate is $0.20 per kWh, how much does it cost to use your computer for a year?

Measuring Electrical Use

How much electrical power do your appliances use?

Procedure

1. Plug the watt meter into an outlet and **observe** the watt meter reading.

2. Plug one of the appliances into the watt meter. **Observe** the watt meter and find the setting that reports "power." **Record** the watt reading and the appliance in your *Science Journal*.

3. Explore different settings of the appliance while plugged into the watt meter. **Record** the watt readings.

4. Estimate how much time you use this appliance at home. **Record** the time in your *Science Journal*.

5. Repeat Steps 1–4 with each of the other appliances.

6. **Use numbers** from your data to calculate the energy consumed for each of the appliances using the equation Energy = Power × Time (kWh = watts/1000 × hours).

Materials
- watt meter
- 5 electrical appliances

Analyze Results

Interpret from your measurements and recorded observations. **Compare** and **infer** from the data you gathered the relationship between various appliances and the amount of electrical power they use.

Create Explanations

1. How much electrical power do your appliances use?

2. Which appliance did you expect to use the most power? Why?

3. Of the appliances tested, which one did you calculate to use the most energy? Is this what you expected? Why or why not?

4. Based on the measurements made in this activity, can you suggest a way that you can conserve energy in your home?

Conserving Electricity Explain

Conservation means saving or preserving. When people talk about conserving electricity, they mean ways of saving—or using less—electricity. You can help conserve electricity in many ways. Read the list of things you can do to save electricity. Think about some of things you are not doing now that you could do to help use less electricity. Conserving electricity saves money and reduces the strain on natural resources. Read through the suggestions on the next page for conserving energy. How much energy does your family use in one month? How can you calculate this? Name three things you can do to reduce your energy consumption.

Lesson Activity

(1) Make an energy audit of your home. First, brainstorm a checklist of things to look for with a partner. Enlist your family's help with the energy audit. Identify what appliances you and your family use each day. Review what you learned in the Structured Inquiry about the power requirements of various appliances. Find ways to make small changes that could reduce your family's overall energy usage.

(2) Next, put together a Conservation Action Plan. Make a list of everything you thought of that would help conserve electricity in your home. Think of ways to spread energy use throughout the day so that you use less energy at peak times.

(3) Compare your list with your partner's list. Add any ideas for conservation actions you did not identify that your partner did.

(4) Review the list of conservation ideas on the next page as well. Then, check on the Internet for additional ideas for ways to conserve electricity. Research how much electricity you can conserve by including each action in your daily life.

(5) Find out your local electric company's rate. Use the results to calculate how much money your family could save by conserving electricity.

How much money would your family save in a month?

> **What You Can Do to Conserve Electricity**
>
> 1. Turn off lights when no one is in the room.
> 2. During the day, open blinds or drapes instead of turning on lights.
> 3. Keep the curtains closed on cold, cloudy days to keep the warm air inside. Also, close them on very hot, sunny days. This helps keep the hot air outside.
> 4. If there is a light that every family member seems to forget to turn off, make a sign and put it next to the switch to remind everyone.
> 5. Replace incandescent bulbs with energy-efficient CFLs or LED lights.
> 6. Turn off televisions, computers, and other electrical devices when not in use.
> 7. Turn down the thermostat on your electric heating systems in cold weather, and turn up the thermostat on air-conditioning in warm weather.
> 8. Close doors and windows when heating or cooling a room.
> 9. Conserve water.

Check for Understanding

How will these measures help conserve the forms of energy you studied earlier?

We use energy resources to generate the electricity we need. Some of these resources are renewable. We can conserve our use of electricity and still have enough to be sufficient for our needs. What are you doing to conserve electricity use?

Concept Check Assess/Reflect

Summary: How is electricity used and conserved? Electricity can be generated from fossil fuels or from renewable sources. Renewable sources include hydroelectric energy, wind energy, solar energy, and geothermal energy. Electric power is the amount of electrical energy used in a certain amount of time. It is measured in watts. Power usage is measured in kilowatts (kWh). There are many small ways to conserve or use less electricity. Conserving electricity also saves money and reduces strain on environmental resources.

1. How is most electricity generated? What are some advantages and disadvantages of using alternative energy sources?
2. If a 150-W light bulb burns for 10 hours a day, what is the bulb's energy use in one day?
3. Describe how dams generate electricity.
4. Identify several ways people can reduce their overall consumption of electricity.

Get to Know
Michael Faraday

Michael Faraday, born in 1791, was the son of a blacksmith. He became an apprentice to a bookbinder when he left school at age 14. It was not unusual for a young man to begin working at that age.

After work, young Michael began reading some of the books he was learning to put together. His interest in science grew. When he was 21, he received tickets to attend a series of lectures given by a famous scientist. Several months after the lectures, he gave the man hundreds of pages of notes about what he had heard. That gesture led to his getting a job at the Royal Institute of Great Britain. He worked there for many years.

During that time, Faraday made it easier to use electricity. He developed the electric motor, which is used today in everything from hair dryers to car motors. He also discovered some chemical elements. Queen Victoria asked him to improve lighthouses so that ships at sea would be safer. Faraday headed the program to install electric lights in lighthouses. This practical application of his work with electromagnetism ensured a constant light for ships at sea. In gratitude for this work, which took nearly 30 years, Prince Albert gave Faraday a house.

Faraday continued sharing his love of science. In 1825, he started a series of Christmas lectures to inspire young people. He spoke at these for 19 years. He died in 1867. The Royal Institute still hosts these lectures every year and now broadcasts them on television.

Called to Serve

Michael Faraday was a faithful member of a branch of the Church of Scotland. He felt that his discoveries showed how God and nature were connected. He held firm in his beliefs as he continued his work.

Concept Check

1. What was unusual about Faraday's interest in and learning about science?
2. What practical application of his work with electromagnetic induction did Faraday investigate?

Energy-Efficient Lighting: Comparing LED Bulbs with CFLs

If you glance up at the ceiling of your classroom, you may see long tubes giving off light. Because they are more energy efficient, these fluorescent tubes are often used in public buildings. In recent years, compact fluorescent bulbs (CFLs) have been made for home use. Developing light-emitting diodes, or LEDs, became the next breakthrough after fluorescent bulbs. The LED sends light in a specific direction. LEDs also remain cool to the touch. They produce little heat and actually re-absorb the heat.

Which type of light bulb saves the most energy? The amount of energy a light bulb uses is measured in watts. A 15-watt CFL uses 75% less energy than a 60-watt incandescent bulb, and it gives about the same amount of light. A 12-watt LED bulb uses 75–80% less energy than an incandescent bulb that puts out the same amount of light.

In general, incandescent bulbs can last between 1000 and 2500 hours. A compact fluorescent light will last about 10,000 hours. If you left a lamp using a CFL burning, it would last 10,400 hours! An LED bulb lasts between 25,000 and 50,000 hours! How could you test these statements?

Both CFLs and LEDs cost more money than incandescent light bulbs. Because they last longer, however, they are a better bargain. Both kinds of lights can be used indoors and outdoors. Many companies are now making holiday lights using LED bulbs because they last so long.

Some people worry about the amount of mercury in CFLs. However, the amount of mercury used in CFLs has decreased over the years. The average CFL bulb now contains less mercury than a mercury thermometer does. The mercury escapes only if the bulb is broken. However, because mercury is highly toxic, most communities have laws regarding the disposal of CFL bulbs.

Concept Check

1. Why is an LED a good choice for a video recorder?
2. Which type of bulb is the best choice in terms of watts being used?

Study Guide

Lesson 1

1. Iron, nickel, and cobalt are magnetic materials. Some magnets are temporary and some are permanent.

2. A magnet has two ends called poles. All magnets are surrounded by a magnetic field. An iron or steel object becomes a temporary magnet when stroked with a magnet or placed in a magnetic field.

3. A magnet is made up of a collection of tiny magnetic domains. The domains of a magnet line up so they point in the same direction.

4. Earth is surrounded by a magnetic field. Compasses work because they align with Earth's magnetic field. The Earth's magnetic field deflects harmful radiation from the Sun.

Lesson 2

1. An electric current can be produced when a magnet moves inside a wire coil or if the coil moves instead of the magnet.

2. A wire carrying an electric current is surrounded by a magnetic field.

3. A coil of wire carrying current makes an iron object within the coil a temporary magnet, producing an electromagnet.

Lesson 3

1. The main parts of a generator are a magnet and a wire coil. A generator changes mechanical energy into electrical energy.

2. The main parts of an electric motor are a rotating wire coil, a power source, a conductor, and a permanent magnet. An electric motor changes electrical energy into mechanical energy.

Lesson 4

1. Electricity can be produced by different sources. Fossil fuels generate most electricity in the United States.

2. Renewable sources of electricity include moving water, solar cells, wind, and geothermal energy.

3. Electric power is measured in watts. The watt rating and the amount of time a device is used determine the amount of electrical energy used.

4. There are many ways to conserve electrical energy, such as using energy-efficient light bulbs, lowering thermostats, and turning off lights when leaving a room.

Show What You Know

Visualize It Complete the following cause-and-effect organizer.

Cause	Effect
The north and south poles of two magnets are placed near each other.	**1.**
	2.
A bar magnet moves in a wire coil.	The electromagnet drops the paperclips clinging to it.
3.	
An iron nail stroked 20 times with a strong magnet.	**4.**

Tell how each pair of terms is related.

5. permanent magnet—temporary magnet
6. generator—electric motor
7. magnetosphere—magnetic field
8. electromagnet—coil
9. magnetic pole—magnetic domains
10. electric power—watt

Multiple Choice

Choose the best answer.

11. What are the glowing lights seen in the sky near Earth's poles called?
    A. solar wind   C. auroras
    B. air molecules   D. magnetic fields

12. Which energy source produces electricity directly from sunlight?
    A. geothermal plants
    B. coal-powered plants
    C. hydroelectric dams
    D. solar panels

13. Which is not an effective electrical conservation strategy?
    A. turn down thermostat when running air conditioners
    B. turn off lights when you leave a room
    C. replace incandescent bulbs with CFLs
    D. turn off computers at the power strip

14. If a laptop computer is rated at 90 W and is used for 7 hours each day for 2 days, what was the computer's daily energy use?
    A. 99 watts   C. 104 watts
    B. 630 watts   D. 1260 watts

15. Which substance is nonmagnetic?
    A. iron   C. nickel
    B. silver   D. cobalt

Check Point

Answer the following questions.

16. Could an electromagnet on a crane be used to move scrap aluminum into a recycling truck? Explain.

17. **Infer** why a magnet attracts a nail to either of its poles, but attracts another magnet to only one of its poles.

18. **Use numbers** to calculate the electrical energy conserved each day by replacing two 100-W incandescent bulbs with 23-W CFLs if the bulbs are each used for 8 hours a day.

19. Examine the device in the picture. (a) How does the domain model explain how it works? (b) What are two ways to increase the strength of the magnetic field it produces?

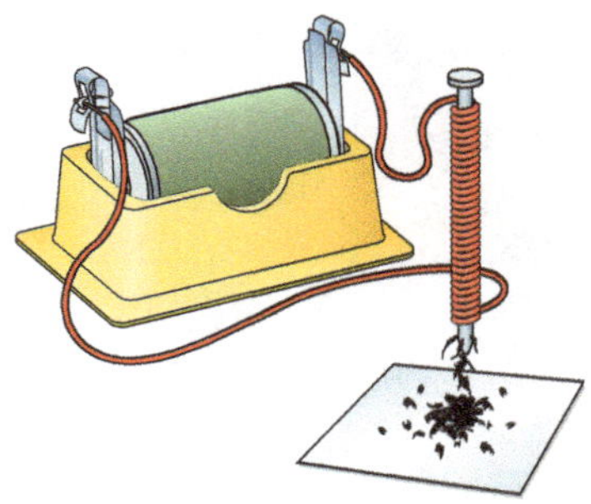

20. **Compare** the way a geothermal plant and a hydroelectric dam produce electricity.

21. A student wants to use a copper rod as the core for an electromagnet. **Predict** what will happen when she tries to use it. Why?

22. If you break a bar magnet in half, what magnetic properties do the halves show?

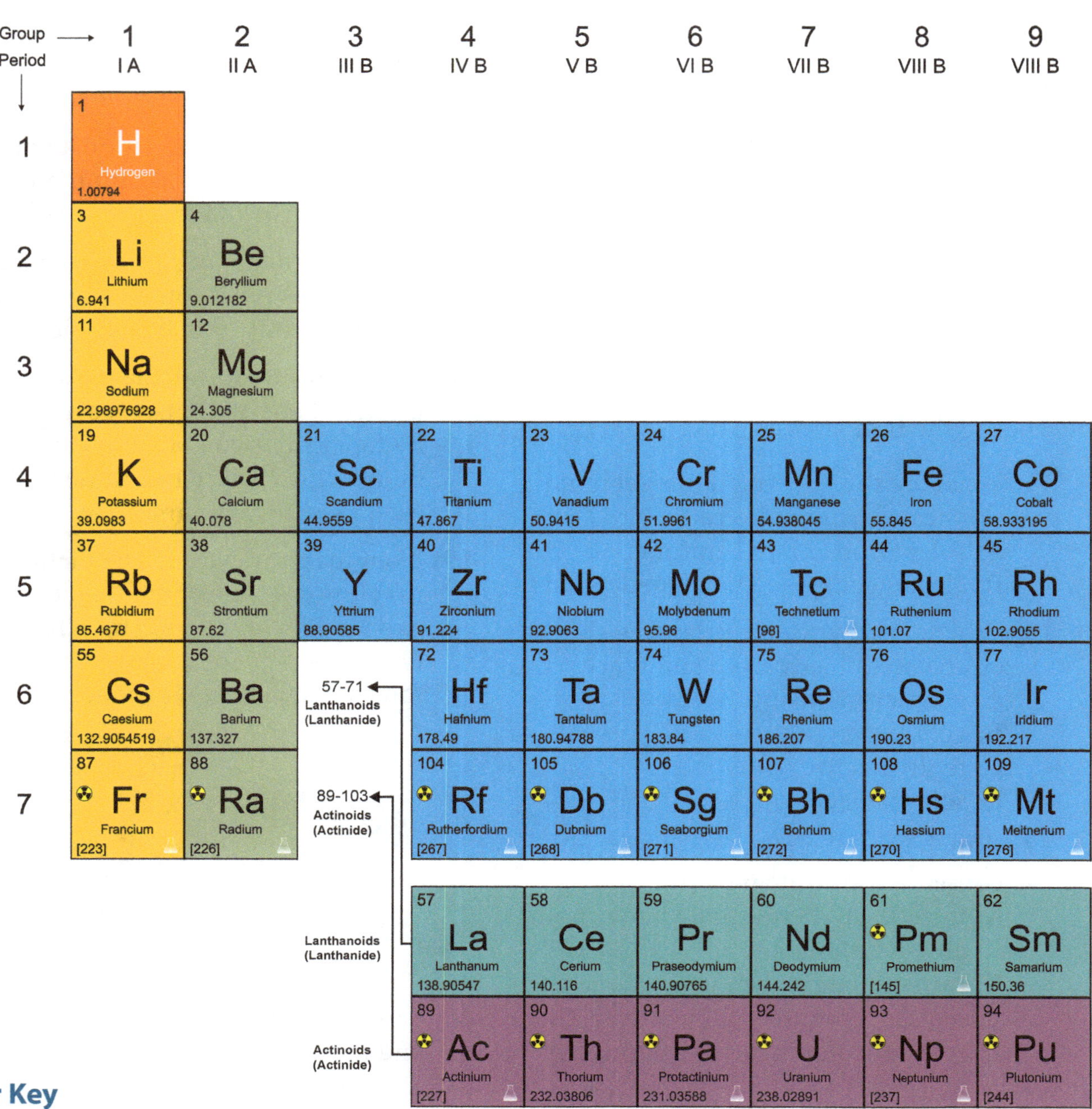

Group
Period
1 I A
2 II A
3 III B
4 IV B
5 V B
6 VI B
7 VII B
8 VIII B
9 VIII B
1 H Hydrogen 1.00794
3 Li Lithium 6.941
4 Be Beryllium 9.012182
11 Na Sodium 22.98976928
12 Mg Magnesium 24.305
19 K Potassium 39.0983
20 Ca Calcium 40.078
21 Sc Scandium 44.9559
22 Ti Titanium 47.867
23 V Vanadium 50.9415
24 Cr Chromium 51.9961
25 Mn Manganese 54.938045
26 Fe Iron 55.845
27 Co Cobalt 58.933195
37 Rb Rubidium 85.4678
38 Sr Strontium 87.62
39 Y Yttrium 88.90585
40 Zr Zirconium 91.224
41 Nb Niobium 92.9063
42 Mo Molybdenum 95.96
43 Tc Technetium [98]
44 Ru Ruthenium 101.07
45 Rh Rhodium 102.9055
55 Cs Caesium 132.9054519
56 Ba Barium 137.327
57-71 Lanthanoids (Lanthanide)
72 Hf Hafnium 178.49
73 Ta Tantalum 180.94788
74 W Tungsten 183.84
75 Re Rhenium 186.207
76 Os Osmium 190.23
77 Ir Iridium 192.217
87 Fr Francium [223]
88 Ra Radium [226]
89-103 Actinoids (Actinide)
104 Rf Rutherfordium [267]
105 Db Dubnium [268]
106 Sg Seaborgium [271]
107 Bh Bohrium [272]
108 Hs Hassium [270]
109 Mt Meitnerium [276]
Lanthanoids (Lanthanide)
57 La Lanthanum 138.90547
58 Ce Cerium 140.116
59 Pr Praseodymium 140.90765
60 Nd Deodymium 144.242
61 Pm Promethium [145]
62 Sm Samarium 150.36
Actinoids (Actinide)
89 Ac Actinium [227]
90 Th Thorium 232.03806
91 Pa Protactinium 231.03588
92 U Uranium 238.02891
93 Np Neptunium [237]
94 Pu Plutonium [244]

Alkali metals
Alkaline earth metals
Transition metals
Post-transition metals
Metaloids
Other nonmetals
Halogens
Noble gases
Lanthanide
Actinide
Radioactive element
Synthetic element
H Gas
Hg Liquid
Li Solid

Color Key

10 VIII B	11 I B	12 II B	13 III A	14 IV A	15 V A	16 VI A	17 VII A	18 VIII A
								2 He Helium 4.002602
			5 B Boron 10.811	6 C Carbon 12.0107	7 N Nitrogen 14.0067	8 O Oxygen 15.9994	9 F Fluorine 18.9984032	10 Ne Neon 20.1797
			13 Al Aluminium 26.9815386	14 Si Silicon 28.0855	15 P Phosphorus 30.973762	16 S Sulfur 32.065	17 Cl Chlorine 35.453	18 Ar Argon 39.948
28 Ni Nickel 58.6934	29 Cu Copper 63.546	30 Zn Zinc 65.38	31 Ga Gallium 69.729	32 Ge Germanium 72.64	33 As Arsenic 74.9216	34 Se Selenium 78.96	35 Br Bromine 79.904	36 Kr Krypton 83.798
46 Pd Palladium 106.42	47 Ag Silver 107.8682	48 Cd Cadmium 112.411	49 In Indium 114.818	50 Sn Tin 118.71	51 Sb Antimony 121.76	52 Te Tellurium 127.6	53 I Iodine 126.90447	54 Xe Xenon 131.293
78 Pt Platinum 195.084	79 Au Gold 196.966569	80 Hg Mercury 200.59	81 Tl Thallium 204.3833	82 Pb Lead 207.2	83 Bi Bismuth 208.9804	84 Po Polonium [209]	85 At Astatine [210]	86 Rn Radon [222]
110 Ds Darmstadtium [281]	111 Rg Roentgenium [280]	112 Cn Copernicium [285]	113 Uut Ununtrium [286]	114 Fl Flerovium [289]	115 Uup Ununpentium [288]	116 Lv Livermorium [293]	117 Uus Ununseptium [294]	118 Uuo Ununoctium [294]

63 Eu Europium 151.964	64 Gd Gadolinium 157.25	65 Tb Terbium 158.9253	66 Dy Dysprosium 162.5	67 Ho Holmium 164.93032	68 Er Erbium 167.259	69 Tm Thulium 168.93421	70 Yb Ytterbium 173.054	71 Lu Lutetium 174.9668
95 Am Americium [243]	96 Cm Curium [247]	97 Bk Berkelium [247]	98 Cf Californium [251]	99 Es Einsteinium [252]	100 Fm Fermium [257]	101 Md Mendelevium [258]	102 No Nobelium [262]	103 Lr Lawrencium [262]

Minerals

Mineral (formula)	Color	Streak	Hardness	Breakage Pattern	Uses and Other Properties
Graphite (C)	black to gray	black to gray	1–1.5	basal cleavage (scales)	pencil lead, lubricants for locks, rods to control some smaller nuclear reactions, battery poles
Galena (PbS)	gray	gray to black	2.5	cubic cleavage perfect	source of lead, used for pipes, shields for X rays, fishing equipment sinkers
Hematite (Fe_2O_3)	black or reddish-brown	reddish-brown	5.5–6.5	irregular fracture	source of iron; converted to pig iron, made into steel
Magnetite (Fe_3O_4)	black	black	6	conchoidal fracture	source of iron, attracts a magnet
Pyrite (FeS_2)	light, brassy, yellow	greenish-black	6–6.5	uneven fracture	fool's gold
Talc ($Mg_3Si_4O_{10}(OH)_2$)	white, greenish	white	1	cleavage in one direction	used for talcum powder, sculptures, paper, and tabletops
Gypsum ($CaSO_4 \cdot 2H_2O$)	colorless, gray, white, brown	white	2	basal cleavage	used in plaster of paris and dry wall for building construction
Sphalerite (ZnS)	brown, reddish-brown, greenish	light to dark brown	3.5—4	cleavage in six directions	main ore of zinc; used in paints, dyes, and medicine
Muscovite ($KAl_3Si_3O_{10}(OH)_2$)	white, light gray, yellow, rose, green	colorless	2–2.5	basal cleavage	occurs in large, flexible plates; used as an insulator in electrical equipment, lubricant
Biotite ($K(Mg,Fe)_3(AlSi_3O_{10})(OH)_2$)	black to dark brown	colorless	2.5–3	basal cleavage	occurs in large, flexible plates
Halite (NaCl)	colorless, red, white, blue	colorless	2.5	cubic cleavage	salt; soluble in water; a preservative

Minerals

Minerals					
Mineral (formula)	**Color**	**Streak**	**Hardness**	**Breakage Pattern**	**Uses and Other Properties**
Calcite ($CaCO_3$)	colorless, white, pale blue	colorless, white	3	cleavage in three directions	fizzes when HCl is added; used in cements and other building materials
Dolomite ($CaMg(CO_3)_2$)	colorless, white, pink, green, gray, black	white	3.5–4	cleavage in three directions	concrete and cement; used as an ornamental building store
Fluorite (CaF_2)	colorless, white, blue, green, red, yellow, purple	colorless	4	cleavage in four directions	used in the manufacture of optical equipment; glows under ultraviolet light
Hornblende ($(CaNa)_{2-3}(Mg,Al,Fe)_5$ $(Al,Si)_2Si_6O_{22}(OH)_2$)	green to black	gray to white	5–6	cleavage in two directions	will transmit light on thin edges; 6-sided cross section
Feldspar ($(KAlSi_3O_8)(NaAlSi_3O_8)$, $(CaAl_2Si_2O_8)$)	colorless, white to gray, green	colorless	6	two cleavage planes meet at 90° angle	used in the manufacture of ceramics
Augite ($(Ca,Na)(Mg,Fe,Al)$ $(Al,Si)_2O_6$)	black	colorless	6	cleavage in two directions	square or 8-sided cross section
Olivine ($(Mg,Fe)_2SiO_4$)	olive, green	none	6.5–7	conchoidal fracture	gemstones, refractory sand
Quartz (SiO_2)	colorless, various colors	none	7	conchoidal fracture	used in glass manufacture, electronic equipment, radios, computers, watches, gemstones

Rocks

Rocks		
Rock Type	**Rock Name**	**Characteristics**
Igneous (intrusive)	Granite	Large mineral gains of quartz, feldspar, hornblende, and mica. Usually light in color.
	Diorite	Large mineral grains of feldspar, hornblende, and mica. Less quartz than granite. Intermediate in color.
	Gabbro	Large mineral gains of feldspar, augite, and olivine. No quartz. Dark in color.
Igneous (extrusive)	Rhyolite	Small mineral grains of quartz, feldspar, hornblende, and mica, or no visible drains. Light in color.
	Andesite	Small mineral grains of feldspar, hornblende, and mica or no visible grains. Intermediate in color.
	Basalt	Small mineral grains of feldspar, augite, and possibly olivine or no visible grains. No quartz. Dark in color.
	Obsidian	Glassy texture. No visible grains. Volcanic glass. Fracture looks like broken glass.
	Pumice	Frothy texture. Floats in water. Usually light in color.
Sedimentary (detrital)	Conglomerate	Coarse grained. Gravel or pebble-size grains.
	Sandstone	Sand-sized grains 1/16 to 2 mm.
	Siltstone	Grains are smaller than sand but larger than clay.
	Shale	Smallest grains. Often dark in color. Usually platy.
Sedimentary (chemical or organic)	Limestone	Major mineral is calcite. Usually forms in oceans and lakes. Often contains fossils.
	Coal	Forms in swampy areas. Compacted layers of organic material, mainly plant remains.
Sedimentary (chemical)	Rock Salt	Commonly forms by the evaporation of seawater.
Metamorphic (foliated)	Gneiss	Banding due to alternate layers of different minerals, of different colors. Parent rock often is granite.
	Schist	Parallel arrangement of sheetlike minerals, mainly micas. Forms from different parent rocks.
	Phyllite	Shiny or silky appearance. May look wrinkled. Common parent rocks are shale and slate.
	Slate	Harder, denser, and shinier than shale. Common parent rock is shale.
Metamorphic (nonfoliated)	Marble	Calcite or dolomite. Common parent rock is limestone.
	Soapstone	Mainly of talc. Soft with greasy feel.
	Quartzite	Hard with interlocking quartz crystals. Common parent rock is sandstone.

Using the Glossary

The Glossary is like a small dictionary. It lists important science terms from your book. The words and terms are listed in alphabetical order. Each word is followed by its meaning. Some words have more than one meaning. This glossary gives the scientific meaning. That is how the word is used in this book. After the meaning is a page number that tells on what page the word is first used.

abiogenesis The idea suggested by Alexander Oparin that living cells arose gradually from nonliving matter millions of years ago. (p. 51)

acquired immunity The protection by which cells "remember" foreign substances so they can respond faster if exposed to them again. (p. 106)

active transport Type of cell transport mechanism that requires energy to function. (p. 54)

ADP (adenosine diphosphate) A special energy molecule that changes light energy into chemical energy. (p. 73)

allergic reaction A reaction by the body to substances such as dust, pollen, food products, or bee stings. An allergic reaction causes immune cells to produce swelling, rashes, hives, sneezing, and itching. (p. 106)

alloy Solid solution made of two or more metals. Bronze is an alloy, a mixture of copper and tin. (p. 430)

alluvial fans Deposits shaped like fans made from rivers that flow down mountain sides and deposit sediment when the water reaches a broad, flat area. (p. 308)

alpine glaciers Glaciers that form high in mountain valleys. (p. 297)

alternating current A current that changes direction because it is produced at a power station using a rotating current generator. (p. 446)

alveoli Small air sacs that are at the end of each bronchiole. (p. 190)

amino acids Organic compounds containing oxygen, hydrogen, and nitrogen that combine to form larger protein molecules. (p. 94)

amperage The rate at which electrons flow through a wire in a given period. (p. 446)

ampere (amp) The unit for measuring the flow of electric current. (p. 446)

anticodon Three bases that pair with a codon using base-pair matching, such as CGU and GCA. (p. 96)

arête A sharp, winding ridge that forms as alpine glaciers erode a cirque. (p. 311)

asthenosphere A partially molten layer of the mantle that lies 60–120 miles below the lithosphere. (p. 265)

atomic mass The total mass of all of the protons and neutrons in the nucleus. (p. 411)

atomic number The number of protons in an element's nucleus. (p. 413)

atoms The tiny particles that make up all matter. Atoms are the basis of atomic structure. (p. 400)

ATP (adenosine triphosphate) A type of high-energy molecule that powers many cell activities. (p. 73)

atrium An upper chamber of the heart. (p. 182)

auroras Different colors of lights in the night sky caused by molecules returning to a natural state and giving off their extra energy. (p. 487)

axons Fibers that carry impulses away from the cell body to other neurons. (p. 139)

batteries A portable source of electrical energy stored in chemical form containing two or more cells. (p. 452)

bedrock Unweathered rock. (p. 302)

binary code A series of 1s and 0s. (p. 468)

biome The largest ecosystems. An ecosystem and geographical community adapted to the temperature and precipitation of an area. (p. 27)

biosphere The part of the Earth in which organisms live. It includes the land, the water, and the atmosphere. (p. 26)

bird-hipped dinosaurs (Ornithischians) Dinosaurs that had a hip structure similar to that of birds. Most Ornithischians ate plants and had bony plates on their skin. (p. 348)

boiling point The temperature at which boiling occurs in a liquid. (p. 375)

brain stem The part of the brain that carries information from the spinal cord and the brain and controls breathing and heartbeat rates, swallowing, and coughing. (p. 142)

C

caldera A large depression formed when a crater and the land around it collapse after an eruption. (p. 283)

capillaries The smallest blood vessels. (p. 185)

carbon cycle A process in which carbon is exchanged between organisms and the environment. (p. 34)

cardiopulmonary resuscitation (CPR) An emergency procedure that involves chest compressions and rescue breathing to keep oxygen and blood flowing in the body. (p. 217)

carrying capacity The largest number of one species that an ecosystem can support without harmful effects. (p. 40)

cartilage Connective tissue that forms part of the skeleton. It is found at the ends of bones and helps cushion them and reduce friction. (p. 167)

catalyst Something that increases the rate of a chemical reaction. (p. 392)

catastrophe A sudden, short-lived, violent event such as a flood, a volcanic eruption, or an earthquake. (pp. 268, 333)

catastrophism A geologic theory that sudden, violent events caused dramatic changes on Earth. (p. 268)

cave An underground hole sometimes formed when groundwater dissolves rock. (p. 308)

cell body The part of the cell that contains the nucleus. (p. 139)

cell cycle The set of events through which a cell moves from one cell division to the next in the life of a cell. (p. 100)

cell membrane The thin "skin" that surrounds and protects the contents of a cell. (p. 53)

cell respiration The process that occurs in the cytoplasm and the mitochondria of cells by which the cell uses oxygen to breakdown sugar into water and carbon dioxide. (p. 75)

cell wall Made of cellulose or silica, it supports the cells, providing structure to help plants hold their shape. (p. 53)

Cenozoic era Refers to "recent life." Fossils from this era include many land animals and birds, such as saber-toothed cats, horses, woolly mammoths, and sloths. (p. 327)

central nervous system Consists of the brain and the spinal cord. (p. 142)

cerebellum (or hindbrain) Controls muscle coordination, balance, and muscle tone. (p. 142)

cerebrum (or forebrain) Largest part of the brain that controls thinking and voluntary movements of the skeletal muscles. (p. 142)

chemical bond Formed when two atoms either share or exchange electrons in a way that links them together. (p. 418)

chemical change The type of change when matter forms new substances with different physical and chemical properties from the original matter. (p. 382)

chemical equations Equations used to describe chemical reactions. (p. 385)

chemical property The behavior of a substance when it undergoes a chemical change or reaction. (p. 380)

chemical reaction When two or more kinds of matter interact. (p. 384)

chemical sedimentary rocks A product of chemical interactions at the atomic level. Form when water evaporates or when water is supersaturated with dissolved ions (i.e., limestone). (p. 241)

chemical symbol A representation of the name of an element using either one capital letter or a combination of one capital and one lowercase letter. (p. 412)

chemical weathering The breakdown of rocks by chemical means. (p. 293)

chlorophyll The green pigment that gives leaves their green color. It traps light during photosynthesis. (p. 72)

chloroplasts Green, oval-shaped organelles in which photosynthesis takes place. (p. 66)

chromosome A threadlike structure of nucleic acids and protein found in the nucleus of most living cells, carrying genetic information in the form of genes. (p. 87)

cinder cone volcano A small, cone-shaped mountain with steep flanks and bowl-shaped crater at the summit that erupts volcanic ash and cinders. (p. 284)

circuit breakers Special switches that are designed to open when too much current passes through them. (p. 467)

circulatory system The group of organs that transports materials from one place in the body to another. It consists of blood, the heart, and a network of blood vessels. (p. 179)

cirque A bowl-shaped landform that forms as an alpine glacier erodes the rocks at the top of its valley. (p. 311)

clastic sedimentary rock A sedimentary rock composed of pieces of other rocks that range in size from clay to boulders; they are named by the size and shape of the clast. (p. 241)

cleavage Minerals that break along smooth planes. (p. 235)

climax community A relatively stable community, reached through ecological succession. (p. 31)

closed circuit An electric circuit in which electrons can flow from one pole of the power source to the other in a continuous path. (p. 458)

cochlea The bony, coiled tube in the ear that is filled with fluid and lined with hair cells. (p. 153)

codon Each of the three bases in RNA. (p. 96)

colloid A liquid mixture containing small particles that do not separate out. (p. 425)

commensalism A relationship between two living things of different species in which one benefits while the other is neither helped nor harmed. (p. 30)

commutator A reversing switch in a motor made to run on direct current. (p. 501)

composite volcano A large mountain with steep sloping flanks that erupts thick (silica-rich) lava, volcanic ash, and cinder. The eruptions of a composite volcano are highly explosive. (p. 284)

compound A molecule that contains at least two different elements. (p. 418)

compression force A pushing force that produces a reverse fault. (p. 279)

concussion A serious brain injury that temporarily changes the way that the brain works. (p. 217)

conductivity A material's ability to allow electrons to flow. (p. 448)

conductor A material that allows electrons to move freely from atom to atom in a continuous path. (p. 448)

connective tissues Connect one part of the body with another and separate one group of cells from another. (p. 119)

conservation The wise use of a natural resource. (p. 255)

continental drift The concept that Earth's continents have slowly drifted by moving across the ocean floor. (p. 270)

continental glaciers Glaciers that form over larger areas of continents. (p. 297)

convergent boundary A type of plate boundary in which the plates come together. (p. 274)

core The inner layer of Earth, composed of a solid inner core and a liquid outer core. (p. 263)

cornea Cornea is the clear covering that protects the eye. It helps to focus light. (p. 150)

corrosion The weakening of metals caused by chemical action. (p. 380)

covalent bond The chemical bond formed when two atoms share electrons. (p. 422)

creep A type of mass wasting in which materials move down a slope over a long time. (p. 315)

crust The surface of Earth, composed of the elements oxygen, silicon, and aluminum. (p. 263)

current electricity The flow of electrons. (p. 445)

cycle A series of events regularly repeated in the same order. (p. 34)

cytoplasm A semi-fluid material that makes up the inside of cells. (p. 53)

D

daughter cells The two new cells that result from cell division. (p. 100)

deductive reasoning The process that begins with general statements to develop a specific conclusion. (p. 147)

delta A fan-shaped deposit formed as rivers or streams empty into larger bodies of water. (p. 308)

dendrites Fibers that carry impulses from other neurons toward the cell body. (p. 139)

denitrifying bacteria A type of bacteria that changes nitrites and nitrates to nitrogen gas, which is returned to the atmosphere. (p. 37)

density A measure of mass per unit of volume, or how closely the particles in an object are packed together. (p. 368)

deposition The dropping of loose materials. (p. 299)

dermis The skin's thick inner layer of loose and dense connective tissue, held together by collagen. (p. 134)

dietitian A person who plans food and nutrition programs, supervises meal preparation, and manages the serving of meals. (p. 204)

diffusion The movement of particles of liquids, gases, or solids from a region of high concentration to one of lower concentration. (p. 54)

diode A semiconductor device with two electrodes that only lets current flow in one direction. (p. 469)

direct current The current used in batteries, which flows in one direction. (p. 446)

displacement The volume of fluid that is pushed aside by an object. (p. 366)

dissolves When a solid spreads and mixes evenly into a liquid. (p. 371)

divergent boundaries A type of plate boundary at which plates separate. (p. 274)

double helix The twisted ladder of a DNA molecule. It is made of two strands of DNA twisted around each other. (p. 88)

drumlins Long, cigar-shaped hills that are usually tens of meters high and no more than 1 kilometer (.6 mile) in length. (p. 312)

dry cell A battery that uses a combination of thick, paste-like chemical for the electrolyte. (p. 455)

ductile A metal's ability to be stretched into a wire. (p. 247)

dunes Sand and silt dropped by the wind that build up to form mounds of sediment. (p. 313)

ecological succession The gradual replacement of one type of biological community by another through a series of changes. (p. 30)

ecologists Scientists who study the organization within the biosphere, including biomes, ecosystems, communities, and populations. (p. 26)

ecology the scientific study of relationships between organisms and their environment. (p. 26)

electric motor A device that uses an electromagnet to change electrical energy into mechanical energy. (p. 499)

electric power The amount of electrical energy used during a certain amount of time. (p. 506)

electric power grid A system connecting power plants across the country. (p. 498)

electrodes Terminals that conduct electricity. (p. 452)

electrolyte A solid or a liquid solution that contains ions from chemical reactions occurring at the electrodes in a battery. (p. 452)

electrolytic cell A device that converts chemical energy to electrical energy. The working part of a battery. (p. 452)

electromagnet A magnet created by an electric current flowing through a coil of wire wrapped around a solid iron core. (p. 492)

electron The cloud of negatively charged particles surrounding the nucleus of every atom. (p. 403)

electronics Devices, such as televisions, radios, and computers, that operate using small electrical parts and the flow of electrons. (p. 468)

electrostatic discharge The sudden release of static electricity between two differently charged objects. (p. 441)

element The simplest form of matter that cannot be broken down into anything simpler without losing its chemical properties. (p. 401)

emergency A situation outside of your control that requires an immediate response, such as calling 911 for assistance. (p. 214)

emergency medical technician (EMT) A health-care professional trained to treat and transport emergency victims. (p. 208)

emulsion A type of mixture that holds two or more liquids, such as mayonnaise and many salad dressings. (p. 425)

endocrine system Helps control body functions through the secretion of hormones. (p. 126)

endoplasmic reticulum (ER) The largest organelle, a network of tube-like channels that winds through the cytoplasm. (p. 63)

endothermic reactions Chemical reactions that absorb energy. (p. 390)

eon The largest division of geologic time. There are two eons, the Precambrian and Phanerozoic. (p. 328)

epidermis The skin's thin outer layer, consisting of five layers of cells. (p. 133)

epithelial tissue Protects all internal and external body surfaces. (p. 116)

era A division of geologic time. There are three eras—Paleozoic, Mesozoic, and Cenozoic. (p. 328)

erosion The movement of weathered material. (p. 296)

eskers Ridges that form within channels of moving water in a glacier. (p. 312)

eukaryotic cell A cell that does have a nucleus and membrane bound organelles. Examples of eukaryotes are plants, animals, fungi, and protists. (p. 56)

exothermic reactions Chemical reactions that release energy. (p. 390)

facilitated diffusion A type of passive transport requiring no energy in which proteins aid in the passing of a substance through the cell membrane. (p. 54)

fermentation The process that allows cells to get energy without the use of oxygen. Fermentation is a common process in some bacteria, yeast, plants, and muscle cells. (p. 75)

flammability The ability to burn or ignite. (p. 380)

floodplain A large, flat area on either side of a river's channel. (p. 307)

follicles Tiny tubes in the skin where hairs grow out of your body. (p. 134)

foot wall The mass of rock lying below a non-vertical fault. (p. 279)

fossils The naturally preserved remains or traces of ancient life that lived in the geologic past. (p. 324)

fracture A break in a bone. (p. 221) Also, minerals that break along rough, uneven surfaces. (p. 235)

frostbite A condition brought on by extreme cold weather conditions that can seriously damage the skin and its underlying tissues. (p. 220)

fuse A device that limits the amount of electrical current flowing through a circuit. (p. 467)

gem A mineral that is valued for its beauty and extensively used in the jewelry industry. (p. 250)

generator A machine that turns mechanical energy into electrical energy. (p. 495)

genes The small sections of DNA that are the basic units of hereditary information. (p. 87)

geologic column A layered sequence of fossils found in rock layers that make up the Earth's crust. (p. 322)

geothermal energy Energy from within Earth. (p. 505)

glands An organ in the body of humans, and other animals, that secretes chemical substances needed by the body. (p. 103)

glucose A simple sugar that is a by-product of photosynthesis. (p. 72)

glycolysis A process that breaks down glucose molecules. (p. 75)

Golgi apparatus The organelle that packages the materials for the endoplasmic reticulum to transport. The Golgi apparatus is made up of a stack of 3 to 20 slightly curved sac-like membranes that float in the cytoplasm. (p. 65)

gradual extinction The disappearance of only one or a few species, usually caused by changes in the species' environment. (p. 342)

group A column in the periodic table of the elements. The elements in a group have similar chemical properties. (p. 413)

hanging wall The mass of rock lying above a non-vertical fault. (p. 279)

heat exhaustion A condition in which a person's body gets too hot. (p. 219)

heatstroke An emergency condition in which a person's body is too hot. (p. 219)

hemoglobin An iron-rich molecule in red blood cells. (p. 180)

historical science Science in which data is provided primarily from past events for which there is no direct experimental data. (p. 332)

homeostasis The ability of an organism to keep its internal environment stable in response to environmental changes. (p. 21)

hormone A chemical substance released by glands. (pp. 103, 126)

horn A formation where three or more adjacent valley glaciers erode adjacent cirques, producing a pyramid-shaped peak. (p. 311)

humus Decayed organic matter that makes up the top layer of soil. (p. 301)

hydroelectricity Electricity that is produced by moving water. Often produced by dams that are built on or near large rivers. (p. 503)

hypothermia A condition in which a person's body loses heat faster than it produces it, causing the person's body temperature to drop. (p. 219)

igneous rocks Formed when molten rocks cools and hardens. (p. 238)

immune system A network of organs that work together to defend the body against disease and infection. (p. 128)

index fossils Preserved remains of animals or plants that occur consistently in the same layers over wide areas. (p. 327)

induction The process of getting electrons to move through a wire by changing the magnetic field. (p. 488)

inductive reasoning Reasoning from a specific case or cases to make a general rule. (p. 147)

inherited traits Characteristics such as height and hair color that are passed from parents to offspring. (p. 87)

insulator A material that resists or blocks the flow of electrons. (p. 448)

integumentary system A group of organs that form a protective outer layer. It is made of the skin, hair, fingernails, and sweat glands. (p. 131)

interdependence The interaction of living things as they rely on each other to live successfully. (p. 21)

ion An atom that gains or loses an electron. (p. 421)

ionic bond The connection formed from the attraction of oppositely charged ions. (p. 421)

iris The colored ring surrounding the pupil that absorbs light and stops reflection. (p. 150)

joint The place where bones meet. Some joints are moveable while others are not moveable. (p. 168)

kilowatt Measurement equal to 1000 watts. (p. 506)

law of conservation of mass The law that states that whenever a chemical change occurs, the total mass of substances in a chemical reaction remains the same throughout the change. (p. 382)

lens The part of the eye that sits behind the pupil and focuses images in the eye. (p. 150)

leukocytes White blood cells that are the parts of the blood that defend the body against disease. (pp. 106, 128)

ligaments Strong bands of connective tissue that hold the bones in the movable joints together. (p. 168)

lightning A natural electrostatic discharge that occurs because of a difference in charge between a cloud and the ground. (p. 442)

limiting factors Things such as disease, parasites, floods, and predation that keep a population from growing any larger. (p. 40)

lithification The way in which sediments become cemented together to form rock layers. (p. 322)

lithosphere The upper layer of Earth's mantle including the crust. (p. 265)

lizard-hipped dinosaurs (Saurischian) The hip is composed of three elements: the ilium, ischium, and pubis with the pubis pointing downward and forward at an angle to the ischium. (p. 349)

load The part of the electric pathway in an electric circuit that uses current, such as a bulb or motor. (p. 458)

luster A property that describes the way a mineral reflects light. (p. 234)

L-wave (surface wave) The last wave to reach a seismic station that travels across the surface of Earth and causes the crust to move side to side and up and down. It causes the damage at Earth's surface. (p. 279)

lymph Clear fluid that surrounds all body cells and saturates them with water and nutrients. (p. 129)

lymph nodes Tiny organs shaped like beans that are found in the neck, armpits, and groin. These nodes filter the lymph and trap foreign materials. (p. 129)

lymphatic system The network of tiny tubes and organs found throughout the body that collects fluid from tissues and returns it to the blood. (p. 129)

lymphocytes The white blood cells that fight harmful bacteria and viruses by releasing antibodies. (p. 130)

lysosomes Specialized vesicles that float freely in the cytoplasm and are responsible for breaking down and digesting materials such as food and worn-out cell parts. (p. 67)

magnetic domains The tiny magnetic regions in magnetic material that must be aligned in order for the material to be magnetic. (p. 483)

magnetic field An invisible area of magnetism all around a magnet. (p. 484)

magnetic poles The places on a magnet where magnetism is the strongest. (p. 481)

magnetosphere A region of Earth's magnetic field that extends 58,000 km (about 36,000 mi) into space. (p. 486)

malleable A metal's ability to be flattened by pounding with a hammer. (p. 247)

mantle The thickest layer of the Earth. It lies between the crust and the core and is made of solid rock that is rich in oxygen, silicone, and magnesium. (p. 263)

marrow The soft fatty substance that fills the cavities of bones and where red and white blood cells and platelets are formed. (p. 166)

mass The amount of matter in an object. (p. 365)

mass extinction The sudden disappearance of large numbers of species, usually caused by major changes on Earth. (p. 342)

mass wasting The downhill movement of rocks and loose material under the influence of gravity. (p. 314)

mechanical weathering The breakdown of rock into smaller pieces by mechanical means, such as freezing and thawing, animal activity, and plant activity. (p. 292)

megafauna A group of very large birds, reptiles, amphibians, and mammals that are now extinct. (p. 350)

melanin A skin pigment. (p. 134)

melting point The temperature at which a solid melts. (p. 373)

Mesozoic era Refers to the middle rock layer. Fossils from this era contain a mixture of marine animals, land animals, and some birds. The most well-known animals in these layers are the dinosaurs. (p. 327)

metalloids One of three groups of elements on the periodic table. Metalloids are elements that transition from metals to nonmetals. They act as conductors and are economically important in the semiconductor industry. (p. 415)

metals One of three groups of elements on the periodic table. All metals are reactive, good conductors, ductile, and malleable. (p. 415)

metamorphic rocks Form when existing rocks are changed by high heat, high pressure, and/or very hot fluids. (p. 242)

microchips Tiny electric circuits printed on thin chips of silicon. (p. 468)

micrometer One-millionth of a meter, represented by the symbol μm. (p. 52)

mitochondria Organelles that release the energy the cell needs to function. (p. 65)

mixture Formed when two or more substances combine without changing the nature of the component substances. (p. 424)

molecule Formed when two or more atoms join together chemically. (p. 418)

multicellular Organisms that contain more than one cell. (p. 52)

muscle tissue Made up of specialized cells or fibers with the ability to contract and produce movement. (p. 120)

muscular system Works with the skeletal system to help a person move. (p. 172)

mutualism A relationship between two living things of different species in which both benefit. (p. 30)

neocatastrophism A geologic theory that retains some of the original assumptions of uniformitarianism while accepting the evidence of catastrophe. (p. 268)

nervous tissue Transmits messages throughout the body and coordinates and controls many of the body's activities. (p. 121)

neuron Nerve cell that is the basic unit of the nervous system. (p. 139)

neutrons Tiny particles contained in the nucleus of an atom with no electrical charge. (p. 403)

nitrogen cycle The process by which nitrogen is cycled between the environment and living things. Nitrogen atoms move slowly between living things, air, soil, and water. (p. 37)

nitrogen fixation The process of changing nitrogen gas into usable nitrogen oxide. (p. 37)

noble gases Colorless and odorless gases that do not react easily with other elements. (p. 416)

nonmetals One of three groups of elements on the periodic table. All nonmetals are poor conductors, brittle, and may be transparent. They may be solids, liquids, or gases at room temperature. (p. 415)

nucleotides The smaller subunits from which DNA is made. Each nucleotide has a sugar molecule, a phosphate group, and one of four bases—*adenine, guanine, thymine,* and *cytosine.* (p. 88)

nucleus The center of an atom; it is often the largest and most visible object inside a cell. (pp. 62, 403)

olfactory Receptors found in the nasal passages that sense chemicals from food as a person chews. (p. 155)

oncologist A physician who studies, diagnoses, and treats cancerous tumors. (p. 201)

open circuit An electric circuit with an open switch so there is a break in the circuit and electrons cannot flow. (p. 458)

optic nerve The part of the eye that carries impulses to the brain, which interprets them as images. (p. 150)

ore Any solid Earth material that can be mined at a profit. (p. 247)

organ A group of tissues that work together to perform a certain task. (p. 122)

organ system A group of organs working together to perform a series of related functions. (p. 122)

organelle The part of a cell that does the jobs needed to keep the cell alive. (p. 57)

organic sedimentary rock Forms from compacted and cemented organic material. (p. 241)

orthodontist A dentist who specializes in straightening teeth with the help of braces and other appliances. (p. 204)

osmosis Type of passive transport requiring no energy in which water molecules pass through the cell membrane from an area of higher concentration to an area of lower concentration to reach equilibrium. (p. 54)

osteocytes The cells within bones that are responsible for growth and repair. (p. 166)

physical property A characteristic of matter that can be observed and measured without changing the matter. (p. 361)

pituitary gland The tiny "master gland" that regulates the body's growth and is located in the brain. (p. 126)

plate tectonics A scientific theory that states that Earth's lithosphere is broken into large slabs of rock, called tectonic plates, that move around on the partly melted section of the mantle. (p. 273)

platelets Small colorless particles that release coagulating chemicals to form clots to stop the flow of blood. (p. 180)

poison Any substance that is taken internally or applied externally that is dangerous to someone's health or that might cause death. (p. 211)

predation The preying (or killing) of one animal on another for food. (p. 30)

primary succession The development of successive plant and animal communities in an area where living things have never existed before. (p. 31)

products New substances produced during a chemical change. (p. 385)

prokaryotic cell A cell that does not have a nucleus. (p. 56)

protein synthesis The process of building proteins that takes place in the ribosomes of the cell. (p. 94)

proteins Large molecules that are the building blocks for cells. (p. 87)

protons Tiny particles contained in the nucleus of an atom with a positive electrical charge. (p. 403)

psychiatrist A physician who specializes in assessing and treating mental illnesses. (p. 201)

pupil The center of the iris that contracts and expands to regulate the amount of light entering the eye. (p. 150)

P-wave (primary wave) The fastest type of earthquake wave that pushes and pulls the rocks through which it travels. (p. 279)

quarry A surface mine from which rock or sediment is removed. (p. 252)

radiometric dating A measurement process that uses the relative proportions of particular radioactive isotopes present in a rock sample to determine its age. (p. 330)

reactants Substances used in chemical reactions that are changed into something else (the products). (p. 385)

reactivity The ability of matter to undergo chemical reactions with other kids of matter. (p. 384)

receptor A specialized cell that receives information from its surroundings and provides it to the brain. (p. 149)

rechargeable battery Batteries that can be used repeatedly. (p. 456)

red blood cells A mature blood cell that is disc shaped and has no nucleus. It contains hemoglobin to carry oxygen to the cells of the body and carries carbon dioxide away from the cells (p. 180)

relative age The age of a rock or formation given relative to the age of other rocks or formations, usually defined as a zone fossil name. (p. 329)

relative dating A dating method that determines the relative order of past events, without necessarily determining their absolute age. It is based on the sequence of rocks and the fossils within the rock strata. (p. 329)

resistance When a material resists the flow of electrons. (p. 448)

respiratory system The body's system that allows a person to inhale oxygen and exhale carbon dioxide. (p. 187)

retina The light-sensitive layer of tissue located in the back of the eye. It collects images formed by the lens and transmits them through the optic nerve to the brain. (p. 150)

RNA (ribonucleic acid) Works with DNA to build proteins. (p. 94)

rock cycle The natural process that changes rocks from one type of rock to another type of rock. (p. 245)

safety engineers Workers who help keep people free from danger, risk, or injury in the workplace. They also work on the designs of buildings and other structures to make sure they are safe. (p. 209)

seafloor spreading A process by which new ocean crust is formed. As new crust forms, it pushes the older crust away from the ridge. It is one of the forces that causes plate movement. (p. 271)

sebaceous glands A microscopic gland found around the bottom of hair follicles that secretes oils to help prevent hair and skin from drying out. (p. 135)

secondary succession The re-growth of a community in an area that has soil and was recently inhabited by living organisms. It occurs after such disturbances as hurricanes, fires, landslides, floods, logging, farming, and grazing. (p. 32)

secretion The process during which proteins are moved out of a cell. (p.102)

sedimentary rocks Rock that forms when sediments are compacted and cemented together. Most sedimentary rocks formed in water or in environments that were once underwater. (p. 241)

selectively permeable Refers to the ability of the cell membrane to control what materials are allowed in and out of the cell. (p. 53)

semiconductors Solid elements, such as silicon or germanium, that have electrical conductivity between that of a conductor and an insulator. (p. 469)

series circuit An electric circuit where the current flows in a single path. (p. 459)

shear force A force acting parallel to a plane that creates a strike-slip fault. (p. 279)

shield volcano A broad mountain with gently sloping flanks. Its eruptions are quiet, with runny lava. (p. 284)

shock A condition caused by loss of blood or other body fluids, trauma, infections, or burns. Shock occurs when the flow of blood throughout the body is reduced. (p. 222)

short-age geology A theory that suggests that much of the geologic record formed in thousands, not millions, of years and that God can intervene in history. (p. 269)

skeletal system Bones and the tissues that connect them. (p. 164)

slump Materials that move down a slope as a unit and along a curved surface. (p. 314)

soil A mixture formed of weathered rock mixed with air, water, and decayed organic matter, or humus. (p. 301)

soil horizons Distinct layers of soil in a soil profile. (p. 302)

soil profile A vertical section of soil from the ground surface to the parent rock. (p. 302)

solar panels Groupings of solar cells used to capture sunlight. (p. 505)

solubility The ability of one substance to dissolve in a solvent. Solubility is a physical property of solutes that can vary depending upon the solvent. (p. 430)

solute A substance that dissolves in another substance. (p. 428)

solution A mixture in which the particles of each substance are mixed evenly. (p. 428)

solvent The substance that a solute is dissolved into. (p. 428)

strata Rock layers formed from sediments deposited by water. (p. 322)

streak The color of a mineral in its powdered form. (p. 234)

subduction The process that causes the more dense crust to push under crust that is less dense. The subducted crust melts to become part of the mantle. (p. 274)

sublimation The change of state when a solid changes directly into a gas. (p. 374)

suffocation A condition that occurs when air is prevented from entering the lungs. Choking is a common cause of suffocation. (p. 210)

superconductors Materials that conduct electricity without any loss of energy due to resistance. (p. 450)

S-wave (secondary wave) A type of earthquake wave that moves up and down through rock. (p. 279)

sweat glands Small glands in the skin that rid the body of heat, excess water, and certain wastes. (p. 135)

synapse The place where an impulse crosses from one neuron's axon to the dendrites of another. (p. 140)

talus Mass wasting that takes place when rocks fall from a higher elevation to a lower elevation. (p. 315)

target cells Cells that respond to hormones. When the hormone binds with DNA in the target cell, the cell responds by directing protein synthesis or controlling specific cell activities. (p. 103)

temporary magnet A magnet made by stroking magnetic material with a permanent magnet. It loses all or most of its magnetism over time as the magnetic domains become misaligned. (p. 481)

tendons Tough bands of tissue that fasten muscles to bones. (p. 173)

tension force A pulling force that produces a normal fault. (p. 279)

texture The size of the mineral grains or the way in which they are arranged in any type of rock. (p. 238)

tissues Groups of similar cells working together to perform specific jobs. (p. 116)

transcription The first step in protein synthesis in which a gene that codes for a protein is copied as an RNA molecule. (p. 95)

transform boundary A type of plate boundary in which the plates move horizontally past one another. (p. 274)

transformers Devices that can alter voltage to push an electric current over long distances. (p. 498)

transistor A small sandwich of two different types of semiconductors. (p. 468)

translation The second step in protein synthesis, which occurs outside the nucleus. During translation, the genetic code on the RNA is read and proteins are built. (p. 96)

turbidite The sedimentary deposit formed by dense underwater or turbidity current. (p. 337)

turbine A machine with moving fan-like blades, a shaft, and a wire that produces an electric current. (p. 498)

unicellular Any life form that consists of just one cell. (p. 52)

uniformitarianism A theory that states that the laws of nature have always been the same and that the natural processes have always happened at the same rate. (p. 268)

vacuole A fluid-filled storage container. Some vacuoles store water, other liquids, or food particles. (p. 65)

valence electron An electron in an atom's outermost shell involved in chemical bonding. (p. 421)

valley A long depression in the surface of the land between uplands, hills and mountains that usually contains a river. (p. 307)

ventricle A lower chamber of the heart. (p. 182)

vesicle Small sac-like structure that forms where part of the smooth ER buds off. It helps transport proteins and other materials that the cells need to be able to do the jobs for which they are designed. (p. 66)

voltage The pressure of an electric current. (p. 446)

volume The amount of space that matter occupies. (p. 365)

watts The unit of measure of power. (p. 506)

weight The measure of the effect of Earth's gravity on an object. (p. 365)

wet cell A battery that contains a liquid electrolyte, which is free to flow and move. (p. 453)

Index

Photo Credits

CVR (C) Ben44/Shutterstock, (Bkgd) Mny-Jhee/Shutterstock, (Banner) (1) Fotorich01/Shutterstock, (2) Janprchal/Shutterstock, (3) Courtesy of Larry Blackmer, (4) Steshkin Yevgeniy/Shutterstock, (5) Lucarelli Temistocle/Shutterstock, (6) NASA Images/NASA, (7) Matej Pavlansky/Shutterstock; **SS 0–SS 1** Beboy/Shutterstock; **SS 2** Bikeriderlondon/Shutterstock; **SS 3** Angela Waye/Shutterstock; **SS 5** Frangipani/Shutterstock; **SS 6, SS 7** TSI Graphics; **SS 8** Jacek Chabraszewski/Shutterstock; **SS 9** Bruce Rolff/Shutterstock; **SS 10** Jakub Krechowicz/Shutterstock; **SS 11** Erhan Dayi/Shutterstock; **12–13** iStockphoto/Thinkstock; **14–15** Zuzule/Shutterstock; **16** iStockphoto/Thinkstock; **17** (br) Pan Xunbin/Shutterstock, (bl) Carolina K. Smith Md/Shutterstock, (tr) Taiga/Shutterstock, (tl) Istockphoto/Thinkstock; **19** Jupiterimages/Thinkstock; **20** Paul Yates/Shutterstock; **21** Ultrashock/Shutterstock; **22** Chepko Danil Vitalevich/Shutterstock; **23** Mj007/Shutterstock; **24–25** Dorling Kindersley RF/Thinkstock; **26** Axz700/Shutterstock; **29** Eric Isselee/Shutterstock; **30** Kirsanov Valeriy Vladimirovich/Shutterstock; **31–32** (all) Dorling Kindersley RF/Thinkstock; **35** Hkannn/Shutterstock; **36** TSI Graphics; **37** (t) Jiang Hongyan/Shutterstock, (b) Hkannn/Shutterstock; **38** Medioimages/Photodisc/Thinkstock; **39** Alila Medical Images/Shutterstock; **40–41** Daniel Herbert/Shutterstock; **42** Karel Gallas/Shutterstock; **43** Kirsanov Valeriy Vladimirovich/Shutterstock; **46–47** Carolina K. Smith Md/Shutterstock; **50** Photos.com/Thinkstock; **51** D. Kucharski K. Kucharska/Shutterstock; **53** Alila Medical Images/Shutterstock; **54** Ducu59us/Shutterstock; **56–57** Lebendkulturen.de/Shutterstock; **59** Steve Schell; **60** Lightspring/Shutterstock; **62–63** Mopic/Shutterstock; **65** Dorling Kindersley RF/Thinkstock; **66** Matthew Cole/Shutterstock; **69** (t) Alila Medical Images/Shutterstock, (b) Zhabska Tetyana/Shutterstock; **70** TSI Graphics; **73** (t) Dorling Kindersley RF/Thinkstock, (b) Steve Schell; **75, 77** TSI Graphics; **78** Oleg Golovnev/Shutterstock; **79** Africa Studio/Shutterstock; **81** (tl) Ducu59us/Shutterstock, (b) Dorling Kindersley RF/Thinkstock; **82–83** Adrian Neal/Thinkstock; **84** Ermess/Shutterstock; **87** Blamb/Shutterstock; **88–89** TSI Graphics; **91** Luka Veselinovic/Shutterstock; **92** Thomas Northcut/Thinkstock; **95, 96** TSI Graphics; **99** Jubal Harshaw/Shutterstock; **100** TSI Graphics; **103** (b) Blamb/Shutterstock, (t) Alila Medical Images/Shutterstock; **104** Stockshoppe/Shutterstock; **107** Alila Medical Images/Shutterstock; **108** Tyler Boyes/Shutterstock; **109** iStockphoto/Thinkstock; **112–113** iStockphoto/Thinkstock; **114–115** Lightspring/Shutterstock; **116–117** Mopic/Shutterstock; **117** (t to b) Wilson's Vision/Shutterstock, Duncan Smith/Photodisk/Thinkstock, Jubal Harshaw/Shutterstock; **119** (t to b) Christopher Meade/Shutterstock, Wilson's Vision/Shutterstock, Wilson's Vision/Shutterstock, Suthep/Shutterstock; **120** (tr) Vetpathologist/Shutterstock, (b) Alila Medical Images/Shutterstock, (tl) Mopic/Shutterstock; **122** TSI Graphics; **123** Matthew Cole/Shutterstock; **127** Alila Medical Images/Shutterstock; **129** (bkgd) Alila Medical Media/Shutterstock, (inset) Alex Luengo/Shutterstock; **131** TSI Graphics; **133** Stockshoppe/Shutterstock; **136** George Doyle/Stockbyte/Thinkstock; **140** Alila Medical Images/Shutterstock; **142** Blueringmedia/Shutterstock; **143** Alex Mit/Shutterstock; **145** Leonello Calvetti/Shutterstock; **147** Fuse/Thinkstock; **149** Blueringmedia/Shutterstock; **150** Zhabska Tetyana/Shutterstock; **152** Alila Medical Images/Shutterstock; **153** Human Ear Anatomy/Shutterstock; **154** Hemera/Thinkstock; **155** Dorling Kindersley RF/Thinkstock; **159** Georgy Shafeev/Shutterstock; **161** Alila Medical Images/Shutterstock; **162–163** Jupiterimages/Thinkstock; **164** Thinkstock; **166** Oleksii Natykach/Shutterstock; **166–167** Grei/Shutterstock; **168** Dorling Kindersley RF/Thinkstock; **169** Alila Medical Images/Shutterstock; **172** Maridav/Shutterstock; **173** Designua/Shutterstock; **175** (l) Leonello Calvetti/Shutterstock, (r) Anatomia3d/Shutterstock; **176** Stihii/Shutterstock; **179** Maxisport/Shutterstock; **180** Alex Mit/Shutterstock; **183** Alila Medical Images/Shutterstock; **184** (l) Designua/Shutterstock, (r) Mmutlu/Shutterstock; **185** Hemera/Thinkstock; **187** Chalabala/Shutterstock; **189** Udaix/Shutterstock; **191** Stockshoppe/Shutterstock; **194** Courtesy of Loma Linda University; **195** iStockphoto/Thinkstock; **198–199** Varuna/Shutterstock; **200** Siri Stafford/Thinkstock; **204** Dean Bertoncelj/Shutterstock; **206** Portokalis/Shutterstock; **208** Comstack/Thinkstock; **211** Ammit Jack/Shutterstock; **213** My Good Images/Shutterstock; **215** Aaron Kohr/Shutterstock; **217** iStockphoto/

Thinkstock; **218** Fivespots/Shutterstock; **219, 221** iStockphoto/Thinkstock; **223** Pav197lin/Shutterstock; **224** Martan/Shutterstock; **225** Keith Brofsky/Thinkstock; **228–229** Stockbyte/Thinkstock; **230–231** Hangingpixels/Shutterstock; **232** Marcelclemens/Shutterstock; **234** iStockphoto/Thinkstock; **235** (l) Sigur/Shutterstock, (r) iStock/Shutterstock; **238** (r) Tyler Boyes/Shutterstock, (l) TSI Graphics; **238–239** Paul B. Moore/Shutterstock; **239** (l) Tyler Boyes/Shutterstock, (r) TSI Graphics; **241** (t to b) iStockphoto/Thinkstock, (all center) TSI Graphics, (b) Sedmi/Shutterstock; **242** (b) Only Fabrizio/Shutterstock, (t) TSI Graphics; **243** (tl) TSI Graphics, (bc) Tyler Boyes/Shutterstock, (tr) Michal812/Shutterstock, (tc) TSI Graphics, (br) TSI Graphics, (bl) TSI Graphics; **245** TSI Graphics; **247** ulisse/Shutterstock; **248** (t) Siim Sepp/Shutterstock, (b) Koya979/Shutterstock; **250** James Jones Jr/ Shutterstock; **251** iStockphoto/Thinkstock; **252–253** iStockphoto/Thinkstock; **254** (l) Jim Parkin/ Shutterstock, (r) iStockphoto/Thinkstock; **256** Photos.com/Thinkstock; **257** iStockphoto/Thinkstock; **259** Siim Sepp/Shutterstock; **260–261** Dr. Morley Read/Shutterstock; **262** Dorling Kindersley Rf/ Thinkstock; **265** Dorling Kindersley RF/Thinkstock; **266** TSI Graphics; **270** USGS; **271** Dorling Kindersley RF/Thinkstock; **272** TSI Graphics; **273** Dorling Kindersley RF/Thinkstock; **274** (b) Dorling Kindersley RF/ Thinkstock, (t) TSI Graphics; **275** Dorling Kindersley RF/Thinkstock; **278** Carlosgardel/Shutterstock; **279** (all) Steve Schell; **282** (t) Lance Cpl. Ethan Johnson, U.S. Marine Corps./NOAA/NGDC; **283** iStockphoto/Thinkstock; **284** Dorling Kindersley RF/Thinkstock; **286** Dorling Kindersley RF/ Thinkstock; **287** Dorling Kindersley RF Thinkstock; **290–291** Archana Bhartia/Shutterstock; **292–293** TTphoto/Shutterstock; **295** Ishbukar Yalilfatar/Shutterstock; **296** Dorling Kindersley/ Thinkstock; **297** iStockphoto/Thinkstock; **298–299** Spirit Of America/Shutterstock; **300** Ollirg/ Shutterstock; **301** Photowings/Shutterstock; **302** Dorling Kindersley RF/Thinkstock; **304** TSI Graphics; **306** Hemera/Thinkstock; **307** 3841128876/Shutterstock; **308** TSI Graphics; **310** Kezza/Shutterstock; **312** TSI Graphics; **313** Denis Burdin/Shutterstock; **314–315** Ajp/Shutterstock; **316** Hecke61/ Shutterstock; **317** Earl D. Walker/Shutterstock; **320–321** Clive Watkins/Shutterstock; **324** (b) Lefteris Papaulakis/Shutterstock, (t) Falk/Shutterstock; **325** (tr) Miguel Garcia Saavedra/Shutterstock, (br) Phaitoon Sutunyawatchai/Shutterstock, (bl) Sascha Burkard/Shutterstock, (cr) Bestimagesevercom/ Shutterstock; **326** (bkgd) TSI Graphics, (t) Denis Barbulat/Shutterstock, (c) Denis Barbulat/Shutterstock, (b) Denis Barbulat/Shutterstock; **329** (l) iStock/Thinkstock, (r) Photo Fun/Shutterstock; **331** Eldad Carin/ Shutterstock; **332** iStock/Thinkstock; **335** iStock /Thinkstock; **336** Bcampbell65/Shutterstock; **337** Kara Jade Quan–Montgomery/Shutterstock; **338** TSI Graphics; **339** Richard Thornton/Shutterstock; **340** TSI Graphics; **342** Phil Pister/Department of Fish and Game, State of California; **344** TSI Graphics; **345** (b) Teguh Tirtaputra/Shutterstock, (t) Miguel Garcia Saavedra/Shutterstock; **346** Lorraine Hudgins/ Shutterstock; **347** Courtesy of Tim Standish; **348** Andreas Meyer/Shutterstock; **349** (l) Hemera/ Thinkstock, (r) Linda Bucklin/Shutterstock; **350** Catmando/Shutterstock; **352** Courtesy of Seventh-day Adventist Church; **353** iStockphoto/Thinkstock; **355** Sascha Burkard/Shutterstock; **356–357** Rimdream/ Shutterstock; **358–359** Creatas/Thinkstock; **360** Aleksandr Kurganov/Shutterstock; **363** TSI Graphics; **364** (l) Fribus Ekaterina/Shutterstock, (r) Karol Kozlowski/Shutterstock; **366** Fedor Bobkov/Shutterstock; **367** (b) TSI Graphics; **370** Rangizzz/Shutterstock; **373** iStockphoto/Thinkstock; **374** Ivo Petkov/ Shutterstock; **375** TSI Graphics; **377** Jupiterimages/Getty Images/Thinkstock; **378** Trevorb/Shutterstock; **380** iStockphoto/Thinkstock; **383** Fotokostic/Shutterstock; **385** Goran Cakmazovic/Shutterstock; **387** Creative Travel Projects/Shutterstock; **389** Anteromite/Shutterstock; **394** Sebastian Kaulitzki/ Shutterstock; **395** Docent/Shutterstock; **397** iStockphoto/Thinkstock; **398–399** Jupiterimages/ Thinkstock; **401** (l) Only Fabrizio/Shutterstock, (r) Fokin Oleg/Shutterstock; **403** Valdis Torms/ Shutterstock; **404** TSI Graphics; **405** (t) Semmick Photo/Shutterstock, (b) TSI Graphics; **406** Baldyrgan/ Shutterstock; **407** TSI Graphics; **408** (l) Fzd.it/Shutterstock, (r) General-Fmv/Shutterstock; **411** Georgios Kollidas/Shutterstock; **412** Concept W/Shutterstock; **413** Photos.com/Thinkstock; **414–415** Concept W/ Shutterstock; **416** Photodisc/Thinkstock; **418** (l) Bobyramone/Shutterstock, (r) Bobyramone/ Shutterstock; **419** (l) Lightspring/Shutterstock, (r) Molekuul.be/Shutterstock; **423** Hemera/Thinkstock; **424** (l) Olga Miltsova/Shutterstock, (r) Jahina_Photography/Shutterstock; **428** Fernando Cortes/